Modern Practice in Stress and Vibration Analysis 2012

(MPSVA 2012)

Journal of Physics: Conference Series Volume 382

Glasgow, Scotland
29-31 August 2012

ISBN: 978-1-62276-849-3
ISSN: 1742-6588

Printed from e-media with permission by:

Curran Associates, Inc.
57 Morehouse Lane
Red Hook, NY 12571

Some format issues inherent in the e-media version may also appear in this print version.

Printed by Curran Associates, Inc. (2013)

For permission requests, please contact the Institute of Physics
at the address below.

Institute of Physics
Dirac House, Temple Back
Bristol BS1 6BE UK

Phone: 44 1 17 929 7481
Fax: 44 1 17 920 0979

techtracking@iop.org

Additional copies of this publication are available from:

Curran Associates, Inc.
57 Morehouse Lane
Red Hook, NY 12571 USA
Phone: 845-758-0400
Fax: 845-758-2634
Email: curran@proceedings.com
Web: www.proceedings.com

TABLE OF CONTENTS

KEYNOTE SPEAKERS

CONTRIBUTED PAPERS

VIBRATION IN STRUCTURAL ELEMENTS

MANUFACTURING APPLICATIONS

STRUCTURAL MECHANICS

ENERGY HARVESTING

MICROMECHANICS AND MATERIALS

NONLINEAR SYSTEM IDENTIFICATION

EXPERIMENTAL STRESS AND STRAIN ANALYSIS

CONTROL OF VIBRATION IN STRUCTURES

SPECIAL SESSION ON IMPACT

BIOMEDICAL APPLICATIONS AND BIOMECHANICS

STRUCTURAL HEALTH MONITORING

MODELLING

NON-DESTRUCTIVE EVALUATION

HIGH POWER ULTRASONIC SYSTEMS

PREFACE

This is the second time that the Modern Practice in Stress and Vibration Analysis conference has come to the University of Glasgow and it is with great pleasure that I write this preface for the event in 2012. The remit of the Modern Practice in Stress and Vibration Analysis conferences is relatively broad and encompasses scientific and technological research in stress analysis, the mechanics of materials, applied dynamics, metrology and instrumentation, system identification, structural health monitoring, nondestructive evaluation, and vibration theory and analysis. Within these relatively traditional subject areas we also see burgeoning new themes emerging, in which new manufacturing technologies, energy harvesting, micro and nano-mechanic applications, biomechanics, and advanced modelling feature very strongly. The conference converges around six keynote addresses over the three days, each one being linked to a central theme for the conference. The first day opens on the morning of Wednesday 29 August 2012 with an address by Professor Walter Lacarbonara of the University of Rome on 'Nonlinear dynamics enabled design and control', in which ideas taken from nonlinear dynamics and once considered to be highly specialised are now informing the design and control of mechanical systems. This is followed by an afternoon address by Professor James R Barber of the University of Michigan on the topic of 'Frictional systems under periodic load – History-dependence, non-uniqueness, and energy dissipation', where fundamental mechanical issues are considered in the performance of loaded mechanical systems in which complicated friction mechanisms play an important role. The second day begins with a morning lecture by Professor Fabrice Pierron of Paris Tech entitled 'A novel photomechanical approach to dynamic testing of materials', and covering the testing of materials, an important theme which has long been central to this conference series. This is followed by the British Society of Strain Measurement's sponsored Measurements Lecture, which also features as the fourth keynote address of the conference, and is given in 2012 by Dr Cathy Holt of the University of Cardiff. The third and final day of the conference opens with a keynote lecture by Professor Wieslaw M Ostachowicz of the Polish Academy of Sciences in Gdansk on another long-standing conference theme, and entitled 'Structural health monitoring by means of elastic wave propagation'. The final keynote lecture takes place in the afternoon of the last day and is given by Professor Jerzy Warminski of the Technical University of Lublin, Poland. The title of the lecture is 'Nonlinear phenomena in mechanical systems dynamics', and is in deliberate juxtaposition to the opening keynote address, emphasising the pervasive nature of modern nonlinear dynamics.

I am delighted to welcome authors and delegates to this Modern Practice in Stress and Vibration conference, run under the auspices of the Institute of Physics Applied Mechanics Group and held at the University of Glasgow. I would like to thank Claire Garland and Dawn Stewart of the Institute of Physics for all their work and assistance, the local organising committee, the scientific committee, and lastly the authors of the papers featured in this conference proceedings. I extend my warmest welcome to all our conference delegates.

Matthew Phillip Cartmell
Conference Organiser

Peer review statement

All papers published in this volume of *Journal of Physics: Conference Series* have been peer reviewed through processes administered by the proceedings Editors. Reviews were conducted by expert referees to the professional and scientific standards expected of a proceedings journal published by IOP Publishing.

Modern Practice in Stress and Vibration Analysis 2012 (MPSVA 2012) IOP Publishing
Journal of Physics: Conference Series **382** (2012) 012001 doi:10.1088/1742-6596/382/1/012001

Nonlinear dynamics enabled systems design and control

Walter Lacarbonara

Department of Structural Engineering, Sapienza University of Rome, via Eudossiana 18. Rome, IT

E-mail: walter.lacarbonara@uniroma1.it

Abstract. There is a growing interest towards design of high-performance structures and devices by seeking ways to exploit advantageously different nonlinearities at different scales rather than constraining operations to avoid nonlinear phenomena. Tools of robust nonlinear modeling and analysis are shown to be turned into design tools for achieving high levels of vibration control authority and synthesis of engineered systems and materials. A brief overview of methods and results on active resonance cancellation and passive nonlinear hysteretic vibration absorbers is illustrated. Recent results on the diffused hysteresis exhibited at the nano-microscale in nanocomposites due to the powerful nonlinear stick-slip mechanism exhibited by carbon nanotubes dispersed in a hosting matrix are discussed. The optimization of the main microstructural parameters is shown to lead to unprecedented levels of damping capacity in next-generation nanostructured materials tailored for wide-band vibrational energy dissipation.

1. Introduction

The field of passive and active nonlinear vibration absorbers has been widely investigated in the last century. Passive autoparametric vibration absorbers (VAs) exploit nonlinear coupling between the main system and the absorber [1] thus overcoming the drawbacks associated with the increase of the system degrees of freedom when the VA is linearly coupled to the main system. Cartmell and Lawson [2] enhanced the performance of the autoparametric VA by using a computerized closed-loop controller. The autoparametric VA is often represented by a pendulum in applications such as for control of the parametric resonance in cantilever beams or suppression of the 1/3-order subharmonic resonance in magnetically levitated (MAGLEV) bodies [3].

Control architectures with passive autoparametric VAs capable of coping with certain nonlinear resonances sometimes turn out to be ineffective for other dynamic instabilities. For example, in the context of a parametrically excited MAGLEV body with an attached autoparametric pendulum, the autoparametric coupling can excite chaotic motions in the main system [4]. In such instances, nonlinear active resonance rejection methods may be necessary. An open-loop control strategy designed via a perturbation procedure turned the pendulum into an active VA achieving a full rejection of the parametric resonance [4].

Different nonlinear mechanisms (e.g., saturation [5], quenching, sub/superharmonic resonances, autoparametric resonances, 0:1 resonances) can be leveraged to produce effective actuator actions for resonance rejection in nonlinear lumped- and distributed-parameter systems. These control laws can be designed to be effective also when excitations and actuations are noncollocated for which classical linear control techniques break down. The nonlinear resonance rejection strategies can be shaped by perturbation techniques [6]. Noncollocated disturbances causing primary and parametric resonance oscillations were addressed via nonlinear actuator actions in a pendulum-type crane architecture [7], in a MAGLEV body [4], in parametrically excited hinged-hinged shallow arches [8], and in simply supported beams subject to parametric-resonance pulsating thrusts [9].

As an alternative to nonlinear active control, passive control achieved by multiple passive viscoelastic VAs linearly coupled with the main system has been extensively investigated. Notwithstanding their desirable fault-tolerance characteristics, viscoelastic (VE) absorbers, often realized as a parallel arrangement of linear springs and dashpots, are rather sensitive to mistuning problems due to heating of the viscoelastic component or due to slight changes of the mass/stiffness distribution. Hysteretic (HYS) absorbers were proposed by Lacarbonara and Vestroni [10, 11] to overcome the limitations of VE absorbers. The rheological hysteretic component can be a device or an assembly of material components with nonlinear (memory-dependent) constitutive behaviors [12] for which unloading occurs on distinct curves different from the so-called virgin loading curve (e.g., a series arrangement of a spring and a friction element, short steel wire ropes, shape memory wires [13, 14]). These nonlinear VAs are proved to be effective for reducing by orders of magnitude the effects of sustained primary resonances in structures such as beams/bridges [15] or for control of self-excited mechanisms such as flutter in aircraft wings [16] and suspension bridges [17, 18].

For applications requiring wide-band dissipation of vibrational energy and noise, the next-generation damping devices will make use of carbon nanotube-based multi-phase composites which exhibit high damping capacities. This is due to the interfacial friction between the nanofillers (CNTs) and the polymer matrix. The combination of extremely large surface area, weak interfacial bonding with the polymer, together with low mass density of the CNTs implies that frictional sliding of nanotubes with the matrix cause significant dissipation of energy with a minimal weight penalty. Recent experimental studies have shown that the damping capacity of polymer nanocomposites was enhanced by over 200% by incorporating a small amount of CNTs [19]. The mechanism behind this enhancement can be explained by shear lag theory and stick-slip theory. A micromechanics-based formulation is proposed [20, 21] to describe the space-wise distributed stick-slip phenomenology and quantify the loss factor and damping ratio. The approach allows to investigate the optimal combination of microstructural parameters to achieve unprecedented increments of damping by over 1000% as in epoxy and PEEK nanocomposites.

2. Passive versus active nonlinear vibration absorbers

An autoparametric VA in the form of a pendulum – in a 1:2 frequency ratio with the main system – was proved [3] to prevent the occurrence of the 1/3-order subharmonic resonance in a MAGLEV body such as the one shown in Fig. 1. The autoparametric VA cannot suppress, however, the principal parametric resonance since large chaotic motions of the main system and VA were found in experiments [4]. This evidence motivated the development of nonlinear control methods aimed at direct rejection of the parametric resonance without using the autoparametric energy transfer.

A simplified model of MAGLEV body [3] is considered in Fig. 1 with the origin O placed at the pendulum pivot point on the body of mass m_1 in its equilibrium under the repulsive magnetic forces in which the gap is denoted by z_o. Let the vertical motion of the body and the pendulum angle be denoted by z and θ, respectively. The pendulum of length r and tip mass m_2 is actuated by the torque $\tau(t)$. The base magnet is subject to the prescribed vertical displacement $z_b = z_{b1} \cos \Omega t$.

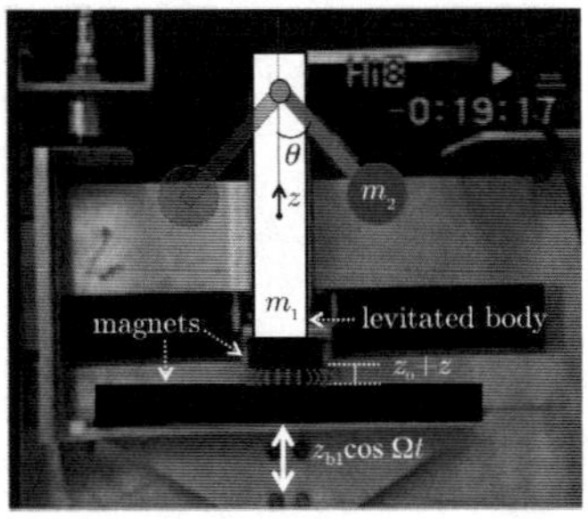

Figure 1. MAGLEV body with the pendulum VA.

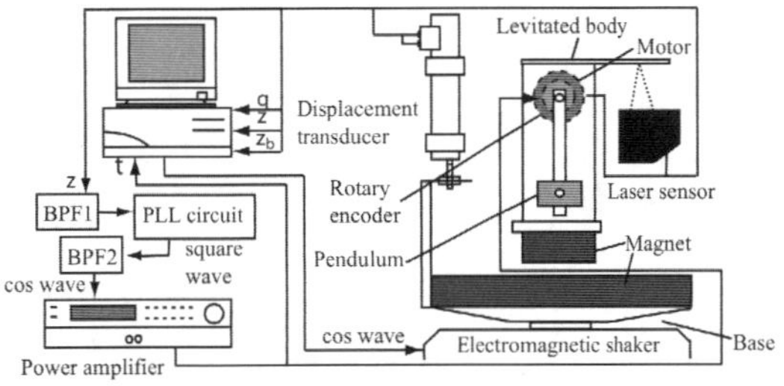

Figure 2. Experimental setup with the block diagram of the phase-lock loop circuit.

The nondimensional equations of motion expanded in Taylor series up to cubic terms are

$$\ddot{z} + (1 - 2\alpha_2\epsilon\cos\nu t)\, z = -\mu_z\dot{z} + \epsilon\cos\nu t - \alpha_2 z^2 - \alpha_3 z^3 + \mu\ddot{z}\theta^2 + \mu r\omega_\theta^2\theta^2$$
$$- \mu r\dot{\theta}^2 - \mu r\omega_\theta^2 b\theta\cos(\nu_\tau t + \phi) + \tfrac{1}{6}\mu r\omega_\theta^2 b\theta^3\cos(\nu_\tau t + \phi), \qquad (1)$$
$$\ddot{\theta} + (\omega_\theta^2 + \ddot{z}/r)\theta = -\mu_\theta\dot{\theta} + \omega_\theta^2 b\cos(\nu_\tau t + \phi) + \tfrac{1}{6}\omega_\theta^2\theta^3 \qquad (2)$$

where $\mu := m_2/(m_1 + m_2)$ is the mass ratio; $\epsilon := z_{b1}/z_o$ is the nondimensional excitation amplitude; (μ_z, μ_θ) are the damping coefficients; α_2 and α_3 are the coefficients of z^2 and z^3, respectively, in the expansion of the magnetic force; and the nondimensional control torque is $\tau = b\cos(\nu_\tau + \phi)$. Time and lengths in Eqs. (1) and (2) have been rescaled by $1/\omega_z$ and z_o, respectively, where ω_z is the natural frequency of the body (when the pendulum is locked).

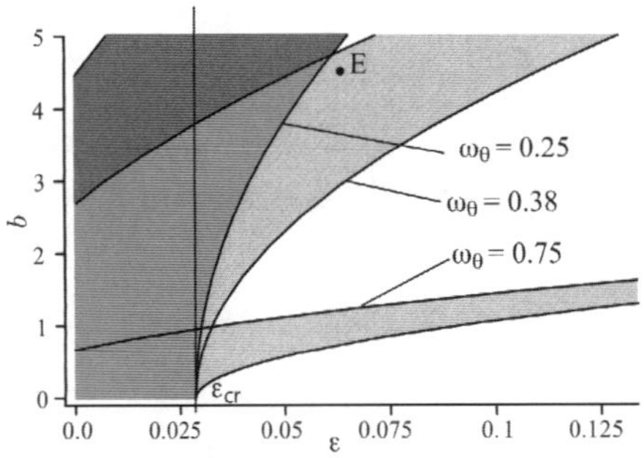

Figure 3. Theoretical range of effective control gains vs. the excitation amplitude for three VAs.

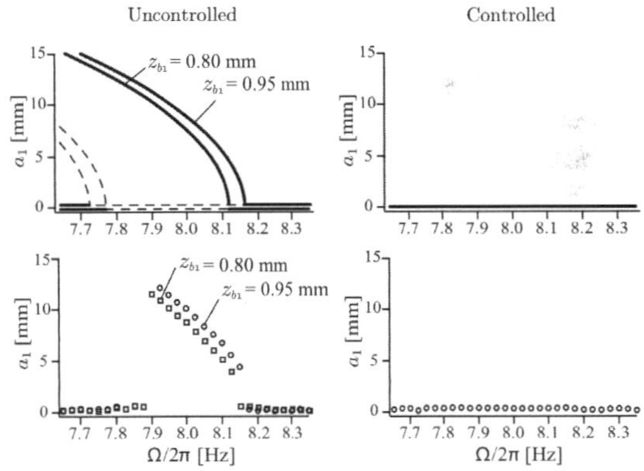

Figure 4. Theoretical (top) and experimental (bottom) frc's of the uncontrolled and controlled system for two excitation levels.

The term proportional to $\theta\ddot{z}$ in Eq. (2) causes the (principal autoparametric resonance) transfer of energy when $\omega_\theta \approx 1/2$. The pendulum torque is a direct excitation term. Setting its frequency to one-half the external excitation frequency $\nu \approx 2$ (i.e., $\nu_\tau = \nu/2$), the resultant pendulum oscillations will exhibit the frequency $\nu/2$ and, by means of part of the vertical reaction force, $\mu\ddot{z}\theta^2$, at the pendulum pivot point, the pendulum will impart to the body a resonant input with frequency equal to 1. This reaction force can counteract the parametric resonance force only if it is of the same order of the parametric resonance term, $2\alpha_2\,\epsilon z\cos\nu t$.

An asymptotic expansion of the dynamic response can be pursued using the method of multiple scales [4]. The nearness of the principal parametric resonance is expressed introducing the detuning parameter $\hat{\sigma}$ such that $\nu = 2 + \epsilon\hat{\sigma}$ and $\nu_\tau = \nu/2 = 1 + \epsilon\hat{\sigma}/2$. By omitting the full details (see [4]), the slow modulation of the complex-valued amplitude $A_z(t) = B_z(t)\exp(i\epsilon\hat{\sigma}t/2)$ of the MAGLEV motion is expressed at the onset of the parametric resonance by

$$\dot{B}_z = -\tfrac{1}{2}\hat{\mu}_z B_z - i(\tfrac{2}{3}\alpha_2 + P_{z1}e^{i2\phi}\hat{b}^2)\bar{B}_z - i(\tfrac{\hat{\sigma}}{2} + P_{z2}\hat{b}^2)B_z \qquad (3)$$

where i is the imaginary unit, the overbar indicates the complex conjugate and P_{z1} is the *effective controller coefficient*. The term proportional to $\bar{B}_z$ is the *effective parametric resonance force* which has to exceed a threshold force for the onset of the parametric resonance. Therefore, by suitably tuning the control phase ϕ and gain b, the effective parametric resonance force can be demoted below the threshold force thus achieving full rejection of the resonance. To this end, the appropriate phase such to minimize the magnitude of the coefficient of $\bar{B}_z$ is $\phi = 0$ if $\mathrm{sgn}(\alpha_2 P_{z1}) = -1$ or $\phi = \pm\pi/2$ if $\mathrm{sgn}(\alpha_2 P_{z1}) = 1$ where sgn denotes the signum function. The stability of the trivial solution is ensured (in which case the parametric resonance is fully rejected) if

$$\left|\tfrac{4}{3}\alpha_2\epsilon + 2P_{z1}\,e^{i2\phi}b^2\right| < \mu_z \qquad (4)$$

which yields in closed form the range of effective control gains shown in Fig. 3.

In the experimental setup shown in Fig. 2, the base ferrite magnet is excited by an electromagnetic shaker while the other magnet is fixed on the lower part of the body. The torque signal was fed to the rotary motor as a sinusoidal signal generated by the (PLL) phase-lock loop circuit (LM565CN) and a power supply. In addition, the phase difference between the body and the pendulum was locked on $\pi/2$ by the PLL device and the phase-shift circuit [4]. Figure 4 portrays the theoretical and experimental frequency-response curves of the uncontrolled and controlled system in which a_1 indicates the amplitude of the levitated body.

The same concepts were applied in theory and experiments to reject principal parametric resonances in structural systems such as shallow arches and beams. A hinged-hinged arch subject to a parametric resonance of the lowest skew-symmetric mode was controlled by a transverse force at the midspan tuned to be in a 1/2-order subharmonic resonance with the excited mode [8]. On the other hand, the principal parametric resonance of a simply supported beam subject to a harmonic end thrust was controlled by noncollocted piezoceramic actuators driven by nonlinear voltage control laws [9].

3. Passive Hysteretic Vibration Absorbers

Passive nonlinear absorbers may in general allow higher vibration attenuation in wider frequency bands due to nonlinear frequency self-tuning. The class of proposed hysteretic VAs comprises rheological elements that exhibit rate-independent hysteresis such as the softening/hardening hysteretic restoring forces possessed by short wire ropes (see Figs. 5a and 6), or the pinched hysteretic forces (see Fig. 5b), or the pseudoelastic forces of shape memory wires (see Fig. 5c).

A hysteretic vibration absorber (HYS VA) (see Fig. 6) was investigated in [10, 11]. The restoring force of the VA mass is provided by short wire ropes under flexure (see Figs. 6c,d). The hysteresis loops of the steel wire ropes, obtained by a Universal Testing Machine, exhibit softening or hardening hysteresis depending on whether the clamps of the wire ropes are laterally unconstrained or constrained when displaced in the vertical direction. The identified loops are represented by solid lines. The constitutive law of the model can be expressed as $N = c\,\dot{y} + k\,y + z$ where y denotes the VA displacement and z is the hysteretic part of the force governed by the Bouc-Wen differential flow law [12].

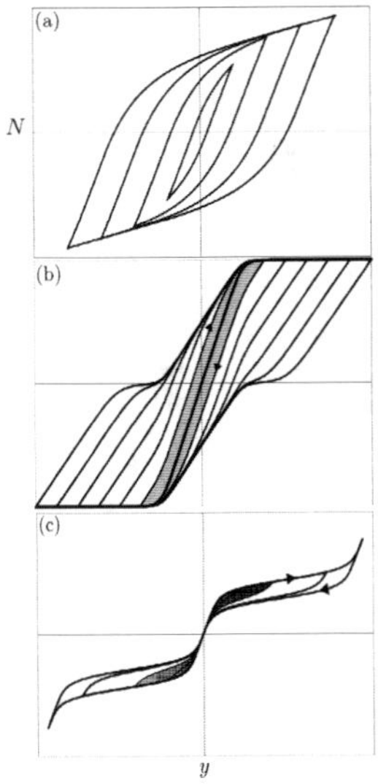

Figure 5. Restoring forces according to the (a) Bouc-Wen model; (b) Masing model [12]; (c) SMA material model [13].

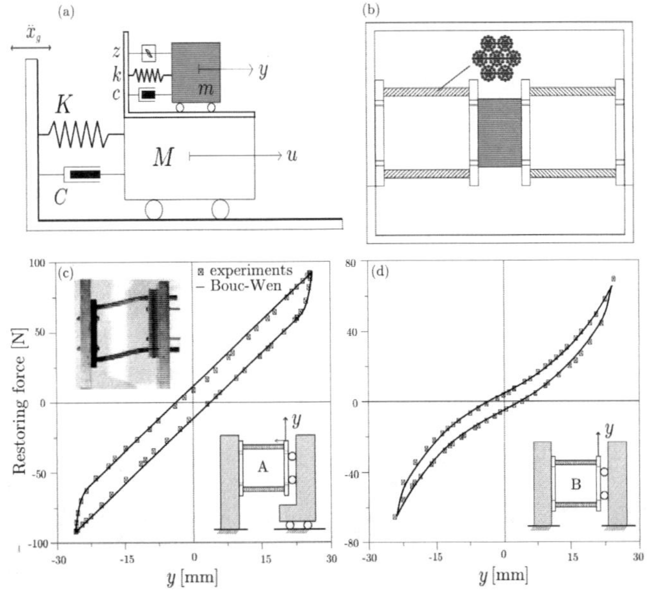

Figure 6. (a) The main system with a HYS VA made of the wire ropes in (b), (c) and (d) hysteresis loops.

The nondimensional equations of motion of the main (spring-mass-damper) system endowed with the HYS VA are:

$$\begin{aligned}
&(1+\mu)\,\ddot{u} + \tilde{C}\,\dot{u} + u + \mu\,\ddot{y} = 0, \\
&\mu\,\ddot{u} + \mu\,\ddot{y} + \tilde{c}\,\dot{y} + \tilde{k}\,y + \tilde{k}_3\,y^3 + z = 0, \\
&\dot{z} = \left[\tilde{k}_z - |z|^n\,(\tilde{\gamma} + \tilde{\beta}\,\mathrm{sgn}(z)\,\mathrm{sgn}(\dot{y}))\right]\dot{y}
\end{aligned} \tag{5}$$

where u and y are the nondimensional displacements of the main mass and VA mass; $\mu := m/M$ is the mass ratio; $(\tilde{k}, \tilde{k}_z, \tilde{k}_3, \beta, n)$ are the nondimensional constitutive parameters of the Bouc-Wen model with the hardening term.

The mass ratio is set to 2 % and the nondimensional amplitude of the base excitation is varied from 0.1 to 1.0 to obtain the frequency-response curves in Fig. 7a. The continuation analysis performed on the 5-dimensional system (in state space) reveals the occurrence of Neimark-Sacher bifurcations in the region of the out-of-phase mode (denoted by points A,B and C,D in Fig. 7a). The bifurcation diagram in the frequency band between the Neimark-Sacher bifurcations shown in Fig. 7b exhibits a scenario of loss of stability to quasiperiodicity, phase-locking, and torus breakdown [12]. The relevance of this finding is that the emergence of these nonlinear attractors is highly beneficial to the HYS VA performance since the amplitude-modulated or low-frequency periodic responses possess lower amplitudes than those exhibited during the softening mode of operation of the VA for which the resonance of the out-of-phase mode gives rise to stable large-amplitude responses [11].

Practical implementations of HYS VAs deal with control of the base-excited primary resonance and random vibrations of beams and footbridges [15], flutter control of aircraft wings [16] and suspension bridges [18]. For suspension bridges, the indicial function approach and a linearized (condensed) model of suspension bridges were employed to study the onset of the flutter condition [17]. The VA parameters were tuned according to an optimization procedure

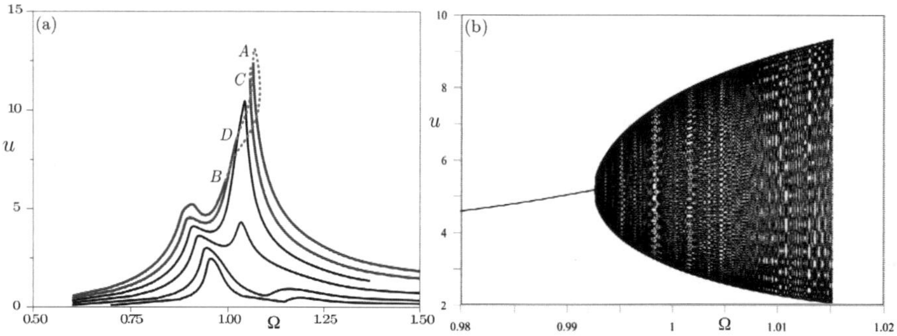

Figure 7. (a) Frequency-response curves of the main mass (see Fig. 6) subject to harmonic base excitations. The dashed lines indicate unstable periodic responses. (b) Bifurcation diagram (Poincaré map) past the Neimark-Sacher bifurcations.

[18]. Multiple HYS VAs (with a mass ratio equal to 0.1 % for an overall mass of 27 tons) distributed on the two sides of the deck of the Runyang Suspension Bridge (span=1490 m) are tuned to suppress flutter of the symmetric torsional mode. The HYS VAs perform better than linear VE VAs since the oscillation amplitudes of the bridge and the HYS devices are an order of magnitude smaller than those achieved with the VE absorbers (see Fig. 8).

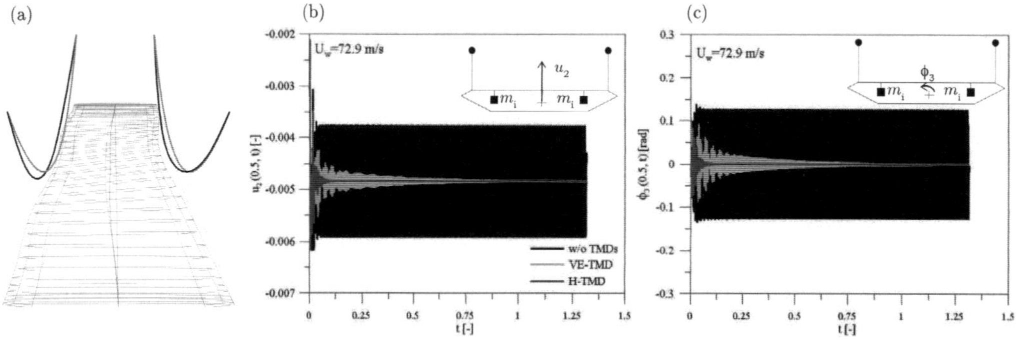

Figure 8. (a) Flutter mode shape, (b) vertical deflection and (c) torsional rotation of the bridge at the flutter wind speed. The uncontrolled response is compared with the response of the structure subject to VE and HYS VAs, here denoted by VE-TMD and H-TMD.

4. Stick-slip hysteresis in carbon nanotube-based composites

During a loading process, a nanocomposite starts elongating and shear stresses are generated between the CNT walls and the surrounding matrix due to the difference in their elastic properties. During the *sticking phase*, the load is transferred to the CNTs which start to elongate with the matrix, if they are well bonded. As the external load exceeds a threshold value that causes a partial debonding of CNTs from the matrix, the CNTs stop elongating together with the matrix and a further increase of the load can only result in further deformation of the matrix. Thus, the polymer starts to flow over the surface of CNTs and deformation energy is dissipated through the frictional slippage between the CNTs and the matrix: this is the *slipping phase*. The CNTs embedded in the hosting matrix are assumed to be aligned with one preferential direction. The domain of the model is given by two sub-domains, $\mathcal{B} = \mathcal{B}_M \cup \mathcal{B}_C$, the matrix and the CNTs (see Figs. 9a,b). To study the mechanism of interfacial slippage between the CNTs and the matrix, a micromechanics-based approach was developed [20, 21] extending the concepts of Eshelby's inclusion theory together with the MOri-Tanaka averaging, and introducing the plastic eigenstrain concept to describe via plastic shear strain increments the distributed slippage of the CNTs. For simplicity, a one-dimensional problem is considered by increasing the shear strain field so that the shear stress along the CNTs walls can be determined using the proposed stick-slip constitutive law and the inelastic shear strain evolutive law characterizing the micromechanics model (see Fig. 9c). Two thermosetting polymers are chosen for the matrix, epoxy resin and PEEK, which exhibit excellent strength and chemical resistance. The constitutive parameters of the adopted model are the CNT volume fraction ϕ_C, the CNT-matrix post-stiking shear stiffness expressed by the exponent M in the plastic flow law, the interfacial shear strength S_0, and the reference plastic strain velocity $\dot{\gamma}_r^p$.

The calculated damping ratios shown in Figs. 9d,e are highly sensitive to the strain amplitude prescribed in the cyclic tests. A peak value is attained at a given strain amplitude depending on the microstructural parameters. The conducted optimization in terms of all microstructural parameters has led to damping ratios as high as 26 % while the baseline matrices are very lightly damped (well below 1 %) thus achieving increments of damping capacity exceeding 1000 %.

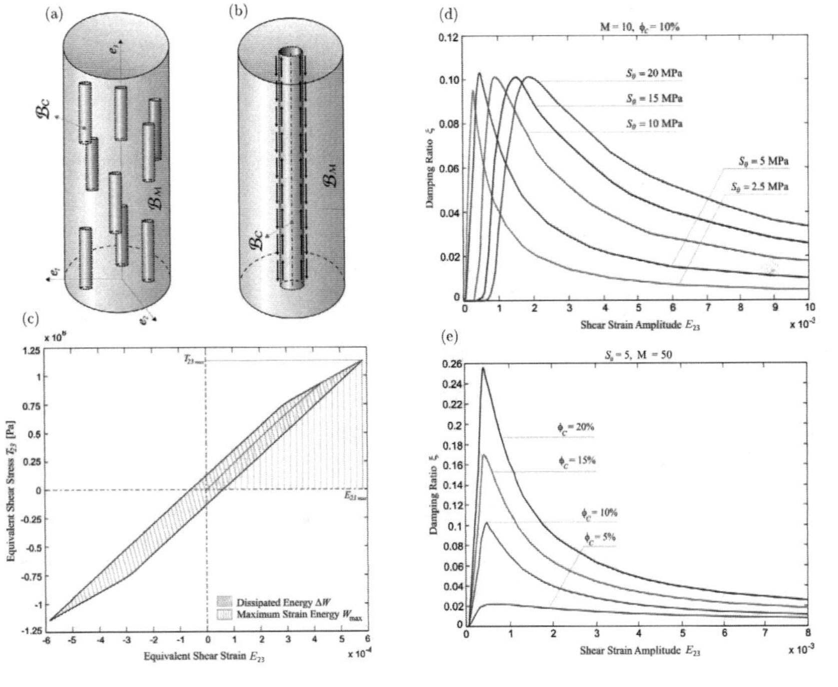

Figure 9. (a) Nanocomposite with CNTs inclusions, (b) shear stress on the CNTs walls, (c) stress-strain curve during stick-slip and associated works, (d) damping ratio ξ vs. shear strain amplitude for various S_0 and (e) for various ϕ_C for the CNT/PEEK composite.

5. Conclusions

An overview of methods and results on nonlinear active resonance cancellation and passive nonlinear hysteretic vibration absorbers has been offered. Recent results on the diffused hysteresis exhibited at the nano-microscale in nanocomposites due to the nonlinear stick-slip mechanism exhibited by carbon nanotubes dispersed in a polymeric hosting matrix have been documented showing the potential to reach unprecedented levels of damping capacity in next-generation nanostructured materials for wide-band vibrational energy dissipation and noise control.

5.1. Acknowledgments

WL acknowledges his colleagues Profs. Fabrizio Vestroni, Hiroshi Yabuno and Giovanni Formica, and his students Andrea Arena, Nicola Carpineto, Arnaldo Casalotti, Michela Taló.

References

[1] Haxton R S and Barr A D S The autoparametric vibration absorber 1972 *J. Eng. Ind-T. ASME*
[2] Cartmell M and Lawson J Performance enhancement of an autoparametric vibration absorber by means of computer control 1994 *J. Sound Vib.* **177**(2) 173-95
[3] Yabuno H, Endo Y and Aoshima N Stabilization of $\frac{1}{3}$-order subharmonic resonance using an autoparametric vibration absorber 1999 *J. Vib. Acoust.* **121** 309-15
[4] Yabuno H, Kanda R, Lacarbonara W and Aoshima N Nonlinear active cancellation of the parametric resonance in a magnetically levitated body 2004 *J. Dyn. Syst-T. ASME* **126** 433-42
[5] Oueini S S , Nayfeh A H and Golnaraghi M F A theoretical and experimental implementation of a control method based on saturation *Nonlinear Dynam.* **13** 189-202
[6] Lacarbonara W, Soper R R and Chin C M Nonlinear vibration control of distributed-parameter systems: a perturbation approach 1999 in Proc. *of 1999 ASME Materials & Mechanics Conference* Blacksburg VA USA June 26-30
[7] Soper R R, Lacarbonara W, Nayfeh A H and Mook D T Open-loop resonance-cancellation control for a base-excited pendulum 1999 *J. Vib. Control* **7** 1265-79
[8] Lacarbonara W, Chin C M and Soper R R Open-loop nonlinear vibration control of shallow arches via perturbation approach 2002 *J. Appl. Mech-T ASME* **69** 325-34
[9] Lacarbonara W, Yabuno H and Hayashi K Nonlinear cancellation of the parametric resonance in elastic beams: theory and experiment 2007 *Int. J. Solids Struct.* **44** 2209-24
[10] Lacarbonara W and Vestroni F Feasibility of a vibration absorber based on hysteresis 2002 in Proc. of *Third World Congress on Structural Control* Como, Apr 7-12
[11] Carpineto N, Vestroni F and Lacarbonara W Vibration mitigation by means of hysteretic tuned mass dampers 2011 *EURODYN 2011 Eighth International Conference on Structural Dynamics* Paper. No. 531 Leuven Belgium July 4-6
[12] Lacarbonara W and Vestroni F Nonclassical responses of oscillators with hysteresis 2003 *Nonlinear Dynam.* **32** 235-58
[13] Lacarbonara W, Vestroni F and Bernardini D Nonlinear thermomechanical oscillations of shape-memory devices 2004 *Int. J. Solids Struct.* **41** 1209-34
[14] Carboni B, Lacarbonara W and Auricchio F Modeling of a shape-memory alloy wire rope 2012 *ICTAM2012 23rd International Congress of Theoretical and Applied Mechanics* Beijing, China August 19-24
[15] Carpineto N, Lacarbonara W and Vestroni F Mitigation of human-induced vibrations in suspension footbridges via multiple tuned mass dampers 2010 *J. Vib. Control* **16** 749-76
[16] Lacarbonara W and Cetraro M Flutter control of a lifting surface via visco-hysteretic vibration absorbers 2011 *International Journal of Aeronautical and Space Sciences* **12**(4) 331-45
[17] Arena A, Lacarbonara W and Marzocca P Unsteady aerodynamic modeling and flutter analysis of long-span suspension bridges 2012 Proc. of *the ASME 2012 International Design Engineering Technical Conferences & Computers and Information in Engineering Conference* Chicago IL USA August 12-15 IDETC/VIB 2012-70289.
[18] Casalotti A Flutter control of long-span suspension bridges by distributed hysteretic tuned mass dampers 2012 MS Thesis Sapienza University of Rome
[19] Surh J and Koratkar N A Energy dissipation in carbon nanotube composites: a review 2008 *J. Mater. Sci.* 43 4370-82
[20] Formica G and Lacarbonara W On the hysteretic properties of debonding nanostructured materials 2011 Proc. of *XX AIMETA Congress of Theoretical and Applied Mechanics* Bologna Sept 12-15
[21] Taló M, Formica G and Lacarbonara W Modeling of hysteresis due carbon nanotube-matrix interaction in nanocomposites 2012 *First International Conference on Mechanics of Nano, Micro and Macro Composite Structures* Politecnico di Torino June 18-20

Frictional systems under periodic loads — History-dependence, non-uniqueness and energy dissipation

J. R. Barber

Department of Mechanical Engineering, University of Michigan, Ann Arbor, MI 48109-2125, U.S.A.

E-mail: jbarber@umich.edu

Abstract. Nominally static contacts such as bolted or shrink-fit joints typically experience regions of microslip when subjected to oscillatory loading. This results in energy dissipation, reflected as apparent hysteretic damping of the system, and also may cause the initiation of fretting fatigue cracks. Early theoretical studies of the Hertzian contact problem by Cattaneo and Mindlin were confirmed experimentally by Johnson, who identified signs of fretting damage in the slip annulus predicted by the theory.

For many years, tribologists assumed that Melan's theorem in plasticity could be extended to frictional systems — i.e. that if there exists a state of residual stress associated with frictional slip that is sufficient to prevent periodic slip in the steady state, then the system will shake down, regardless of the initial condition. However, we now know that this is true only if there is no coupling between the normal and tangential loading problems, as will be the case notably when contact occurs on a symmetry plane.

For all other cases, periodic loading scenarios can be devised such that shakedown occurs for some initial conditions and not for others. The initial condition here might be determined by the assembly protocol — e.g. the order in which a set of bolts is tightened — or by the exact loading path before the steady cycle is attained. This non-uniqueness of the steady state persists at load amplitudes above the shakedown limit, in which case there is always some dissipation, but the dissipation per cycle (and hence both the effective damping and the susceptibility to fretting damage) depends on the initial conditions. This implies that fretting fatigue experiments need to follow a well-defined assembly protocol if reproducible results are to be obtained. We shall also present results showing that when both normal and tangential forces vary in time, the energy dissipation is very sensitive to the relative phase of the oscillatory components, being greatest when they are out of phase.

With sufficient clamping force, 'complete' contacts (i.e. those in which the contact area is independent of the normal load) can theoretically be prevented from slipping, but on the microscale, all contacts are incomplete because of surface roughness and some microslip is inevitable. In this case, the local energy dissipation density can be estimated from relatively coarse-scale roughness models, based on a solution of the corresponding 'full stick' problem.

1. Introduction

Many engineering structures comprise a number of separate elastic components connected by nominally rigid connections such as bolted joints, shrink fits etc., which rely on friction to prevent relative motion. Typically these connections transmit a substantial normal 'clamping' force that remains constant over time, and also time-varying forces that depend on the operation of the machine or the environmental conditions. The latter are often periodic in nature, and may result (for example) from machine vibration, operating cycles, or diurnal changes in environmental temperature. In this paper, we shall restrict attention to cases in which elastodynamic effects can be neglected and the contact problem regarded as quasi-static. This is not a serious restriction, since even in problems of impact, the quasi-static Hertzian contact theory has been shown to give good results [1].

Contact problems can be characterized as 'complete' or 'incomplete', depending on whether the contact area is determined by the geometry of the contacting components [complete], or whether it varies depending on the applied loads [incomplete]. In the latter case, the normal contact pressure usually tends to zero at the edge of the contact area, and a relatively small tangential force is sufficient to generate a region of 'microslip' — i.e. a region in which the deformability of the materials permits local tangential motion, even though the greater part of the contact area remains stuck. Under periodic loading, the microslip region will experience small amplitude reversed slip and work will be done against the frictional forces. If the Coulomb friction law applies, this will be reflected as apparent non-linear hysteretic (i.e. rate independent) damping in the overall behaviour of the structure [2]. Indeed, it has been estimated that in typical engineering structures, the energy dissipation in microslip at joints far exceeds that due to internal damping in the material [3]. Also, the extremely localized plastic deformation associated with this energy dissipation creates an environment favourable to the initiation and propagation of fretting fatigue failures that are a critical determining factor in the life of aeroengines [4, 5].

By contrast, if the contact is complete, it is theoretically possible to apply a sufficiently large clamping force to prevent microslip under any given tangential loading. However, we note that all real engineering surfaces are rough on the microscale, and hence all contact problems are incomplete in the sense that the 'actual contact area' (i.e. the area in which the surfaces are within the range of interatomic forces) always increases with increasing normal load. Thus, even in nominally complete contacts, we must anticipate some frictional energy loss and potential fretting damage [6].

2. Shakedown and Melan's theorem

The slip that occurs during the first few cycles of loading causes a state of residual stress that generally tends to reduce the amplitude of subsequent periodic slip [7]. This is analogous with the behaviour of elastic-plastic structures under periodic loading, and for many years researchers in contact mechanics assumed that there should exist a frictional equivalent of Melan's theorem [8] which states essentially that if an elastic-plastic structure can shake down (meaning that there exists an allowable state of residual stress that would be sufficient to prevent further plastic deformation under the given periodic loading), then it will do so.

The equivalent frictional theorem can be enunciated as *"If a set of time-independent tangential displacements at the interface can be identified such that the corresponding residual stresses when superposed on the time-varying stresses due to the applied loads cause the interface tractions to satisfy the conditions for frictional stick throughout the contact area at all times, then the system will eventually shake down to a state involving no slip, though not necessarily to the state so identified."* This theorem was eventually proved by Klarbring *et al.*, both for discrete (e.g. finite element) systems [9] and continuous systems [10], but only under the rather restrictive condition that the normal and tangential contact problems should be uncoupled. In other words, the normal contact tractions must be unaffected if a tangential slip is imposed on the system at any part of the contact interface. The basis of both proofs is to define a scalar norm representing the difference between the instantaneous state and that which would occur at the same point in the cycle in the identified 'safe shakedown state'. In the discrete case, this is expressed as a sum over the set of contact nodes, whereas in the continuum case, it is represented by an integral over the contact area. It can then be shown that *if the system is uncoupled,* this norm changes only during periods (and at locations) where there is slip and that any such changes in the norm are necessarily negative by virtue of the conditions defining the Coulomb friction law.

For all other (i.e. coupled) cases, counter-examples to the frictional Melan's theorem can be found, in which there exist loading scenarios for which the system may shake down or not shake down, depending on the initial conditions [9, 10]. A simple heuristic argument for this conclusion is that when the system is coupled, any point that remains stuck throughout the loading cycle could be subjected to a slip displacement sufficiently small to avoid violating the friction law, and if the system is coupled, this may affect the normal tractions at other points sufficiently to permit or inhibit cyclic slip.

3. Behaviour of coupled systems

The frictional Melan theorem shows that we must anticipate history-dependence in the steady-state response of coupled systems. In fact, the above heuristic argument suggests that the theorem is a special case of a more general result (as yet unproven) that the steady-state frictional response to periodic loading in an elastic structure can depend on the initial conditions if and only if there exists coupling between the normal and tangential contact problems [13].

There is some anecdotal evidence for this conjecture, notably:

(i) Fretting fatigue tests are very consistent for smooth 'Hertzian-like' contact geometries, but much more erratic when a rectangular indenter is pressed against a plane surface. The former geometry is reasonably approximated by two half planes, which involves no coupling when the materials are similar, whereas the latter involves significant normal-tangential coupling.

(ii) Experimental measurements of the effective damping of nominally identical bolted joints exhibit significantly different behaviour, and even the same system, if disassembled and then reassembled, can give different results. Depending on the assembly protocol employed, this might be equivalent to a change in initial conditions.

To understand the reasons for this history-dependence, it is helpful to consider a two-dimensional discrete system, which might comprise a finite element solution of a two dimensional frictional elastic contact problem. The internal nodal displacements can be eliminated by standard static reduction procedures, leaving a system with two degrees of freedom (displacement components) at each contact node. The nodal forces (contact reactions) can then be expressed as

$$\begin{aligned} q_j &= q_j^w + A_{ji}v_i + B_{ij}w_i \\ p_j &= p_j^w + B_{ji}v_i + C_{ji}w_i \ , \end{aligned} \tag{1}$$

where p_j, q_j are the normal and tangential reactions respectively at node j, v_i, w_i are the tangential (slip) and normal (separation) displacements at node i, p_j^w, q_j^w are the nodal reactions that would be obtained if all the nodes were constrained in the position $v_i = w_i = 0$, and the summation convention is implied. Notice that the matrices $\boldsymbol{A}, \boldsymbol{B}, \boldsymbol{C}$ represent a partitioning of the stiffness matrix reduced to the contact nodes and hence $\boldsymbol{A}$ and $\boldsymbol{C}$ must be positive definite, but no such restriction applies to $\boldsymbol{B}$.

We consider the case where the time-varying external loads $\boldsymbol{p}^w(t), \boldsymbol{q}^w(t)$ can be written in the form

$$\boldsymbol{p}^w(t) = \boldsymbol{p}_0 + \lambda \boldsymbol{p}_1(t) \ ; \quad \boldsymbol{q}^w(t) = \boldsymbol{q}_0 + \lambda \boldsymbol{q}_1(t) \ , \tag{2}$$

where $\boldsymbol{p}_0, \boldsymbol{q}_0$ represent the time-dindependent mean loads, $\boldsymbol{p}_1(t), \boldsymbol{q}_1(t)$ are periodic vectors with mean value zero and λ is a non-negative scalar load factor characterizing the magnitude of the periodic loading.

In the simple case where there are only two contact nodes and they are both in contact, $w_1 = w_2 = 0$ and the Coulomb friction inequalities $|q_j| \leq fp_i$ with equations (1) demand that

$$\begin{array}{rcll}
(A_{11} - fB_{11})v_1 + (A_{12} - fB_{12})v_2 & \leq & fp_1^w - q_1^w & \text{I} \\
(A_{11} + fB_{11})v_1 + (A_{12} + fB_{12})v_2 & \geq & -fp_1^w - q_1^w & \text{II} \\
(A_{21} - fB_{21})v_1 + (A_{22} - fB_{22})v_2 & \leq & fp_2^w - q_2^w & \text{III} \\
(A_{21} + fB_{21})v_1 + (A_{22} + fB_{22})v_2 & \geq & -fp_2^w - q_2^w \;, & \text{IV}
\end{array} \qquad (3)$$

where f is the coefficient of friction and the equalities in these relations control the motions

$$\text{I:} \quad \dot{v}_1 < 0\,; \quad \text{II:} \quad \dot{v}_1 > 0\,; \quad \text{III:} \quad \dot{v}_2 < 0\,; \quad \text{IV:} \quad \dot{v}_2 > 0\,, \qquad (4)$$

with the dot indicating the derivative with respect to time. We restrict attention to cases where the coefficient of friction is sufficiently small for the rate problem so defined to be well-posed. The limiting coefficient imposed by this condition can be determined from Klarbing's P-matrix condition [11]. For higher coefficients of friction, conditions can be reached where the system becomes wedged, or where sudden elastodynamic state changes are preciptated even when the loading rate is extremely small [12].

Each of the constraints (3) excludes the region on one side of a certain straight line in v_i-space, suggesting that tracking the motion of the instantaneous state in this space might be a useful way of characterizing the evolution of the system [14]. If the nodes are to remain in contact, there must at all times t exist a region that is not excluded by any of the constraints. Furthermore, as the external loads p_i^w, q_i^w vary in time, the constraint lines (3) move whilst retaining the same slope, and hence cause the instantaneous state P to move either vertically or horizontally in accordance with (4), as illustrated in Figure 1.

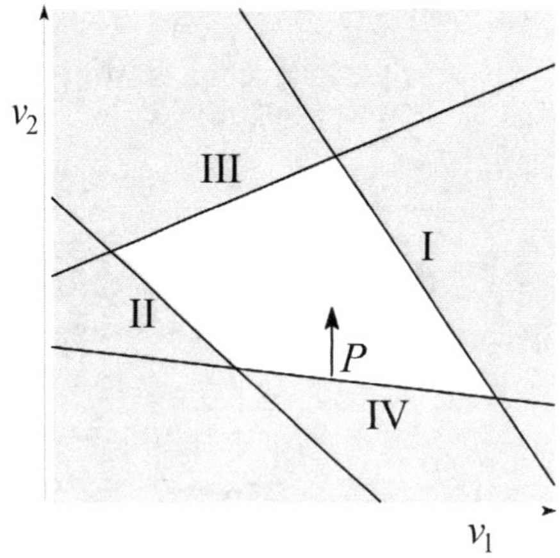

Figure 1: Motion of the instantaneous state P due to the advance of constraint IV ($\dot{v}_2 > 0$).

If shakedown is to be possible there must exist a safe shakedown region that is not excluded by any of the constraints *at any time t in the periodic loading cycle.* Ahn *et al.* [14] showed that the system would always shake down if the safe shakedown region so defined was a quadrilateral, but that the occurrence of shakedown depends on the initial conditions if it is a triangle.

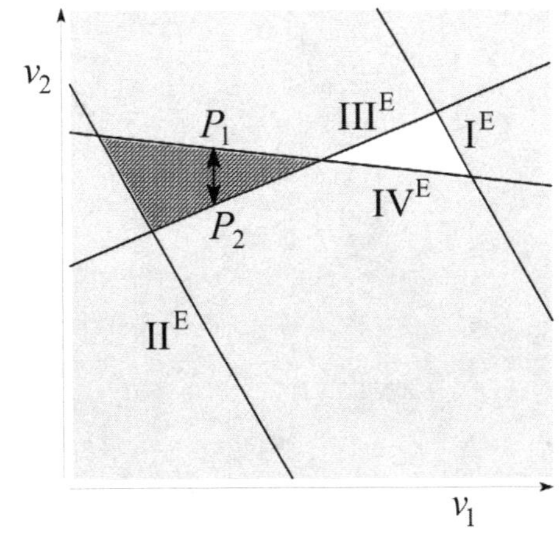

Figure 2: Cyclic slip limit cycle in the case where the safe shakedown region is triangular..

The reason for this is apparent from Figure 2, which shows the extreme positions I^{E}, IIE, $\mathrm{III}^{\mathrm{E}}$, IVE of the four constraints — i.e. the positions that exclude the maximum amount of space in the diagram. These extreme positions are not generally reached at the same time, in which case it is possible for the system to oscillate along the line P_1P_2 implying cyclic slip at node 2, because the constraint II never moves far enough to the right to push P into the safe (unshaded) triangle. On the other hand, if we start from a position in the safe triangle, the constraints will never push us out of it and shakedown will be assured. This criterion is easily extended to the N-node system and is equivalent to the requirement that if all $2N$ extremal frictional constraints are active in defining the safe region in v_i-space, then the safe region is always reached. If one constraint or more is inactive in defining this region, then the steady state depends on the initial conditions and may comprise shakedown or cyclic slip with the node(s) associated with the inactive constraint(s) permanently stuck.

3.1. System memory

For the steady-state behaviour under periodic loading to depend on the initial conditions, the system must in some sense posess a 'memory' of these initial conditions. Now at any node that is instantaneously separated or slipping, we have two equations to determine the degrees of freedom v_i, w_i, namely $p_i = q_i = 0$ for separation and $w_i = 0, q_i = \pm f p_i$ for slip (the sign taken depending on the direction of slip). Thus, if an N-node system experiences a period when all nodes are simultaneously either slipping or separated, there will be $2N$ equations for $2N$ unknowns and the state is uniquely determined by the *instantaneous* values of $p_i^w(t), q_i^w(t)$. In this case, the subsequent evolution of the system cannot depend on the initial conditions. By contrast, if a node is instantaneously stuck, we have only one equation, $w_i = 0$, and the friction *inequality* $-fp_i < q_i < fp_i$. It follows that at any given time, the system memory must reside in the slip displacements v_i at those nodes that are instantaneously stuck.

We conclude that if all the nodes slip *at some time during the cycle,* but not all at the same time, the memory must somehow be exchanged between nodes during each cycle and we would expect this to lead to a degradation of memory and hence the attraction of the system to a unique steady state. Behaviour of this kind is illustrated in Figure 3, where the constraints are assumed to advance and recede in the sequence I, III, II, IV. Starting from two distinct points A, B, we find that the trajectory is attracted to a unique rectangular orbit.

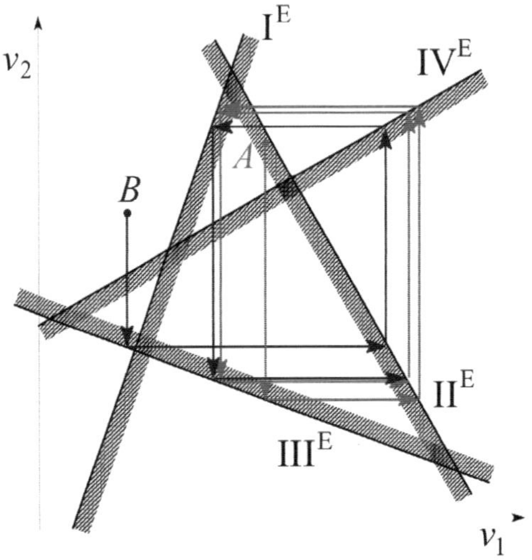

Figure 3: Attraction of a coupled system to a unique periodic steady state.

This issue has been explored for the two-node system by Andersson [15], who shows that for more complex loading patterns, conditions can arise where a steady periodic state is *unstable,* so that small perturbations from it grow with successive cycles until the sequence of stick/slip/separation states is altered, at which point we anticipate a qualitatively distinct stable (attractive) steady state. In some many cases, two such attractive orbits exist and the state eventually reached depends on which side of the unstable orbit the initial condition is located. However, in other cases, the partition of the initial condition space into catchment areas for the two stable orbits becomes fractal in the neighbourhood of the unstable orbit.

3.2. Effect of load factor

Consider now a system where the loading is defined by equation (2). If the scalar load factor λ is sufficiently small, the system will shake down for all initial conditions and we can identify a critical value of $\lambda = \lambda_1$ at which one of the extreme constraints ceases to be active and a larger value (λ_2) at which the safe shakedown region becomes null. In the range $\lambda_1 < \lambda < \lambda_2$ the system may or may not shake down depending on the initial conditions. For $\lambda > \lambda_2$, shakedown is impossible, but the steady state (and hence the frictional dissipation per cycle) still exhibits some dependence on initial conditions as long as a subset of nodes remains permanently stuck. However, at some even larger value λ_3, we anticipate reaching a condition where all the nodes slip at least once during the cycle, so the system is attracted asymptotically to a unique steady state. Above a still larger value λ_4 there will exist at least one point in the cycle at which all the nodes either slip or separate, in which case the unique steady state is reached in the first one or two cycles. This kind of behaviour under gradually increasing periodic loading was documented by Jang and Barber [16] for a system comprising a two-dimensional elastic body containing a set of plane cracks. Of course, for systems that are nominally stuck, such as a bolted joint, there will usually be a permanently stuck region, implying that $\lambda < \lambda_3$.

4. Uncoupled systems

Conditions are considerably simpler if the system is uncoupled, and the most practically useful class of systems for which this condition is satisfied comprise those for which the two contacting bodies can be approximated by half spaces of similar materials [17].

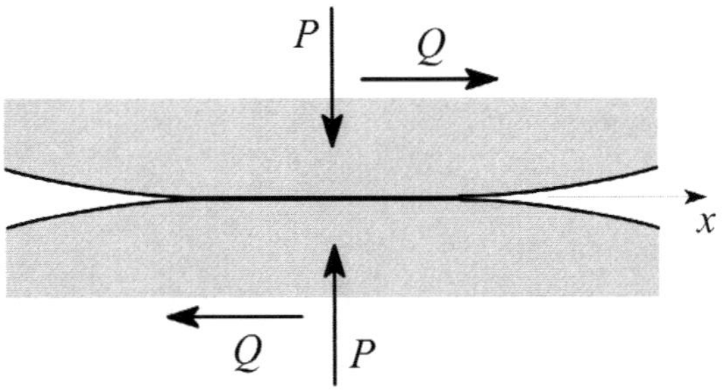

Figure 4: The Cattaneo-Mindlin problem.

4.1. Cattaneo and Mindlin's problem

The earliest theoretical study of microslip in a nominally static contact is that due to Cattaneo [18], who considered the problem of the Hertzian contact of Figure 4, comprising two quadratic surfaces loaded first by a normal compressive force P, which is then held constant whilst a tangential force Q is applied.

This loading scenario is illustrated in Figure 5, where we also identify the limiting lines $Q = \pm fP$ beyond which gross slip (sliding) would occur. Cattaneo showed that the tangential contact tractions $q(x, y)$ can be written as the superposition

$$q(x, y) = fp(x, y) - fp_S(x, y) \; , \tag{5}$$

where $p(x, y)$ are the corresponding (Hertzian) normal tractions and $p_S(x, y)$ are the normal tractions at a smaller fictitious load P_S defined by

$$P_S = P - \frac{Q}{f} \; . \tag{6}$$

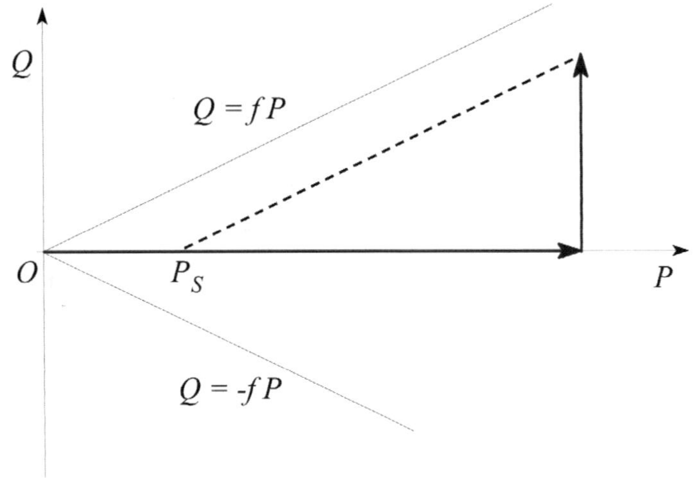

Figure 5: Load path for the Cattaneo-Mindlin problem.

Figure 5 shows a geometrical construction from which the load P_S can be determined. We also note that if $\mathcal{A}(P)$ denotes the contact area at a normal load P, microslip will occur in the region $\mathcal{A}(P) - \mathcal{A}(P_S)$ and stick in the region $\mathcal{A}(P_S)$.

It should be noted that this form of superposition was to some extent prefigured in a much earlier solution by Carter [19] to the related problem of a cylinder rolling on a plane. Cattaneo's result was 'rediscovered' apparently independently by Mindlin [20], who also considered the case where the tangential force varies periodically in time [21, 22]. For the contact of two spheres, they predicted that microslip would occur in an annulus at the edge of the circular contact area, a result that was verified experimentally by Johnson [23] by observing the fretting damage to the surfaces in this region after a period of oscillation.

4.2. The Ciavarella-Jäger theorem

The superposition defined by equations (5,6) applies to a much broader class of contact problems, as was shown, again apparently independently, by Ciavarella [24, 25] and Jäger [26]. If the loading is as shown in Figure 5, it applies exactly to all two-dimensional contact problems between similar materials for which the bodies may be approximated by half planes, and in an approximate sense to corresponding three-dimensional problems. The approximation involved in the latter case is that the slip directions predicted deviate somewhat from the direction of the local frictional traction, particularly near the stick-slip boundary. This approximation is also inherent in the Cattaneo and Mindlin solutions and its magnitude was assessed by Munisamy *et al.* [27] and found to be extremely small. We shall see that Ciavarella and Jäger's result is extremely useful in estimating the frictional energy dissipation in fairly general contacts, including those involving rough surfaces.

4.3. Periodic loading

The simplest form of periodic loading for the system of Figure 4 can be written

$$P(t) = P_0 + P_1 \cos(\omega t) \; ; \quad Q(t) = Q_0 + Q_1 \cos(\omega t - \phi) \; ,$$

which comprises a constant mean load (P_0, Q_0) and a sinusoidal periodic load in which the normal and tangential components (P_1, Q_1) may not necessarily be in phase. The resulting trajectory would then be elliptical as shown in Figure 6. The periodic cycle must be reached by some preliminary loading phase, which we here identify by the segment OA. In other words, the contact must be 'assembled' in some way.

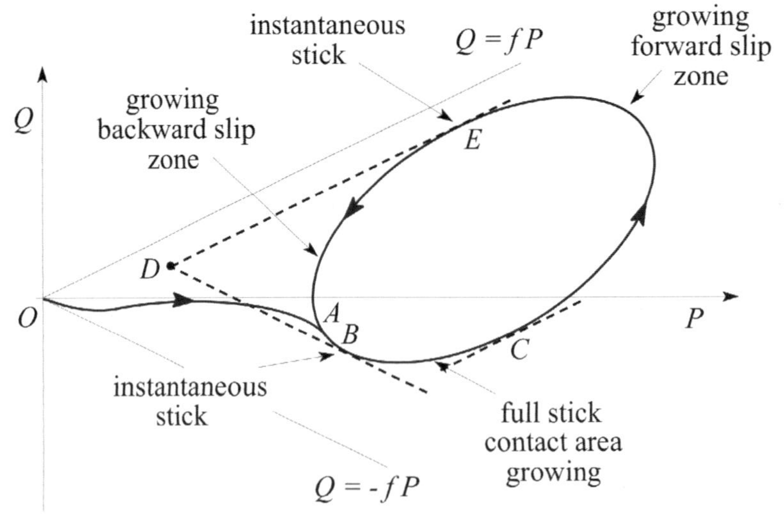

Figure 6: Periodic loading path.

The evolution of the traction distribution can be tracked by considering the incremental problem in which the normal and tangential loads $P(t), Q(t)$ increase by $\Delta P, \Delta Q$ during the small time increment Δt. If the (compressive) normal load is increasing $(dP/dt > 0)$, and if $|dQ/dP| < f$, this incremental problem involves complete stick, so there is no microslip during the phase BC in Figure 6. Once we pass C, a microslip zone begins to develop at the edge of the contact area and elementary calculations [28] show that the change in the traction distribution can be described by a superposition similar to that of Ciavarella and Jäger.

This state continues until we reach the point E, where the derivative dQ/dP is again of magnitude f, but the normal load is decreasing ($dP/dt < 0$). At this point the contact sticks everywhere instantaneously and a reverse microslip zone starts to grow from the edges of the contact area. As before, the resulting incremental tractions can be defined by superposed Ciavarella-Jäger distributions. One result of some interest is that the minimum extent of the stick region occurs just before the points E, B are reached and is equal to the contact area at a reduced load P_D, where the point D is identified by the geometrical construction shown in Figure 6. The tractions in the permanently stuck region $\mathcal{A}(P_D)$ depend on the initial load path OA, and this constitutes the memory of the system. However, these stored tractions do not influence those in the rest of the contact area, and hence they do not affect the frictional energy dissipation or the damage due to fretting, which are therefore independent of the initial assembly protocol.

4.4. In-phase loading

Figure 6 is presented for the case where the tangential and normal loads are out of phase. In the special case where the phase lag $\phi = 0$, the ellipse will condense to a straight line which is traversed back and forth during the cycle. The behaviour then depends critically on the magnitude of the ratio Q_1/fP_1. If $Q_1/fP_1 < 1$, the entire loading phase ($dP/dt > 0$) occurs in a state of stick. Furthermore, during unloading ($dP/dt < 0$) the system passes through the same states as during loading, but in reverse order, so there is no tendency for microslip and no energy is dissipated at any point in the cycle. By contrast, if $Q_1/fP_1 > 1$, microslip occurs during both the loading and unloading phases and the tractions are given by Ciavarella-Jäger distributions, with the entire contact area becoming instantaneously stuck at the two extreme points.

4.5. Frictional energy dissipation

The frictional energy dissipation during the loading cycle can be calculated either in a local sense, using the instantaneous local tractions and slip displacements, or in a global sense, using the applied tangential force $Q(t)$ and the corresponding relative rigid-body displacement of the contacting bodies. In the former case, for two-dimensional geometries, we can obtain the dissipation *per unit area* as

$$W(x) = \oint q(x,t)\dot{v}(x,t)dt \ ,$$

where $v(x,t)$ is the local slip displacement, and the integral is performed around one complete loading cycle ($t = 0..2\pi/\omega$). The distribution of $W(x)$ through the contact area provides a good measure of the severity of fretting and its maximum is likely to correlate with the location of the first fretting fatigue cracks. For a Hertzian geometry, this maximum is found to be about half way between the edge of the contact area and the edge of the permanent stick zone [29]. The total dissipation per cycle $\mathcal{W}$ can then be obtained by integrating $W(x)$ over the maximum extent $\mathcal{A}$ of the interface — i.e.

$$\mathcal{W} = \int_{\mathcal{A}} W(x)dx \ .$$

However, a more direct approach is to evaluate the integral

$$\mathcal{W} = \oint Q(t)\dot{V}(t)dt \ , \tag{7}$$

where V represents the relative tangential rigid-body displacement of the two contacting bodies. The integrand in this case represents the incremental work done by the load $Q(t)$ during a time increment dt, but not all this work constitutes frictional dissipation — there is also a contribution to the elastic strain energy of the bodies. However, if the integral is performed over a complete cycle, the strain energy at the two limits will be identical and the integral will therefore represent the frictional dissipation per cycle.

The integral (7) can also be written

$$\mathcal{W} = \oint Q(t)\frac{dV}{dQ}dQ \; , \tag{8}$$

and hence can be evaluated if we can determine the *incremental tangential compliance* dV/dQ. Also, since the traction distribution corresponding to dQ is at all points in the cycle the derivative of one or more expressions of the form of equation (5), dV/dQ can be written down by a similar superposition associated with the contact areas $\mathcal{A}(P)$ and $\mathcal{A}(P_S)$. In the two dimensional problem, it is necessary to define a finite reference point in the elastic bodies to avoid mathematical problems associated with logarithmically unbounded displacements at infinity, but the choice of this reference point does not affect the final result and the compliances are found to be identical with the corresponding normal compliances dU/dP, where U is the rigid-body approach. Thus, the integral (8) can be determined provided we know the relation between normal force P and the corresponding normal approach for the particular profiles of the contacting bodies. In three-dimensional problems, this equivalence is not exact except for certain contact geometries such as the ellipse, but the ratio between normal and tangential compliance is quite tightly bounded, and for a wide range of three-dimensional contact geometries is closely approximated by

$$\frac{dV}{dQ} = \lambda\frac{dU}{dP} \quad \text{where} \quad \lambda = \frac{(2-\nu)}{2(1-\nu)} \tag{9}$$

and ν is Poisson's ratio [30, 31].

4.6. Effect of surface roughness

We have already remarked that most engineering surfaces are rough and hence the contact problem is incomplete, even in cases where the large scale geometry makes it appear complete. Thus, for example, if two coextensive rectangular blocks are pressed together, the contact area $\mathcal{A}(P)$ will comprise a set of 'actual contact areas' at the peaks of the asperities on the two surfaces, and these contact areas will grow with increasing normal load. This implies that the stick zones in the corresponding Ciavarella-Jäger distributions are defined by the set of actual contact areas $\mathcal{A}(P_S)$ at the fictitious load P_S. It also implies that there will be dissipation and the potential for fretting damage even in contacts where the solution of the 'smooth' contact problem predicts complete stick.

Roughness wavelengths are typically short compared with the dimensions of the nominal contact area, so a convenient way of incorporating the effect of roughness is first to solve the equivalent smooth problem, find the spatial distribution of the mean and alternating tractions $p(x,y,t), q(x,y,t)$, and then approximate the local dissipation per unit area due to roughness as being that obtained in a nominally uniform case where two plane surfaces with the given roughness are subjected to the same nominal traction cycle.

Many theories have been advanced to describe the contact of rough surfaces, the most widely used being those of Greenwood and Williamson [32] and Persson [33]. Though approaching the problem in very different ways, both authors conclude that the relation between macroscale quantities such as the normal force P and the incremental stiffness dP/dU is approximately linear because the population of microscale contact events retains approximately the same distribution over a wide range of load levels. As P is increased, the number of actual contact areas increases. Small areas get larger, but a sufficient number of new smaller areas are established to preserve an approximately similar size distribution. Using Persson's theory, Akarapu *et al.* [34] argue that

$$\frac{dP}{dU} = \frac{P}{\gamma\sigma} , \tag{10}$$

where σ is the RMS roughness height and γ is a numerical constant of order unity. More generally, the incremental normal stiffness dP/dU due to roughness can be bounded, based only on the peak-to-valley height [35], and the bounding procedure implies that a relatively coarse numerical model of the surface is sufficient to give good estimates. In other words, only the long wavelength terms in the surface profile spectrum make significant contributions to the incremental stiffness.

Using (9, 10), we then have

$$\frac{dV}{dQ} = \frac{\lambda\gamma\sigma}{P}$$

and the integral in (8) can be performed numerically for the Ciavarella-Jäger distributions developed during the load cycle of Figure 6. Putigniano *et al.* [36] showed that the resulting expression can be presented dimensionlessly in the form

$$\hat{\mathcal{W}} \equiv \frac{\mathcal{W}}{\lambda\gamma\sigma f^2 P_0} = f\left(\hat{P}_1, \hat{Q}_1, \phi\right) \quad \text{where} \quad \hat{P}_1 = \frac{P_1}{P_0} ; \quad \hat{Q}_1 = \frac{Q_1}{fP_0} ,$$

and for $\hat{Q}_1 < 0.5$ they found that $\hat{\mathcal{W}} \sim \hat{Q}_1^3$. Notice that if the mean force P_0 is replaced by the mean normal traction p_0, the corresponding expression for $\mathcal{W}$ will then define the dissipation per unit area of the nominal contact interface. The expressions $\hat{P}_1, \hat{Q}_1$ are already normalized and will remain unchanged if forces are replaced by tractions.

We recall that for in-phase loading the dissipation is zero for $Q_1/fP_1 < 1$, which here corresponds to $\hat{Q}_1 < \hat{P}_1$. For larger values of $\hat{Q}_1$, dissipation occurs for all phase angles, but is significantly larger when $\phi = \pm\pi/2$, as shown in Figure 7.

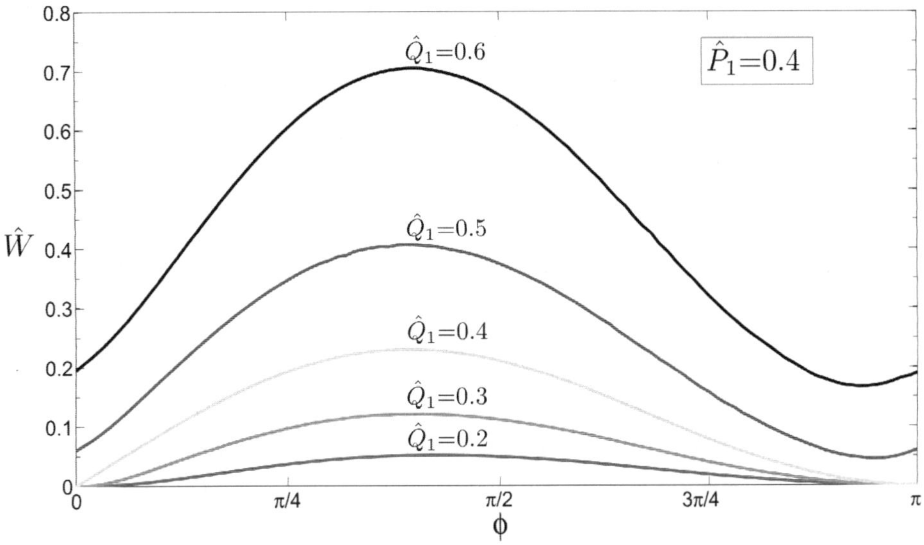

Fig.7. Dissipation as a function of relative phase ϕ for $\hat{P}_1 = 0.4$ and various values of $\hat{Q}_1$.

5. Conclusions

The behaviour of elastic systems with frictional interfaces is qualitatively influenced by elastic coupling between the normal and tangential tractions and displacements. When coupling is present, the behaviour under cyclic loading, including frictional dissipation and susceptibility to fretting damage, will generally depend on the initial conditions, which might include the intial loading or assembly protocol. By contrast, for uncoupled systems, including the contact of bodies of similar materials that can be modelled as half spaces, the steady-state response is independent of initial conditions, and simple procedures, based on a theorem due to Ciavarella and Jäger, can be used to calculate the dissipation. In particular, it is possible to estimate the dissipation associated with surface roughness in nominally tight joints, and this is found to depend only on the long wavelength components in the surface roughness.

References

[1] Johnson, K L 1985 *Contact Mechanics* Cambridge University Press, Cambridge
[2] Wentzel H 2006 Modelling of frictional joints in dynamically loaded structures: a review Technical report KTH Solid mechanics Royal Institute of Technology (2006) http://www-old.hallf.kth.se/forskning/publikationer/rapport_419.pdf
[3] Gaul L and Lenz J 1997 Nonlinear dynamics of structures assembled by bolted joints *Acta Mechanica* **125** 169–181 doi:101007/BF01177306
[4] Murthy H Harish G and Farris T N 2004 Efficient modeling of fretting of blade/disk contacts including load history effects *ASME Journal of Tribology* **126** 56–64
[5] Nowell D Dini D and Hills D A 2006 Recent developments in the understanding of fretting fatigue *Engineering Fracture Mechanics* **73** 207–222
[6] Björkland S 1997 A random model for micro-slip between nominally flat surfaces *ASME Journal of Tribology* **119** 726–732
[7] Churchman C M and Hills D A 2006 General results for complete contacts subject to oscillatory shear *Journal of the Mechanics and Physics of Solids* **54** 1186–1205
[8] Melan E 1936 Theorie statisch unbestimmter Systeme aus ideal-plastichem Baustoff *Sitzungsber d Akad d Wiss Wien* **2A(145)** 195–218
[9] Klarbring A Ciavarella M and Barber J R 2007 Shakedown in elastic contact problems with Coulomb friction *International Journal of Solids and Structures* **44** 8355–65
[10] Barber J R Klarbring A and Ciavarella M 2008 Shakedown in frictional contact problems for the continuum *Comptes Rendus Mecanique* **336** 34–41.
[11] Klarbring A 1999 Contact friction discrete mechanical structures and mathematical programming In: Wriggers P Panagiotopoulos P (Eds) *New Developments in Contact Problems* Springer 55–100
[12] Ahn Y J 2010 Discontinuity of quasi-static solution in the two-node Coulomb frictional system *International Journal of Solids and Structures* **47** 2866–71
[13] Barber J R 2011 Frictional systems subjected to oscillating loads *Annals of Solid and Structural Mechanics* **2** 45–55
[14] Ahn Y J Bertocchi E and Barber J R 2008 Shakedown of coupled two-dimensional discrete frictional systems *Journal of the Mechanics and Physics of Solids* **56** 3433– 40
[15] Andersson L-E 2012 Private communication
[16] Jang Y H and Barber J R 2011 Frictional energy dissipation in materials containing cracks *Journal of the Mechanics and Physics of Solids* **59** 583–594
[17] Hills D A Nowell D and Sackfield A 1993 *Mechanics of elastic contact* Butterworth-Heinemann Oxford
[18] Cattaneo C 1938 Sul contatto di due corpi elastici: distribuzione locale degli sforzi *Rendiconti dell'Accademia Nazionale dei Lincei* **27** 342–348 434–436 474–478 (In Italian)
[19] Carter F C 1926 On the action of a locomotive driving wheel *Proceedings of the Royal Society of London* **A112** 151–157
[20] Mindlin R D 1949 Compliance of elastic bodies in contact *ASME Journal of Applied Mechanics* **16** 259–268
[21] Mindlin R D Mason W P Osmer T F and Deresiewicz H 1952 Effects of an oscillating tangential force on the contact surfaces of elastic spheres Proc 1st US Nat Cong Appl Mech 1951 203–208
[22] Mindlin R D and Deresiewicz H 1953 Elastic spheres in contact under varying oblique forces *ASME Journal of Applied Mechanics* **75** 327–344
[23] Johnson K L 1955 Surface interaction between elastically loaded bodies under tangential forces *Proceedings of the Royal Society of London* **A230** 531–548

[24] Ciavarella M 1998a The generalized Cattaneo partial slip plane contact problem I-Theory II-Examples *International Journal of Solids and Structures* **35** 2349–78

[25] Ciavarella M 1998b Tangential loading of general three-dimensional contacts *ASME Journal of Applied Mechanics* **65** 998–1003

[26] Jäger J 1998 A new principle in contact mechanics *ASME Journal of Tribology* **120** 677–684

[27] Munisamy R L Hills D A and Nowell D 1994 Static axisymmetrical Hertzian contacts subject to shearing forces *ASME Journal of Applied Mechanics* ASME JAppl Mech **61** 278–283

[28] Barber J R Davies M and Hills D A 2011 Frictional elastic contact with periodic loading *International Journal of Solids and Structures* **48** 2041–47

[29] Davies M Barber J R and Hills D A 2012 Energy dissipation in a frictional incomplete contact with varying normal load *International Journal of Mechanical Sciences* in press

[30] Sevostianov I and Kachanov M 2008 Normal and tangential compliances of interface of rough surfaces with contacts of elliptic shape *International Journal of Solids and Structures* **45** 2723–2736

[31] Campañá C Persson B N J and Müser M H 2011 Transverse and normal interfacial stiffness of solids with randomly rough surfaces Journal of Physics: Condensed Matter **23** 085001 [DOI: 10.1088/0953-8984/23/8/085001].

[32] Greenwood J A and Williamson J B P 1966 The contact of nominally flat surfaces *Proceedings of the Royal Society of London* **A295** 300–319

[33] Persson B N J 2001 Theory of rubber friction and contact mechanics *Journal of Chemical Physics* **115** 3840–61

[34] Akarapu S Sharp T and Robbins M O 2011 Stiffness of contacts between rough surfaces *Physics Review Letters* **106** Art 204301

[35] Barber J R 2003 Bounds on the electrical resistance between contacting elastic rough bodies *Proceedings of the Royal Society (London)* **459** 53–66

[36] Putignano C Ciavarella M and Barber J R 2011 Frictional energy dissipation in contact of nominally flat rough surfaces under harmonically varying loads *Journal of the Mechanics and Physics of Solids* **59** 2442–54

Journal of Physics: Conference Series **382** (2012) 012003 doi:10.1088/1742-6596/382/1/012003

Structural health monitoring by means of elastic wave propagation

Ostachowicz Wiesław[1,2]*, **Radzieński Maciej**[1,3]

[1] Academy of Sciences, Gdańsk, Poland

[2] Gdynia Maritime University, Faculty of Navigation, Gdynia, Poland

[3] Gdansk University of Technology, Faculty of Electrical and Control Engineering, Gdansk, Poland

*wieslaw@imp.gda.pl

Abstract. This paper presents numerical and experimental approach for damage detection in structures. Presented methods are based on the phenomenon of elastic wave propagation. The results are provided for isotropic and anisotropic plates with damage in form of fixed mass or notch cut.

1. Introduction

The scope of Structural Health Monitoring (SHM) includes constant monitoring of the structure's material condition (in real–time), for the elements of the structure as well as for the whole structure during its useful lifetime. The main goal of the SHM technique is to conduct research aimed at developing advanced non–destructive diagnostic methods of assessing integrity of diverse composite structures. Determining location, size and type of damage together with estimating the remaining operational lifespan of investigated structures is very important for assessing technical condition of many safety–critical machines and devices, as well as of means of transport (e.g. aircraft, helicopters, land and sea engineering structures).

Developing such methods and then applying them during construction and operation of composite structures allows for evaluating degradation, and therefore remaining life, of such structures. Application of the SHM methods is doubtlessly enable avoiding dangerous situations, and even more importantly, serious catastrophic scenarios.

Among various techniques available, a health monitoring system based on piezoelectric transducers and Lamb wave propagation seems to be a promising method for inspection of metallic and composite plate-like structures.

Guided waves induced by piezoelectric transducers are extensively used for damage detection purpose. A numerical model based on time–domain spectral element method has been developed to simulate elastic wave propagation in metallic and composite structures induced by the piezoelectric transducers. Moreover, numerical models constitute the foundation for developing perfect tools for designing and verifying new concepts of signal processing and testing methodology. Visualisations of elastic waves excited by the piezoelectric actuator in selected structures have been performed. Apart from visualisation of propagating waves also the interaction of guided waves with various types of damage have been investigated.

Wide range of numerical simulations and experimental validations of analysed structures provide helpful information about dispersion, mode conversion, thermo–mechanical processes and wave scattering from stiffeners and boundaries. It can allow one to optimise excitation signal parameters and sensor placement, as well as enable analysis of signals reflected from damage.

2. Lamb Waves

Lamb waves are specific type of elastic waves that propagate in solid media bounded by two parallel surfaces. They were discovered and formulated by Horace Lamb in 1917 [1]. In non-destructive testing or SHM those waves are often called guided Lamb waves or just guided waves, because of finite dimensions of a real test structure.

Two forms of Lamb waves exist in infinite solutions: symmetric (S0,S1....) and anti-symmetric (A0,A1,...) which depends on the product of excited frequency and plate thickness. For an aluminium alloy structure only two basic modes coexist up to 1.8 MHz·mm which is presented by the means of dispersion curves in figure 1 and figure 2. Those fundamental modes are found to be used in damage detection methods.

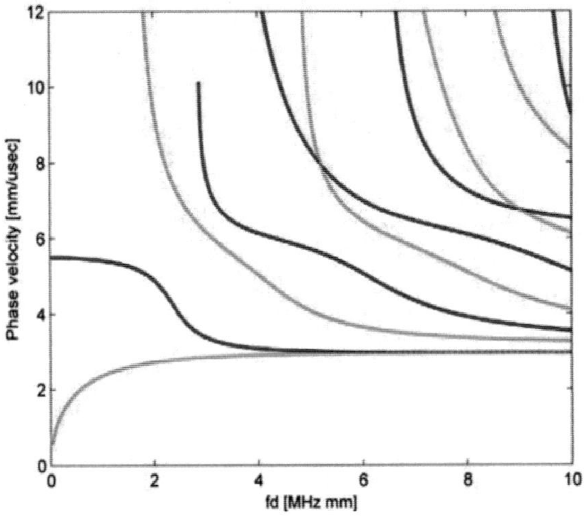

Figure 1: Phase velocities of symmetric (blue) and anti-symmetric (green) modes of Lamb waves.

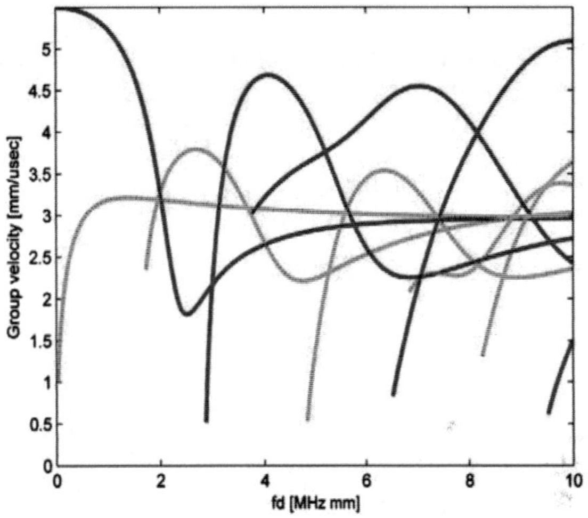

Figure 2: Group velocities of symmetric (blue) and anti-symmetric (green) modes of Lamb waves.

In comparison to isotropic plates, dispersive relations in composite material are more complex. The solution must satisfy Christofel's equation for each layer, the continuity condition at the interfaces and the traction–free boundary conditions at the plate surfaces [2]. Base on Mindlin's plate theory approximate solution can be found [3]. This approach shows that group velocity of the transverse wave (which approximates A0 Lamb wave mode) is a function of a direction of propagation and relative volume fraction of fibres. Some examples for GFRC and GFRC are presented in figure 3, where green line stands for 0% and 100% volume fraction of fibres and blue line stands for 20, 40, 60, 80% volume fraction of fibres composite material.

Journal of Physics: Conference Series **382** (2012) 012003 doi:10.1088/1742-6596/382/1/012003

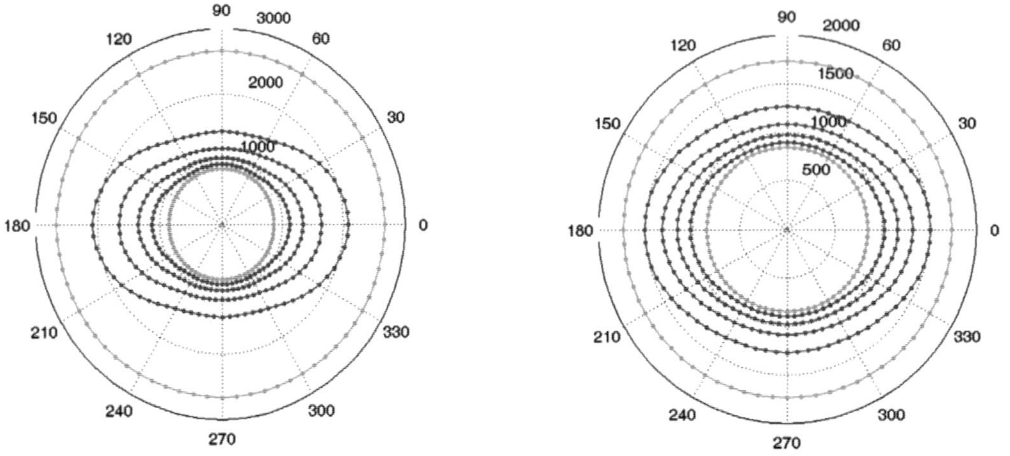

Figure 3: Group velocities surfaces a) glass-epoxy composite, b) graphite epoxy composite.

3. Wave propagation modelling

Modelling guided waves propagation in real structures is challenging task. An efficient method was proposed by Doyle [4] in form of FFT-based Spectral Element Method. However, it can not be applied for 3D geometry. Both Finite Element Method and Finite Difference Method may be used for complex geometies [5] but they are not efficent enough for elastic wave propagation phenomenon modelling.

Spectral Element Method developed by IFFM research group overcame those problems [3, 6]. Additionally authors propose spectral elements for electromechanical coupling [7] which is important in case of piezoelectric transducer modelling. Some exemplary results for guided wave propagation in 2 mm thick aluminium alloy halfpipe are presented in figure 4.

Figure 4: Guided wave propagation in halfpipe structure at three consecutive time steps.

4. Damage imaging techniques

Due to different approaches to guide wave registration two main groups of damage detection methods may be defined. First one of them is based on number of sensors mounted on the specimen surface to excite and/or register time signals. Second group utilizes full wavefield of propagating guided waves which may be obtained by shearography or Doppler vibrometry.

4.1. Fixed sensors

Registered signals by the piezoelectric transducer may be used for damage localization by various inverse techniques. However, this approach is very computationally intense. Utilizing information about wave propagation velocity and the position of the sensors, direct methods may be applied in order to create damage maps. Among direct damage detection method two main approaches may be distinguished: pulse-echo or pitch-catch. The former is based on assumption that discontinuity of a structure is causing elastic wave reflections. Additional wave packages in registered signal allow one to determine the distance between excitation, damage and sensor. The latter group utilized information about direct patch of elastic wave between excitation and receivers. Any changes due to damage within amplitude, velocity, type of mode may be recognized as a material discontinuity [8, 9, 10].

4.2. Full wavefield

Another approach for elastic wave-based damage localization utilizes non-contact measurements of full wavefield, which is possible to register using shearography or scanning Doppler vibrometry. Due to information about full wavefield of propagating guided waves new damage detection possibilities arise. One of the first damage detection method incorporating full wavefield obtained by Doppler vibrometer was presented by Ruzzene et al. [11]. This approach was based on visualization of the energy of propagating wave distribution over the measured surface. Any local change in energy maps may be recognized as geometry feature of specimen or damage.

Radzieński et. al [12] developed RMS method by normalizing amplitude of propagating waves spread on surface which facilitate analysis of created damage maps and was called weighted RMS. In paper [13] Ruzzene proposed utilizing 3D Fourier Transform for filtering incident waves leaving only reflections from potential damages. The calculated energy of filtered signal should have the biggest values in the position of damage.

The studies presented in this paper covers comparison between three of those methods (RMS, weighted RMS, incident wave filtering) as well as new method developed by the authors. Proposed approach utilizes 2D Fourier Transform for wave image filtering. Spectral pattern of propagating guided waves in specimen is determined by averaging first few time snapshots in wavenumber domain. By taking advantage of this information it is possible to define specific filter which is successively used for eliminating main component of propagating guided waves from each registered time frame. In results the filtered signals contains mostly information about changes in wave propagation pattern with its positions. Map of distribution of energy of this filtered signal in expected to have highest values in position where propagating guided waves are affected by material discontinuity.

5. Results

First tested structure was 1 x 1 x 0.002 m aluminum alloy plate. One quarter of this specimen was considered in numerical and experimental researches, as indicated by a grey rectangle at figure 5. Simulation of damage in form of fixed 1 g mass was introduced. Exemplary results of abovementioned methods for both numerical and experimental studies are presented in figure 7 and figure 8 respectively.

Second specimen used in investigations was 0.44 x 0.44 x 0.0032 m glass-epoxy composite plate composed of 4 layers of glass fibre mats with vertically oriented fibres. In this specimen damage was introduced in form of narrow notch cuts with length 4 mm in positions indicated at figure 6a and 6b. In numerical model notch cut was simulated by separation of nodes between elements. In experiment notch cuts were about two thirds of the specimen thickness deep. Results obtained for this structure are presented in figures 9 and 10 for numerical and experimental data respectively.

In all numerical and experimental studies burst sine excitation at carrier frequency 35 kHz applied in the geometrical centre of specimens were used.

It should be noted that experimental data was filtered prior to applying damage localization techniques, using median filtering in space domain. This provided elimination of several sparsely distributed measurement points with low signal to noise ratio, which is typical for scanning laser vibrometry measurements.

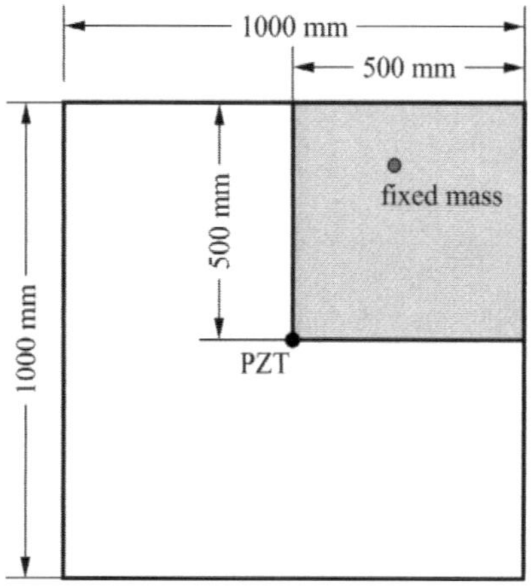

Figure 5: Scheme of aluminium alloy plate.

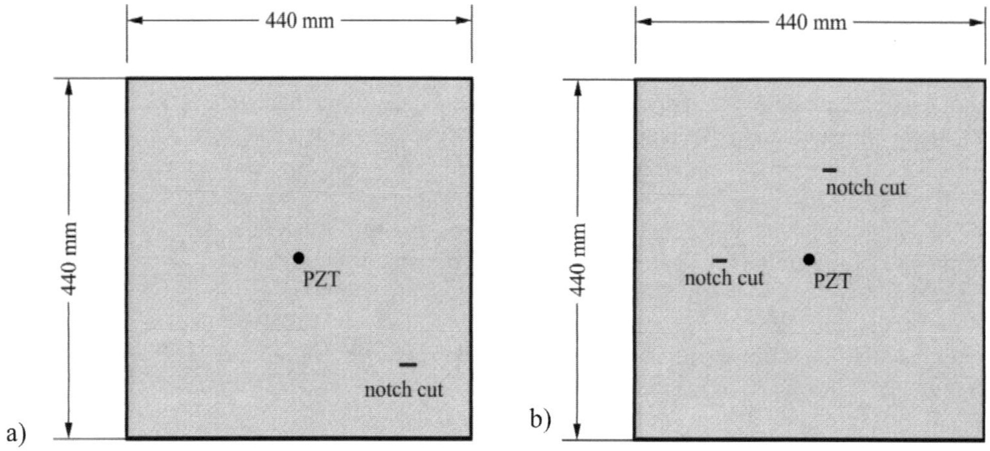

Figure 6: Scheme of composite plate a) numerical, b) experimental.

In case of aluminum alloy plate damage may be localized using all tested methods. The best results were obtained using incident wave filtering. However, it should be stated that signals used for creation of damage maps in this particular approach were limited to time step in which guided waves reached the boundaries of structure. In this manner top right corner was not covered by the signal and may be considered as so called 'dead zone'. This is only problem in case when investigated area is in proximity or includes specimens boundaries.

In numerical results obtained for composite plate all four tested techniques were able to correctly localize position of damage. Incident wave filtering method was prepared separately for every quarter of specimen and then merged into one map (figure 9c and 10c).

Nevertheless, experimental results were not so encouraging. Top notch cut was barely visible in RMS based methods and incident wave filtering method (figure 10a, 10b, 10c). Second notch cut cannot be pinpointed by those three methods. Due to the fact that this crack was oriented perpendicular to excited wave front, incident wave was almost not affected by damage. However, proposed by the authors method is capable of excluding information about wave pattern changes also from reflected by the structure boundaries guided waves. In case of notch cut of composite plate situated on the left side of excitation, guided waves reflected from the top and the bottom boundaries of specimen were perpendicular to this damage. Therefore localization of both cuts by proposed method was possible, what is presented in figure 10d.

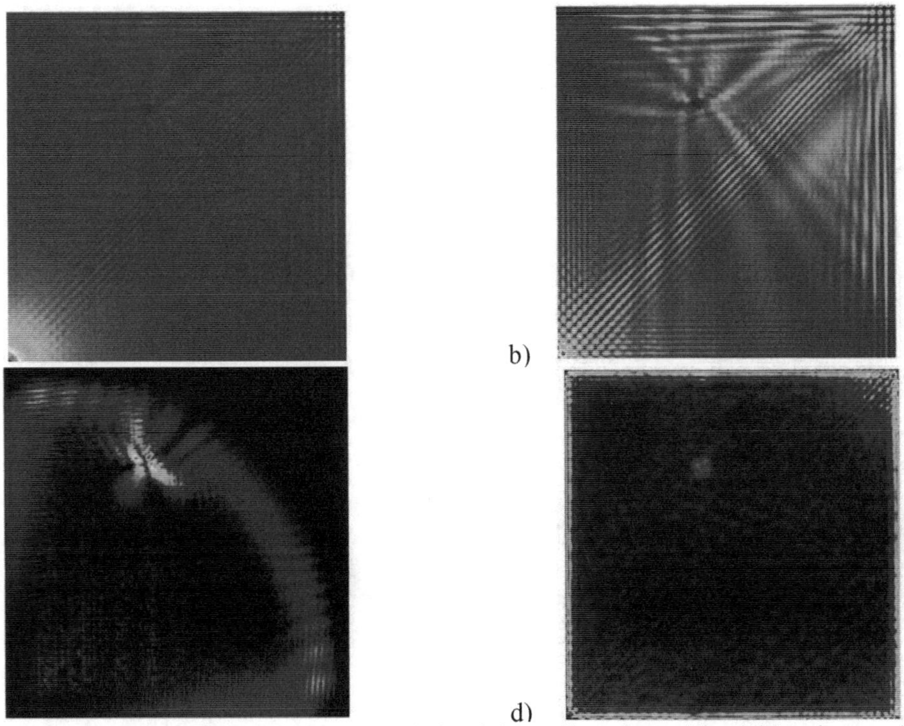

Figure 7: Damage detection maps for numerical aluminium alloy plate with additional mass a) RMS, b) Weighted RMS, c) Incident wave filtering, d) Adaptive wave image filtering.

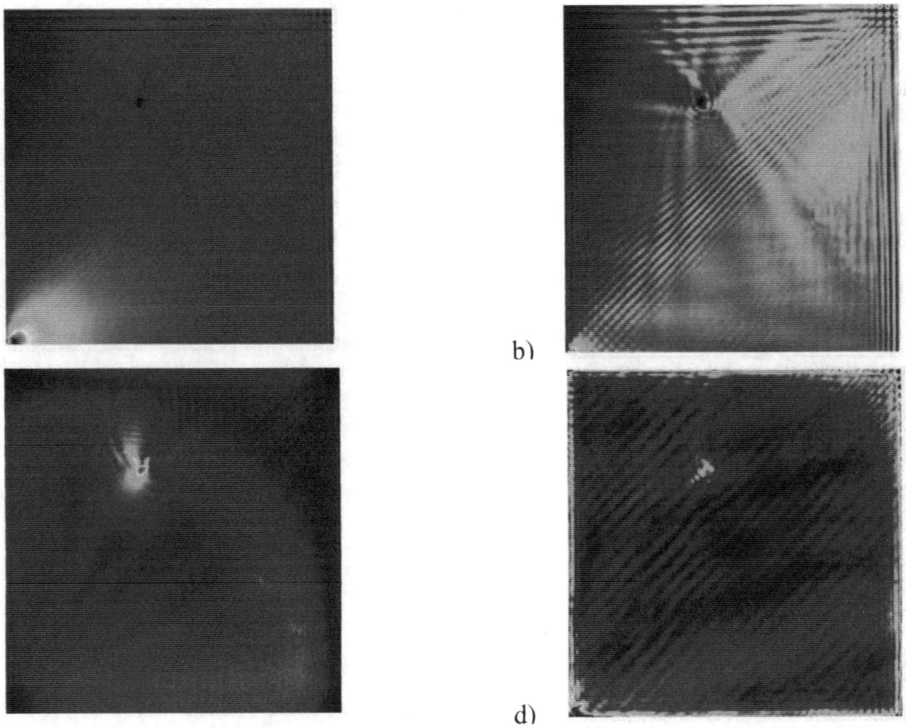

Figure 8: Damage detection maps for experimental aluminium alloy plate with additional mass a) RMS, b) Weighted RMS, c) Incident wave filtering, d) Adaptive wave image filtering

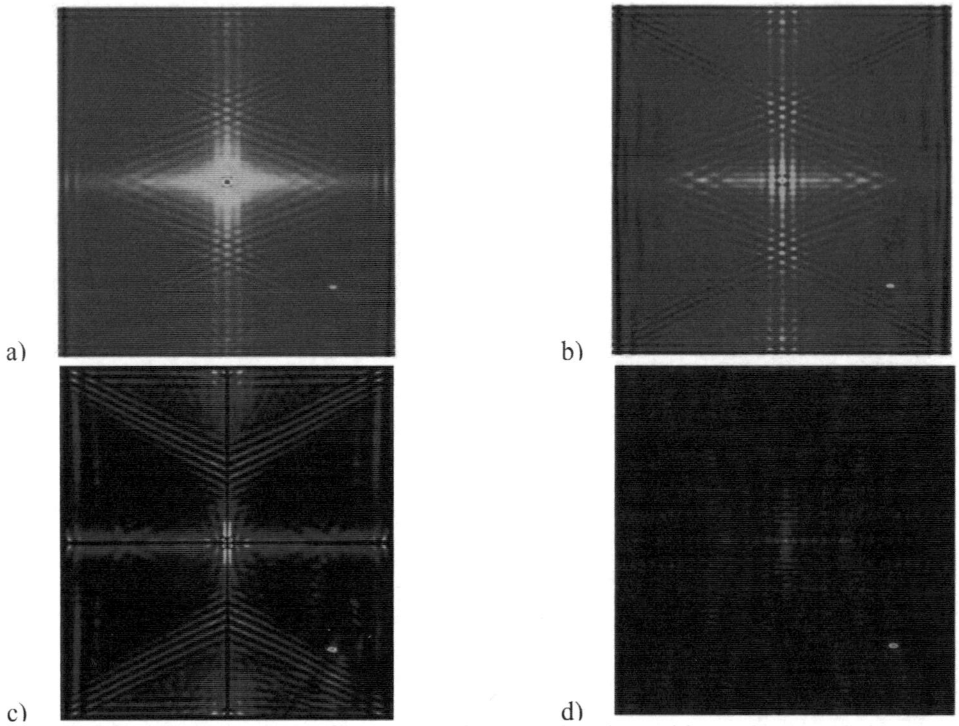

Figure 9: Damage detection maps for numerical composite plate with crack a) RMS, b) Weighted RMS, c) Incident wave filtering, d) Adaptive wave image filtering.

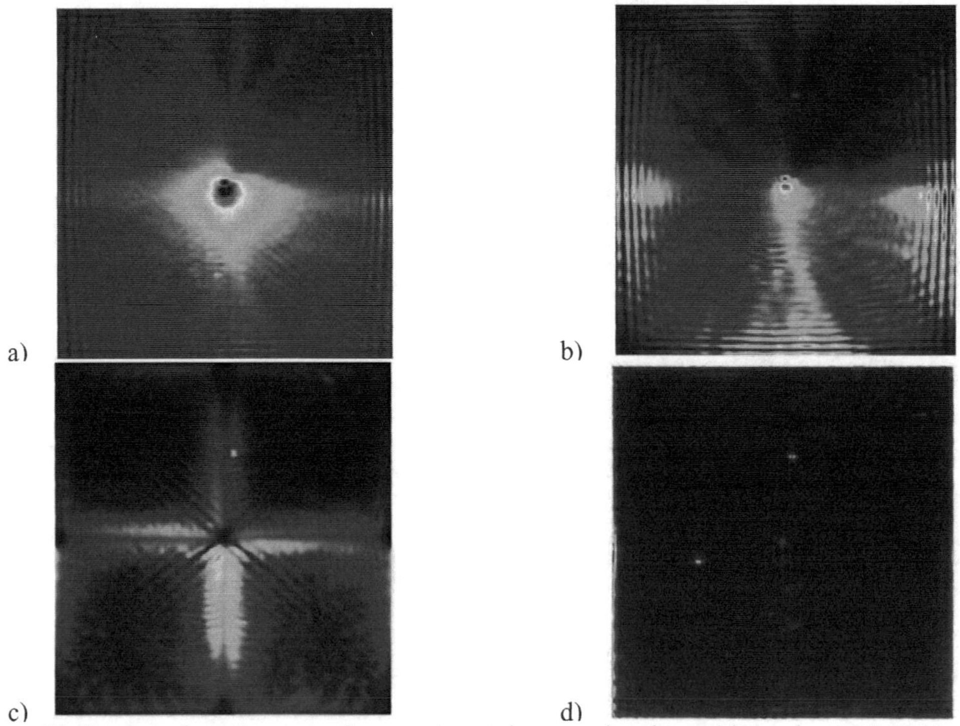

Figure 10: Damage detection maps for experimental composite plate with cracks a) RMS, b) Weighted RMS, c) Incident wave filtering, d) Adaptive wave image filtering.

6. Conclusions

Spectral Element Method was found to be very helpful in designing effective Structural Health Monitoring systems. Numerical results verified by the experimental research confirm its capability of modelling and testing damage detection algorithms.

Adaptive wave image filtering method for damage detection was proposed. Numerical and experimental results proved that this technique can be used as an effective tool for localization of cracks in composite plate.

It has been shown that in spite of the fact that guided wave propagation is very complex phenomenon it can be successfully utilized for damage detection and localisation in both isotropic and anisotropic plates. There are many various approaches for this process and depending on the type of structure, implementation possibilities, exploitation environment, accessibility and time constraints one should satisfy its needs. The capability of employing more than one SHM techniques simultaneously in order to improve the overall performance of structure technical condition evaluation should be also considered.

References

[1] H. Lamb, On waves in an elastic plate. Proc. of the Royal Society of London, 93 (1917), 293-312

[2] J. Rose, Ultrasonic Waves in Solid Media, Cambridge University Press, 1999.

[3] P. Kudela, A. Zak, M. Krawczuk, W. Ostachowicz, Modelling of wave propagation in composite plates using the time domain spectral element metod, J. Sound Vib., 302 (2007) 728–745.

[4] J.F. Doyle, Wave Propagation in Structures, Springer–Verlag, 1997.

[5] P.P. Delsanto, R.S. Schechter, R.B. Mignogna, 1997. Connection machine simulation of ultrasonic wave propagation in materials III: The three dimensional case, Wave Motion, 26 (1997) 329–339.

[6] W. Ostachowicz, P. Kudela, Spectral Element Method for wave propagation modeling in 2D and 3D solids, Proc. of the 7th Int. Conf. on SHM, 2 (2009) 2213–2221, Stanford, USA.

[7] P. Kudela, W. Ostachowicz W., 3D time–domain spectral elements for stress waves modelling," 7th Int. Conf. MPSVA, Cambridge, UK, J. Physics: Conf. Series, 181 (2009), paper no 012091

[8] P. Kudela, W. Ostachowicz, A. Zak, Damage detection in composite plates with embedded PZT transducers, Mech. Syst. Signal Pr. 22 (2008) 1327–1335.

[9] F. Schubert: A Conceptual Study on Guided Wave Based Imaging Techniques for SHM with Distributed Transducer Array, Proc. Structural Health Monitoring (2008) 748–757.

[10] W. Ostachowicz, P. Kudela, P. Malinowski, T. Wandowski, Damage localisation in plate–like structures based on PZT sensors, Mech. Syst. Signal Pr. 23 (2009) 1805–1829.

[11] M. Ruzzene, S.M. Jeong, T. Michaels, J. Michaels, B. Mi, Simulation and measurement of ultrasonic waves in elastic plates using laser vibrometry, Review of Quntitative Nondestructive Evaluation, 24 (2005), 172-179.

[12] M. Radzieński, Ł. Doliński, M. Krawczuk, A. Żak, W Ostachowicz, Application of RMS for damage detection by guided elastic waves, J. Phys.: Conf. Ser. 305 (2011), 012085 1-10.

[13] M. Ruzzene, Frequency-wavenumber domain filtering for improved damage visualization, Smart Materials and Structures, 16s (2007), 2116-2129.

Journal of Physics: Conference Series **382** (2012) 012004 doi:10.1088/1742-6596/382/1/012004

Nonlinear phenomena in mechanical system dynamics

J Warminski[1], K Kecik[1], A Mitura[1] and M Bochenski[1]

[1]Department of Applied Mechanics, Lublin University of Technology, Nadbystrzycka 36, 20–618 Lublin, POLAND

E-mail: `j.warminski@pollub.pl`

Abstract. The goal of the paper is to present selected, untypical, and intuitively unexpected phenomena from nonlinear mechanics. Particular attention is paid to the dynamics of self-, parametric and external excited systems. Interactions between these various vibration types lead mainly to quasi-periodic responses. However, in the selected domains of system parameters, the effect of frequency locking is observed. Furthermore, external harmonic force imposed on such a system produces a specific internal loop inside a resonance zone. As an example of nonlinear autoparametric systems, a structure (oscillator) with an attached pendulum is presented. The nonlinear terms introduced by pendulum motion cause instabilities in the resonance region. This instability transits the pendulum to rotation or chaotic motion. An application of nonlinear couplings for the reduction of unwanted vibrations is also studied. In order to reduce vibrations, the main structure is coupled to an electrical oscillator by a quadratic term. It has been shown that such a coupling leads to the amplitude saturation phenomenon which can then be used to design a nonlinear control strategy.

1. Introduction

In many dynamical systems nonlinear phenomena may completely alter intuitively expected behaviour and can drastically change their dynamical responses [1]. Mechanical vibrations can be caused by different reasons and, depending on the method of excitations, they are divided into three main types. We may distinguish between forced vibration, parametric vibrations and self-excited vibrations. The two former categories belong to mathematical models represented by nonautonomous equations. All the above systems are well known and deeply investigated in the literature, however mainly separately.

Paper [2] is one of the first that considered interactions between two different vibration types. There is only a preliminary conclusion offered on possible interactions between parametric and self-excitations. More extensive research is presented in [3], and advanced analysis is delivered in monograph [4] and papers [5, 6, 7, 8]. Coupled van der Pol – Mathieu oscillators are studied there for one and many degree of freedom systems. The influence of different nonlinear terms on the localisation of the frequency locking zones which follow the second kind of Hopf bifurcation, is studied using numerical and analytical methods in [9, 10].

The dynamics of parametrically and self-excited vibrations taking into account the influence of external force is presented in paper [11]. The authors conclude that five solutions may appear in the studied one–degree–of–freedom system. In fact this phenomenon was shown in former works such as [12, 13], where an additional external force influence, acting on a self- and parametrically excited system, was studied in detail for one and two degree of freedom models. It was shown that if the excitation frequency of the parametric and external terms was tuned in the ratio 2:1, then essential changes in the main parametric resonance region occurred. For small values of excitation amplitude the phenomenon of an internal resonance loop was discovered. Comprehensive research considering interactions between parametric, external and self- excitations, as well as for regular, chaotic and hyperchaotic motion is published in [14]. The discussion of non-ideal systems is delivered in that monograph and in paper [15] as well.

Pendulum-like systems are commonly used in many practical applications, including special dynamical dampers [16, 1]. We may take advantage of the pendulum's dynamics to suppress vibrations of high buildings, bridges or helicopter blades under flutter conditions [17]. But, a harmonically excited pendulum may also undergo complicated dynamics. Particular behaviour may occur if the pendulum and the main structure such as an oscillator are coupled by an inertial resonance condition. Such a system generates various motions, from simple periodic oscillation to complex chaos [18]. The presence of the coupling terms can lead to a certain type of instability which is referred to as autoparametric resonance. This kind of phenomenon takes place when the external resonance and the internal resonance meet, due to the coupling terms. Small parametric excitations, or a small change of initial conditions, then produce large response [1]. Similar dynamics governs the behaviour of coupled beam structures where the L-shaped beam is a good example [19]. The concept of the use of a combination of the magnetorheological damper together with the nonlinear spring applied in the autoparametric system suspension is presented in [16, 1]. By activating magnetorheological damping or varying nonlinear stiffness of the supporting spring, the unwanted dynamics can be eliminated or moved away. The solution with the MR damper gives reliable control possibilities and can help to react properly in critical situations.

Sometimes we may take advantage from nonlinear coupling. As is shown in [20] nonlinear coupling can be used for vibration suppression. Introducing quadratic terms on purpose, which couple the cantilever beam and a controller, leads to the saturation phenomenon. The system's response depends on the controller damping and gains. The same control algorithm for the reduction of flexural and torsional vibrations of a plate is presented in [21]. To increase the effectiveness of the control the measurement of the frequency of excitation is added to the system. Analytical, numerical and experimental results for Nonlinear Saturation Control (NSC) and the linear position feedback algorithm (PPF) are studied in [22, 23, 24]. Those two algorithms are combined and then applied in experimental tests [22]. Advantages of hybrid, proportional linear and saturation nonlinear, controllers are presented in that paper. The influence of non-uniformities in the beam properties for a system with a saturation controller is presented in [25], where the additional effect of modal coupling for system response is studied. The effectiveness of the NSC method for vibration suppression of a light composite beam with Macro Fibre Composites (MFC) actuators is also confirmed in [26]. Advantages of MFC actuators in comparison with traditional PZT elements are stated in [27]. The proposed nonlinear controllers may work very effectively, under the condition that their nonlinear behaviour is known and well designed.

The main purpose of this paper is to point out nonlinear interactions which, in consequence, can lead to very complex dynamics, including chaotic motion, even for models having a very simple structure. Advantages and disadvantages of the nonlinear couplings will be demonstrated for selected examples.

2. A loop phenomenon in the dynamics of a self and parametrically exited system driven by an external force

Interactions between self and parametric vibrations can lead to very complex behaviour, even for very simple models [12, 13]. Near the parametric resonance regions the frequency locking phenomenon takes place (this is sometimes called frequency synchronisation). Outside these regions the response is quasi-periodic, represented by quasi-periodic limit cycles on Poincaré maps.

The lumped mass dynamics of the considered system is described by the following differential equations of motion which take the matrix form

$$\mathbf{m}\ddot{\mathbf{x}} + \mathbf{f_d}(\mathbf{x}, \dot{\mathbf{x}}) + \mathbf{k}\mathbf{x} + \mathbf{f_p}(\mathbf{x}, \omega_p, t) + \mathbf{f_x}(\mathbf{x}) = \mathbf{f_e}(\omega_e, t), \tag{1}$$

where the subscripts used in the notation denote, d – self (damping), p – parametric and e – external excitation terms and $\mathbf{x}$ is a general coordinates vector.

Let us consider, as an example, a two degree of freedom system as presented in Fig.1. In this example we assume that the parametric resonance frequency is 2ω while the external excitation frequency is ω. This case corresponds to practical engineering applications [29]. The differential equations of motion of the model in Fig. 1 have the following form

$$\begin{aligned} m_1\ddot{X}_1 + f_{d1}\left(X_1, \dot{X}_1\right) + \delta_1 X_1 + \gamma_1 {X_1}^3 + (\delta_{12} - \mu\cos 2\omega t)(X_1 - X_2) &= q\cos\omega t \\ m_2\ddot{X}_2 + f_{d2}\left(X_2, \dot{X}_2\right) + \delta_2 X_2 + \gamma_2 {X_2}^3 - (\delta_{12} - \mu\cos 2\omega t)(X_1 - X_2) &= 0, \end{aligned} \tag{2}$$

where, $f_{d1}(X_1, \dot{X}_1)$, $f_{d2}(X_2, \dot{X}_2)$ are nonlinear self-excitation damping functions. They can be modelled by van der Pol or Rayleigh type terms. For the Rayleigh model considered for further analysis, these functions are defined as $f_{d1} = -\alpha_1\dot{X}_1 + \beta_1\dot{X}_1^3$, $f_{d2} = -\alpha_2\dot{X}_2 + \beta_2\dot{X}_2^3$.

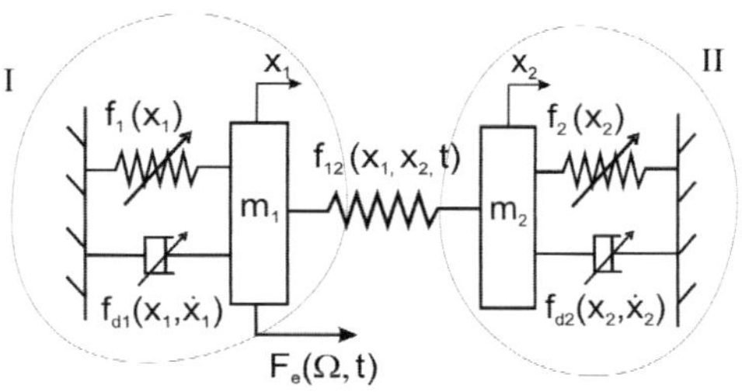

Figure 1. Model of coupled self, parametrically and externally excited oscillators with 2 DOF.

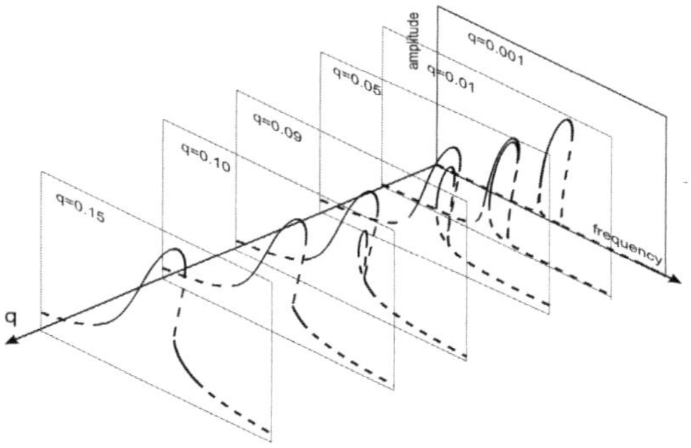

Figure 2. Resonance curves versus external excitation amplitude q.

At first, to check the possible dynamics and interactions between self and parametric vibrations driven by an external harmonic force, we consider a one degree of freedom system. This case corresponds to a single oscillator (No. I in Fig. 1) described by Eq. $(2)_1$. The resonance curves obtained for $m_1 = 1$, $\alpha_1 = 0.01$, $\beta_1 = 0.05$, $\delta_1 = 1$, $\gamma_1 = 0.1$, $\delta_{12} = 1$, $\mu = 0.2$ are presented in Fig.2. We see that the resonance curves have an unexpected shape. For a small amplitude of external force (parameter q) the internal loop occurs inside the resonance curve. This loop may exist only below a certain threshold of excitation amplitude q. It means that even for a single degree of freedom system five solutions, which represent steady states, exist. However, stability analysis shows that only two of them are stable (solid lines of the resonance curve). For large values of q (above the threshold) the loop disappears and the curve has a classical form. The loop occurrence phenomenon is studied in detail in [12].

We may expect that for many degrees of freedom system the loop should also appear inside the resonance curve. In order to verify this the set of coupled nonlinear equations (2) is transformed to normal coordinates Y_j, $(j = 1, 2)$, by a transformation $\mathbf{X} = \mathbf{uY}$, where $\mathbf{u}$ is a modal matrix. Taking into account that around resonances mainly one Y_j normal coordinate in involved, the problem is solved analytically, assuming weakly nonlinear coupling [28]. This

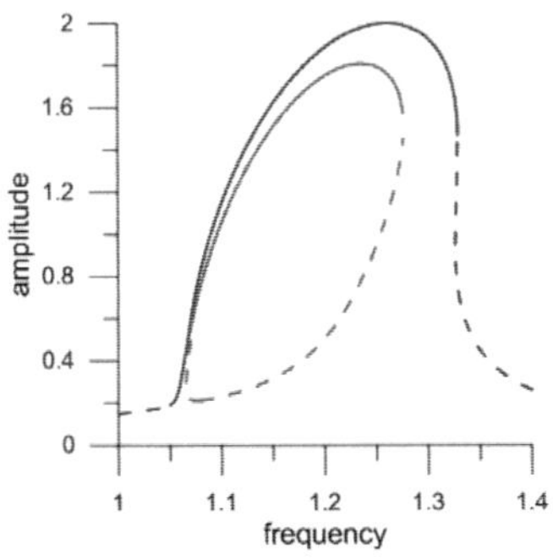

Figure 3. Resonance curve around the second natural frequency; linear normal mode (LNM) projection.

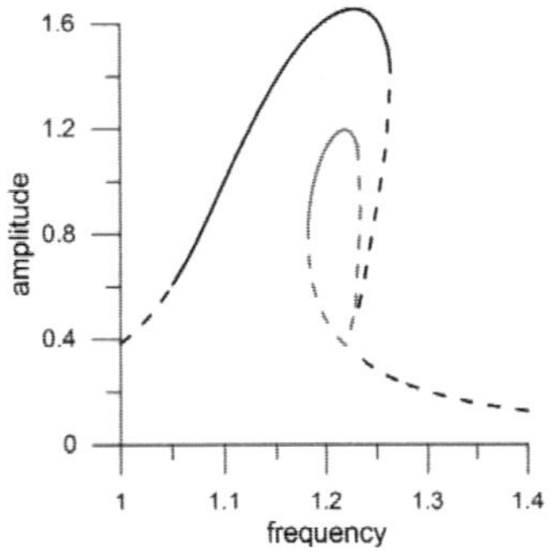

Figure 4. Resonance curve around the second natural frequency; nonlinear normal mode (NNM) projection.

approach cannot be accepted if the system is strongly nonlinear or various types of nonlinear terms are involved in its dynamics. Therefore, as an alternative, the nonlinear normal modes (NNMs) formulation is proposed. The modes are determined on the basis of free vibrations of a nonlinear system. Details of that formulation are presented in [29]. The NNMs include geometric nonlinear terms therefore these terms vanish after the coordinate transformation. However, self- and parametric excitations still exist. After the NNM projection the system is uncoupled into two single nonlinear oscillators

$$M_j\ddot{Y}_j + M_j\omega_{0j}^2 Y_j - C_{\alpha j}\dot{Y}_j + C_{\beta j}\dot{Y}_j^3 - C_{\mu j}Y_j\cos 2\omega t = C_{q_j}\cos\omega t \ , \tag{3}$$

where, M_j is modal mass and ω_{0j} natural frequency $(j = 1, 2)$. Because the NNMs are amplitude dependent, the modal mass, natural frequency and modal matrix are also amplitude dependent. Equation (3) is solved by the multiple time scales method together with the constraint equation for NMNs. The resonance curves obtained for $j = 2$ and $M_2 = 1.089$, $\omega_{02} = 1.16752$, $C_{\alpha 2} = 0.01044$, $C_{\beta 2} = 0.0501$, $C_{\mu 2} = 0.292988$, are presented in Figs.3 and 4. We see that for the two degree of freedom system the loop exists as well. However, the loop found by the LNM transformation (Fig.3) is located on the left part of the resonance curve. Contrary to this the NNM transformation gives a much smaller loop, located on the right side of the curve. The solution obtained by NNMs is much closer to real system dynamics and is in a good agreement with direct numerical simulations.

3. Regular and chaotic dynamics of an autoparametric system with magnetorheological damping

As has been mentioned in the introduction systems with pendulums are very sensitive and often, due to their strong nonlinearity, may behave unpredictably. The autoparametric pendulum-like system presented in Fig.5 is composed of the main structure i.e. a nonlinear oscillator and the pendulum. The oscillator's suspension is considered as a classical linear suspension with viscous damping or a nonlinear suspension with a nonlinear spring and a magnetorheological (MR) damper. The nonlinear spring is introduced by a structural modification while the MR damper behaviour can be modified on-line from viscous effects, if the system is not activated, to a mixed mode system with viscous and dry friction components, when the damper is activated. The differential equations of coupled motion of the oscillator and the pendulum have the form

$$\begin{aligned} \ddot{X} + F_d\left(X, \dot{X}\right) + F_s\left(X\right) + \mu\lambda(\ddot{\varphi}\sin\varphi + \dot{\varphi}^2\cos\varphi) &= q\cos\omega t \\ \ddot{\varphi} + \alpha_2\dot{\varphi} + \lambda\left(\ddot{X} + 1\right)\sin\varphi &= 0\,. \end{aligned} \tag{4}$$

Functions F_d and F_s represent the nonlinear damping and stiffness and they are defined respectively as, $F_d = \alpha_1 \dot{X} + \alpha_3 \tanh\left(e\dot{X}\right)$, $F_s = X + \gamma X^3$.

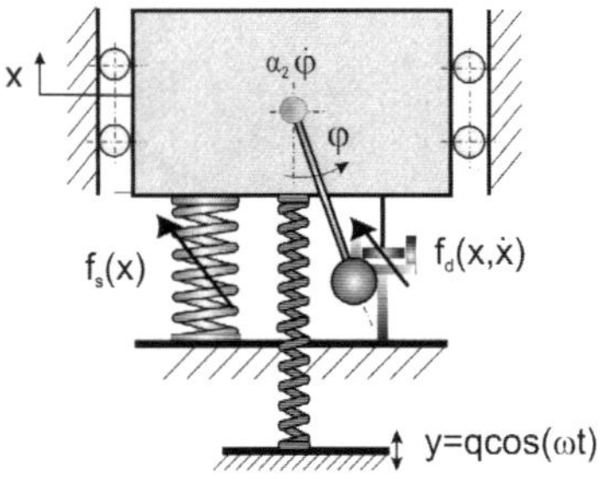

Figure 5. Model of an autoparametric system with a pendulum and a nonlinear semi-active suspension.

A sample resonance curve of the pendulum obtained for data as follows: $\alpha_1 = 0.261354$, $\alpha_2 = 0.1$, $\alpha_3 = 0$, $\mu = 17.2279$, $\lambda = 0.127213$, $q = 2.45094$ and $\gamma = 0$ is presented in Fig.6(a).

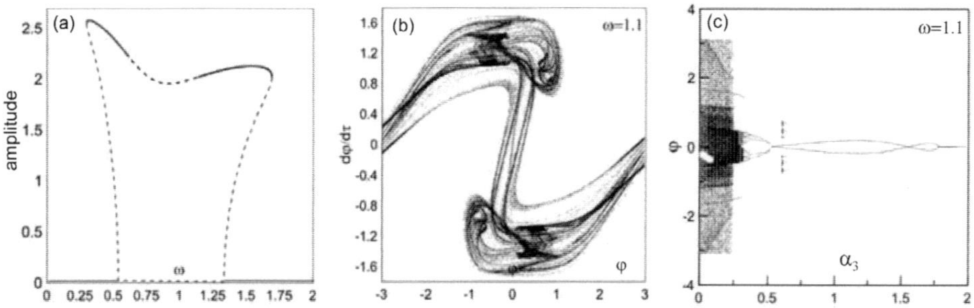

Figure 6. Resonance curve of the pendulum with marked unstable regions (dashed line) (a), chaotic attractor for $\omega = 1.1$, $\alpha_3 = 0$ (b), and bifurcation diagram of the chaotic attractor versus the MR damping parameter α_3 (c); $\alpha_1 = 0.2613$, $\alpha_2 = 0.1$.

Analysis of Eqs.4 exhibits instability zones in the pendulum motion. In Fig.6 (a) we see an instability near the middle of the resonance curve. In this region the pendulum may go to chaotic motion or to regular rotations [16, 18, 14]. A strange chaotic attractor obtained for frequency $\omega = 1.1$ is presented in Fig.6 (b). The attractor has been determined for parameter $\alpha_3 = 0$, which means that the MR damper is not activated. The bifurcation diagram in Fig.6 (c) shows the influence of MR damping. We see that an increase of MR damping effectively suppresses rotations or oscillations of the pendulum. By applying a semi-active MR device we may control the dynamics of the pendulum and the structure treated as a whole.

4. Nonlinear control

We may take some advantages from strongly nonlinear couplings which may be applied to get special effects in a control strategy. Let us consider a light composite cantilever beam with active elements. Embedded Macro Fiber Composite (MFC) actuators are used for the real implementation of the control strategy. A sketch of a real system used for the validation of the theoretical results is presented in Fig.7.

The model of the beam (the plant) is assumed to be nonlinear, including geometrical and inertia nonlinear terms. The beam is forced kinematically by vertical motion of the clamped end. The basic structure is then coupled with an additional oscillator realised by the electrical part. In order to suppress vibrations, the beam is coupled with the controller by means of quadratic

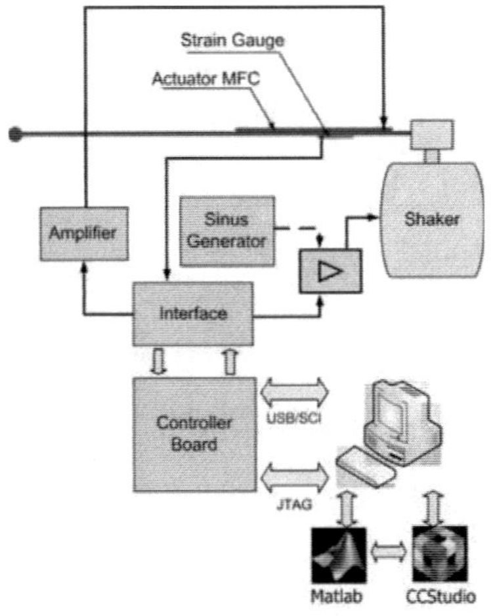

Figure 7. A sketch of the experimental setup used for nonlinear control of a composite beam with embedded MFC actuators.

terms. The mathematical model of this nonlinear control problem, taking into account the first mode response, has the form

$$\begin{aligned} \ddot{v} + 2\mu\omega_s\dot{v} + \omega_s^2 v + \beta v^3 + \delta\left(v\dot{v}^2 + v^2\ddot{v}\right) &= f\,\sin(\Omega t) + \gamma u^2 \\ \ddot{u} + 2\zeta\omega_c\dot{u} + \omega_c^2 u &= \alpha u v\,, \end{aligned} \tag{5}$$

where $f = y_0\xi\Omega^2$ and the excitation frequency Ω is close to ω_s. Due to the quadratic coupling the effect of amplitude saturation can be achieved. But this requires tuning of the natural frequency of the controller to the natural frequency of the beam in the ratio $\omega_c = \frac{1}{2}\omega_s$. The physical parameters of the beam taken for numerical and experimental tests have values, Young's modulus: E=25.5 $\times$ 10^9 Pa, density: ρ=2100 kg/m^3, length: L=236 mm, width: b=12.8 mm. The dimensionless parameters calculated for various thickness h and tip mass M values are presented in Table 1. They allow a study of the dynamics of the beam with hard ($\beta > 0$) and soft ($\beta < 0$) elasticity terms. However, the global characteristic depends also on the nonlinear inertia terms (parameter δ), which may overturn the slope of the resonance characteristic. The other parameters for all cases take these values: $\mu = 0.01$, $\zeta = 0.001$, $\alpha = 2$, $\gamma = 0.01$. The amplitude of excitation y_0 is varied.

Table 1. Physical and dimensionless parameters of the nonlinear control system.

Physical	parameters	Dimensionless	parameters		
h (mm)	M (gram)	ω_s^2	β	δ	ξ
1.0	0.5	9.3825	14.4108	3.2746	0.89663
2.1	0.5	10.7430	25.2913	3.8922	0.83866
2.1	5.0	4.8883	-7.1401	1.0827	1.24173
2.1	15	2.2012	-7.5974	-0.32638	1.85326

As we may observe in Fig. 8, for a well tuned structure and properly selected parameters around a relative frequency $\Omega/\omega_s \approx 1$, the response of the system is very well suppressed. Moreover, the system's response is well suppressed even if the excitation amplitude is increased. The diagrams

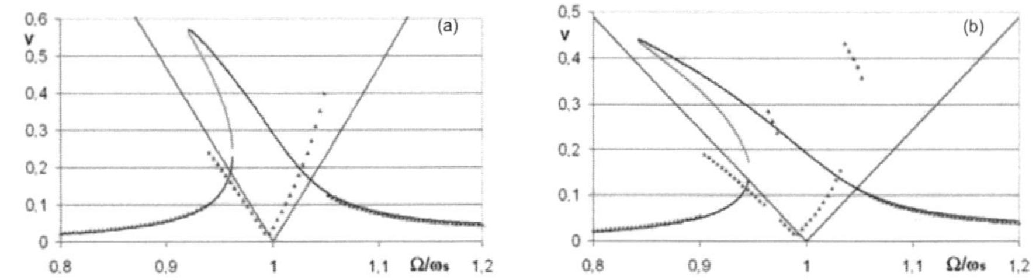

Figure 8. Resonance curve of the beam without and with the activated controller; (a) case 1 (b) case 3 from Table 1.

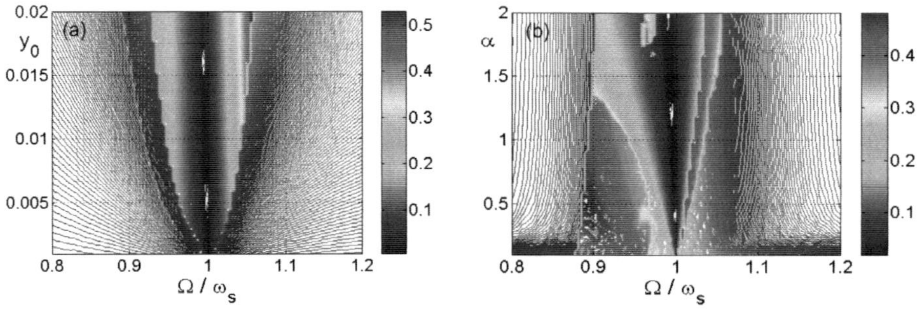

Figure 9. Two parameter diagrams (a) amplitude versus excitation frequency - case 1 (b) amplitude versus gain α - case 3 from Table 1.

frequency-amplitude and frequency-gain α (Fig. 9) present regions of low amplitude vibration (dark-blue colour). The red colour represents vibrations of high amplitude. This vibration suppression is caused by nonlinear coupling, leading to the saturation phenomenon [28, 30].

5. Conclusions and remarks

A few selected phenomena of nonlinear dynamics are presented in this paper. It has been shown that the interaction between self, parametric and external vibrations leads to the occurrence of an internal loop inside the resonance curve. Then, five solutions, even for a simple one-degree-of-freedom system are found. Nonlinear couplings in the autoparametric pendulum-like structures result in instabilities in the resonance regions. Then the pendulum may go to full rotations or chaotic motion. It has been shown that the application of a semi-active MR damper may successfully reduce the pendulum's vibration and allow control of the system behaviour. The quadratic nonlinearity, coupling the light composite beam and the MFC active element, may effectively suppress the structure vibrations. The saturation phenomenon maintains the system response at the design level even if the amplitude of excitation increases.

6. References

[1] Warminski J, Lenci S, Carttmell P M, Rega G and Wiercigroch M 2012 *Nonlinear Dynamic Phenomena in Mechanical Engineering Systems* in Solids Mechanics and Its Applications **181** (Springer)

[2] Minorski N 1962 *Nonlinear oscillations* (v. Nostrand)

[3] Tondl A 1978 *On the interaction between self-excited and parametric vibrations* Monographs and Memoranda **25** (National Research Institute for Machine Design, Prague)

[4] Abadi O 2003 *Nonlinear Dynamics of Self-Excitation in Autoparametric Systems* (University of Utrecht Netherlands).

[5] Warminski J 2003 Regular, chaotic and hyperchaotic vibrations of nonlinear systems with self, parametric and external excitations *Scientific Journal FACTA UNIVERSITATIS* **3** 14 891-905
[6] Bakri T, Nabergoj R, Tondl A and Verhurlst F 2004 Parametric excitation in non-linear dynamics *International Journal of Non-Linear Mechanics*, **39** 2, 311-329
[7] Verhulst F 2005 Quenchening of self-excited vibrations *Journal of Engineering Mathematics* **53** 349-358.
[8] Pust L and Tondl A 2008 System with a non-linear negative self-excitation *International Journal of Non-Linear Mechanics* **43** 497 - 503
[9] Rand R, Barcilon A and Morrison T 2005 Parametric Resonance of Hopf Bifurcation *Nonlinear Dynamics* **39** 411-421
[10] Tondl A and Nabergoj R 2004 The effect of parametric excitation on a self-excited three-mass system *International Journal of Non-Linear Mechanics* **39** 821-832.
[11] Pandey M, Rand R and Zehnder A 2008 Frequency Locking in a Forced Mathieu-van der Pol-Duffing System *Nonlinear Dynamics* **54** 3-12
[12] Szabelski K and Warminski J 1995 The self-excited system vibrations with the parametric and external excitations *Journal of Sound and Vibration*, **187** (4) 595-607.
[13] Szabelski K and Warminski J 1997 Vibrations of a non-linear self-excited system with two degrees of freedom under external and parametric excitation *Nonlinear Dynamics* **14** 23-36
[14] Warminski J 2001, *Regular and Chaotic Vibrations of Parametrically and Self-Excited Systems with Ideal and Non-Ideal Energy Sources* (Lublin University of Technology Publisher, in Polish)
[15] Warminski J, Balthazar J M and Brasil R M L R F 2001 Vibrations of non-ideal parametrically and self-excited model *Journal of Sound and Vibration* **245** (2) 363-374
[16] Warminski J and Kecik K 2006 Autoparametric vibrations of a non-linear system with pendulum *Mathematical Problems in Engineering* Article ID 80705, Hindawii Publishing Corporation.
[17] Nagasaka I, Ishida Y, Ishii T, Okada T and Koyoma T 2007 Vibration suppression of helicopter blades by pendulum absorbers. Analytical and experimental investigations in case of rigid -body mode *Transactions of the Japan Society of Mechanical Engineers* **73** 725 129–137.
[18] Warminski J and Kecik K 2009 Instabilities in the main parametric resonance area of a mechanical system with a pendulum *Journal of Sound and Vibration* 322 (3):612–628
[19] Warminski J, Cartmell M P, Bochenski M and Ivanov I 2008 Analytical and experimental investigations of an autoparametric beam structure *Journal of Sound and Vibration* **315** 3 486-508
[20] Oueini S S, Nayfeh A H and Pratt J R 1998 A Nonlinear Vibration Absorber for Flexible Structures *Nonlinear Dynamics* **15** 259-282
[21] Ashour O N and Nayfeh A H 2002 Adaptive control of flexible structures using a nonlinear vibration absorber *Nonlinear Dynamics* **28** 309-322
[22] Pai P F, Wen B, Naser A S and Schulz M J 1988 Structural vibration control using PZT patches and non-linear phenomena *Journal of Sound and Vibration* **215** (2) 273-296
[23] Kwak M K and Heo S 2007 Active vibration control of smart grid structure by multi-input and multi-output positive position feedback controller *Journal of Sound and Vibration* **304** 230-245
[24] Shan J, Liu H and Sun D 2005 Slewing and vibration control of a single-link flexible manipulator by positive position feedback (PPF) *Mechatronics* **15** 487-503
[25] Saguranrum S, Kunz DL and Omar HM 2003 Numerical simulations of cantilever beam response with saturation control and full modal coupling. *Comput. Struct.* **81** 1499–1510.
[26] Warminski J, Bochenski M, Jarzyna W, Filipek P and Augustyniak M 2011 Active Suppression of Nonlinear Composite Beam Vibrations by Selected Control Algorithms *Communications Nonlinear Science and Numerical Simulation* **16** 5 486-508
[27] Sodano H A, Park G and Inman D J, 2004 An investigation into the performance of macro-fiber composites for sensing and structural vibration applications *Mechanical Systems and Signal Processing* **18**, 683-697
[28] Nayfeh A, 2000 *Nonlinear Interactions, Analytical, Computational and Experimental Methods* (Wiley)
[29] Warminski J 2010 Nonlinear normal modes of a self-excited system driven by parametric and external excitations *Nonlinear Dynamics* **61**(4) 677–689
[30] Mitura A, Kazmir T, Warminski J, Augustyniak M and Jarzyna W 2010 Vibration suppression of a composite plate with MFC active elements *Machine Dynamics Problems* **34** 2 86-92

Acknowledgments

The research leading to these results has received funding from the European Union Seventh Framework Programme (FP7/2007-2013), FP7-REGPOT-2009-1, under grant agreement No: 245479. The support by the Polish Ministry of Science and Higher Education-grant no 1471-1/7.PR UE/2010/7-is also acknowledged.

Numerical Study of a Jeffcott Rotor Model with a Snubber Ring

J. Páez Chávez[1], M. Wiercigroch[2] and H. Demrdash[3]

[1,2] Centre for Applied Dynamics Research, School of Engineering, University of Aberdeen, Aberdeen AB24 3UE, UK

[1] Instituto de Ciencias Matemáticas, Escuela Superior Politécnica del Litoral, P.O. Box 09-01-5863, Guayaquil, Ecuador

[3] Faculty of Engineering, The University of Nottingham Malaysia Campus, 43500 Semenyih, Selangor Darul Ehsan, Malaysia

E-mails: [1] jpaez@abdn.ac.uk, [2] m.wiercigroch@abdn.ac.uk,
[3] hazem.demrdash@nottingham.edu.my

Abstract. In this article we study a two-degrees-of-freedom model of a rotor system with a bearing clearance. During operation the rotor makes intermittent contact with an outer snubber ring, which results in complex dynamical behaviour. Specifically, the system will be analyzed numerically by a path following method, where we will use the toolbox TC-HAT, which is a module for modeling non-smooth systems by AUTO 97.

1. Introduction

Rotor systems have attracted much attention due to their rich dynamics and also because of their wide industrial applications, such as aero engines, gas turbines, compressors, generators, and others. There are different sources of nonlinearities for this type of systems. One of the most important of them are the bearing clearances, which allow rotors to make intermittent contact between their components (e.g. rotor-stator dynamics), which leads to non-smooth dynamics. This is often produced by centrifugal forces due to rotor mass imbalance. Even if the rotor is well balanced at the beginning of its operating life, the balance generally deteriorate with use, which may produce high amplitude vibrations that can lead to catastrophic failures. Therefore it is important to understand the dynamics of rotor systems in order to design suitable control methods in order to avoid unwanted behaviour.

Rotor systems with bearing clearances have been considered in the past, putting special attention on Jeffcott rotors. For instance, rub interactions in rotating machines have been studied in [1], [2–5]. Stability analysis has been performed in [6], while the spontaneous sidebanding was considered in [7]. On the other hand, subharmonic and synchronous responses of rotor systems have been investigated in [3], [8–10]. Various numerical studies have revealed the rich dynamics behind this system (see [11], [12]), which includes the existence of period-doubling bifurcations and fractal basins of attraction. Apart from this, the theoretical models have been compared with experimental data, finding good correlations with the theory, see [13], [14]. The next phase of this research is the bifurcation analysis of the periodic orbits of the rotor system, which is the main concern of this article. To this end, we will use the toolbox TC-HAT (see [15]) which encompasses a set of basic tools for the bifurcation analysis of periodic

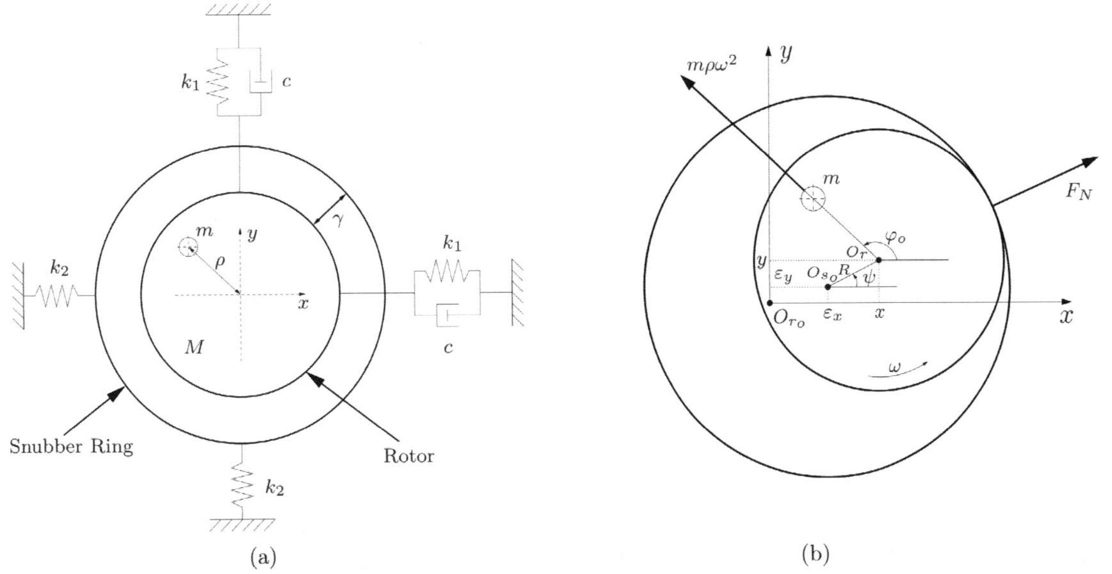

Figure 1. (a) Physical model and (b) geometrical representation of the Jeffcott rotor with a snubber ring as proposed in [14].

trajectories of non-smooth dynamical systems. It functions as a driver to a modified version of AUTO 97 (cf. [16]).

2. Physical model and equations of motion

Consider a two-degrees-of-freedom model of a Jeffcott rotor system with a snubber ring, as shown in Figure 1. A rotating mass M is excited by an out-of-balance ρm, within a massless elastic snubber ring. During operation the rotor makes intermittent contact with the snubber ring, resulting in a complex dynamical behaviour. The equations of motion are obtained in the coordinate system shown in Figure 1(b), centered at the static position of the rotor O_{r_0}. The vector $(\varepsilon_x, \varepsilon_y)$ represents the displacement of O_{r_0} from the center of the resting position of the snubber ring O_{s_0}, which is separated from the center of the rotor O_r by R. A detailed description of the variables and parameters of the model can be found in Appendix A. Under this notation, the dimensionless equations of motion are[1] (cf. [11])

$$\widetilde{x}'' + 2\nu\widetilde{x}' + \widetilde{x} + \widetilde{F}_{Nx} = \eta_m \widetilde{\rho}\eta^2 \cos(\eta\tau + \varphi_0), \tag{1}$$

$$\widetilde{y}'' + 2\nu\widetilde{y}' + \widetilde{y} + \widetilde{F}_{Ny} = \eta_m \widetilde{\rho}\eta^2 \sin(\eta\tau + \varphi_0), \tag{2}$$

where

$$\begin{pmatrix} \widetilde{F}_{Nx} \\ \widetilde{F}_{Ny} \end{pmatrix} = \frac{1}{k_1\gamma} F_N = \begin{cases} \dfrac{1}{k_1\gamma} \begin{pmatrix} k_2(R-\gamma)\cos(\psi) \\ k_2(R-\gamma)\sin(\psi) \end{pmatrix}, & R \geq \gamma, \\ 0, & R < \gamma, \end{cases} \tag{3}$$

$$= \begin{cases} \begin{pmatrix} \widetilde{K}(\widetilde{x} - \widetilde{\varepsilon}_x)\left(1 - \dfrac{1}{\widetilde{z}}\right) \\ \widetilde{K}(\widetilde{y} - \widetilde{\varepsilon}_y)\left(1 - \dfrac{1}{\widetilde{z}}\right) \end{pmatrix}, & \widetilde{z} \geq 1, \\ 0, & \widetilde{z} < 1, \end{cases} \tag{4}$$

[1] In what follows, $\widetilde{x}'$ means $\frac{d\widetilde{x}}{d\tau}$ and so on.

represents the dimensionless normal force produced when the rotor is in contact with the snubber ring.

3. Bifurcation analysis with TC-HAT

For the numerical implementation of the rotor system in TC-HAT, we need first to write the equations (1)–(2) as a set of first-order ODEs with discontinuous right-hand side:

$$v' = \begin{cases} f_{\text{imp}}(v,\alpha,\eta), & \widetilde{z} \geq 1 \text{ (i.e. the rotor is in contact with the snubber ring)}, \\ f_{\text{ins}}(v,\alpha,\eta), & \widetilde{z} < 1 \text{ (i.e. no contact with the snubber ring)}, \end{cases} \tag{5}$$

where $v = (v_1, v_2, v_3, v_4, s)$ represents the state variables of the model with $(v_1, v_2) = (\widetilde{x}, \widetilde{y})$, $s = \eta\tau \bmod 2\pi$ and $\alpha = (\eta_m, \widetilde{\rho}, \nu, \widetilde{K}, \varphi_0, \widetilde{\varepsilon}_x, \widetilde{\varepsilon}_y)$. Furthermore, the vector fields are defined as follows

$$f_{\text{ins}}(v,\alpha,\eta) = \begin{pmatrix} v_3 \\ v_4 \\ \eta_m\widetilde{\rho}\eta^2\cos(s+\varphi_0) - 2\nu v_3 - v_1 \\ \eta_m\widetilde{\rho}\eta^2\sin(s+\varphi_0) - 2\nu v_4 - v_2 \\ \eta \end{pmatrix},$$

$$f_{\text{imp}}(v,\alpha,\eta) = \begin{pmatrix} v_3 \\ v_4 \\ \eta_m\widetilde{\rho}\eta^2\cos(s+\varphi_0) - 2\nu v_3 - v_1 - \widetilde{K}(v_1 - \widetilde{\varepsilon}_x)\left(1 - \dfrac{1}{\widetilde{z}}\right) \\ \eta_m\widetilde{\rho}\eta^2\sin(s+\varphi_0) - 2\nu v_4 - v_2 - \widetilde{K}(v_2 - \widetilde{\varepsilon}_y)\left(1 - \dfrac{1}{\widetilde{z}}\right) \\ \eta \end{pmatrix}.$$

Moreover, in order to describe the trajectories of the piecewise-smooth system (5), we have introduced several segments and event functions as shown in Table 1 (cf. [15]). In this table

Index	Segment	Vector Field	Event Function	Jump Function
I_1	Inside	f_{ins}	h_{imp}	g_{id}
I_2	Impact	f_{imp}	h_{imp}	g_{id}
I_3	Phase-inside	f_{ins}	h_{ph}	g_{ph}
I_4	Phase-impact	f_{imp}	h_{ph}	g_{ph}
I_5	Turning	f_{ins}	h_{tur}	g_{id}

Table 1. Segments and event functions for the numerical analysis.

the event and jump functions are defined as follows: $h_{\text{imp}}(v,\alpha) = (v_1 - \widetilde{\varepsilon}_x)^2 + (v_2 - \widetilde{\varepsilon}_y)^2 - 1$, $h_{\text{ph}}(v,\alpha) = s - 2\pi$, $h_{\text{tur}}(v,\alpha) = v_3$, $g_{\text{id}}(v) = v$ and $g_{\text{ph}}(v) = (v_1, v_2, v_3, v_4, s - 2\pi)$. The first row of Table 1 indicates that the segment I_1 corresponds to the situation when the rotor is not in contact with the snubber ring, so the dynamics of the system is governed by the vector field f_{ins}, and during this segment we monitor the condition $h_{\text{imp}}(v,\alpha) = 0$ in order to detect an impact with the snubber ring, in which case the jump function g_{id} gives the initial point for the next segment, and so on.

To begin with the numerical analysis we choose the initial non-impacting periodic orbit shown in Figure 2(a), for the parameter values $\eta = 0.85$, $\eta_m = 0.0017$, $\widetilde{\rho} = 70$, $\nu = 0.125$, $\widetilde{K} = 30$, $\varphi_0 = 0$, $\widetilde{\varepsilon}_x = 0.75$ and $\widetilde{\varepsilon}_y = 0$. This orbit has the cyclic signature $\{I_5, I_3\}$, which is marked by the colors black (I_5) and green (I_3) in Figure 2(a). The red curve represents the impact condition

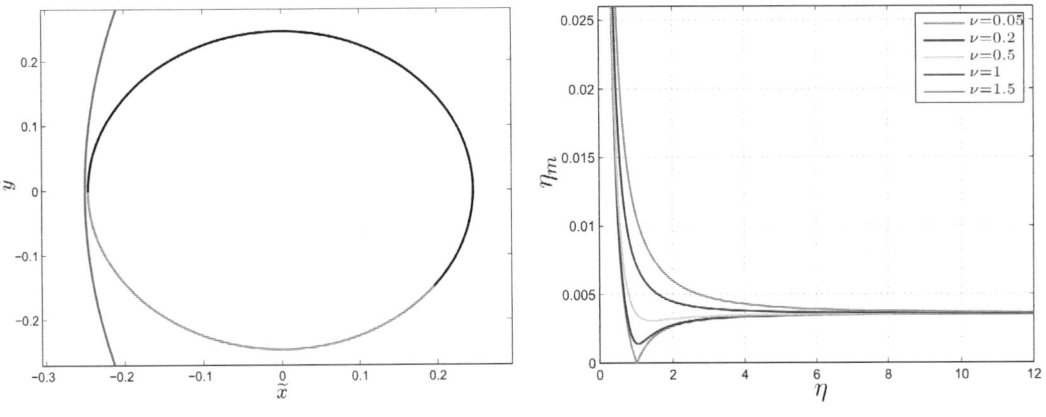

(a) (b)

Figure 2. (a) Initial periodic orbit and (b) grazing curve for different viscous damping coefficients of the rotor.

$h_{\text{imp}}(v, \alpha) = 0$, which means that in this regime the rotor is close to impacting the snubber ring. Indeed, when performing the continuation of this orbit with TC-HAT by increasing the frequency ratio η, we find a grazing bifurcation at $\eta \approx 0.85281$. With this value we can find a numerical approximation of the relation between the mass imbalance of the rotor (η_m) and the frequency of rotation (η) that produces the impact with the snubber ring (grazing curve). This two-parameter continuation is shown in Figure 2(b), for different values of viscous damping ν. Here we can identify a region between the horizonal axis and the grazing curves, where the rotor does not impact the snubber ring no matter how high the frequency is. This information is important for the analysis of forced vibrations in rotor systems, where the main concern is to determine the relation between the mass imbalance of the rotor and the frequency of rotation in order to avoid impacts with the outer snubber ring, as this phenomenon may cause serious malfunctions and damage in the rotor system.

After grazing, the dynamics of the rotor becomes nonlinear, which gives rise to a more complex behaviour. When we perform the one-parameter continuation for frequency values above the grazing point $\eta \approx 0.85281$, the following bifurcations are found: two period-doubling bifurcations (PD) at $\eta \approx 0.85395$ and $\eta \approx 0.877789$; and two fold bifurcations at $\eta \approx 1.52393$ and $\eta \approx 1.3331$. In Figure 3 we present a phase diagram of the rotor system for $\eta = 0.878$ (after the PD bifurcations) where the black and brown curves mark the co-existing period-1 and -2 orbits of (5), respectively.

4. Conclusions

We have studied numerically the dynamics of a two-degrees-of-freedom rotor model with a bearing clearance described by a non-smooth system of ODEs. During operation the rotor makes intermittent contact with an outer snubber ring, which produces a complex dynamical behaviour. In the past, similar systems have been investigated by brute force numerical methods. Here we implemented a path following method by using the toolbox TC-HAT, a module allowing to model non-smooth systems by AUTO 97. First, by varying the rotational frequency we determined period-1 orbits which make zero velocity impact with the snubber ring (grazing) for different values of the viscous damping, keeping the other system parameters fixed. Next, by applying the continuation method, we have found two period-doubling and two fold bifurcations. This reveals some similarity to grazing induced bifurcations (e.g. [17]), which are typical smooth

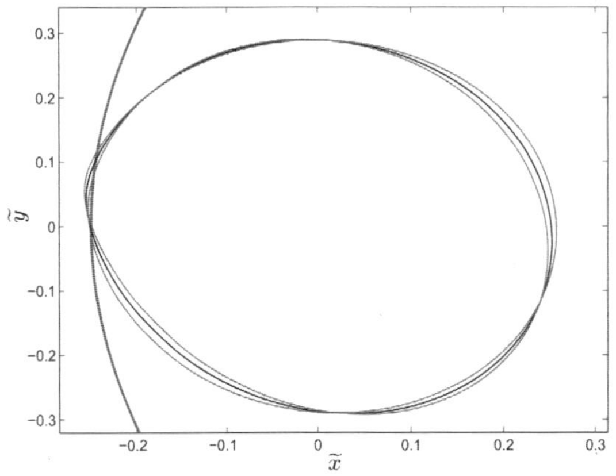

Figure 3. Periodic orbits of the rotor system after grazing.

bifurcation scenarios.

References

[1] Choy F K and Padovan J 1987 *Journal of Sound and Vibration* **113** 529–545

[2] Muszynska A and Goldman P 1995 *Chaos, Solitons and Fractals* **5** 1683–1704

[3] Childs D W 1982 *ASME Journal of Engineering Power* **104** 533–541

[4] Chu F and Zhang Z 1998 *Journal of Sound and Vibration* **210** 1–18

[5] Chu F and Zhang Z 1997 *Int. J. Eng. Sci.* **35** 963–973

[6] Ganesan R 1996 *Mech. Mach. Theory* **31** 781–798

[7] Ehrich F F 1992 *J. Vibr. Acoust.* **114** 498–505

[8] von Groll G and Ewins D J 2002 *J. Vibr. Acoust.* **124** 350–358

[9] Diken H 2001 *Journal of Sound and Vibration* **243** 117–125

[10] Ding Q 2002 *Journal of Sound and Vibration* **252** 817–833

[11] Karpenko E V, Wiercigroch M and Cartmell M P 2002 *Chaos, Solitons and Fractals* **13** 1231–1242

[12] Karpenko E V, Pavlovskaia E E and Wiercigroch M 2003 *Chaos, Solitons and Fractals* **15** 407–416

[13] Karpenko E V, Wiercigroch M, Pavlovskaia E E and Neilson R D 2006 *Journal of Sound and Vibration* **298** 907–917

[14] Gonsalves D H, Neilson R D and Barr A D S 1995 *Nonlinear Dynamics* **7** 451–470

[15] Thota P and Dankowicz H 2008 *SIAM J. Appl. Dyn. Sys.* **7** 1283–1322

[16] Doedel E J, Champneys A R, Fairgrieve T F, Kuznetsov Y A, Sandstede B and Wang X J 1997 *Auto97: Continuation and bifurcation software for ordinary differential equations (with HomCont)* Computer Science, Concordia University, Montreal, Canada available at `http://cmvl.cs.concordia.ca`

[17] Ing J, Pavlovskaia E E, Wiercigroch M and Banerjee S 2008 *Philos. Trans. R. Soc. Lond., Ser. A, Math. Phys. Eng. Sci.* **366** 679–704

Appendix A. Variables and parameters of the model

Notation	Description
c	viscous damping coefficient
k_1	rotor stiffness
k_2	snubber ring stiffness
M	mass of the rotor
ρm	out-of-balance
t	time
x	displacement of the rotor in the horizontal direction
y	displacement of the rotor in the vertical direction
R	radial displacement of the rotor relative to the equilibrium position of the snubber ring
ε_x	eccentricity of the rotor in the x-direction
ε_y	eccentricity of the rotor in the y-direction
F_N	normal force produced when the rotor hits the snubber ring
γ	radial clearance between the rotor and the snubber ring
ω_n	natural frequency of the rotor $\sqrt{\frac{k_1}{M}}$
φ_0	initial phase shift
ω	shaft rotational velocity
$\widetilde{K}$	stiffness ratio $\frac{k_2}{k_1}$
$\widetilde{x}$	displacement ratio of the rotor in the x-direction $\frac{x}{\gamma}$
$\widetilde{y}$	displacement ratio of the rotor in the y-direction $\frac{y}{\gamma}$
$\widetilde{z}$	radial displacement ratio $\frac{R}{\gamma}$
$\widetilde{\varepsilon}_x$	dimensionless eccentricity of the rotor in the x-direction $\frac{\varepsilon_x}{\gamma}$
$\widetilde{\varepsilon}_y$	dimensionless eccentricity of the rotor in the y-direction $\frac{\varepsilon_y}{\gamma}$
η	frequency ratio $\frac{\omega}{\omega_n}$
η_m	mass ratio $\frac{m}{M}$
ν	damping ratio of the rotor $\frac{c}{2\sqrt{k_1 M}}$
τ	dimensionless time $\omega_n t$
ψ	angle between the radial displacement of the rotor and the horizonal axis
$\widetilde{\rho}$	dimensionless radius $\frac{\rho}{\gamma}$

Modern Practice in Stress and Vibration Analysis 2012 (MPSVA 2012) IOP Publishing
Journal of Physics: Conference Series **382** (2012) 012006 doi:10.1088/1742-6596/382/1/012006

Linear Modal Analysis of L-shaped Beam Structures - Parametric Studies

Fotios Georgiades[1], Jerzy Warminski[1], Matthew P. Cartmell[2]

[1]Department of Applied Mechanics, Lublin University of Technology, Lublin, Poland

[2]School of Engineering, University of Glasgow, Glasgow, UK

f.georgiadis@pollub.pl

Abstract. Linear modal analysis of L-shaped beam structures indicates that there are two independent motions, these are in-plane bending and out of plane motions including bending and torsion. Natural frequencies of the structure can be determined by finding the roots of two transcendental equations which correspond to in-plane and out-of-plane motions. Due to the complexity of the equations of motion the natural frequencies cannot be determined explicitly. In this article we nondimensionalise the equations of motion in the space and time domains, and then we solve the transcendental equations for selected values of the L-shaped beam parameters in order to determine their natural frequencies. We use a numerical continuation scheme to perform the parametric solutions of the considered transcendental equations. Using plots of the solutions we can determine the natural frequencies for a specific L-shape beam configuration.

1. Introduction

There has been continuous interest in the dynamics of L-shaped coupled structures since the 1960s. Roberts and Cartmell in [1,2] studied autoparametric resonances within an L-shape beam structure. Balachandran and Nayfeh performed nonlinear dynamic analysis considering only in-plane motions [3]. Warminski *et al* formulated in [4] the third order partial differential nonlinear equations of an L-shaped beam structure with different flexibilities in the two orthogonal directions, without taking into account rotary inertia effects. Ozonato *et al.* studied post-buckled chaotic vibrations of an L-shaped beam structure considering nonlinear motions but only in plane bending [5]. In [7] Georgiades *et al.* extracted the linear equations of motion of an L-shaped beam structure, considering the inextensionality conditions and rotary inertia terms and in [8] they performed a linear modal analysis of an L-shaped beam structure. The natural frequencies of a given L-shaped beam structure can be determined by the roots of two transcendental equations which correspond to in-plane bending and out-of-plane motions.

In this article we start from the equations of motion, which we nondimensionalise in space and time. Then the nondimensional frequencies can be determined by the roots of the two transcendental equations. Using a numerical continuation technique we perform a parametric study to determine the nondimensional natural frequency of the first mode for in-plane and out-of-plane motions for various sets of parameters.

2. Theory

2.1 Equations of motion

Let us consider beams made of isotropic material and constant cross section with respect to the longitudinal direction. Initially, we nondimensionalise the equations of motion in space and time, using,

$$t = a_i\tau \text{ and } s = l_1 S. \tag{1a,b}$$

noting that the scaling in time is different for the two different motions, with i = 1,2 for in-plane motion and out-of-plane motion respectively. For in-plane bending we use the following nondimensionlised parameters,

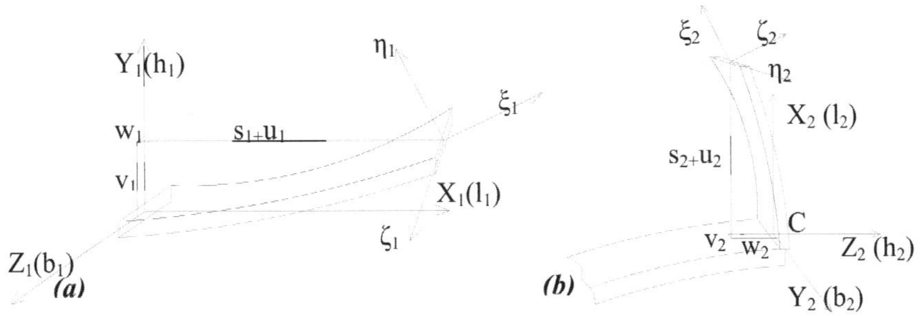

Figure 1. Indication of axis orientations, displacements and dimensions for (a) the primary beam, and (b) the secondary beam.

$$a_1 = \sqrt{\frac{\bar{D}_{\zeta_1} m_1 l_1^4}{D_{\zeta_1}}},\ \bar{m}_2 = \frac{m_2}{m_1},\ \bar{l}_2 = \frac{l_2}{l_1},\ \bar{D}_{\eta_2} = \frac{D_{\eta_2}\bar{D}_{\zeta_1}}{D_{\zeta_1}}, \tag{2a-d}$$

for which $\bar{D}_{\zeta_1}$ can have any arbitrary value.

For the out-of-plane motion we use the following nondimensionlised parameters,

$$\begin{gathered} a_2 = \sqrt{\frac{\bar{D}_{\eta_1} m_1 l_1^4}{D_{\eta_1}}},\ \bar{m}_2 = \frac{m_2}{m_1},\ \bar{l}_2 = \frac{l_2}{l_1},\ \bar{D}_{\zeta_2} = \frac{D_{\zeta_2}\bar{D}_{\eta_1}}{D_{\eta_1}},\ \bar{I}_{\eta_1} = \frac{I_{\eta_1}}{m_1 l_1^2}, \\ \bar{I}_{\xi_1} = \frac{I_{\xi_1}}{m_1 l_1^2},\ \bar{D}_{\xi_1} = \frac{D_{\xi_1}\bar{D}_{\eta_1}}{D_{\eta_1}},\ \bar{I}_{\xi_2} = \frac{I_{\xi_2}}{m_1 l_1^2},\ \bar{D}_{\xi_2} = \frac{D_{\xi_2}\bar{D}_{\eta_1}}{D_{\eta_1}}, \end{gathered} \tag{3a-i}$$

where $\bar{D}_{\eta_1}$ can have any arbitrary value. Before proceeding with the parametric study we define the following dimensionless ratios,

$$\beta_1 = \frac{b_2}{b_1},\ \beta_2 = \frac{b_2}{h_1},\ \beta_3 = \frac{h_1}{l_1},\ \beta_4 = \frac{h_2}{l_1}. \tag{4a-d}$$

Therefore using these ratios (eq. 4a-d) and considering that both beams are made of the same material (aluminium with $\nu = 0.33$) and considering the transcendental equations given by [8] we determine the nondimensional natural frequencies.

3. Results of Parametric Study

The roots of the transcendental equations have been determined using the optimisation toolbox within Matlab. In both cases, we used a numerical continuation technique. According to this technique the initial input value for the optimisation technique was the one from the already determined nearest parameter value, but using a very fine mesh for the variation of each parameter in a specific range. In this way we tried to avoid jumps from the roots of the transcendental equations which would correspond to the first mode to other roots which correspond to higher modes. In order to scale the system of equations appropriately, and to avoid singular values in parts of the transcendental equations, we used the following nondimensional rescaling value for the stiffnesses:

$$\bar{D}_{\zeta_1} = \bar{D}_{\eta_1} = 1000.$$

Therefore, for a specific configuration of the L-shaped beam structure, and considering the results from the parametric study, the natural frequencies in rad/sec are given by,

$$\omega_{in,0} = \frac{\omega_{in}}{a_1},\ \text{and}\ \omega_{out,0} = \frac{\omega_{out}}{a_2}. \tag{26a,b}$$

3.1 In-plane motion

In the case of in-plane motion we performed a parametric study considering variations of $\bar{m}_2$, $\bar{D}_{\eta_2}$ and $\bar{l}_2$. Figures 2a-d depict the nondimensional natural frequencies which correspond to the first natural frequency for the in-plane bending. There are two characteristics in Fig. 2a-d which clearly demonstrate: a) the stiffening effect with the increase of bending stiffness, or decrease of mass, and b) that for low length dimensions ($\bar{l}_2$) the secondary beam behaves like a tip-mass, and the bending stiffness is not playing any essential role.

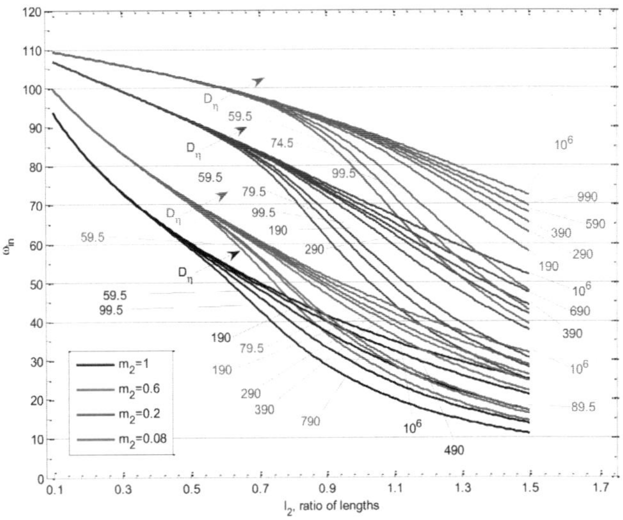

Figure 2a. Nondimensional natural frequencies of the first mode for in-plane motion.

3.2 Out-of-plane motion

In Figure 3-5 the nondimensional natural frequencies of the first mode are depicted for out of plane motion for β_1=9,1,0.1 respectively. In these Figures the left side corresponds to $\bar{l}_2 \in [0.1,0.8]$ and the right side Figures correspond to $\bar{l}_2 \in [0.8,1.5]$, noting that the parametric values of the pair (β_2, β_4) are also indicated. The convergence of many curves to one with a decrease in $\bar{l}_2$, for which the secondary beam behaves like a tip mass, is evident only in Figs. 3-5 and in those cases of very small dimensions for the secondary beam in comparison to the primary beam dimensions.

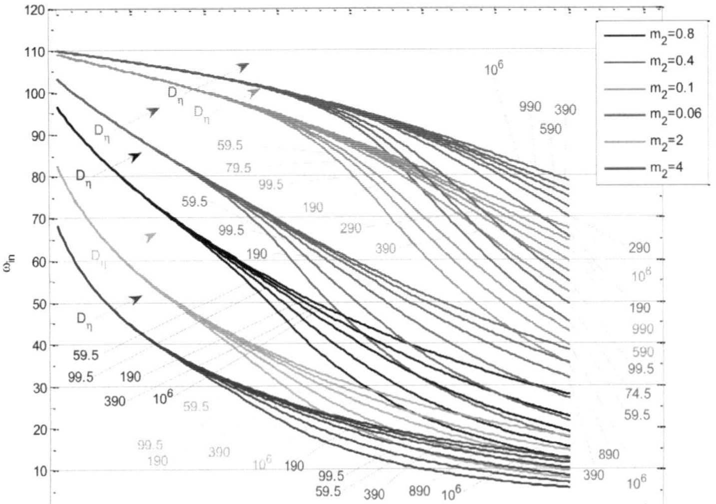

Figure 2b. Nondimensional natural frequencies of the first mode for in-plane motion.

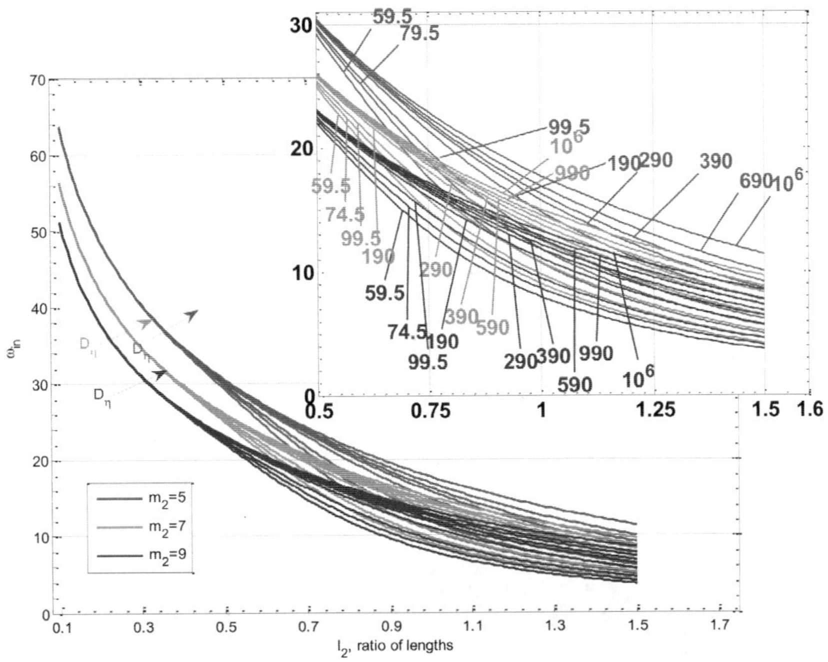

Figure 2c. Nondimensional natural frequencies of the first mode for in-plane motion.

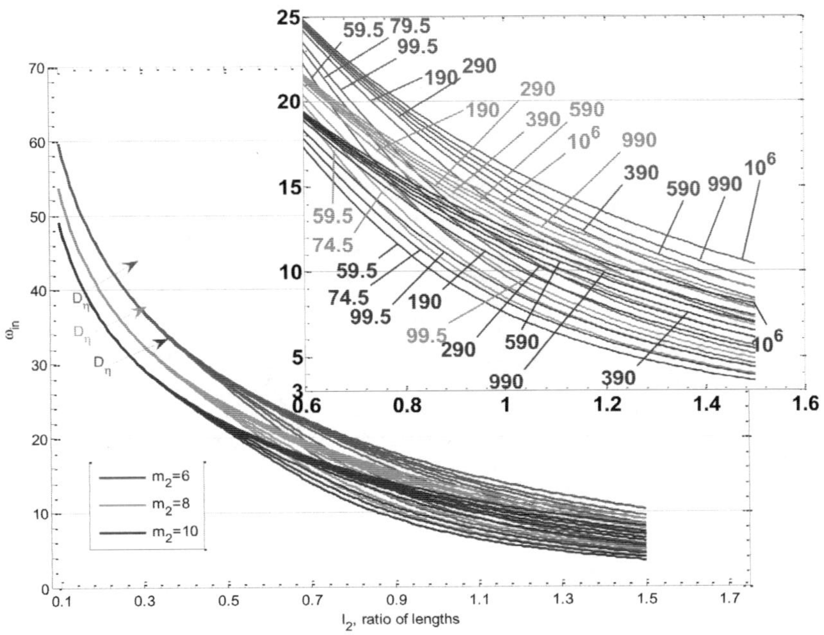

Figure 2d. Nondimensional natural frequencies of the first mode for in-plane motion.

4. Conclusions

In this article a parametric study was performed for the determination of the nondimensional natural frequency of the first system mode for in-plane and out-of-plane bending, for various sets of parameters. For the case of in-plane bending the effects related to an increase in stiffness or a decrease in mass were demonstrated. When the dimensions of the secondary beam are much smaller than those of the primary beam then the system behaves like a cantilever beam with a tip mass.

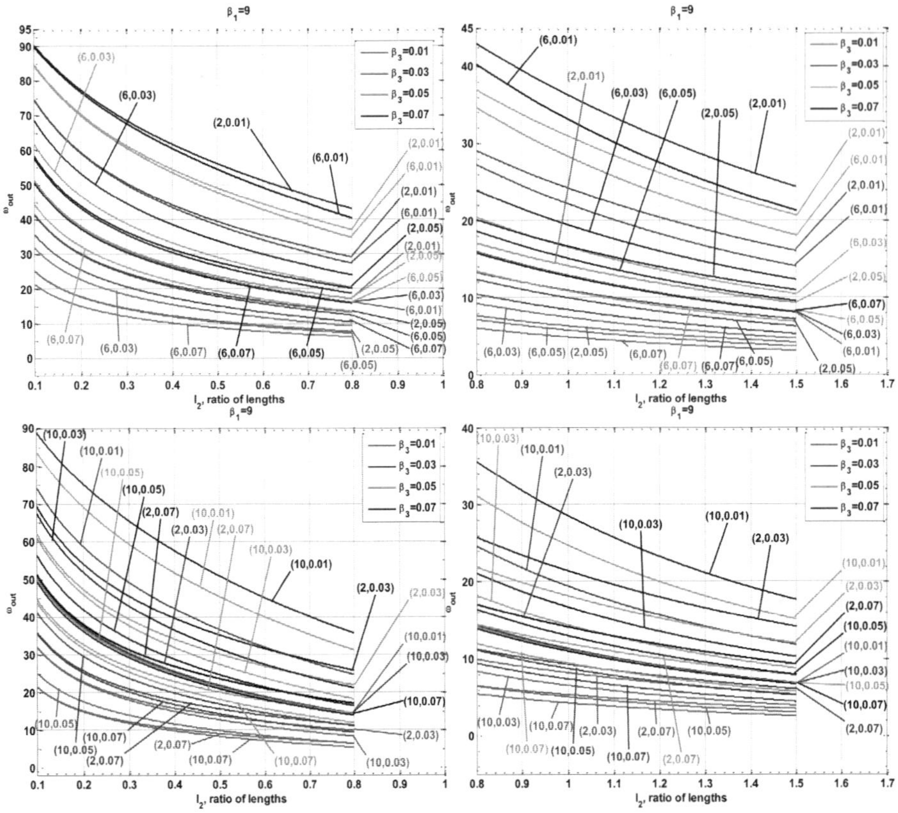

Figure 3. Nondimensional frequencies for the first mode of out-of-plane motions with β_1=9.

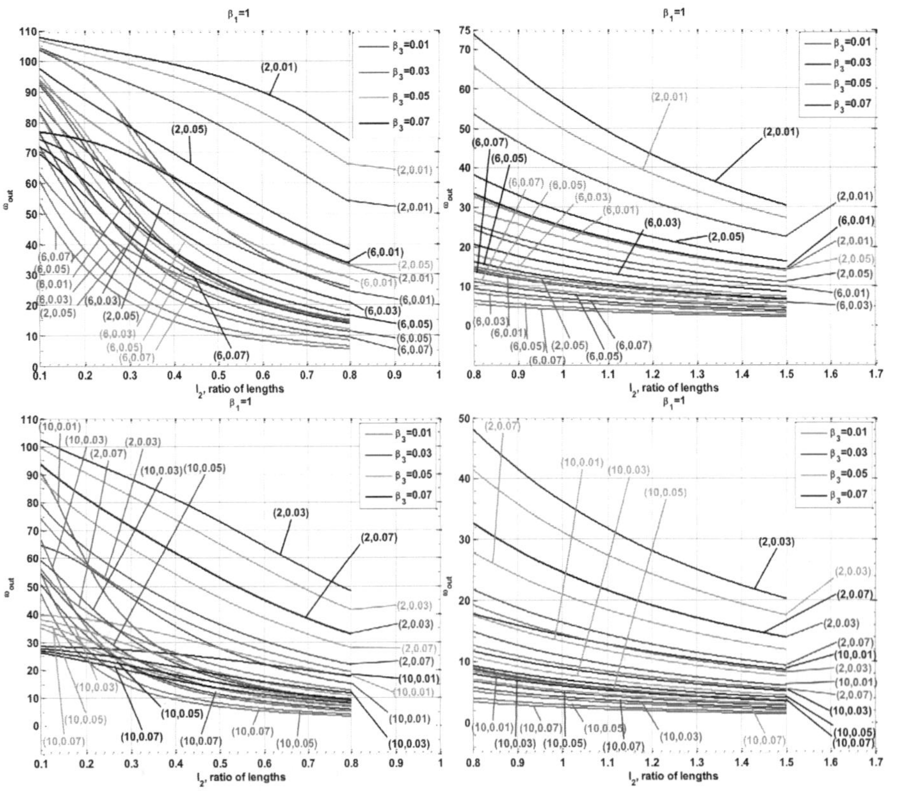

Figure 4. Nondimensional frequencies for the first mode of out-of-plane motions with β_1=1.

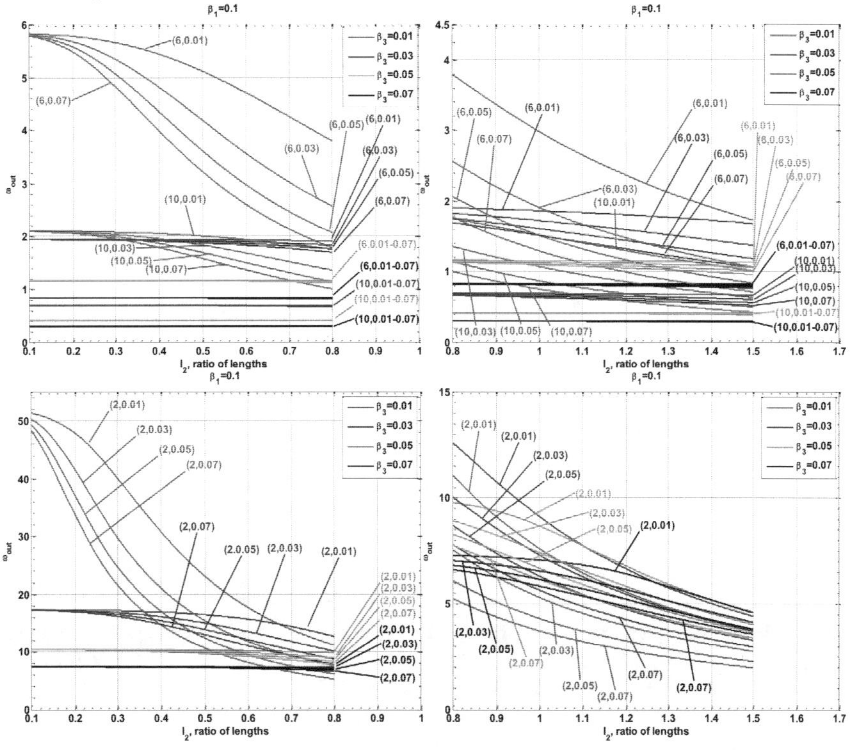

Figure 5. Nondimensional frequencies for the first mode of out-of-plane motions with β_1=0.1.

5. Acknowledgements
The research leading to these results has received funding from the European Union Seventh Framework Programme (FP7/2007-2013), FP7 - REGPOT - 2009 - 1, under grant agreement No:245479. The support by Polish Ministry of Science and Higher Education – Grant No. 1471-1/7.PRUE/2010/7 – is also acknowledged by the second author.

6. References
[1] Roberts, J.W., Cartmell, M.P., 1984, Forced vibration of a beam system with autoparametric coupling effects, Strain, 20, 123-131.
[2] Cartmell, M.P., Roberts, J.W., 1987, Simultaneous combination resonances in a parametrically excited cantilever beam, Strain, 23, 117–126.
[3] Balachandran, B., Nayfeh A.H., 1990, Nonlinear motion of beam-mass structure, Nonlinear Dynamics, 1, 39–61.
[4] Warminski, J., Cartmell, M.P., Bochenski, M., Ivanov, I., 2008, Analytical and experimental investigations of an autoparametric beam structure, Journal of Sound and Vibration, 315, 486-508.
[5] Ozonato, N., Nagai, K.-I., Maruyama, S., Yamaguchi, T., 2012, Chaotic Vibrations of A Post-Buckled L-Shape Beam with an Axial Constraint, Nonlinear Dynamics, 67, 2363-2379.
[6] Nayfeh, A.H., Pai, F., 2004, Linear and Nonlinear Structural Mechanics, John Willey, New Jersey.
[7] Georgiades, F., Warminski, J., Cartmell, M.P., Towards Linear Modal Analysis for an L-shaped Beam-Equations of Motion, Mechanics Research Communications, under review.
[8] Georgiades, F., Warminski, J., Cartmell, M.P., Linear Modal Analysis of L-shaped Beam, Mechanical systems and signal processing, under review.
[9] Ewins, D.J., 2000, Modal Testing (theory, practice and application), Research Studies Press Ltd.

Journal of Physics: Conference Series **382** (2012) 012007 doi:10.1088/1742-6596/382/1/012007

An analysis of the effects of the orientation angle of a surface crack on the vibration of an isotropic plate

Rainah Ismail and M P Cartmell

School of Engineering, University of Glasgow, Glasgow G12 8QQ, Scotland, UK

Email: rismail@eng.gla.ac.uk

Abstract. This paper proposes a vibration analysis for an isotropic plate containing an arbitrarily orientated surface crack as an enhancement to previous work on cracked plates for which the orientation of the crack angle was not included. The governing equation of motion of the plate with this enhanced crack modelling represents the vibrational response based on classical plate theory into which a developed crack model has been assimilated. The formulation of the angled crack is based on a simplified line-spring model and the cracked plate is subjected to transverse harmonic excitation with arbitrarily chosen boundary conditions. It is found that the vibrational characteristics of the plate structure can be affected significantly by the orientation of the crack in the surface plate. For reasons of comparison and validation a finite element model is used for a further modal analysis in order to corroborate the effect of crack length and crack orientation angle on the modal parameters i.e. the natural frequency and also the vibrational amplitude, as predicted by the analysis. The results show excellent agreement between the two methods.

1. Introduction

A comprehensive background on plates has been provided by Timoshenko and Woinowsky-Krieger [1], and the research of this paper is focused on thin plate structures in order to develop enhanced modelling for reliable, light and efficient structures. This type of structure can lead to unwanted instances of high vibration and then may lead to damaging effects on the structure. Cracks can form and propagate catastrophically with very little warning and then can create a complete unbalance of the structure which leads to ultimate failure. Failure of a structure can result in terrible consequences, economically and most probably and importantly in terms of loss of life. The dynamic responses of rectangular plates with cracks, or minor irregularities under different loading conditions, have been investigated in the past by many researchers for different boundaries conditions [2]. The length, position, and orientation of a crack will affect the vibration characteristics of any host structure, in addition to other effects caused by material properties, plate geometry, and boundary conditions.

Most studies of vibration analysis in cracked plates have investigated cracks which have tended to be located at the centre or edge of the plate, and parallel to one side of the plate. Only a few papers [3-6] have investigated the vibration analysis of a plate with a crack which is not parallel along one side of the plate. However, no literature contains any substantial references to the study of a plate with cracks of variable orientation undergoing forced vibration. The aim of this research is to extend the forced vibration analysis of the cracked plate discussed by Israr [2] and Israr *et al.* [7] by considering an alternative geometry whereby the crack orientation is variable.

2. Modelling of a plate with a crack of variable angular orientation

An analytical approach is presented for the forced vibration analysis of a plate containing an arbitrarily orientated surface crack, based on three different boundary conditions. The method was based on classical plate theory. Firstly, the equation of motion is derived for an isotropic plate containing the angled surface crack which is arbitrarily orientated at an angle β with respect to the x axis of the plate, and this plate is loaded by an arbitrarily located concentrated force, $\bar{q}_z$ at some point (x_0, y_0) [8]. The crack formulation representing the surface crack of variable angular orientation with length $2a$ is based on a simplified line-spring model [8-9]. The final form of the equation of motion for the forced vibration in a thin plate with a variably oriented surface crack emerges, as follows,

$$D\left(\frac{\partial^4 w}{\partial x^4}+2\frac{\partial^4 w}{\partial x^2\partial y^2}+\frac{\partial^4 w}{\partial y^4}\right)=-\rho h\frac{\partial^2 w}{\partial t^2}+n_x\frac{\partial^2 w}{\partial x^2}+q_z+\frac{aD\left(1+\cos 2\beta\right)\left(\frac{\partial^4 w}{\partial y^4}+\nu\frac{\partial^4 w}{\partial x^2\partial y^2}\right)}{3\left(\frac{\alpha_{bt}}{6}+\alpha_{bb}\right)(3+\nu)(1-\nu)h+2a}$$
$$-\frac{a(1+\cos 2\beta)}{(6\alpha_{tb}+\alpha_{tt})(1-\nu^2)h+2a}n_o\frac{\partial^2 w}{\partial y^2}+\frac{2aD\sin 2\beta\left(\frac{\partial^4 w}{\partial x\partial y^3}+\nu\frac{\partial^4 w}{\partial x^3\partial y}\right)}{3\left(\frac{C_{bt}}{6}+C_{bb}\right)(1+\nu)h+2a}-\frac{2a\sin 2\beta}{(6C_{tb}+C_{tt})(1+\nu)h+2a}n_o\frac{\partial^2 w}{\partial x\partial y} \tag{1}$$

where D is the flexural rigidity of the plate, represented by $D=Eh^3/12(1-\nu^2)$, with E, h and ν representing the modulus of elasticity, plate thickness, and Poisson ratio, respectively, and where $\rho h(\partial^2 w/\partial t^2)$ is the inertia force, with ρ defining the plate density. n_x and n_0 are the in-plane forces in the x and y directions, respectively. The non-dimensional compliance coefficients α_{tt}, α_{bb} and $\alpha_{bt}=\alpha_{tb}$ can be found in [10], and the compliance coefficients C_{tt}, C_{bb}, and $C_{bt}=C_{tb}$ can be seen in [11] and [12]. Next Galerkin's method is applied in the usual manner to discretise the partial differential equation in (1) and transform the transverse deflection coordinate $w(x, y, t)$ into time dependent modal coordinates. Subsequently, by employing the Berger formulation, the derived governing equation of motion of the cracked plate is converted into a nonlinear ordinary differential equation model in terms of a chosen modal coordinate, and then by considering the system to be under the influence of weak classical linear viscous damping μ, and the load to be harmonic [2], the final equation of motion for the system takes the form of a specialised Duffing equation,

$$\ddot{\psi}_{ij}(t)+2\mu\dot{\psi}_{ij}(t)+\omega_{ij}^2\psi_{ij}(t)+\gamma_{ij}\psi_{ij}^3(t)=\frac{\eta_{ij}}{D}q\cos\Omega_{ij}t \tag{2}$$

where $\omega_{ij}^2=\frac{K_{ij}}{M_{ij}}$, $\gamma_{ij}=\frac{G_{ij}}{M_{ij}}$, and $\eta_{ij}=\frac{Q_{ij}}{M_{ij}}$ (3) and $M_{ij}=\frac{\rho h}{D}\sum_{i=1}^{\infty}\sum_{j=1}^{\infty}A_{ij}\int_0^{l_1}\int_0^{l_2}X_i^2Y_j^2\,\mathrm{d}x\mathrm{d}y$ (4)

$$K_{ij}=\sum_{i=1}^{\infty}\sum_{j=1}^{\infty}A_{ij}\int_0^{l_1}\int_0^{l_2}\Bigg[X_i^{iv}Y_j+2X_i^{"}Y_j^{"}+Y_j^{iv}X_i-\frac{a(1+\cos 2\beta)}{3\left(\frac{\alpha_{bt}}{6}+\alpha_{bb}\right)(3+\nu)(1-\nu)h+2a}\left(Y_j^{iv}X_i+\nu X_i^{"}Y_j^{"}\right)$$
$$-\frac{2a\sin 2\beta}{3\left(\frac{C_{bt}}{6}+C_{bb}\right)(1+\nu)h+2a}\left(X_i^{'}Y_j^{'''}+\nu X_i^{'''}Y_j^{'}\right)\Bigg]X_iY_j\,\mathrm{d}x\mathrm{d}y \tag{5}$$

$$G_{ij}=\sum_{i=1}^{\infty}\sum_{j=1}^{\infty}A_{ij}^3\int_0^{l_1}\int_0^{l_2}\left[-P_{1ij}X_i^{"}Y_j+\frac{a(1+\cos 2\beta)}{(6\alpha_{tb}+\alpha_{tt})(1-\nu^2)h+2a}P_{2ij}X_iY_j^{"}+\frac{2a\sin 2\beta}{(6C_{tb}+C_{tt})(1+\nu)h+2a}P_{2ij}X_i^{'}Y_j^{'}\right]X_iY_j\,\mathrm{d}x\mathrm{d}y \tag{6}$$

$$P_{1ij}=\frac{6}{h^2l_1l_2}\sum_{i=1}^{\infty}\sum_{j=1}^{\infty}\int_0^{l_1}\int_0^{l_2}\left[\left(\frac{\partial X_i}{\partial x}\right)^2Y_j^2+\nu\left(\frac{\partial Y_i}{\partial y}\right)^2X_j^2\right]\mathrm{d}x\mathrm{d}y,\quad P_{2ij}=\frac{D}{h^2l_1l_2}\sum_{i=1}^{\infty}\sum_{j=1}^{\infty}\int_0^{l_1}\int_0^{l_2}\left[\left(\frac{\partial Y_j}{\partial y}\right)^2X_i^2+\nu\left(\frac{\partial X_i}{\partial x}\right)^2Y_j^2\right]\mathrm{d}x\mathrm{d}y \tag{7}$$

and $q_{ij} = \frac{q_o(t)}{D} Q_{ij}$ where $Q_{ij} = X_i(x_0)\, Y_j(y_0)$ is the integral of the delta function given by Israr *et al.* [7] in the form,

$$\int_{-\infty}^{\infty} X_i(x)\delta(x - x_o)dx = X_i(x_o) \tag{8}$$

also noting that Ω_{ij} is the excitation frequency, A_{ij} is the arbitrary amplitude, ψ_{ij} is the time dependent modal coordinate for the system, and, X_i and Y_j represent the characteristic or modal functions in the x and y directions of the cracked rectangular plate, respectively, with i, j being the plate mode designators.

3. Numerical results

In this section, simulation results are presented for the intact plate and the enhanced cracked plate model with arbitrarily chosen CCFF type boundary conditions. The type of material used in this investigation is an aluminium alloy of 5083 grade, with the modulus of elasticity $E = 7.03 \times 10^{10}$ Nm^{-2}, plate density ρ = 2660 kgm^{-3}, Poisson's ratio ν = 0.33, and measured damping ratio of μ = 0.08. A convergence study was firstly carried out with [8] in order to verify the correctness of the model within an analytical model of the plate for three different types of boundary conditions [9]. In section 3.1 simulation results are presented for investigation of the natural frequency of the first mode of the intact plate and enhanced cracked plate model for various aspect ratios, and factors that influence changes in the trend of the natural frequency for this boundary condition are discussed in section 3.2.

Table 1. Natural frequencies for the clamped-clamped free-free (CCFF) boundary condition

Crack angle, β (deg)	First Mode Natural Frequency, ω_{ij} (rad/s)					
	Intact Plate	Cracked Plate a = 3mm	Cracked Plate a = 7.5mm	Intact Plate	Cracked Plate a = 3mm	Cracked Plate a = 7.5mm
	Square plate (l_1=0.3, l_2=0.3)			Square plate (l_1=0.15, l_2=0.15)		
	268.22	-	-	1072.89	-	-
0°		247.29	223.78		989.15	895.10
20°		265.40	247.30		1061.58	989.21
40°		279.11	268.65		1116.42	1074.59
60°		282.94	279.22		1131.74	1116.86
80°		275.47	275.35		1101.90	1101.42
90°		268.22	268.22		1072.89	1072.89
	Rectangular plate (l_1=0.15, l_2=0.3)			Rectangular plate (l_1=0.3, l_2=0.15)		
	770.27	-	-	770.27	-	-
0°		743.88	715.78		650.48	498.48
20°		765.52	742.67		703.75	586.80
40°		782.78	769.00		757.28	689.93
60°		788.05	783.03		788.12	762.82
80°		779.10	778.91		784.84	782.59
90°		770.27	770.27		770.27	770.27

3.1. A plate with a surface crack of variable angular orientation

Dimensions of the plate considered in this investigation are l_1= 150 mm, l_2 = 300 mm, and plate thickness h = 3mm. A load of 10 N was applied at a distance of 225 mm from the fixed edge on the x-axis and 115 mm from the fixed edge on the y-axis. Table 1 shows the results for the first mode natural frequencies ω_{ij} at different lengths of half-crack and for different values of crack orientation angle, β. The orientation angle is arbitrarily chosen from 0° to 80°, in 20° steps and finishes at 90°. The crack is also located at the centre of the plate with 3mm and 7.5mm half-crack lengths, respectively. The results generally show that the natural frequency reduces with an increase in half-crack length. Furthermore, in terms of the crack orientation angle effect it can clearly be seen that the natural frequency increases with the increase in the crack angle, up to 60° and then decreases when β exceeds 60°. This similar trend in the crack orientation effect was also found by [3] and [4].

3.2. Factors which influence changes in the natural frequency

In this section a parametric study is performed on the natural frequency equation of the cracked plate for the case of the CCFF boundary condition. This type of boundary condition is selected because the results showed interesting changes in the trend of the natural frequency values at an angle of 60°, and this trend is different to those for SSSS and CCSS. This study is intended to examine configurations to determine the physical parameters that influence the changeover of the maximum value of the crack orientation angle. Here, *maximum* means that value of crack orientation angle for which the natural frequency reaches a maximum value before it decreases. It is found that all the factors of Figure 1 influence the natural frequency, for example when the half-crack length is 1 mm the natural frequency increases for the crack angle, β up to 50.77° and reduces when β exceeds 50.77°. When the half-crack length increases to 30 mm, the natural frequency increases up to 75.06° and then decreases when β exceeds 75.06°. The results, given in Figure 1(a)-(c) show that the crack orientation angle increases with the crack length and plate aspect ratio, however it decreases with an increase in the plate thickness (Figure 1(b)). For the Poisson ratio it can be seen that in Figure 1(d) the crack angle decreases up to 0.42, and then increases when the ratio exceeds 0.42 while the results of density and modulus of elasticity show a very slight difference for the crack orientation angle for a maximum natural frequency value in which significant changes cannot be observed from Figures 1(e)-(f).

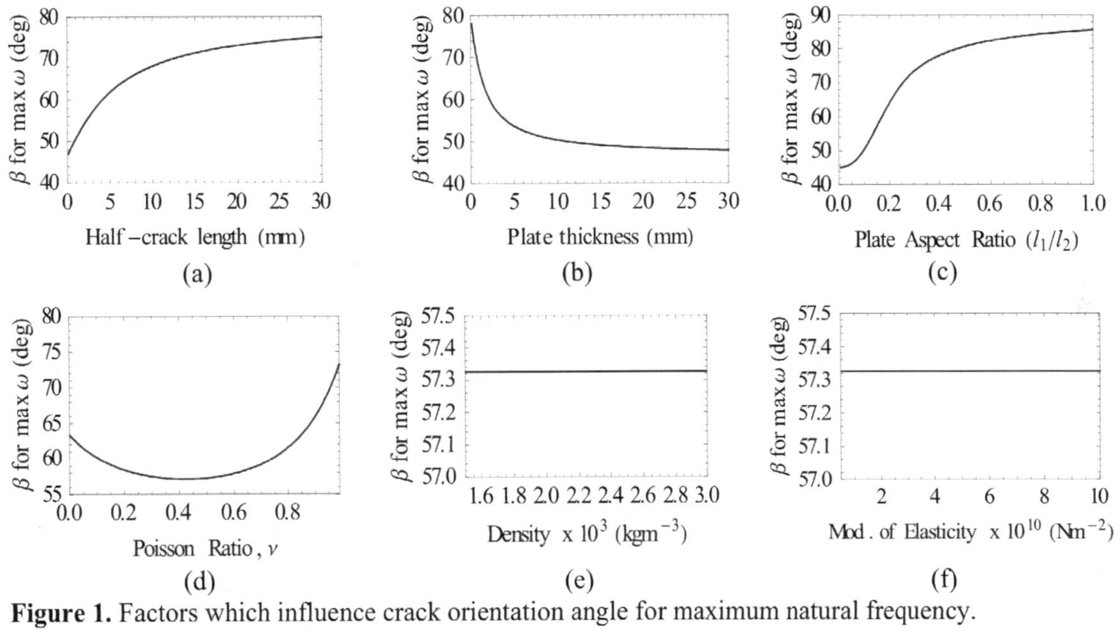

Figure 1. Factors which influence crack orientation angle for maximum natural frequency.

4. Finite Element Method (ABAQUS/CAE 6.9.1)

In this study the modal characteristics of an intact plate and twelve cracked plates, namely the natural frequencies, mode shapes and displacements, were all evaluated using Abaqus/CAE. In Table 2 the frequency values from frequency extraction analysis for the three first modes of the cracked plate for orientation angle β are shown together with the results for an intact plate. It can be seen that the large crack shifts the frequency values of the 1st, 2nd and 3rd modes downwards, as expected, and is due to reduced plate stiffness. For both crack lengths of 3 mm and 7.5 mm, the frequency values increase monotonously from 0° up to 60°, and then decrease when β is more than 60°. Table 2 also shows the results of the amplitude response of the simulations, and, as expected, the amplitude values increase due to the small crack in the plate. These amplitude responses are also affected by the crack orientation angle where the amplitude decreases from 0° up to 60° when the frequency value increases, and then increases again when the frequency value decreases. Figure 2 shows the first three mode shapes of the plates for a crack length of 3 mm and orientation angle of 60°. The dark blue areas in these figures indicate nodal displacements for the first three modes of vibration, representing the areas where the displacement is close to zero.

Table 2. Frequency extraction analysis and amplitude response

			FEA Results			
		Crack Orientation angle, β	Frequency (Hz)			Amplitude (mm)
			Mode I	Mode II	Mode III	Mode I only
Intact Plate		-	122.94	259.80	525.16	7.372
Cracked Plate	3 mm	0°	122.73	259.32	524.18	7.483
		20°	122.77	259.49	524.37	7.379
		40°	122.84	259.82	525.00	7.263
		60°	123.05	260.79	527.47	7.200
		80°	122.82	259.54	524.38	7.727
		90°	122.73	259.36	524.22	7.739
Cracked Plate	7.5 mm	0°	122.69	259.13	524.00	7.564
		20°	122.75	259.37	524.18	7.227
		40°	122.81	259.74	524.79	6.969
		60°	123.04	260.72	527.34	6.796
		80°	122.81	259.57	524.29	7.309
		90°	122.69	259.33	524.12	7.614

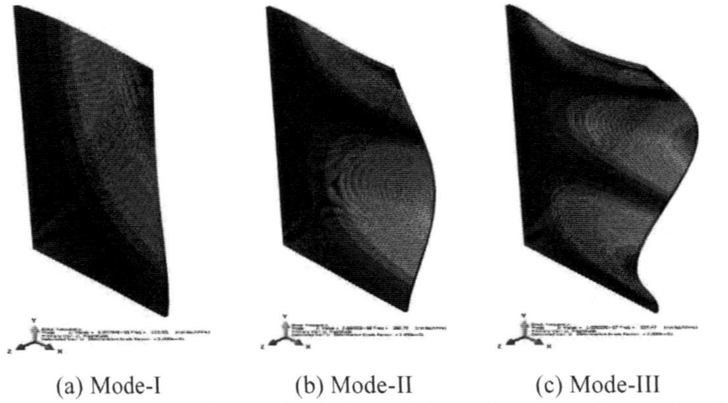

(a) Mode-I (b) Mode-II (c) Mode-III

Figure 2. Vibration mode shapes of a cracked plate with crack orientation angle, $\beta = 60°$.

5. Comparative study

A comparative study of the theoretical modelling and FE approaches is presented here. Both sets of results in Table 3 show a significant change in natural frequency and response amplitude for the different lengths of the crack and the varying orientation angle of the crack. The natural frequencies obtained for both results decrease slightly with an increase in the crack length for every crack inclination angle. In addition, it is apparent that the FE predicted frequency and amplitude trends are similar to the analytical results for which the frequency values increase from 0° up to 60° and then decrease when β is more than 60°, while the amplitude responses behave conversely.

6. Conclusions

The results show overall that the cracked plate model is very sensitive to the crack length and orientation angle and that it is able to predict well for the cases tested using the CCFF boundary conditions. Orientation angles for which the natural frequency is maximum are affected by crack

Table 3. Comparison results between theoretical and FE analysis for Mode I only.

		Crack Orientation angle, β	Frequency (Hz) Analy.	Frequency (Hz) FEA	Error (%)	Amplitude (mm) Analy.	Amplitude (mm) FEA	Error (%)
Intact Plate		-	122.58	122.94	0.29	7.979	7.372	7.61
Cracked Plate	3 mm	0°	118.38	122.73	3.67	8.019	7.483	6.68
		20°	121.82	122.77	0.78	7.220	7.379	2.20
		40°	124.57	122.84	1.39	6.993	7.263	3.86
		60°	125.41	123.05	1.88	6.898	7.200	4.38
		80°	123.98	122.82	0.94	7.524	7.727	2.70
		90°	122.58	122.73	0.12	7.979	7.739	3.01
Cracked Plate	7.5 mm	0°	113.91	122.69	7.71	8.066	7.564	6.22
		20°	118.18	122.75	3.87	6.951	7.227	3.97
		40°	122.37	122.81	0.36	6.659	6.969	4.66
		60°	124.61	123.04	1.26	6.548	6.796	3.79
		80°	123.95	122.81	0.92	7.336	7.309	0.37
		90°	122.58	122.69	0.09	7.979	7.614	4.57

length, plate thickness, plate aspect ratio, Poisson ratio, plate density and modulus of elasticity. Simulation with standard parameter properties showed changes in the natural frequency values at an angle of approximately 57° which is reasonably close to 60°. Very close agreement was obtained between the analytical and FE results with the maximum error in the prediction of the frequency value at about 7.71% and the amplitude response at 7.61%.

References

[1] S.P.Timoshenko, S.Woinowsky-Krieger, Theory of plates and shells, McGraw-Hill, New York (1959).

[2] A.Israr, Vibration analysis of cracked aluminium plates, PhD Thesis, University of Glasgow (2008).

[3] K.Maruyama, O.Ichinomiya, Experimental study of free vibration of clamped rectangular plates with straight narrow slits, JSME International Journal, Ser. 3, Vibration, control engineering, engineering for industry 32 (2) (1989) 187-193.

[4] D.Wu, S.S.Law, Anisotropic damage model for an inclined crack in thick plate and sensitivity study for its detection, International Journal of Solids and Structures 41 (2004) 4321-4336.

[5] C.S.Huang, A.W.Leissa, Vibration analysis of rectangular plates with side cracks via the Ritz Method, Journal of Sound and Vibration 323 (2009) 974-988.

[6] C.S.Huang, A.W.Leissa, C.W.Chan, Vibrations of rectangular plates with internal cracks or slits, International Journal of Mechanical Sciences 53 (2011) 436-445.Another reference

[7] A.Israr, M.P.Cartmell, E.Manoach, I.Trendafilova, W.M.Ostachowicz, M.Krawczuk, A.Zak, Analytical modelling and vibration analysis of partially cracked rectangular plates with different boundary conditions and loading, Journal of Applied Mechanics 76 (2009) 1-9.

[8] Ismail, R., and Cartmell, M.P., An investigation into the vibration analysis of a plate with a surface crack of variable angular orientation. Journal of Sound and Vibration, 331 (2012) 2929-2948.

[9] Z.J.Zeng, S.H.Dai, Stress intensity factors for an inclined surface crack under biaxial stress state, Engineering Fracture Mechanics 47 (2) (1994) 281-289.

[10] J.R.Rice, N.Levy, The part-through surface crack in an elastic plate, Journal of Applied Mechanics 39 (1972) 185-194.

[11] P.F.Joseph, F.Erdogan, Surface crack in a plate under antisymmetric loading conditions, International Journal Solids Structures 27 (6) (1991) 725-750.

[12] Y.C.Lu, Y.G.Xu, Line-spring model for a surface crack loaded antisymmetrically, National University of Defense Technology Technical paper (1986).

The Dynamic Response of an Euler-Bernoulli Beam on an Elastic Foundation by Finite Element Analysis using the Exact Stiffness Matrix

Jeong Soo Kim[1], **Moon Kyum Kim**[2]
Researcher, School of Engineering, Yonsei University, 134 Shinchon-dong, Seodaemun-gu, Seoul 120-749, Korea

E-mail: coffee1210@yonsei.ac.kr

Abstract. In this study, finite element analysis of beam on elastic foundation, which received great attention of researchers due to its wide applications in engineering, is performed for estimating dynamic responses of shallow foundation using exact stiffness matrix. First, element stiffness matrix based on the closed solution of beam on elastic foundation is derived. Then, we performed static finite element analysis included exact stiffness matrix numerically, comparing results from the analysis with some exact analysis solutions well known for verification. Finally, dynamic finite element analysis is performed for a shallow foundation structure under rectangular pulse loading using trapezoidal method. The dynamic analysis results exist in the reasonable range comparing solution of single degree of freedom problem under a similar condition. The results show that finite element analysis using exact stiffness matrix is evaluated as a good tool of estimating the dynamic response of structures on elastic foundation.

1. Introduction

There are lots of methods for estimating the dynamic behaviour of structures. Especially because of simplicity and convenience, beam on elastic foundation analysis is used to analyse structures such as footings, piles, buried structures, slabs in mechanical and civil engineering. Beam on elastic foundations have been studied extensively over the years [1]. Many research of beam on foundation model, e.g. Winkler foundation, Pasternak foundation, have been combined with the finite element method for considering efficiently lots of loading conditions, shape, dynamic behaviours etc. However, there is not the approach of the finite element method using stiffness matrix, which is derived from the exact solution of beam on elastic foundation, in the dynamic problems.

The objective of this study is estimating the adaptability of exact stiffness matrix in the dynamic finite element analysis of beam on elastic foundation problems. The exact element stiffness of beam on foundation is formulated from the Winkler model [2], which is assumed that the foundation reaction is proportional to the deflection of the footing at the every point [3]. Then, simple structures resting on elastic foundation are analysed by static and dynamic finite element analysis using exact stiffness matrix. Though the number of elements is small and the length of elements is arbitrary, the finite element analysis results using exact stiffness matrix show a good agreement with theoretical or

---

[1] Ph.D. candidate, E-mail address : coffee1210@yonsei.ac.kr
[2] Professor, E-mail address : applymkk@yonsei.ac.kr

reference results. Thus, the use of elements based on exact shape function can be expected to reduce total elements in the analysis.

2. Dynamic FEM Formulation

2.1. Formulation of shape function matrix to the Winkler model

The solution of Euler-Bernoulli beam on Winkler foundation can be described the equation (1) in the form of matrix [2, 4].

$$u(\mathrm{x}) = \boldsymbol{\phi}\,\mathbf{c} \tag{1}$$

Where

$$\boldsymbol{\phi} = [\mathrm{e}^{\beta x}\cos\beta x \quad \mathrm{e}^{\beta x}\sin\beta x \quad \mathrm{e}^{-\beta x}\cos\beta x \quad \mathrm{e}^{-\beta x}\sin\beta x], \quad \mathbf{c} = [\mathrm{c}_1 \quad \mathrm{c}_2 \quad \mathrm{c}_3 \quad \mathrm{c}_4]^{\mathbf{T}}$$

The equation (1) is expressed as the equation (2) in terms of a nodal displacement vector by applying element boundary conditions, $u(0) = u_1, u'(0) = u'_1, u(L) = u_2$, and $u'(L) = u'_2$.

$$u(\mathrm{x}) = \boldsymbol{\phi}\,\mathbf{c} = \boldsymbol{\phi}\mathbf{A}^{-1}\boldsymbol{d} = \boldsymbol{N}\boldsymbol{d} \tag{2}$$

Where

$$\mathbf{A} = \begin{bmatrix} 1 & 0 & 1 & 0 \\ \beta & \beta & -\beta & \beta \\ \mathrm{e}^{\beta L}\cos\beta L & \mathrm{e}^{\beta L}\sin\beta L & \mathrm{e}^{-\beta L}\cos\beta L & \mathrm{e}^{-\beta L}\sin\beta L \\ \beta\mathrm{e}^{\beta L}(\cos\beta L - \sin\beta L) & \beta\mathrm{e}^{\beta L}(\sin\beta L + \cos\beta L) & -\beta\mathrm{e}^{-\beta L}(\cos\beta L + \sin\beta L) & \beta\mathrm{e}^{-\beta L}(-\sin\beta L + \cos\beta L) \end{bmatrix}$$

$$\boldsymbol{N} = \boldsymbol{\phi}\mathbf{A}^{-1} \tag{3}$$

2.2. Dynamic governing equation and exact stiffness matrix for finite element method

As the equation (3) is substituted into the weak form of governing equation of beam on elastic foundation [3, 5], the equation of motion can be derived as the (4) equation.

$$\boldsymbol{M}\ddot{\boldsymbol{d}} + \boldsymbol{C}\dot{\boldsymbol{d}} + \boldsymbol{K}\boldsymbol{d} = \boldsymbol{f}_{ext} \tag{4}$$

Where $\boldsymbol{M}$, $\boldsymbol{C}$, $\boldsymbol{K}$, and $\boldsymbol{f}_{ext}$ is mass, damping, stiffness matrix, and nodal force vector, respectively. The stiffness matrix is composed with two parts, beam stiffness $\boldsymbol{K}_b$ and ground stiffness $\boldsymbol{K}_s$ as the equation (5).

$$\boldsymbol{K} = \boldsymbol{K}_b + \boldsymbol{K}_s \tag{5}$$

Where

$$\boldsymbol{K}_b = \int_0^L \boldsymbol{B}^T EI\boldsymbol{B}\,dx, \qquad \boldsymbol{K}_s = \int_0^L \boldsymbol{N}^T k\boldsymbol{N}\,dx\,, \qquad \boldsymbol{B} = \frac{\partial \boldsymbol{N}}{\partial x}$$

As substituting the shape function $\boldsymbol{N}$ derived in section 2.1 into equation (4), not the 3rd polynomials which is used ordinarily as the shape function, we can obtain the stiffness matrix based on exact solution of the Winkler foundation.

3. Numerical examples

3.1. Problem statement

Dynamic finite element code using the formulation of exact stiffness matrix derived in section 2 is made in MATLAB language. To verify the code and analyze the characteristics of finite element method by using exact stiffness matrix, some numerical examples are analyzed. The reference values are obtained from theoretical solution of Winkler model and results of finite element analysis using 3rd order polynomial shape function [6]. All the examples are assumed the following conditions:

- The material properties of beam and ground are linear-elastic.
- The beam is slender and prismatic, not containing shear deformation.
- Except beams on elastic foundation, other members are frame elements. Flexural and Axial behaviour of the element is described by 3rd polynomials and linear function, respectively.
- The ground region which is not below the flexible beam has small flexural stiffness only.

3.2. Static finite element analysis

To verify our program codes, finite element models of infinite beam on foundation are created. The results are compared with the corresponding theoretical values [2]. The concentric load $p = 1$ kN or moment $M = 10$ kN $\cdot$ m are loading on the center of infinite beam with the flexural stiffness $EI = 132.533 \times 10^{-6}$ m^4, respectively. The beam is also supported by ground with $k = 43$ MPa. The models compose 3 finite elements and use symmetric boundary conditions in the each case. Each result is shown in Figure 1. (a) and 1. (b), respectively.

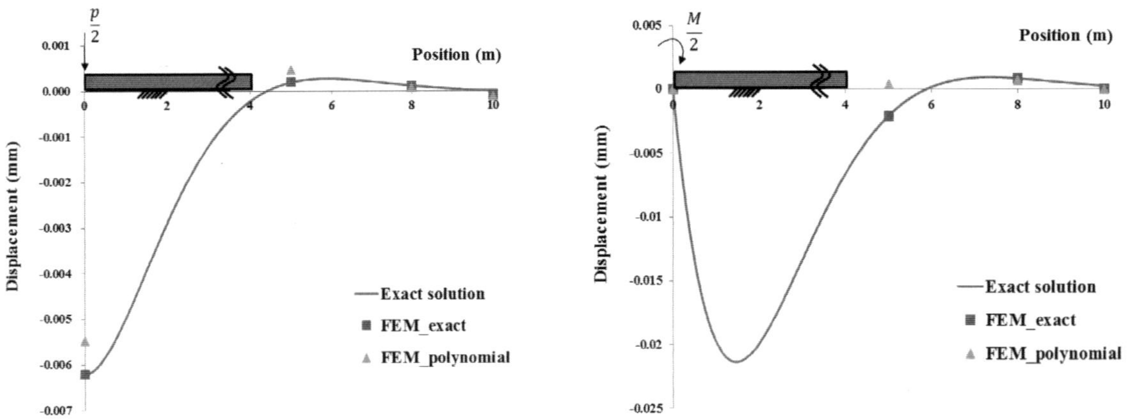

(a) Case 1: Concentric load **(B)** Case 2: Concentric moment

Figure 1. Comparison with FEM results of infinite beam on elastic foundation

The results of using the exact stiffness matrix show that the variation of beam of deflection is not sensitive to the element length, ratio to flexural and ground stiffness than the case of the ordinary stiffness matrix. However, the results of two cases converge on the theoretical values as increasing the number of elements. Therefore, the number of elements and length is important than the shape function type. The variation of results according to the relative stiffness and element length are listed in the table 1 and 2.

Table 1. Comparisons with the midpoint deflection according to the relative stiffness (case 1)

Type of Stiffness Element	Relative Stiffness (β)	Deflection ($\times 10^{-6}$,m)	Error [b] (%)	Relative Stiffness (β)	Deflection ($\times 10^{-6}$,m)	Error [b] (%)
Exact	0.1688	2.139	9.0010	0.5337	6.206	0.0193
Ordinary [a]		2.135	8.8009		5.477	11.7325
Exact	0.3001	3.546	1.6047	0.9490	11.035	0.0001
Ordinary [a]		3.475	0.4184		7.247	34.3271

[a] 3rd polynomial
[b] referred to theoretical results

Table 2. Comparisons with the midpoint deflection according to the element length [a] (case 1)

Element length type	Stiffness type	Displacement ($\times 10^{-6}$,m) Node #1	Node #2	Node #3	Node #4	Node #5	Node #6
	Extra	6.206	2.899	0.225	0.276	0.116	0.070
Uniform [b]	Ordinary	6.174	2.877	0.219	0.275	0.114	0.070
	Theoretical	6.205	2.899	0.228	0.267	0.116	0.070
	Extra	6.206	4.985	2.899	1.212	0.225	0.007
Non-uniform [c]	Ordinary	6.192	4.973	2.897	1.231	0.281	0.116
	Theoretical	6.205	4.984	2.899	1.213	0.228	0.007
	Extra	6.206	0.276	0.214	0.116	0.020	0.070
Non-uniform [d]	Ordinary	5.104	0.679	0.347	0.099	0.073	0.216
	Theoretical	6.205	0.267	0.206	0.116	0.046	0.007

[a] Finite element models have 5 elements. All elements relative stiffness β is 0.5337.
[b] element dimension : 2/2/2/2/2 (m) [b] element dimension : 1/1/1/1/6 (m) [c] element dimension : 6/1/1/1/1 (m)

3.3. Dynamic finite element analysis
We test our code by applying the static example and check the sensitivity of the element type in section 3.2. To examine the characteristic of finite element model using exact stiffness matrix, a simple frame structure resting on elastic foundation, as shown in the Figure 2, is considered.

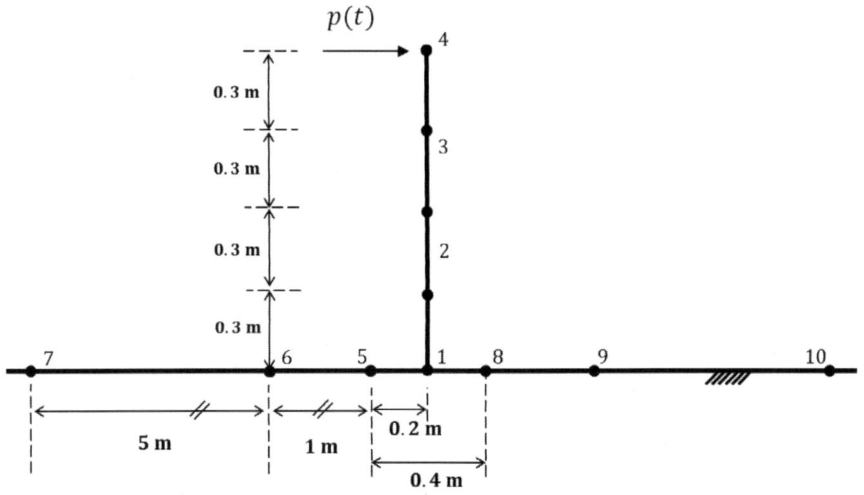

Figure 2. Simple frame structure supported elastic foundation

Journal of Physics: Conference Series 382 (2012) 012008 doi:10.1088/1742-6596/382/1/012008

The horizontal point load is 0.45kN rectangular pulse during 0.01 second. There are properties of the finite element model in the table 3. The time step is $\Delta t = 1 \times 10^{-6}$ second and the observation time is the same the duration of the pulse. To observe how the type of stiffness matrix influences the dynamic response, 3 cases numerical analysis are performed.

- Case 1: Single degree of freedom model with equivalent stiffness and mass (FEM_SDOF)
- Case 2: Finite element model with the exact stiffness matrix (FEM_exact)
- Case 3: Finite element model with the stiffness matrix derived 3rd polynomial (FEM_polynomial)

Table 3. Materials and geometric parameters

Elastic modulus (GPa)	Unit weight (kN/m^3)	Area (m^2)	Moment of inertia of Area (m^4)	Ground reaction coefficient (MPa)
210	76	0.025	0.004019	14

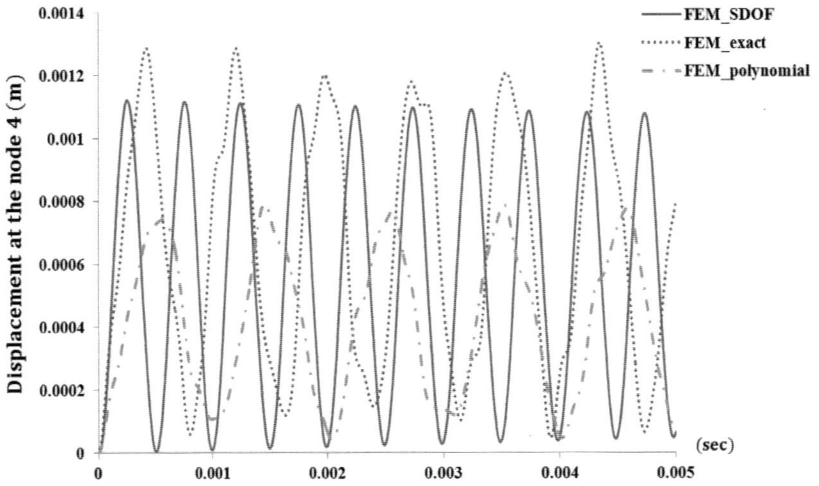

Figure 3. Comparisons with the numerical results

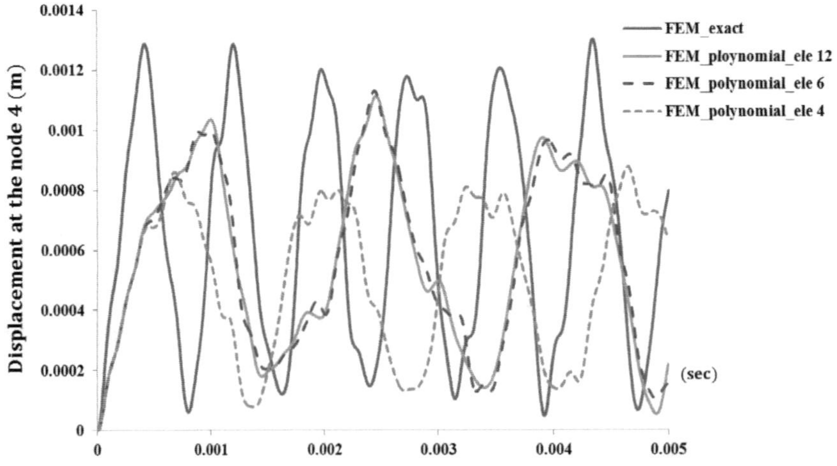

Figure 4. Displacement according to the number of elements

In the Figure 3, the horizontal displacement of the node 4 at in the model obtained by using the exact stiffness matrix is plotted with the results of other cases. The case 1 can be considered as theoretical values because it is able to be determined analytically. Though it has a few elements in the model, the results of case 2 appear to exist in the reasonable range. Because the both stiffness of case 1 and 2 are derived from the solution of Winker model, this result is seemed to be natural. On the other hands, the results of case 3 have difference with the case 1. It can be considered because the ground stiffness is approximated by 3rd order polynomial shape function, in despite of theoretical vertical and flexural stiffness of beam on foundation is described as combination of exponential functions.

If the model is more discretized, the result of case 3 is close to the results of other cases. In the figure 4, the result of case 3 is improved with the increase of the number of elements of the model, 4, 6, and 12. However, the results of using element stiffness based on the exact solution are not sensitive to the number of elements of model and element length. It can be considered because the exact solution contain exponential function of βx, especially the relative stiffness ratio β is small.

4. Conclusions

We perform the finite element analysis using the stiffness matrix derived from the exact solution of beam on elastic foundation. The stiffness and mass matrix are formulated to make the program code. The code is verified by performing the static finite element analysis of infinite beam on elastic foundation. The results show a good agreement with the theoretical solution of Winker model although our models have a few numbers of elements. The results obtained by using exact stiffness matrix are not sensitive to the number of elements and element length then the results of stiffness matrix using 3rd polynomial in the static and dynamic problems. Therefore, the application of exact stiffness matrix in the dynamic problem can be expected to reduce computational costs, time, and sensitivity of the number of element and size.

References

[1] Salih N Akour 2010 Dynamics of nonlinear beam on elastic foundation *World Congress on engineering proceeding*: vol 2
[2] Moshe Eisenberger and David Z. Yankelevsky 1985 Exact stiffness matrix for beams on elastic foundation *Computers & structures* vol 21 pp 1355-1359
[3] Ugural A.C. 1981 Advanced strength and applied elasticity *Prentice Hall* chapter 9 pp 349-359
[4] Bing Y. Ting. et al 1984 Beam on elastic foundation finite element *Journal of Structural Engineering* vol 110 pp 2324-2339
[5] I. M. Smith 2004 Programming the finite element method *Wiley* chapter 3 pp 100-106
[6] Yi Wang et al 2011 Transient responses of beam with elastic foundation supports under moving wave load excitation *International journal of engineering and technology* vol 1 pp 137-143

Crush Can Behaviour as an Energy Absorber in a Frontal Impact

Atanu Bhuyan, Olga Ganilova
School of Engineering, University of Glasgow, Glasgow, G12 8QQ, UK

E-mail: olga.ganilova@glasgow.ac.uk

Abstract. The work presented is devoted to the investigation of a state-of-the-art technological solution for the design of a crush-can characterized by optimal energy absorbing properties. The work is focused on the theoretical background of the square tubes, circular tubes and inverbucktube performance under impact with the purpose of design of a novel optimized structure. The main system under consideration is based on the patent US 2008/0185851 A1 and includes a base flange with elongated crush boxes and back straps for stabilization of the crush boxes with the purpose of improvement of the energy-absorbing functionality. The modelling of this system is carried out applying both a theoretical approach and finite element analysis concentrating on the energy absorbing abilities of the crumple zones. The optimization process is validated under dynamic and quasi-static loading conditions whilst considering various modes of deformation and stress distribution along the tubular components. Energy absorbing behaviour of the crush-cans is studied concentrating on their geometrical properties and their diamond or concertina modes of deformation. Moreover, structures made of different materials, steel, aluminium and polymer composites are considered for the material effect analysis and optimization through their combination. Optimization of the crush-can behaviour is done within the limits of the frontal impact scenario with the purpose of improvement of the structural performance in the Euro NCAP tests.

1. Introduction

The automotive industry gives the priority to crashworthiness and occupant safety during the design and manufacture of vehicles. Occupant safety is primarily characterised by vehicle deceleration pulses and deformation of the occupant compartment. Introduction of crumple zones in a vehicle is a trusted engineering solution to this problem. Thin-walled structures also known as crush cans are connected to the vehicle bumpers which absorb sufficient energy to ensure minimum damage to the passenger and critical vehicle parts during frontal impact, thereby contributing to the automobile body crash safety. The majority of research in this area is devoted to the investigation of thin-walled tubular structures with varieties of cross-sections and their behaviour [1-6].

The key concepts have been developed in the area of energy absorbers for crash safety of automobiles, as well as a number of inventions which have been made along the years. A lot of these patents have found their application in the manufacturing process. For example, Evans et al. [13] invented an energy absorber that includes a base flange and elongated crush boxes forming a continuous face surface on the energy absorber. Back straps are integrated across the back of the crush boxes to improve the energy absorbing functionality by keeping a check on unacceptable spreading of side walls of the crush boxes during an impact.

Common shapes of collapsible energy absorbers include circular tubes, square tubes, frusta, struts, honeycombs and sandwich plates [1]. Quite often the modes of deformation of these structures are studied by axial crush, lateral indentation, lateral flattening, inversion and splitting. In [7] the authors represent a study on energy absorption values of thin-walled tubes with different geometric dimensions using finite element simulation. In this case a nonlinear finite-element model is developed and analysed using LS-DYNA. A comparative study, in terms of the energy absorption properties, is performed for square, circular and elliptic tubes of steel and aluminium. The analysis concluded that the elliptic tubes absorb more energy during collisions. It was also demonstrated that the energy absorption can be increased with the width increase. Energy absorption per unit weight of steel and aluminium is also calculated demonstrating that steel tubes have a better ability to absorb comparing to aluminium. Different material and geometry combinations thus were found to show interesting variations in performances. Another work concentrating on the design optimisation [9] introduces tapered thin-walled tubes with axisymmetric indentations. The present work attempts to go along the lines of the research mentioned in order to design a novel optimised structure for a crush-can.

Due to the superior behaviour a number of works are devoted to circular tube structures [2-6]. The buckling of a circular cylindrical shell under axial loading is a classical problem in solid mechanics. Circular tubes prove to have one of the best energy absorbing abilities under axial compression, which explains their high frequency of occurrence in automotive crash safety applications. Plastic energy can be dissipated in thin metallic (or composite) tubes in several modes of deformation, viz. inversion, splitting, lateral indentation, lateral flattening and axial crushing. Thin circular tubes thus represent the most widespread shape of collapsible impact energy absorbers, owing to their high frequency of occurrence as structural elements.

An extensive review is done on a number of related experimental investigations of crush structures under quasi-static and dynamic conditions [10]. Discussion of some innovative theoretical and analytical models can be found in the reviews similar to [1]. As a rule finite element analysis is performed using ABAQUS and LS-DYNA [14] to study certain energy absorption characteristics and crashing parameters [8].

To make a viable analysis on crush-can behaviour it is important to set all the required load and boundary conditions. In a frontal crash, the impact load is experienced by the bumper assembly under different angles based on the barrier offset, and the varieties of oblique load. Reyes et al. [10] perform quasi-static experiments and numerical simulations to study the behaviour of aluminium columns in alloy AA6060 subjected to oblique loading. The force was applied under three different angles to the centreline of the column thereby being the oblique load [11,12]. In this case LS-DYNA was used to validate the numerical model and describe the force-displacement curve. It was demonstrated that the deformation modes are dependant on both load angle and thickness of the structure. The energy absorption was proved to drop with an increase of the loading angle after a drastic drop under the load angle of 5°.

The main objective of the present work is to investigate the state-of-the-art technological solution for the design of a crush-can characterised by optimal energy absorbing properties and to propose a novel design of a circular crush-can optimising its shape and behaviour. The main system under consideration is based on the patent US 2008/0185851 A1 [13] and includes a base flange with elongated crush boxes and back straps for stabilization of the crush boxes with a purpose of improvement of the energy-absorbing functionality. The modelling of this system is carried out applying both theoretical approach and finite element analysis concentrating on the energy absorbing abilities of the circular crush-cans.

2. Mathematical formulation

The crush-cans geometry can be represented as a thin-walled tube with a circular or square cross-section. In our case, a cylindrical tube of diameter D, length L and thickness t is considered, as shown in Figure 1. The longitudinal axis of the geometry is set along the z axis. Going along the lines of the pre-existing patent [13], one end of the crush-can is clamped and the other end is fixed to the frontal bumper cover. This makes the non-clamped end free to respond to the impact but constrained in terms of rotation.

Alexander in his work [2] assumes a simple mode of collapse of the tube and determines the work required to achieve this mode. This theoretical analysis was presented for an axisymmetric crush of a thin-walled cylindrical shell subjected to a static axial loading. The final solution for the mean crush load for this mode was derived empirically:

$$P_{av} = K\sigma_Y t(Dt)^{1/2} \tag{1}$$

where P_{av} is the mean crushing load , σ_Y is the yield stress and K is a mathematically and experimentally derived constant and is also a function of the material's Poisson's ratio υ ($K \approx 6.08$ for υ=0.25).

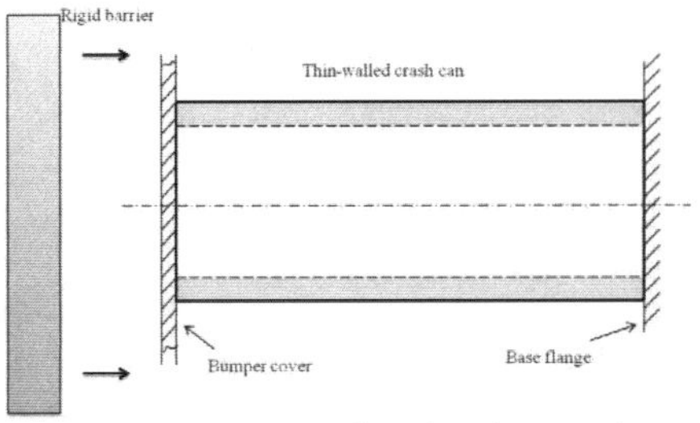

Figure 1. Geometry of a circular crush-can

The study for an energy absorber is incomplete without a consideration of the energy analysis. In a front vehicle crash, to be specific, there are two types of impact: elastic and plastic; necessarily a combination in the actual case. The elastic energy ($E_{Plastic}$) primarily includes the energy absorbed by the crush cans while a small fraction of it is dissipated as heat to the surrounding (the latter can be neglected for simulation purposes). Hence,

$$E_{plastic}=E_{Deformation}=E_a \tag{2}$$

where the term E_a is introduced as the energy absorbed by the deforming structures. This energy is a consequence of elastic-plastic deformation of the structures.

With this basic information, the absorption characteristics are to be assessed and for this purpose we lay down some crashing parameters that have been accepted as standard in similar studies. The parameters under consideration are average failure load/mean crush load (P_{av}), Specific Energy Absorption (SEA) and Crush Force Efficiency (CFE). The mean crush load is used as the response parameter for energy absorption capability and can be defined as:

$$P_{av} = \frac{E_a}{\delta} \tag{3}$$

where E_a is the energy absorbed during a collapse and δ represents displacement.

The SEA characterises the ability of the structure to absorb more energy with the least possible weight. It is therefore the energy absorbed per unit mass, and can be represented as:

$$SEA = \frac{E_a}{m} \tag{4}$$

where m is the mass of the individual crush can/thin-walled structure.

CFE which is defined as the ratio of the mean crush load to the peak crush load (P_{max}) with the most desirable value being unity, can be written as

$$CFE = \frac{P_{av}}{P_{\max}} \tag{5}$$

3. Novel design discussion

To study the behaviour of a crush-can in a front impact, under axial impact loading an energy absorption comparison is done for circular tubes with different geometric dimensions. For investigation of the effect of geometry and material properties, parameters presented in [7], are chosen for the thin-walled tube under consideration. Load conditions are the 50% frontal offset impact at 55 km/h [15]. The total crush mass of the vehicle (curb mass set at 1350 kg) is calculated according to Table 1.

Journal of Physics: Conference Series **382** (2012) 012009 doi:10.1088/1742-6596/382/1/012009

Table 1. Crash mass of vehicle [15]

Curb mass	1350 kg
Luggage	113 kg
Dummies	149 kg
Total Crash Mass	1612 kg

With regard to the information presented for various front crash impact tests, the impact is considered to be lasting for 100 milliseconds. This is an approximated value chosen to maintain simplicity in the calculations. The crash impact load is thus applied over the unclamped end of the structure and is treated as a distributed force over the section area of the crush tube.

3.1. Numerical parameters and pre-simulation calculations

The input load (P_i) is calculated using the concept of Newton's second law which relates to the change of momentum of the vehicle during the crush. The crush-cans undergoing deformation don't change the overall mass of the vehicle as the pre-crash speed of 55 km/h is brought to a stop within the typically assumed 100 ms duration. Thus the impact force is calculated as:

$$P_i = TCM \frac{V_A - 0}{\Delta t} \tag{6}$$

where *TCM* is the total crash mass from Table 1, V_A is the equivalent pre-crash speed and Δt is the crash duration. Therefore $P_i = 2.4627 \times 10^5$ N. This can be considered as the peak crush load P_{max}. The tube dimensions under consideration are presented in Table 2.

Table 2. Tube parameters

Thickness (mm)	Length (mm)	Width (mm)
1.25	150	30
1.5	150	30
1.25	150	50
1.5	150	50

The metals used in the current work are steel (AISI 1045) and aluminium (alloy; Al 6063), the properties of which are listed in Table 3.

Table 3. Mechanical properties

	E (GPa)	σ_Y (MPa)	υ	ρ (kg/m^3)
Steel	205	530	0.30	7850
Aluminium	99	214	0.33	2700

3.2. Comparative geometrical analysis

A parametric study of tube geometry facilitates the model verification as FEM is introduced into the analysis. According to the Table 2, we have four cases of circular tube parameters, i.e. two values of section diameters, 30 mm and 50 mm, and two values of tube section thickness, 1.25 mm and 1.5 mm. Four different geometries are investigated on their energy absorbing performance studied by reviewing the stress distribution and performance curves as load vs. displacement and total energy vs. time graphs. The crush-can behaviour is analysed with regard to the work presented in [7]. For a steel circular tube of geometrical dimensions presented in Table 1, the mean crush load P_{av} is calculated using Eq(2), CFE is obtained according to Eq(5), (Table 4).

Table 4. Calculating P_{av} and *CFE* for different tube geometry

Thickness (mm)	Length (mm)	Width (mm)	P_{av} (N)	CFE (%)
1.25	150	30	2.46×10^4	10.01
1.5	150	30	3.24×10^4	13.17
1.25	150	50	3.18×10^4	12.93
1.5	150	50	4.18×10^4	16.99

3.3. Comparative analysis of material performance

From Table 4, it is obvious that the last geometry shows the best *CFE*, thus demostrating optimum energy absorbing characteristics which is in agreement with the work [7].

At this stage, Aluminium is introduced for a comparative material study. Since Aluminium has lower mass density, it may be worthwhile to explore its energy absorbing capability per weight performance. Thus circular tubes of 50 mm wide, 150 mm long and 1.5 mm thick are investigated with Aluminium as the structure material and compared with the corresponding steel model. Results of the structural performance of Aluminium and Steel, for tubes of t = 1.5 mm, L = 150 mm and D = 50 mm are presented in Table 5.

Table 5. Calculating P_{av} and *CFE* for different materials

Material	P_{av} (N)	CFE (%)
Steel	3.18×10^4	12.93
Aluminium	1.69×10^4	6.86
Aluminium (t=2.25 mm)	3.10×10^4	12.60

Since steel tubes are almost twice as strong as aluminium, a higher section thickness of 2.25 mm was considered for the *CFE* calculations. Considering the low weight advantages of the Aluminium, different combinations of the two metals are attempted in the modelling. This contributes to proposing an optimised crush-can design as a novel energy absorbing solution.

4. Design verification applying FEM

An impact energy absorption performance of a thin-walled tube is investigated using FEM software ABAQUS/Explicit version 6.10 to study the modes of deformation of tubes under quasi-static loading conditions. The three dimensional model generated for the circular tube in ABAQUS is shown in Figure 2. The simulation is carried out using the direct matrix solver with minimum increment size of 1×10^{-5}, with structured hex-dominated mesh elements of type C3D8R.

Figure 2. 3D model of a circular crush-can

From the observed behaviour established in Tables 4 and 5, three novel designs are proposed and investigated. These are the combinations of steel and aluminium distributed as

1) Half a tube (clamped end) is made of aluminium and the other half is made of steel;

2) Half a tube (clamped end) is made of steel and the other half is made of aluminium;

3) The stress concentration region is reinforced with steel and the rest of the tube is made of aluminium. The models and results for simulation of the crush are presented in Figures 3, 4 and 5 respectively.

Figure 3. Combination 1 model, stress distribution before impact and stress distribution after impact. Impact load $=2.46 \times 10^5$ N, left end clamped and right end free-rigid.

Figure 4. Combination 2 model, stress distribution before impact and stress distribution after impact.Impact load =2.46x10^5 N, left end clamped and right end free-rigid.

Figure 5. Combination 3 model, stress distribution before impact and stress distribution after impact. Impact load =2.46x10^5 N, left end clamped and right end free-rigid.

For the third model, the steel material zone is proposed arbitrarily but with an instinctive decision based on the observed stress concentration pattern, and is subject to experimental verification. For simulation purposes, however the region starts 10 mm away from the base and the thickness is assigned to be of 30 mm. The three models considered are analysed based on the same crushing parameters and P_{av} is calculated independently from L. Using the query tool, the mass and relative displacements are obtained, which is used to calculate the *SEA* from Eq(3) and Eq(4) (Table 6).

Table 6. Comparative assessment of the proposed models

Parameter (all in SI units)	Combination 1	Combination 2	Combination 3
Mass	0.181	0.181	0.128
Relative displacement (end to end); δ	19.2 x 10^{-3}	9.05 x 10^{-3}	19.9 x 10^{-3}
Relative displacement (mid to base)	0.62 x 10^{-3}	0.88 x 10^{-3}	0.65 x 10^{-3}
P_{av}	2.44 x 10^4	2.44 x 10^4	1.56 x 10^4
CFE (%)	9.91	9.91	6.34
E_a	468.48	220.82	310.44
SEA	2588.29	1220	2425.31

The total energy for each of the three models is presented in Figure 6. This assists in a better understanding of the performance of each design suggested.

5. Conclusions

The behaviour of three novel designs for circular crush-cans (Figures. 3, 4 and 5) is studied. The modelling of these systems is carried out applying both theoretical approach and FEM concentrating on the energy absorbing abilities. The optimization process is validated under quasi-static loading conditions whilst considering various modes of deformation and stress distribution along the tubular components. Moreover, structures made of different materials, steel and aluminium, are considered for the material effect analysis and optimization through their combination. Steel, being stiffer, demonstrated a better value of *CFE* and aluminium being lighter and more prone to crumpling results in a better *SEA*. Considering overall characteristics for an effective energy absorber the first model (Combination 1, Figure 3) is the one exhibiting a better performance and can be recommended as an optimal option for a crush-can design. In terms of the cost, Combination 1 is also cheaper to be manufactured than Combination 3, due to the larger proportion of aluminium which is more than three times cheaper than steel. In future Combination 3 can be further optimised using a more detailed simulation or series of experiments, or a combination of both. The approach and designs presented can be employed for further optimisation and development of other innovative design solutions. Optimization of the crush-can behaviour is done within the limits of the front impact scenario with the purpose of improvement of the structural performance in the Euro NCAP tests.

References

[1] Alghamdi A A A 2000 *Thin-Walled Structures* **39** 189-213.
[2] Alexander J M 1960 *Quart. J. Mech. Appl. Math* **13(1)** 10–5.
[3] Abramnowicz W and Jones N 1984 *Int. J. Impact Engng.* **2(3)** 263-281.
[4] Abramowicz W and Jones N 1984 *Int. J. Impact Engng* **2(2)** 179–20.
[5] Pugsley A G and Macaulay M 1960 *Quart. J. Mech.Appl. Math* **13(1)** 1–9.
[6] Pugsley A G 1979 *Quart. J. Mech. Appl. Math* **32(1)** 1–7.
[7] Marzbanrad J, Mehdikhanlo M and Pour A S 2009 *Turkish J. Eng. Env. Sci.* **33** 159–166.
[8] Aljibori H S S 2010 *European J. of Scientific Research* **39(1)**, 111-121.
[9] Acar E *et al* 2011 *Thin-Walled Structures* **49** 94-105.
[10] Reyes A, Hopperstad O S, Langseth M 2004 *Int. J. of Solids and Structures* **41(5-6)** 1645-1675.
[11] Han D C and Park S H 1999 *Thin-Walled Structures* **35** 167-184.
[12] Park J S, Park S H and Han D C 2000 Seoul 2000 FISITA World Automotive Congress, 2000G291.
[13] Evans D, Burke J T, Bastien K R , Holdren K E, Ralston D D and Gorman D 2008 Energy absorber with crash boxes and back straps. U.S. Patent No. 2008/0185851 A1.
[14] Choi W M *et al* 2011 *World Academy of Science, Engineering and Technology* **76**
[15] Ultra Light Steel Auto Body – Engineering Report, Phase 2 Findings, Version 1.0.2, ULSAB Consortium, March 1998, Porsche Engineering Services Inc., accessed April 6, 2012, http://www.autosteel.org/~/media/Files/Autosteel/Programs/ULSAB/ulsab_eng_rpt_complete.ashx.

Vibration analysis of a trimorph plate as a precursor model for smart automotive bodywork

A Big-Alabo[1] and M P Cartmell

Systems, Power and Energy Research Division, School of Engineering, University of Glasgow, G12 8QQ, Scotland, UK

Email: a.big-alabo.1@research.gla.ac.uk; Matthew.Cartmell@glasgow.ac.uk

Abstract
This study investigates the vibration characteristics of a proposed candidate structure for smarter car bodies. The material is conceived as a three-layer laminated structure in the form of a trimorph plate. The vibration response of the plate is investigated for large deflections by considering the effects of geometric nonlinearity. First, the governing equation for the mid-point deflection of the plate is developed based on classical laminate plate theory (CLPT). The governing equation is solved, and a simulation is run for different possible layer-stacking sequences. Comparisons are made between the nonlinear vibration response of this trimorph plate both with and without the effects of the von Kármán geometric nonlinearity. The results show that for the same material properties the different layer-stacking sequences produce different vibration responses, and from there it is concluded that layer-stacking sequencing is a basis for the definition of a suitable material configuration for high performance automotive applications.

1. Introduction

Passive safety mechanisms in automotive bodywork appear in the form of integrated tubes and plates used to fortify the frontal, rear, and side members. These applications are structurally unintelligent as they do not respond to impact in a precisely predictable and controllable manner. Little has been done to investigate structural intelligence within automotive bodywork for passive safety applications. Big-Alabo and Cartmell [1] have proposed a multi-layered trimorph plate with sensing and actuation layers for a structural intelligence application for automotive bodywork. They studied the vibration characteristics of the trimorph plate for small displacements taking into account the effect of membrane forces, which introduced a quadratic nonlinearity in the model. Their results show that the nonlinear effect was not seen to be generally significant, and this was justifiably so because small displacements are predominately characterised by linear phenomenon. This study also shows that the vibration characteristics of the trimorph plate are dependent on the stacking sequence of the host/sensor/actuator layers. However, real automotive accident situations require that investigations are carried out for large displacements of the bodywork material, due to different forms of impact.

[1] To whom any correspondence should be addressed.

Large displacements of a plate are characterised by membrane force stretching, which introduces geometric nonlinearity. Manoach and Trendifilova [2] studied the large displacement of a rectangular plate for damage detection. They developed a model of plate vibration based on Mindlin plate theory and von Kármán geometric nonlinearity, and solved the reduced model using a finite difference method. Kazanci and Mecitoğlu [3] studied the nonlinear vibration of a laminated composite plate based on the von Kármán large deflection theory of thin plates. They used a combined Galerkin and finite difference method for the solution of the nonlinear equations of the plate. Other studies of large displacements of plates based on the von Kármán geometric nonlinearity can be found in [4 – 7] etc. This paper investigates the vibration characteristics of a trimorph plate for large displacements. Geometric nonlinearity of the von Kármán type was considered in developing the governing equation, and comparisons were made for the nonlinear response characteristics of the trimorph plate both with and without the von Kármán geometric nonlinearity. The Navier and Galerkin solution methods have been used to obtain the reduced order time-dependent models, and bespoke *Mathematica* codes were used to integrate numerically the equations of motion.

2. Governing Equation

The trimorph plate is modelled as a thin plate and therefore all the appropriate modelling assumptions apply. Also, perfect bonding between the layers is assumed. The plate is simply supported at this stage of investigation and the applied load is considered as a distributed pressure function. By introducing linear viscous damping into the equation of motion for the transverse displacement of a thin rectangular plate subject to membrane forces, we have:

$$\frac{\partial^2 M_x}{\partial x^2} + 2\frac{\partial^2 M_{xy}}{\partial x \partial y} + \frac{\partial^2 M_y}{\partial y^2} = \rho h \frac{\partial^2 w}{\partial t^2} + c\frac{\partial w}{\partial t} - q(x,y,t) + N_x \frac{\partial^2 w}{\partial x^2} + 2N_{xy}\frac{\partial^2 w}{\partial x \partial y} + N_y \frac{\partial^2 w}{\partial y^2} \quad (1)$$

The moment effects and membrane forces in equation (1) for an unsymmetric laminated plate with isotropic layers are given as [1]:

$$\begin{Bmatrix} M_x \\ M_y \\ M_{xy} \end{Bmatrix} = \begin{bmatrix} B_{11} & B_{12} & 0 \\ B_{12} & B_{11} & 0 \\ 0 & 0 & B_{66} \end{bmatrix} \begin{Bmatrix} \varepsilon_x^0 \\ \varepsilon_y^0 \\ \varepsilon_{xy}^0 \end{Bmatrix} + \begin{bmatrix} D_{11} & D_{12} & 0 \\ D_{12} & D_{11} & 0 \\ 0 & 0 & D_{66} \end{bmatrix} \begin{Bmatrix} k_x^0 \\ k_y^0 \\ k_{xy}^0 \end{Bmatrix} \quad (2)$$

$$\begin{Bmatrix} N_x \\ N_y \\ N_{xy} \end{Bmatrix} = \begin{bmatrix} A_{11} & A_{12} & 0 \\ A_{12} & A_{11} & 0 \\ 0 & 0 & A_{66} \end{bmatrix} \begin{Bmatrix} \varepsilon_x^0 \\ \varepsilon_y^0 \\ \varepsilon_{xy}^0 \end{Bmatrix} + \begin{bmatrix} B_{11} & B_{12} & 0 \\ B_{12} & B_{11} & 0 \\ 0 & 0 & B_{66} \end{bmatrix} \begin{Bmatrix} k_x^0 \\ k_y^0 \\ k_{xy}^0 \end{Bmatrix} \quad (3)$$

Equations (2) and (3) are based on the classical laminate theory. The elements of the stiffness matrices $[A_{ij}], [B_{ij}]$ *and* $[D_{ij}]$ are determined according to the expressions derived in [1]. The strain-displacement relationships for the mid-plane deflection of a rectangular plate based on the von Kármán geometric nonlinearity are given as:

$$\left.\begin{aligned} \varepsilon_x^0 = \frac{\partial u^0}{\partial x} + \frac{1}{2}\left(\frac{\partial w}{\partial x}\right)^2; \varepsilon_y^0 = \frac{\partial v^0}{\partial y} + \frac{1}{2}\left(\frac{\partial w}{\partial y}\right)^2; \varepsilon_{xy}^0 = \frac{\partial u^0}{\partial y} + \frac{\partial v^0}{\partial x} + \frac{\partial w}{\partial x}\cdot\frac{\partial w}{\partial y} \\ k_x^0 = -\frac{\partial^2 w}{\partial x^2};\ k_y^0 = -\frac{\partial^2 w}{\partial y^2};\ k_{xy}^0 = -2\frac{\partial^2 w}{\partial x \partial y} \end{aligned}\right\} \quad (4)$$

By substituting equations (2), (3), and (4) into equation (1), and then neglecting the in-plane displacements, the governing equation for the transverse displacement of the trimorph plate is derived as:

$$\begin{aligned}
&\rho h\frac{\partial^2 w}{\partial t^2}+c\frac{\partial w}{\partial t}+D_{11}\frac{\partial^4 w}{\partial x^4}++2(D_{12}+2D_{66})\frac{\partial^4 w}{\partial x^2\partial y^2}+D_{11}\frac{\partial^4 w}{\partial y^4}\\
&-\left\{B_{11}\left[\left(\frac{\partial^2 w}{\partial y^2}\right)^2+\left(\frac{\partial^2 w}{\partial x^2}\right)^2\right]+2B_{12}\frac{\partial^2 w}{\partial x^2}\cdot\frac{\partial^2 w}{\partial y^2}+4B_{66}\left(\frac{\partial^2 w}{\partial x\partial y}\right)^2\right.\\
&+\frac{1}{2}\frac{\partial^2}{\partial x^2}\left[B_{11}\left(\frac{\partial w}{\partial x}\right)^2+B_{12}\left(\frac{\partial w}{\partial y}\right)^2\right]+\frac{1}{2}\frac{\partial^2}{\partial y^2}\left[B_{12}\left(\frac{\partial w}{\partial x}\right)^2+B_{11}\left(\frac{\partial w}{\partial y}\right)^2\right]\\
&\left.+2B_{66}\frac{\partial^2}{\partial x\partial y}\left[\frac{\partial w}{\partial x}\cdot\frac{\partial w}{\partial y}\right]\right\}+\frac{1}{2}\left[A_{11}\left(\frac{\partial w}{\partial x}\right)^2+A_{12}\left(\frac{\partial w}{\partial y}\right)^2\right]\frac{\partial^2 w}{\partial x^2}\\
&+\frac{1}{2}\left[A_{12}\left(\frac{\partial w}{\partial x}\right)^2+A\left(\frac{\partial w}{\partial y}\right)^2\right]\frac{\partial^2 w}{\partial y^2}+2A_{66}\frac{\partial w}{\partial x}\cdot\frac{\partial w}{\partial y}\cdot\frac{\partial^2 w}{\partial x\partial y}\\
&=q(x,y,t)
\end{aligned}\tag{5}$$

In this study, $q(x,y,t)$ is taken as a sinusoidally distributed and varying load. Equation (5) was solved using the Navier [1] and the Galerkin [8] solution methods, and the following ODE models were respectively obtained.

$$\ddot{w}+\bar{c}\dot{w}+\bar{D}w-\bar{B}w^2=\bar{Q}\sin\omega t \tag{6}$$

$$\ddot{w}+\bar{c}\dot{w}+\bar{D}w-\bar{B}w^2-\bar{A}w^3=\bar{Q}\sin\omega t \tag{7}$$

where

$\bar{c}=\frac{c}{\rho h}=2\mu\sqrt{\bar{D}}$; $\bar{Q}=\frac{q}{\rho h}$ are the damping constant per unit mass and excitation magnitude per unit mass respectively;

$\bar{D}=\frac{\pi^4}{\rho h}\left[D_{11}\left(\frac{1}{a^4}+\frac{1}{b^4}\right)+\frac{2(D_{12}+2D_{66})}{a^2b^2}\right]$ is the bending elastic stiffness per unit mass;

$\bar{B}=\frac{2\pi^4}{\rho h}\left[B_{11}\left(\frac{1}{a^4}+\frac{1}{b^4}\right)+\frac{B_{12}+B_{66}}{a^2b^2}\right]$ *for Navier solution*

$\bar{B}=\frac{32\pi^4}{9\rho h}\left[B_{11}\left(\frac{1}{a^4}+\frac{1}{b^4}\right)+\frac{5B_{12}+3B_{66}}{a^2b^2}\right]$ *for Galerkin solution*

} is the bending-extension elastic stiffness per unit mass; and

$\bar{A}=\frac{\pi^4}{32\rho h}\left[3A_{11}\left(\frac{1}{a^4}+\frac{1}{b^4}\right)+\frac{6A_{12}-4A_{66}}{a^2b^2}\right]$ is the extensional elastic stiffness per unit mass.

3. Numerical Simulations and Discussion of Results

A trimorph plate made up of Al, PVDF, and PZT layers has been investigated. The plate is a $1m\times 1m$ square plate and the thicknesses of the layers are $1.3mm$, $1.0mm$, and $1.7mm$ respectively. A damping ratio of 0.005 and excitation amplitude of $10N/m^2$ has been used, and the material properties of the layers can be found in [1]. Since the trimorph plate comprises three different layers, the stacking of the layers can be in six different permutations defined by: Al/PVDF/PZT, Al/PZT/PVDF, PVDF/Al/PZT, PVDF/PZT/Al, PZT/PVDF/Al, and PZT/Al/PVDF. Tables 1 and 2 summarise the response characteristics of the nonlinear models without, and with, the von Kármán geometric nonlinearity respectively. The response characteristics show that the bending-extension stiffness for the response with the von Kármán geometric nonlinearity is approximately twice that of the response without the von Kármán geometric nonlinearity based on the Navier solution method. The effect of the membrane stretching is such that at an excitation frequency equal to the undamped linear natural frequency, the transverse displacement of the plate is reduced. The results show that this membrane stretching effect is more pronounced when the von Kármán geometric nonlinearity is considered. In simulating the results, the transverse displacements have been normalised with respect to the total plate thickness.

Table 1: *Response characteristics for Navier solution model without von Kármán nonlinearity.*

S/N	Layer Arrangement from top to down	Plate stiffness ($\bar{D}$) [N/mkg]	Plate stiffness ($\bar{B}$) [N/m^{2}kg]	Excitation freq. [rad/s]	Damping const. [Ns/mkg]	Normalised Steady-state response (w/h) [-]	Steady state set-in time [s]
1	Al/PVDF/PZT	27387	6.79902×10^6	165.49	1.6549	0.188247	~ 30
2	Al/PZT/PVDF	14062.7	4.59591×10^6	118.586	1.18586	0.202384	~ 30
3	PVDF/PZT/Al	34284.9	7.8617×10^6	185.162	1.85162	0.181128	~ 20
4	PVDF/Al/PZT	32336	7.78368×10^6	179.822	1.79822	0.179962	~ 20
5	PZT/Al/PVDF	15232	4.90664×10^6	123.418	1.23418	0.198429	~ 30
6	PZT/PVDF/Al	27497	6.74457×10^6	165.822	1.65822	0.189366	~ 30

Table 2: *Response characteristics for Navier solution model with von Kármán nonlinearity.*

S/N	Layer Arrangement from top to down	Plate stiffness ($\bar{D}$) [N/mkg]	Plate stiffness ($\bar{B}$) [N/m^{2}kg]	Excitation freq. [rad/s]	Damping const. [Ns/mkg]	Normalised Steady-state response (w/h) [-]	Steady state set-in time [s]
1	Al/PVDF/PZT	27387	1.3234×10^7	165.49	1.6549	0.123921	~ 10
2	Al/PZT/PVDF	14062.7	9.33216×10^6	118.586	1.18586	0.128723	~ 10
3	PVDF/PZT/Al	34284.9	1.59692×10^7	185.162	1.85162	0.116823	~ 10
4	PVDF/Al/PZT	32336	1.5656×10^7	179.822	1.79822	0.116397	~ 10
5	PZT/Al/PVDF	15232	9.87807×10^6	123.418	1.23418	0.126893	~ 20
6	PZT/PVDF/Al	27497	1.31533×10^7	165.822	1.65822	0.124668	~ 20

Table 3: *Natural frequencies of the trimorph plate for the different layer-stacking.*

S/N	Layer Arrangement from top to down	Linear model	Nonlinear model without von Kármán GN [1]	Nonlinear model with von Kármán GN (Navier Solution)	Nonlinear model with von Kármán GN (Galerkin Solution)
1	Al/PVDF/PZT	165.49	161.00	158.00	162.00
2	Al/PZT/PVDF	118.59	112.00	107.50	113.50
3	PVDF/PZT/Al	185.16	181.50	178.50	182.50
4	PVDF/Al/PZT	179.82	176.00	172.50	177.00
5	PZT/Al/PVDF	123.42	117.00	112.50	118.50
6	PZT/PVDF/Al	165.82	161.50	158.50	162.50

Note: 'GN' means geometric nonlinearity and '~' means approximately.

Table 3 compares the natural frequencies of the different nonlinear models with that of the linear model. For the nonlinear models, the natural frequencies are obtained from the frequency-response plots (Figures 1, 2, and 3).

It is important to mention that the nonlinear effect due to membrane forces can only be observed very close to the undamped linear natural frequency of the trimorph plate. This explains the reason for investigating the frequency-domain responses in this region. The reduction in the natural frequency of the nonlinear responses results from the nonlinear softening effect introduced by the membrane forces. This softening effect is least pronounced in the Galerkin solution model, and this is so because unlike the Navier solution model, the Galerkin solution model accounts for the pure extensional stiffness due to membrane stretching. Hence, the Navier solution model overestimates this softening effect and the Galerkin solution model is a more accurate representation the trimorph plate response.

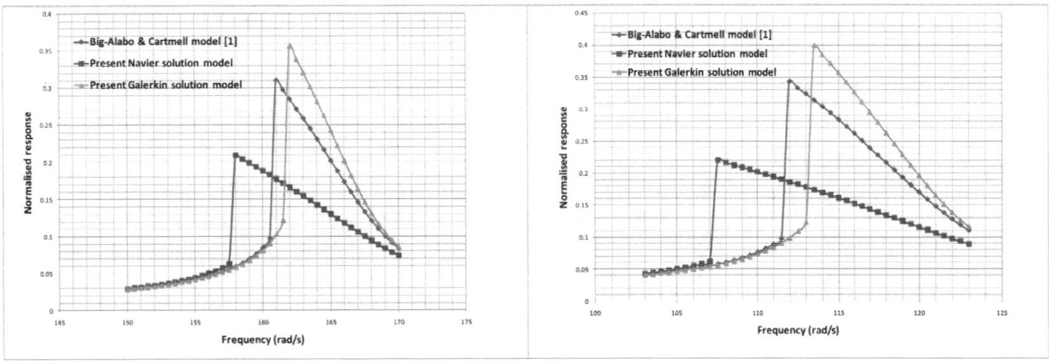

Figure 1: *Frequency-domain response for Al/PVDF/PZT and Al/PZT/PVDF respectively.*

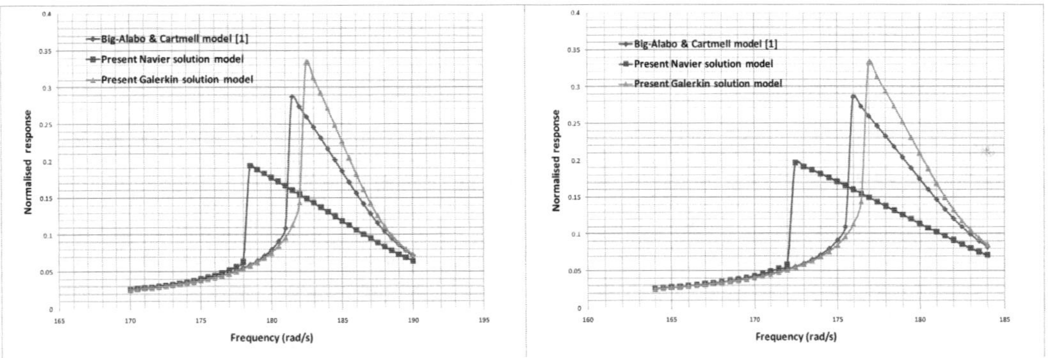

Figure 2: *Frequency-domain response for PVDF/PZT/Al and PVDF/Al/PZT respectively.*

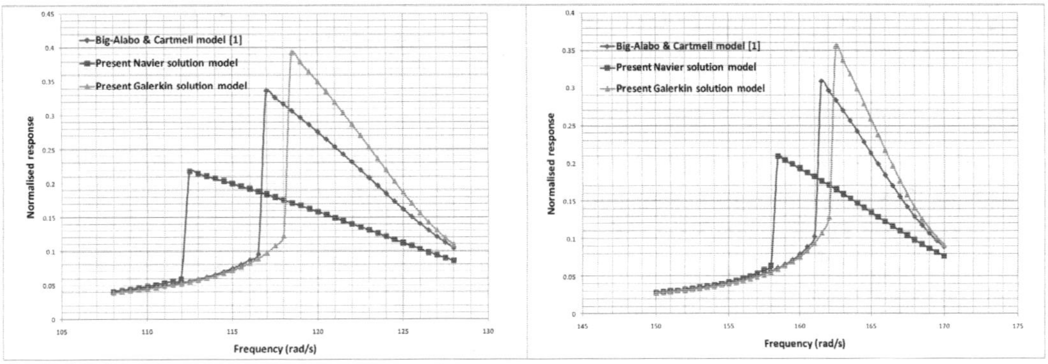

Figure 3: *Frequency-domain response for PZT/Al/PVDF and PZT/PVDF/Al respectively.*

For the purpose of comparing the response qualities of the different layer-stacking, a trimorph configuration is sought with the highest natural frequency, lowest response amplitude, and lowest steady-state response time. A higher natural frequency translates into higher bending resistance, a lower response amplitude translates into lower deformation, and a lower steady-state response time translates into a shorter time to stable response. The PVDF/Al/PZT or PVDF /PZT/Al layer-stacking meets these conditions better than the other layer-stacking sequences.

4. Conclusion

The governing equation for the large displacements of the trimorph plate with the von Kármán geometric nonlinearity has been developed. Two reduced order nonlinear models were arrived at based on the Navier and Galerkin solution methods. The results show that the steady-state displacements at the undamped linear natural frequency and natural frequency of the trimorph plate are further reduced as a result of geometric nonlinearity. It was concluded that the Galerkin solution model with quadratic and cubic nonlinearities represents the large displacements of the trimorph plate more accurately because it accounts for extensional stiffness. The models developed here are precursor models that will be further developed for impulse loading, material nonlinearity, and other boundary conditions, in order to investigate automotive accident scenarios with a view to developing a better material for optimised active damage mitigation.

Acknowledgement

The first author wishes to thank the Commonwealth Scholarship Commission, UK for granting him the scholarship for his PhD research, of which this work is a part.

References

[1] Big-Alabo A and Cartmell M P 2011 Vibration analysis of a trimorph plate for optimised damage mitigation *Journal of Theoretical and Applied Mechanics* **49 (3)** pp 641 – 664

[2] Manoach E and Trendifilova I 2008 Large amplitude vibrations and damage detection of rectangular plates *Journal of Sound and Vibration* **315** pp 591 – 606

[3] Kazanci Z and Mecitoğlu Z 2008 Nonlinear dynamic behavior of simply supported laminated composite plates subjected to blast load *Journal of Sound and Vibration* **317** pp 883 – 897

[4] Chen J, Dawe D J and Wang S 2000 Nonlinear transient analysis of rectangular composite laminated plates *Composite Structures* **49** pp 129 – 139

[5] Jayakumar K, Yadav D and Nageswara Rao B 2009 Nonlinear free vibration analysis of simply supported piezo-laminated plates with random actuation electric potential difference and material properties *Communication in Nonlinear Science and Numerical Simulation* **14** pp 1646 – 63

[6] Kerur S B and Ghosh A 2011 Active control of geometrically non-linear transient response of smart laminated composite plate integrated with AFC actuator and PVDF sensor *Journal of Intelligent Material Systems and Structures* **22** pp 1149 – 60

[7] Kazanci Z 2011 Dynamic response of composite sandwich plates subjected to time-dependent pressure pulses *International Journal of Non-Linear Mechanics* **46** pp 807 – 817

[8] Szilard R 2004 *Theory and Applications of Plate Analysis: Classical, Numerical and Engineering Methods* (John Wiley & Sons Inc.: New Jersey)

Free vibration of a rod undergoing finite strain

S.M. Mousavi*[1], S.J. Fariborz[2]
[1]Department of Civil and Structural Engineering, Aalto University, Finland
[2]Department of Mechanical Engineering, Amirkabir University of Technology, Iran

Abstract
The finite strain longitudinal free vibration of a rod is studied. Utilizing second Piola-Kirchhoff stress and Green strain tensors, the equation of motion is written in terms of displacement in reference configuration. Three different types of homogenous boundary conditions may be considered for the rod, leading to three nonlinear eigenvalue problems. The series solutions with three terms satisfying the boundary conditions are utilized and the relationships between amplitudes of vibration are obtained by means of the Galerkin method. The backbone curves are drawn and the internal resonance between different modes of vibration is analyzed.

Keywords: Finite strain; Rod; Backbone curve; Internal resonance

1. Introduction

The nonlinear vibration of continuous structures has been the subject of study by numerous investigators. To mention a few articles most pertinent to the present study, Lewandowski [1] used an analytical approach to analyze free vibration of beams with large displacement. In another article Lewandowski [2], analyzed the free vibration of beams, membranes and plates with moderately large deformation. The Galerkin procedure in time and space variables was used to obtain systems of nonlinear equations for the amplitude and frequency of vibration. The backbone curves were drawn and internal resonance between modes of vibration was investigated. The longitudinal vibration of a rod and lateral vibration of a beam subjected to nonlinear response of the surrounding medium, posing weakly nonlinear equations, were studied by Andrianov et al. [3]. In all of these works, structures sustained large deformation but infinitesimal strain. Therefore, nonlinear terms appeared only in strain-displacement relationships and the terms with the highest order of derivative were linear in the ensuing differential equations.

In this work, we consider free vibration of a rod with finite strain. The equation of motion in terms of displacement is strongly nonlinear with quadratic and cubic nonlinear terms. The analysis is carried out for rods with three types of boundary conditions. The application of the Galerkin method results in systems of nonlinear algebraic equations specifying the backbone curves. Moreover, the interaction between modes of vibration i.e., the phenomenon of internal resonance is investigated.

2. Formulation

The deformation gradient tensor is defined as, Malvern [4]

$$F = u\overleftarrow{\nabla} + I \tag{1}$$

where u is the displacement vector and I signifies the unit matrix. The equations of motion in the absence of body force, for a medium undergoing finite strain in material coordinates, are

$$\nabla.[SF^T] = \rho_0 \frac{\partial^2 u}{\partial t^2} \tag{2}$$

* Corresponding author: E-mail address: mahmoud.mousavi@aalto.fi

where ρ_0 is the density of the medium in the undeformed configuration and S denotes the second Piola-Kirchhoff stress tensor. The second Piola-Kirchhoff stress tensor is energetically conjugate to the Green strain tensor. Therefore, Hooke's law may be used as the constitutive equation of the medium, Malvern [4].

$$S = cE \tag{3}$$

where c is a fourth order tensor representing the elastic behavior of material and E is the Green strain tensor defined as

$$E = \frac{1}{2}\left[F^T F - I\right] \tag{4}$$

Let the lateral boundary of the rod be constrained such that only longitudinal deformation $U(x,t)$ occurs. Consequently, the gradient deformation tensor in Cartesian coordinates becomes a diagonal matrix with components

$$F_{xx} = 1 + \frac{\partial U}{\partial x} \quad , \quad F_{yy} = 1 \quad , \quad F_{zz} = 1 \tag{5}$$

Employing (4) the only non-zero component of Green strain tensor is

$$E_{xx} = (1 + \frac{1}{2}\frac{\partial U}{\partial x})\frac{\partial U}{\partial x} \tag{6}$$

For isotropic materials with modulus of elasticity Y and Poisson's ratio ν, the stress-strain relationships (3) become

$$S_{ij} = \frac{Y}{1+\nu}\left[E_{ij} + \frac{\nu}{1-2\nu}E_{kk}\delta_{ij}\right] \tag{7}$$

where δ_{ij} is the Kronecker delta. Substituting (6) into (7), we observe that the shear stresses vanish and the normal stress components yield

$$S_{xx} = \frac{Y(1-\nu)}{(1+\nu)(1-2\nu)}(1 + \frac{1}{2}\frac{\partial U}{\partial x})\frac{\partial U}{\partial x}$$

$$S_{yy} = S_{zz} = \frac{Y\nu}{(1+\nu)(1-2\nu)}(1 + \frac{1}{2}\frac{\partial U}{\partial x})\frac{\partial U}{\partial x} \tag{8}$$

Equation (2) by virtue of Eqs. (5) and (8) becomes

$$\left[(\frac{\partial U}{\partial x})^2 + 2\frac{\partial U}{\partial x} + \frac{2}{3}\right]\frac{\partial^2 U}{\partial x^2} = \frac{2\rho_0(1+\nu)(1-2\nu)}{3Y(1-\nu)}\frac{\partial^2 U}{\partial t^2} \tag{9}$$

In the particular case of infinitesimal deformation of medium, the nonlinear terms in Eq. (9) become insignificant and we arrive at

$$\frac{\partial^2 U}{\partial x^2} = \frac{\rho_0(1+\nu)(1-2\nu)}{\mu(1-\nu)}\frac{\partial^2 U}{\partial t^2} \tag{10}$$

To render the equations dimensionless, the following non-dimensional variables are introduced

$$\Psi = \frac{U}{l} \quad , \quad \zeta = \frac{x}{l} \tag{11}$$

where l is the undeformed length of the rod. Equations (9) and (10) in view of (11) become

$$\left[(\frac{\partial \Psi}{\partial \zeta})^2 + 2\frac{\partial \Psi}{\partial \zeta} + \frac{2}{3}\right]\frac{\partial^2 \Psi}{\partial \zeta^2} = \frac{2}{3}\delta\frac{\partial^2 \Psi}{\partial t^2} \tag{12}$$

and

$$\frac{\partial^2\Psi}{\partial\zeta^2}=\delta\frac{\partial^2\Psi}{\partial t^2} \tag{13}$$

where $\delta=\rho_0 l^2(1+v)(1-2v)/\mu(1-v)$.

3. Solution of eigenvalue problem

The solution to nonlinear equation (12) is accomplished by means of Galerkin method in time and space variables by taking series solution with three terms readily satisfying the boundary conditions. It is worth noting that for the analysis of internal resonance a series with three terms representing the first three modes of vibration is sufficient. The ends of a rod are either fixed or free. Consequently, three different cases depending upon the type of boundary conditions may be considered. These are analyzed separately, in what follows.

3.1. Rod with fixed-fixed ends

The boundary conditions in this case are $\Psi(0,t)=0$ and $\Psi(1,t)=0$. The one-harmonic Fourier series expansion of solution of Eq. (12) satisfying the boundary conditions is

$$\Psi(\zeta,t)=\sum_{n=1}^{3}a_n\sin(n\pi\zeta)\cos(\lambda t) \tag{14}$$

The application of Galerkin method yields

$$\frac{2}{T}\int_0^T\int_0^1 R(\zeta,t)\sin(n\pi\zeta)\cos(\lambda t)d\zeta dt=0 \qquad \text{n=1,2,3} \tag{15}$$

where $T=2\pi/\lambda$ and the residue $R(\zeta,t)$ is determined by substituting Eq. (14) into (12) leading to

$$\begin{aligned}R(\zeta,t)=\pi^2[&(a_1\pi\cos(\pi\zeta)+2a_2\pi\cos(2\pi\zeta)+3a_3\pi\cos(3\pi\zeta))^2(\cos(\lambda t))^2\\&+2(a_1\pi\cos(\pi\zeta)+2a_2\pi\cos(2\pi\zeta)+3a_3\pi\cos(3\pi\zeta))(\cos(\lambda t))+2/3][a_1\sin(\pi\zeta)\\&+4a_2\sin(2\pi\zeta)+9a_3\sin(3\pi\zeta)]\cos(\lambda t)\\&-2/3\,\delta\lambda^2[a_1\sin(\pi\zeta)+a_2\sin(2\pi\zeta)+a_3\sin(3\pi\zeta)]\cos(\lambda t)\end{aligned} \tag{16}$$

Equation (15) after carrying out the necessary manipulations leads to three algebraic equations for the amplitudes and frequenceis of vibration as

$$\begin{aligned}&\frac{3\pi^4}{4}\left[\frac{1}{8}a_1^3+a_2^2a_1+\frac{9}{4}a_3^2a_1+\frac{3}{8}a_1^2a_3+\frac{3}{2}a_2^2a_3\right]+\frac{1}{3}a_1\left(\pi^2-\delta\lambda^2\right)=0\\&a_2\left[\frac{3\pi^4}{4}\left(2a_2^2+a_1^2+9a_3^2+3a_1a_3\right)+\frac{1}{3}\left(4\pi^2-\delta\lambda^2\right)\right]=0\\&\frac{3\pi^4}{4}\left[\frac{1}{8}a_1^3+\frac{81}{8}a_3^3+\frac{9}{4}a_1^2a_3+9a_2^2a_3+\frac{3}{2}a_1a_2^2\right]+\frac{1}{3}a_3\left(9\pi^2-\delta\lambda^2\right)=0\end{aligned} \tag{17}$$

The following nontrivial solutions may be obtained for the system of Eqs. (17):

$$i)\ a_2=a_3=0,\quad a_1=\pm\frac{4\sqrt{2}}{3\pi^2}\sqrt{\delta\lambda^2-\pi^2},\quad \lambda_1\geq\frac{\pi}{\sqrt{\delta}} \tag{18}$$

ii) $a_1 = a_3 = 0\,,\quad a_2 = \pm\frac{\sqrt{2}}{3\pi^2}\sqrt{\delta\lambda^2 - 4\pi^2}\,,\quad \lambda_2 \geq \frac{2\pi}{\sqrt{\delta}}$ (19)

iii) $a_1 = a_2 = 0\,,\quad a_3 = \pm\frac{4}{9\sqrt{6}\pi^2}\sqrt{\delta\lambda^2 - 9\pi^2}\,,\quad \lambda_3 \geq \frac{3\pi}{\sqrt{\delta}}$ (20)

The backbone curve, i.e., the amplitude-frequency relationship for the above cases are depicted in Fig. 1.

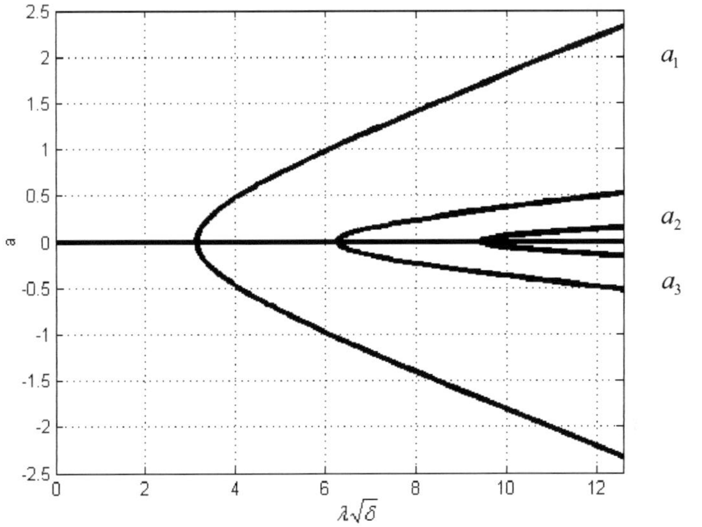

Fig. 1: Backbone curves for rods with fixed-fixed and free-free ends

The bifurcation point for each mode is equal to the linear natural frequency of the rod obtained from Eq. (13). The frequency at bifurcation points for second and third modes are integral multiples of the first mode, $\lambda_2 = 2\lambda_1$ and $\lambda_3 = 3\lambda_1$, thus we expect the phenomenon of internal resonance. First, the possibility of internal resonance between the first and third modes of vibration is investigated.

iv) $a_2 = 0\,,\quad a_1 \neq 0\,,\quad a_3 \neq 0$

Applying the above conditions to Eqs. (17) and eliminating $\delta\lambda^2$, we arrive at the equation specifying the interaction between the first and the third modes of vibration

$$a_1^3 + 63a_3^3 + 17a_1^2a_3 - 3a_1a_3^2 + \frac{256}{9\pi^2}a_3 = 0 \quad (21)$$

The internal resonance occurs when the values of a_1 and a_3 satisfying Eq. (21) result in real values for the frequency from the following equation

$$\lambda = \frac{\pi}{\sqrt{\delta}}\left[1 + \frac{9\pi^2}{16}\left(\frac{1}{2}a_1^2 + 9a_3^2 + \frac{3}{2}a_1a_3\right)\right]^{1/2} \quad (22)$$

The plot of a_1 versus a_3 for the internal resonance is shown in Fig. 2, and in Fig. 3, the backbone curves of internal resonance are plotted. It is worth noting that for $a_1 \neq 0$, $a_2 \neq 0$, $a_3 = 0$, no real frequency can be found. Thus, the internal resonance between the first and second modes does not exist.

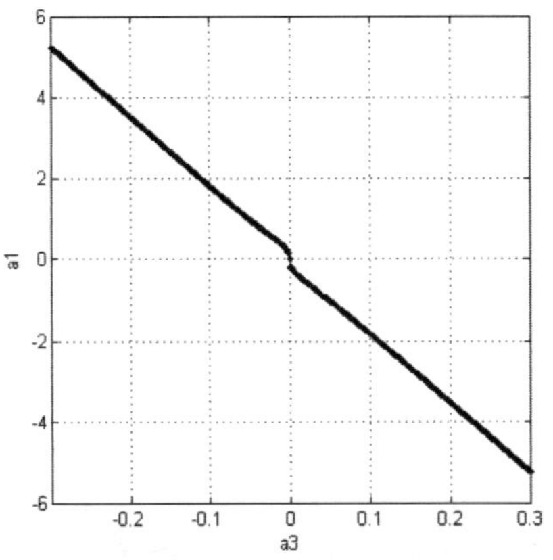

Fig. 2: Internal resonance for rods with fixed-fixed ends

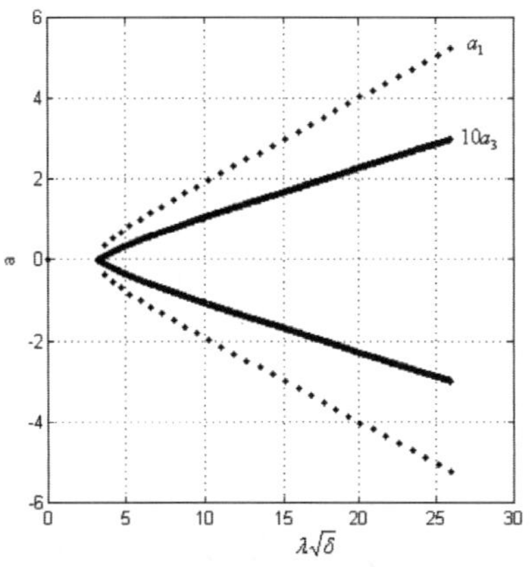

Fig. 3: Backbone curve for a_1 and a_3 due to internal resonance for rod with fixed-fixed ends

3.2. Rod with free-free ends

The boundary conditions in this case are $\partial\Psi(0,t)/\partial x = 0$ and $\partial\Psi(l,t)/\partial x = 0$. The solution of Eq. (12) is considered as

$$\Psi(\zeta,t) = \sum_{n=1}^{3} a_n \cos(n\pi\zeta)\cos(\lambda t) \tag{23}$$

Application of Galerkin method leads to the system of algebraic equations

$$\begin{gathered}
\frac{3\pi^4}{4}\left[\frac{1}{8}a_1^3 + a_2^2 a_1 + \frac{9}{4}a_3^2 a_1 - \frac{3}{8}a_1^2 a_3 + \frac{3}{2}a_2^2 a_3\right] + \frac{1}{3}a_1\left(\pi^2 - \delta\lambda^2\right) = 0 \\
a_2\left[\frac{3\pi^4}{4}\left(2a_2^2 + a_1^2 + 9a_3^2 + 3a_1a_3\right) + \frac{1}{3}\left(4\pi^2 - \delta\lambda^2\right)\right] = 0 \\
\frac{3\pi^4}{4}\left[-\frac{1}{8}a_1^3 + \frac{81}{8}a_3^3 + \frac{9}{4}a_1^2 a_3 + 9a_2^2 a_3 + \frac{3}{2}a_1 a_2^2\right] + \frac{1}{3}a_3\left(9\pi^2 - \delta\lambda^2\right) = 0
\end{gathered} \tag{24}$$

The nontrivial solutions of the above equations are identical with those of Eqs. (17). Therefore, Fig. 1 shows the bifurcation points and backbone curves. Internal resonance only between the first and third modes, where $\lambda_3 = 3\lambda_1$ exists and the equation for amplitudes of vibration for internal resonance is

$$a_1^3 - 63a_3^3 - 17a_1^2 a_3 - 3a_1 a_3^2 - \frac{256}{9\pi^2}a_3 = 0 \tag{25}$$

The frequency for internal resonance between the first and third modes may be obtained from the following equation

$$\lambda = \frac{\pi}{\sqrt{\delta}}\left[1 + \frac{9\pi^2}{16}\left(\frac{1}{2}a_1^2 + 9a_3^2 - \frac{3}{2}a_1 a_3\right)\right]^{1/2} \tag{26}$$

Fig. 4, shows the plot of a_1 against a_3 for internal resonance. The backbone curves for internal resonance are shown in Fig. 5.

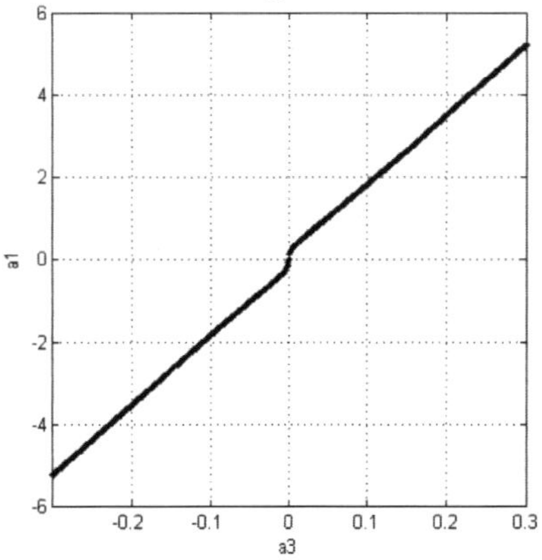

Fig. 4: Internal resonance for rods with free-free ends

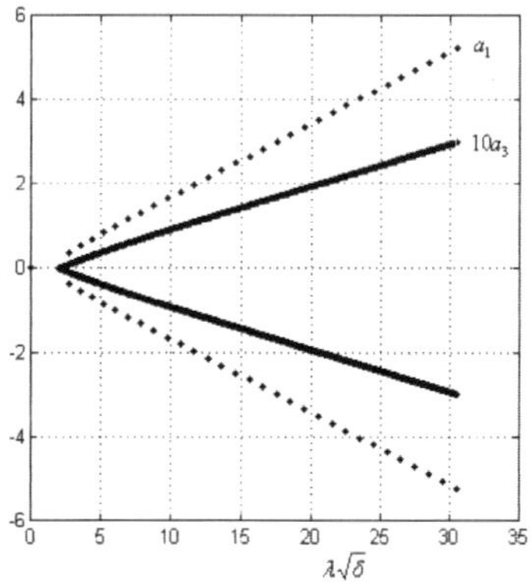

Fig. 5: Backbone curve for a_1 and a_3 due to internal resonance for rod with free-free ends

3.3. Rod with fixed -free ends

The solution for a rod fixed at $\zeta = 0$ and free at $\zeta = 1$ is taken as

$$\Psi(\zeta,t) = \sum_{n=1}^{3} a_n \sin[(2n-1)\pi\zeta / 2]\cos(\lambda t) \tag{27}$$

Similar to part (3.1), the system of algebraic equations yield

$$\begin{aligned}
&\frac{3\pi^4}{64}\left[\frac{1}{8}a_1^3 + \frac{9}{4}a_2^2 a_1 + \frac{25}{4}a_3^2 a_1 + \frac{3}{8}a_1^2 a_2 + \frac{15}{4}a_1 a_2 a_3 + \frac{45}{8}a_2^2 a_3\right] + \frac{1}{3}a_1\left(\frac{\pi^2}{4} - \delta\lambda^2\right) = 0 \\
&\frac{3\pi^4}{64}\left[\frac{1}{8}a_1^3 - \frac{9}{8}a_1^2 a_3 + \frac{45}{4}a_1 a_2 a_3 + \frac{9}{4}a_1^2 a_2 + \frac{81}{8}a_2^3 + \frac{225}{4}a_2 a_3^2\right] + \frac{1}{3}a_2\left(\frac{9\pi^2}{4} - \delta\lambda^2\right) = 0 \\
&\frac{3\pi^4}{64}\left[\frac{45}{8}a_1 a_2^2 + \frac{15}{8}a_1^2 a_2 + \frac{25}{4}a_3 a_1^2 + \frac{225}{4}a_3 a_2^2 + \frac{625}{8}a_3^3\right] + \frac{1}{3}a_3\left(\frac{25\pi^2}{4} - \delta\lambda^2\right) = 0
\end{aligned} \tag{28}$$

The nontrivial solutions to the above equations are:

$$i)\ a_2 = a_3 = 0,\quad a_1 = \pm\frac{16\sqrt{2}}{3\pi^2}\sqrt{\delta\lambda^2 - \frac{\pi^2}{4}},\quad \lambda_1 \ge \frac{\pi}{2\sqrt{\delta}} \tag{29}$$

$$ii)\ a_1 = a_3 = 0,\quad a_2 = \pm\frac{16\sqrt{2}}{27\pi^2}\sqrt{\delta\lambda^2 - \frac{9\pi^2}{4}},\quad \lambda_2 \ge \frac{3\pi}{2\sqrt{\delta}} \tag{30}$$

$$iii)\ a_1 = a_2 = 0,\quad a_3 = \pm\frac{16\sqrt{2}}{75\pi^2}\sqrt{\delta\lambda^2 - \frac{25\pi^2}{4}},\quad \lambda_3 \ge \frac{5\pi}{2\sqrt{\delta}} \tag{31}$$

The bifurcation points coincide with the natural frequencies of linear vibration obtained from Eq. (13), and backbone curves are depicted in Fig. 6.

The internal resonance between the first and second modes, $\lambda_2 = 3\lambda_1$, leads to the following case.

$iv)\ a_3 = 0,\quad a_1 \ne 0,\quad a_2 \ne 0$

From the system of Eqs. (28), after eliminating $\delta\lambda^2$, we arrive at the equation specifying the interaction between the first and the second modes of vibration as

$$3a_1^3 + 189a_2^3 + 51a_1^2a_2 - 9a_1a_2^2 + \frac{1024}{3\pi^2}a_2 = 0 \tag{32}$$

The internal resonance occurs when the values of a_1 and a_2 satisfying Eq. (32) give real value for the frequency

$$\lambda = \frac{\pi}{2\sqrt{\delta}}\left[1 + \frac{9\pi^2}{64}\left(\frac{1}{2}a_1^2 + 9a_2^2 + \frac{3}{2}a_1a_2\right)\right]^{1/2} \tag{33}$$

The plot of a_1 against a_2 is shown in Fig. 7 and the backbone curves are depicted in Fig. 8. It is noteworthy to mention that internal resonance is confined to this case. In Figs. 2, 4 and 7 a saddle point may be observed at the origin. Moreover, since internal resonance occurs between modes of vibration with ratio of frequencies 3 at bifurcation points, Eq. (12) may be considered as an equation with cubic nonlinearity.

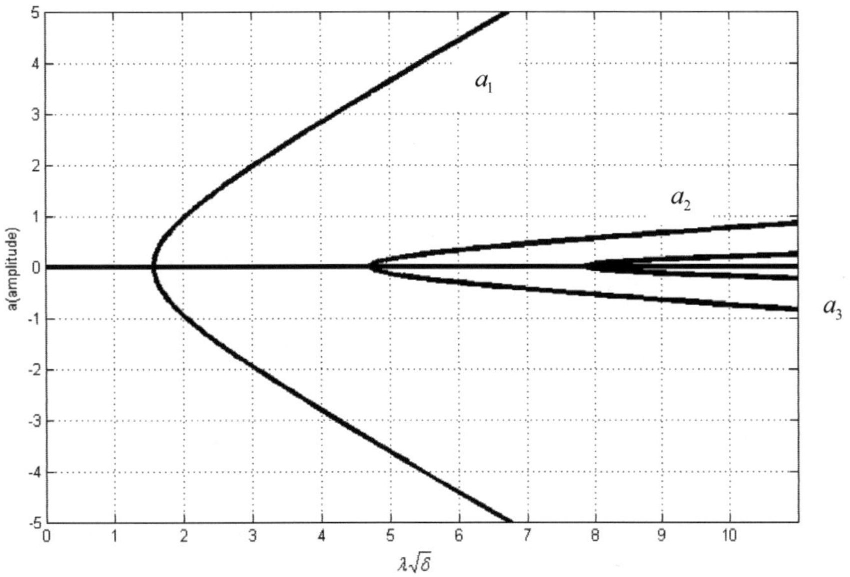

Fig. 6: Backbone curves for rods with fixed-free ends

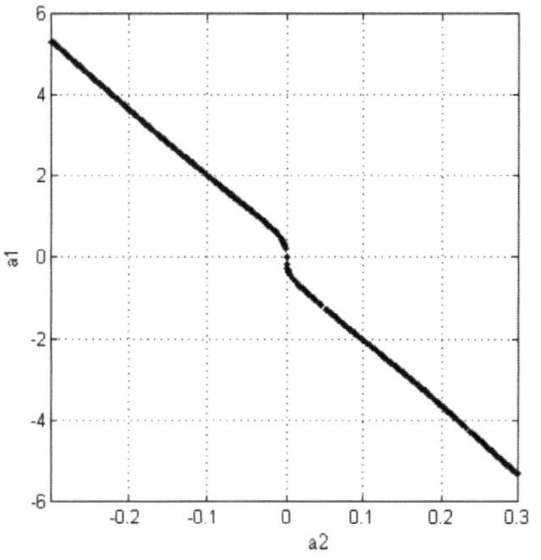

Fig. 7: Internal resonance for rods with fixed-free ends

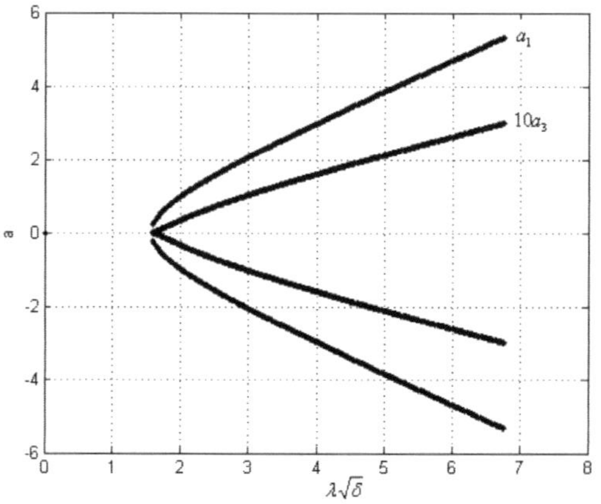

Fig. 8: Backbone curve for a_1 and a_2 due to internal resonance for rod with fixed-free ends

Conclusion:

In this article, the free vibration of a rod is investigated. By solving an eigenvalue problem, the phenomenon of internal resonance is observed between certain modes. This analysis conveys that for the rod with fixed-fixed ends, internal resonance occurs between first and third modes, while no interaction between first and second or second and third modes are observed. This behavior has been reported previously. The same behavior is seen in the rod with free ends. On the other hand, for the rod with a free end and a fixed end, it is seen that there is interaction between first and second modes and no interaction exists between the first and third or second and third modes.

References

[1] Lewandowski, R., 1994, Solution with bifurcation points for free vibration of beams: an analytical approach, J. Sound Vib., 177(2), 239-249.

[2] Lewandowski, R., 1996, On beams membranes and plates vibration backbone curves in cases of internal resonance, Meccanica, 31(3), 323 – 346.

[3] Andrianov, V., Danishevs'kyy, V.V., 2002, Asymptotic approach for non-linear periodical vibrations of continuous structures, J. Sound Vib., 249(3), 465-481.

[4] Malvern, L.E., 1969, Introduction to the Mechanics of a Continuous Medium, Prentice Hall, Englewood Cliffs, New Jersey.

Chatter control in the milling process of composite materials

K Kecik[1], R Rusinek, J Warminski and A Weremczuk

Lublin University of Technology, Department of Applied Mechanics

k.kecik@pollub.pl

Abstract. In this paper, a model of the milling process of fibre reinforced composite material is shown. This classical one degree of freedom model of the milling process is adjusted for composite materials by variable specific cutting forces, which describe the fibre resistance. The stability lobe diagrams are determined numerically. Additionally, to eliminate the chatter vibration, small relative oscillations between the workpiece and the tool are introduced. Basing on numerical simulations the range of amplitude and the frequency of excitation is found for chatter reduction.

1. Introduction

A composite material consist of two or more materials that results in better properties than when the individual components are used alone [1]. The aerospace and aeronautical industries have been major users of composite technology in the last decades. Composite materials provide distinctive advantages in the manufacture of advanced products because of attractive features such as simultaneous high strength and light weight. There are a large number of various composite materials used in aircraft, automotive, aerospace and many other industries. In modern engineering high demands are being placed on components made of composites in relation to their dimensional precision as well as to their surface quality. As a result of potential applications there is a great need to understand the questions associated with the machining of composite materials. They are more difficult to machine than classical materials mainly because they are anisotropic, nonhomogeneous and are reinforced by hard glass or carbon fibres [3]-[4].That leads to fast tool wear [5]. Anisotropic and inhomogeneous material properties create problems during machining of composite materials. The cutting forces depend on the fibre direction of the laminate. Machining forces and finish surface quality may change depending on the direction of cutting. In addition, the rubbing effect of the deboned particles on the machined surface degrades the surface quality of the product [6], [7]. Unwanted self-excited vibrations between the tool and workpiece, called chatter, are a primary cause of surface finish roughness in cutting both classical and composite materials. During the machining of composite materials, additional problems may arise especially when the matrix is carbon or glass fibre reinforced.

The primary aim of this article is the reduction of chatter vibrations during composite material milling. The paper is based on a classical one degree of freedom milling model with a chatter control system (CCS) which generates small relative vibrations between the workpiece and the tool in order to improve cutting conditions.

2. Modelling of the milling process

2.1. Modelling of milling process with active workpiece vibration

In the milling process, material is removed from a workpiece by a cutting tool, which rotates with speed Ω (in rpm). A schematic representation of the milling process is shown in *Fig. 1*. The cutting force F_j acting on jth tooth ($j=0,1, .., z$) in the x direction is represented by tangential (F_{tj}) and normal (F_{nj}) component

$$F_j = \left(F_{tj} \sin\varphi_j - F_{nj} \cos\varphi_j\right) g_j, \tag{1}$$

where z represents the number of tool teeth, and g_j defines when j-th tooth is active. Tangential and radial cutting force actions on the tool are generally proportional to the axial depth of cut (a_p) and chip thickness (h_j)

$$F_{tj} = K_t a_p h_j (t)^{\kappa},\ F_{nj} = K_n a_p h_j (t)^{\kappa}, \tag{2}$$

where K_t and K_n are specific cutting forces which depend on the cutting material properties, depth of cut a_p and fibre orientation. The coefficient κ also depends on the material, and is usually estimated about 0.8-1[8].

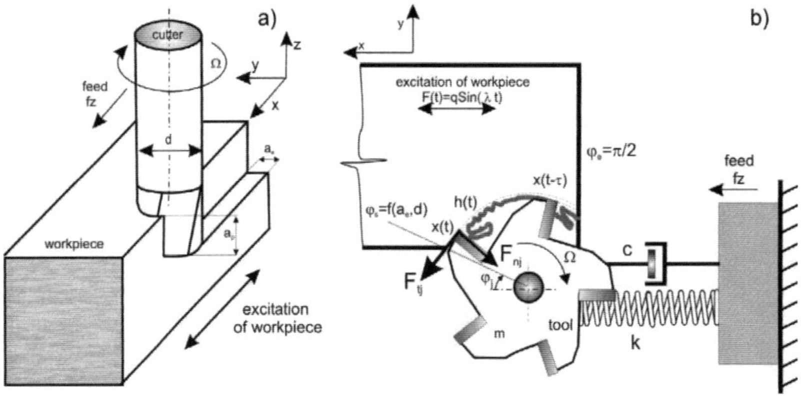

Fig.1. Model of down milling process with excitation of workpiece.

The chip thickness $h_j(t)$ is a function of the feed of the cutter f_z, tool vibrations (*regeneration effect*) and the active excitation of workpiece $F(t)= q\sin(\lambda t)$. The actual chip thickness is

$$h_j = \left[f_z + x(t) - x(t-\tau)\right]\cos\varphi_j - q\sin(\lambda t), \tag{3}$$

where the tooth passing period is $\tau=60/z\Omega$. Parameters q and λ are responsible for active vibration control. This workpiece excitation in the real system can be caused by piezoelectric stack actuators.
The step function g_j is defined in order to check whether the tool is in cut or not:

$$g_j(\varphi_j) = \begin{Bmatrix} 1, & \varphi_s \le \varphi_j \le \varphi_e \ and\ h_j > 0 \\ 0 & elsewhere \end{Bmatrix}. \tag{4}$$

Usually, the conditions described in Eq. (4) are realized by Heaviside functions. In order to calculate the total milling forces, the number of teeth (z) on the cutter and the radial depth of cut (a_e) and diameter of cutter (d) must also be known. Then, for down milling the entry (φ_s) and exit (φ_e) angles are defined

$$\varphi_s = \arcsin\left(\frac{d-2a_e}{d}\right),\quad \varphi_e = \pi/2. \tag{5}$$

The equation of motion, for one degree of freedom milling model can be written as

$$m\ddot{x} + c\dot{x} + kx = a_p h(t)^{\kappa}\left(K_t \sin\varphi - K_n \cos\varphi\right) g_j, \tag{6}$$

where k, c and m are stiffness, damping and mass obtained by modal tests [8].

2.2. Fibre orientation effect

Because of the anisotropic nature of fibre composites, distinctly different cutting mechanisms have been observed depending on the fibre orientation with respect to the direction of tool motion [9]-[11]. The peculiar aspect of cutting with a rotating tool such as in milling is that the fibre orientation angle θ, is not constant, but varies continuously with cutting edge position around the cutter axis [12]. *Fig. 2* shows the definition of fibre orientation with respect to the angle of tool rotation.

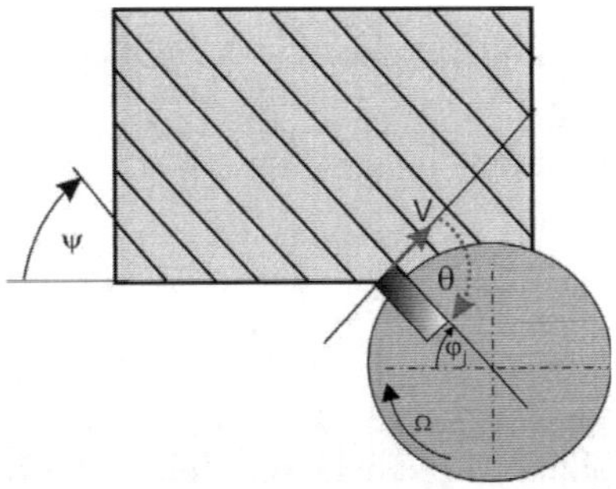

Fig.2. Fibre orientation of unidirectional laminate for down milling process.

In our example, for down milling of a unidirectional laminate the fibre direction angle is defined by $\psi=45^0$. The temporary angle θ, is measured clockwise from the cutting velocity vector (V) to the location of the teeth. For the case shown in *Fig. 2*, it can be written that

$$\theta = \pi / 2 - \left(\varphi_j - \psi\right). \tag{7}$$

The specific cutting forces K_n and K_t corresponding to these fibre orientations are shown in *Fig. 3a* [12]. The specific cutting forces rise gradually with an increase in fibre orientation and reach a maximum between 60^0 and 120^0. Interestingly, for the fibre direction angle $\Psi=45^0$ the specific cutting force is higher if the edge is near the exit (then the chip thickness is smallest –down milling, *Figure 3b*).

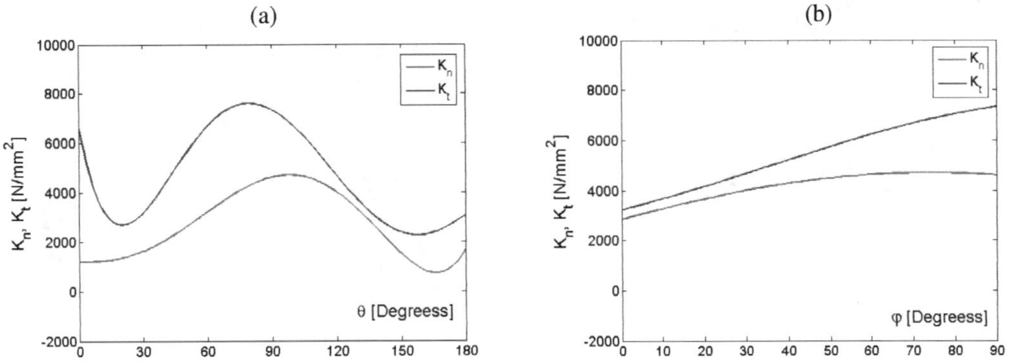

Fig. 3. Variation of specific cutting forces K_t and K_n depending on fibre orientation Θ (a), and tool position φ for $\Psi=45^0$ (b) [12].

Table 1. Coefficients of equation (8).

Coefficient	Kn [N/mm^2]	Kt [N/mm^2]
a_0	30.5840	137.1059
a_1	0	−536.5678
a_2	0	1115.3522
a_3	109.6388	−786.1277
a_4	−76.3806	224.6042
a_5	13.2425	−22.5474
n	−0.6937	−0.7329

The mathematical function which characterises the specific cutting forces is formulated on the basis of [12] in the form

$$K_{n,t} = a_p^n \left(a_0 + a_1\theta + a^2\theta^2 + a_3\theta^3 + a^4\theta^4 + a_5\theta^5 \right), \tag{8}$$

where a_0, a_1, a_2, a_3, a_4, a_5, and n are regression model coefficients and are given in Table 1. The angle θ in the equation (8) is in radians. The model is verified by experiment in paper [12].

3. Results and discussion

3.1. The influence of fibre orientation on milling stability

During the machining the problem of dynamic stability is very important from the practical point of view[13]. Therefore, section a stability analysis is performed for high speed machining (HSM) conditions (100 rpm<Ω<20 krpm). The governing equation (6) with (7) and (8) is solved with the help of the fourth order Runge-Kutta method with a fixed time step of *0.0001*. Stability lobe diagrams (SLD) have been generated using parameters obtained from experimental modal tests of a real spindle - tool system (CNC milling machine *Blue Bird MG6037PKK*). Based on the modal analysis performed on the milling machine with the one flute PCD end cutter (Cuttech, radius of *6mm*), the modal parameters are: $m=0.8kg$, $k=2.22\cdot10^7$ N/m, $c=57,3Ns/m$. The radial depth of cut $a_e=6mm$, feed $f_z=0.03mm/tooth$, number of teeth z=1 and exponent $\kappa=0.9$. To identify the region of process stability, on the basis of numerical simulations, an amplitude criterion is introduced, namely if the amplitude is greater than *1mm* we classify the process as unstable. A similar criterion has been used in paper [13], where SLD obtained by simulation is compared with the results obtained by commercial software *CutPro9*. Additionally, in papers [13] and [14] new indices to detect unstable motion are proposed. In *Figure 4a* the stability lobe diagram (SLD) for a classical model with constant specific cutting forces is shown. The black area denotes unstable milling parameter ranges, but the white colour indicates ranges of parameters for stable machining.

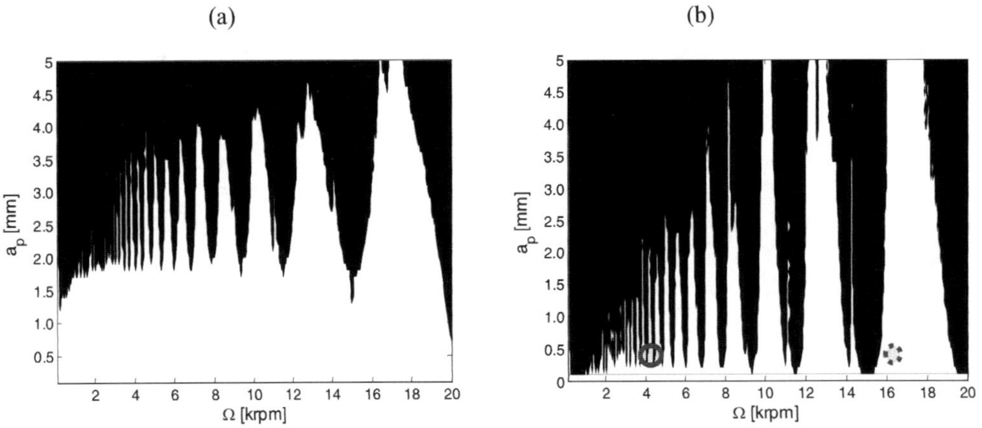

Fig.4. The classical SLD obtained for constant values of $K_t=7500\ N/mm^2$ and $K_n=4000\ N/mm^2$ (a), SLD obtained for composite model described by eqs. (7) and (8) for $\Psi=45^0$ (b)

A numerical stability lobes diagram has been done by point-by-point investigation of a gird in the parameter space consisting of the spindle speed resolution *50rpm* and depth of cut resolution *0.1mm*. The SLD presented in *Fig. 4a* and *4b* are much different. The chatter frequency seems the same, however the critical depth of cut for classical model equal about *1mm (Fig.4b)*, but for composite materials is shifted down and equal *0.1mm*. The results suggest that the classical approach to determine the stability of cutting process fails. Interestingly, that for composite material exist regions, where the cutting process with large depth of cut is possible.

Journal of Physics: Conference Series **382** (2012) 012012 doi:10.1088/1742-6596/382/1/012012

3.2. Control of chatter vibration

In the literature several methods are proposed to counteract chatter. Generally, components based on the integration of mechatronic systems into the machine structure are the most promising [15]. Martinez et al. [16] demonstrated the feasibility of piezoelectric stack actuators in chatter reduction.

(a) (b)

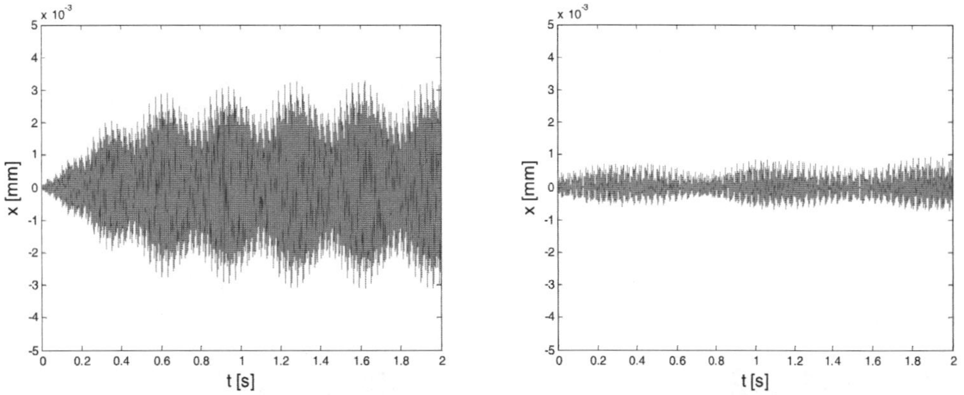

Fig.6.Time history of tool vibration without activation of workpiece (a) and with activation $q=0.5mm$ and $\lambda=20000rad/s$ (b) for $a_p=0.5mm$ and $\Omega=6krpm$ and fibre orientation $\Psi=45^0$.

(a) (b)

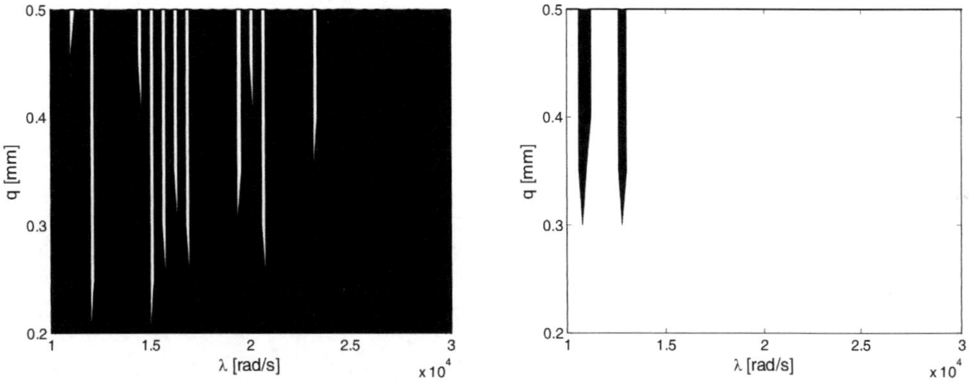

Fig.7.Two parameters space plot: amplitude and frequency of workpiece excitation, $\Omega=6krpm$ (a) and $\Omega=18krpm$ (b), for $a_p=0.5mm$ cutting parameters for fibre orientation $\Psi=45^0$.

In *Fig.6*, the time history of tool vibrations, for two cases, is presented. The first time course (*Fig. 6a*) is obtained for $\Omega=6krpm$ and depth of cut $a_p=0.5mm$ (circled point by solid line in *Fig. 4b*). The amplitude of vibration without the chatter control system (CCS) reaches a value of about $x=3mm$, but if we activate the control system, the vibrations amplitude is below *1mm* (*Fig.6b*). This denotes that the machining process is being stabilized. The main problem is how to choose parameters of CCS. Therefore, *Fig. 7*, presents the two parameter space plot: amplitude and frequency of excitation (λ-q). For the unstable point from *Fig.4b* ($a_p=0.5mm$, $\Omega=18krpm$) CCS works correctly if parameters λ and q are selected from white region *Fig. 7a*, while λ-q from the black regions does not stabilize the milling process. An activation of CCS for stable cutting parameters $a_p=0.5mm$, $\Omega=6krpm$ (*Fig.4b*) can destabilize the process if λ and q are selected from black regions (*Fig. 7b*). This is very important from the engineering point of view because CCS can eliminate chatter vibration or on the contrary – generate it. Therefore, the use of CCS must be carefully applied.

4. Conclusion and final remarks

Determination of the cutting process stability of typical commonly used materials can be calculated quite easy. The problem arises when the workpiece is made of material which has heterogeneous properties e.g. composites. Then the classical approach fails, and commercial software does not have properties for heterogeneous materials. Therefore, analysis of composite cutting materials is important from the practical point of view. The presented model can be used to determine the stability chart of the machining process.

To improve the milling process and eliminate chatter vibrations, the CCS is proposed for composites materials. This method should be carefully applied, because the specific parameters of CCS can generate vibrations. Therefore, this method of stabilizing requires detailed verification.

Acknowledgements

The research leading to these results has received funding from the European Union Seventh Framework Programme (FP7/2007-2013), FP7-REGPOT-2009-1, under grant agreement No: 245479. The support by the Polish Ministry of Science and Higher Education-grant no 1471-1/7.PR UE/2010/7-is also acknowledged.

References

[1] Peters ST. *Hanbook of composites.* Chapman&Hall, London 1998.

[2] Chinmaya RD, Yung CS. Modeling of machining of composite materials: A review. *International Journal of Machine Tools & Manufacture*, 2012 (in print).

[3] Komanduri R. Machining of fiber-reinforcedcomposites, *MachiningScience an Technology*, 1(1), 113–152, 1997.

[4] Nayak D, Singh I, Bhatnagar N, Mahajan P. Ananalysis of machining Induced damagesin FRP composites a micromechanics finite element approach, AIP Conference Proceedings, 712, 327–331, 2004.

[5] Komanduri R, Machining of fiber-reinforced composites, *Machining Science and Technology*, 1 (1), 1997, 113–152.

[6] Arola D, Sultan MB, Ramulu M. Finite element modeling of edge trimming fiber reinforced plastics. *Transactions of the ASME, Journal of Manufacturing Science and Engineering*, 124 (1), 32–41, 2002.

[7] Monaghnan J, Brazil D. Modeling the flow processes of a particle reinforced metal matrix composite during machining, *Composites Part A* 29(A), 87–99, 1998.

[8] Insperger T, Stepan T, Bayly PV, Mann BP. Multiple chatter frequency in milling processes, *Journal of Sound and Vibration*, 262, 2003, 333-345.

[9] Koplev A, Lystrup A, Vorm T. The cutting process, chips, and cutting forces in machining CFRP, Composites 14 (4), 1983, 371–376.

[10] Kalla D, Sheikh-Ahmad J. Twomy J. Prediction of cutting forces in helical end milling fiber reinforced polymers, *International Journal of Machine Tools and Manufacture*, 50, 882–891, 2010.

[11] Zhang LC. Cutting composites: A discussion on mechanics modelling. *Journal of Materials Processing Technology*, 209, 2009, 4548–4552.

[12] Sheikh-Ahmad JY. Machining of Polymer Composites.Springer 2009.

[13] Kecik K, Rusinek R, Warminski J. Modelling of high speed milling process with frictional effect. *Proceedings of the Institution of Mechanical Engineers, Part K, Journal of Multi-body Dynamics*, 2012 (submitted).

[14] Kecik K, Rusinek R, Warminski J. Stability lobes analysis of nickel superalloys milling. *International Journal Bifurcation and Chaos*, 21(10), 2011, 1-12.

[15] Neugebauer R, Denkena B, Wegener K. Mechatronic Systems for Machine Tool. *Annals of the CIRP* 56(2), 2007, 657-686.

[16] Martinez DR, Hinnerichs TD, Redmond JM. Vibration control for PrecisionManufacturing using Piezoelectric Actuators. *Journal of Intelligent Material Systems and Structures*, **7**, 1996, 182-191.

A study of weld quality in ultrasonic spot welding of similar and dissimilar metals

Z Al-Sarraf, M Lucas
School of Engineering, University of Glasgow, Glasgow, G12 8QQ, UK

E-mail: z.sarraf.1@research.gla.ac.uk

Abstract. Several difficulties are faced in joining thinner sheets of similar and dissimilar materials from fusion welding processes such as resistance welding and laser welding. Ultrasonic metal welding overcomes many of these difficulties by using high frequency vibration and applied pressure to create a solid-state weld. Ultrasonic metal welding is an effective technique in joining small components, such as in wire bonding, but is also capable of joining thicker sheet, depending on the control of welding conditions. This study presents the design, characterisation and test of a lateral-drive ultrasonic metal welding device. The ultrasonic welding horn is modelled using finite element analysis and its vibration behaviour is characterised experimentally to ensure ultrasonic energy is delivered to the weld coupon. The welding stack and fixtures are then designed and mounted on a test machine to allow a series of experiments to be conducted for various welding and ultrasonic parameters. Weld strength is subsequently analysed using tensile-shear tests. Control of the vibration amplitude profile through the weld cycle is used to enhance weld strength and quality, providing an opportunity to reduce part marking. Optical microscopic examination and scanning electron microscopy (SEM) were employed to investigate the weld quality. The results show how the weld quality is particularly sensitive to the combination of clamping force and vibration amplitude of the welding tip.

1. Introduction

Ultrasonic welding can be defined as a solid-state joining process in which materials are held together by a normal force whilst a high-frequency shear vibration is applied. During this process the vibration deforms, shears, and flattens surface asperities, scatters interstitial oxides and contaminants, and increases the contact area of the parts being welded [1]. The first demonstration of ultrasonic welding was in the early 1950s and was limited to grain refinement and soldering [2], but now the technique can be applied to various softer and harder metals [3]. Ultrasonic welding has become an efficient joining technique for many industrial and scientific applications, using lighter and more versatile equipment to produce a stronger, smaller, and more precise weld [2]. Furthermore, ultrasonic welding does not require any solder or filler and therefore has some associated environmental and economic benefits [4]. Ultrasonic welding systems consist of a power supply, transducer, booster, horn and anvil. The horn is tuned to operate in the longitudinal mode but precise design and characterisation are required to ensure the required dynamic characteristics of the device are achieved and the subsequent weld is of high strength and quality.

In this study, two investigations have been carried out to improve weld strength and quality; numerical design and subsequent experimental characterisation of an integrated ultrasonic spot welding horn to enhance the vibration characteristics at the welding surface, and secondly an experimental study of the effects of process parameters on the weld itself, considering issues such as tool/workpiece adhesion and weld quality. The weld strength is characterised experimentally in terms of the results of repeated tensile shear tests. Weld quality is also investigated by using microscopic observations of the deformed surface areas under different welding conditions. The ability to use

Published under licence by IOP Publishing Ltd

Journal of Physics: Conference Series **382** (2012) 012013 doi:10.1088/1742-6596/382/1/012013

amplitude profiling to increase weld strength, enhance weld quality and reduce part marking of joining similar and dissimilar materials is also demonstrated.

2. Design and characterisation of the welding system

2.1 Numerical design of the ultrasonic spot welding horn

A design of a lateral-drive ultrasonic welding system has been carried out to meet the requirement for high ultrasonic amplitude and low clamping force that are necessary to weld thinner parts [5], Fig. 1.

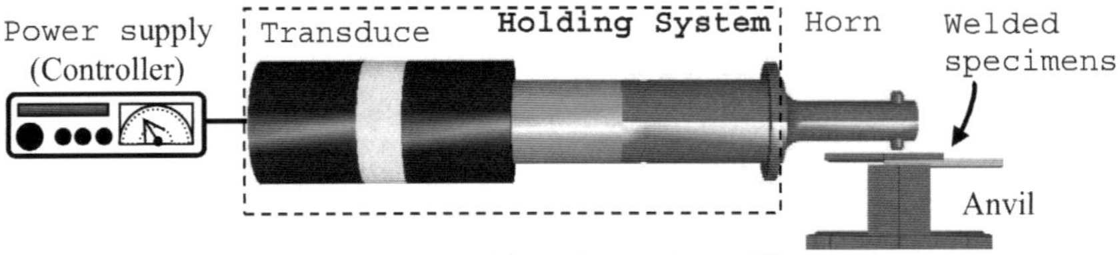

Figure 1. Lateral-drive ultrasonic welding device.

Several factors are considered in the design of such an ultrasonic spot welding horn including resonant frequency, frequency separation, amplitude amplification, amplitude uniformity and stress concentration. The horn is made from steel to give good acoustic properties and acceptable wear characteristics. An integrated spot welding tip with knurled flat surface was designed. High amplitude amplification and low stress are obtained from adopting a catenoidal horn profile which is clamped at a nodal plane. Finite element (FE) analysis has been carried out to extract the shape of the operating vibration mode and those of some of the surrounding modes, as shown in Fig. 2.

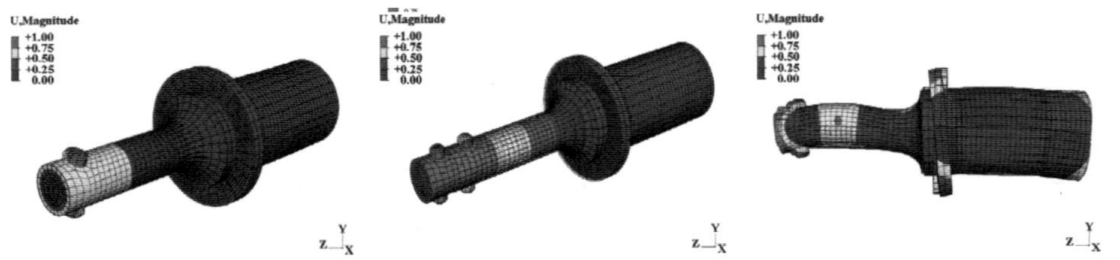

Figure 2. FEA predicted torsional (13.2 kHz), longitudinal (20.8 kHz) and bending (22.6 kHz) modes

2.2 Experimental characterisation of the ultrasonic spot welding horn

The numerical results are compared to the results of an Experimental Modal Analysis (EMA), shown in Fig. 3, carried out using a 3D laser Doppler vibrometer (LDV) and random excitation over a frequency range from 0 to 40 kHz. The amplifications, resonant frequencies and mode shapes of the horn are found to be in agreement and the system is therefore considered to operate as designed.

Figure 3. EMA measured torsional (13.7 kHz), longitudinal (20.8 kHz) and bending (23.5 kHz) modes

3. Welding Experiments

3.1 Experimental set up

The welding stack (horn connected with transducer) is driven by a 1 kW ultrasonic generator (Sonic Systems), to provide a mechanical vibration at 20.8 kHz. A tensile test machine (Zwick Roell) is employed to secure the welding stack by holding it at the nodal plane, as shown at left in Fig. 4(a). The upper moving cross-head of the test machine is equipped with a 2 kN load cell to measure the clamping force, which is recorded as a function of time. The knurled flat welding tip surface contacts the upper specimen of the weld coupon during welding, while the lower specimen is held by a knurled anvil as shown in Fig. 4(b). In this study, the weld samples were cut from aluminium and copper strips, and their properties were obtained from ASTM [6] and BSI [7] codes. The specimens were prepared to 50 x 10 mm size from roll sheets with 0.1 mm and 0.5 mm thickness, and the overlap between specimens was set at 10 mm to accommodate the constant spot diameter of 6 mm of the welding tip, as shown in Fig. 4(c). Dummy plates were added to avoid bending due to structural misalignment of specimens during the test. Five tensile tests at each set of weld conditions were carried out, the results were averaged, and the standard deviation for the tests was obtained. Welding of similar and dissimilar metals was performed with different arrangement of stacking order such as: (Al-Al), (Cu-Cu), (Al-Cu) and (Cu-Al). Welds were obtained (Fig. 4(c)), for different sheet thicknesses and several tests were carried out to ameliorate the problem of horn sticking. However, a few tests failed either due to insufficient power or applied load.

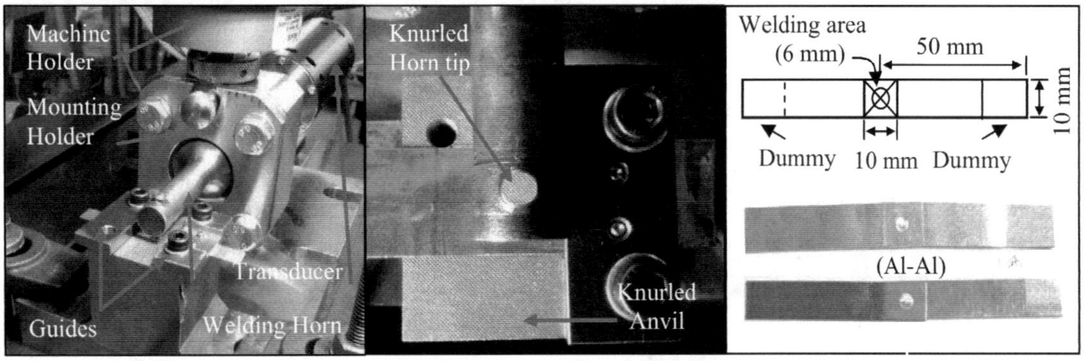

(a) (b) (c)

Figure 4. Experimental set-up of welding tools, knurled surfaces, specimen layout and welded coupons.

3.2 Stepped amplitude

Most welding systems operate with constant ultrasonic amplitude. An alternative, namely amplitude stepping, performs a weld using two different ultrasonic amplitudes. The trigger between the transitions of the amplitude setting can be made by time, energy level or power value [1]. A stepped amplitude weld starts with a higher amplitude value that the weld interface requires to provide the higher velocities for surface scrubbing to create a solid-state weld at intimate surfaces. Then the amplitude is dropped to reduce frictional heating and softening of the sample, reducing shear as the weld forms and thus minimising damage. A series of experiments have been conducted using stepped amplitude to investigate opportunities for enhanced weld strength and weld quality.

3.3 The effect of constant and stepped amplitude on weld strength in similar and dissimilar materials

Various welds of different specimen thicknesses were obtained from applying different constant and stepped vibrational amplitudes. Fig. 5 plots the weld strength against clamping force, in joining of similar materials of thickness 0.1 mm. The error bars represent one standard deviation of the five tests for each parameter set. Welding strength is seen to increase with clamping force, although it is noted that excessive clamping force may generate high friction and suppress the relative motion of the surfaces, resulting in a reduced weld strength [9]. Further, the higher ultrasonic amplitude, in this case 40 μm, although resulting in increased weld strength, also resulted in higher standard deviation and increased tool/workpiece adhesion with significant part marking. In Fig. 5, the benefit of using stepped amplitude, 40-17 μm, is seen. When stepped amplitude is applied the result is a reduction in the size of the error bards (lower standard deviations) and increased weld strength up to a critical value of clamping force. Slightly higher weld strength is recorded in joining of Al-Al than Cu-Cu under identical process parameters.

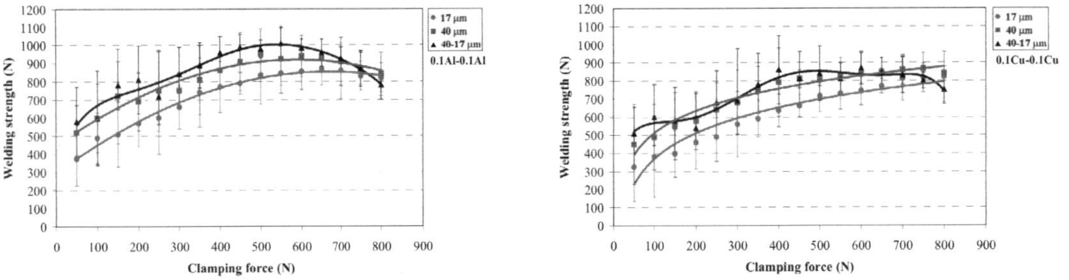

Figure 5. Variation of weld strength vs. clamping force for joining (Al-Al) and (Cu-Cu).

In welding dissimilar materials it is important to determine which specimen is placed on top, as the results in Fig. 6 suggest a superior bond is created when the horn bears against aluminium for the two metals tested in this study. It is also noted that sticking becomes more prevalent at higher clamping forces, particularly when the clamping force is above 500-600 N, and the weld strength for constant and stepped amplitude becomes progressively lower. It was also observed that the material hardness and surface roughness significantly affect the weld strength.

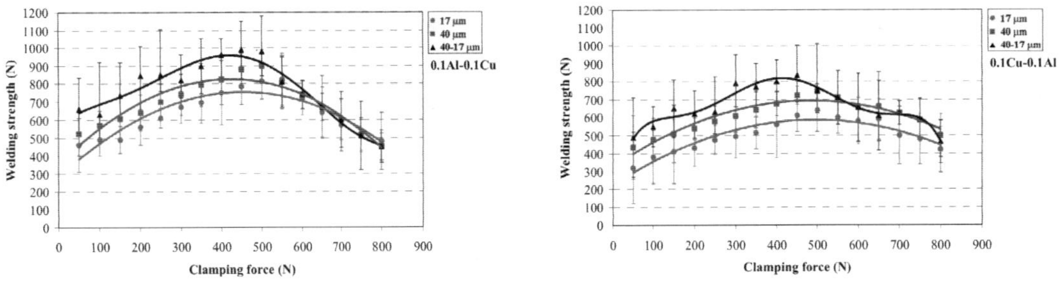

Figure 6. Variation of weld strength vs. clamping force for joining (Al-Cu) and (Cu-Al).

3.3 Microscopic observation and scanning electron microscopy (SEM) investigation

The fracture surfaces of the weld coupons after tensile tests were observed using an optical microscope. Fig. 7 show the weld zone of the Al-Cu specimens, observed on the Al side and Cu side in Fig.7 (a) and (c) respectively, with their respective enlargements shown in Fig. 7 (b) and (d). These results show the microscopic features, in the weld zone, of the plastic flow of the metal due to an applied force of 500 N and ultrasonic amplitude of 17 μm. The fracture surfaces of the broken weld show much clearer evidence of deposition of Al on Cu the side, while there is little Cu located on the Al side. Furthermore, the difference in hardness and surface roughness between the Al and Cu specimens strongly affects metal deformation during ultrasonic welding, with the harder metal (Cu) undergoing less deformation, producing no increase in welding area that could improve weld quality. Figure 8 shows higher magnifications, using a scanning electron microscope, of the fracture surfaces for the Al side and Cu side of the Al-Cu weld specimens, for the welding conditions of 500N clamping force, 17 μm ultrasonic amplitude and 1 sec welding time. The x-ray diffraction reveals a chemical composition for Al of 79.48 (wt %), which confirms the high deposition of Al on the Cu side. Little

Cu is present on the fracture surface of the Al side, and the spectrum analysis of reaction surface records a chemical composition for Cu of 11.29 (wt %).

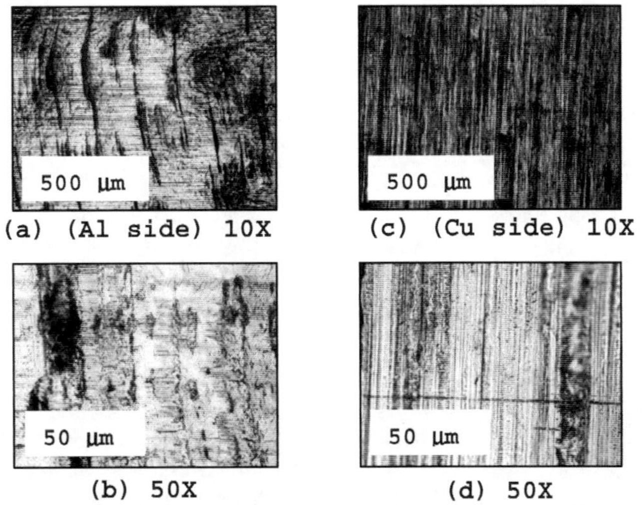

Figure 7. Microscopic photograph for Al-Cu joint from Al side (a) and (b), and from Cu side (c) and (d).

These results suggest that the weld coupon metal susceptibility to plastic deformation under ultrasonic vibration and static clamping load is related to the metal hardness and surface condition such as roughness, oxides and contaminants. Al is a softer metal than Cu and is therefore more readily deformed plastically during ultrasonic welding. The higher surface roughness of the harder metal also strongly affects the bond strength and can cause lower weld quality. The oxide layer of the metal has to be removed to create intimate surfaces for high quality welds.

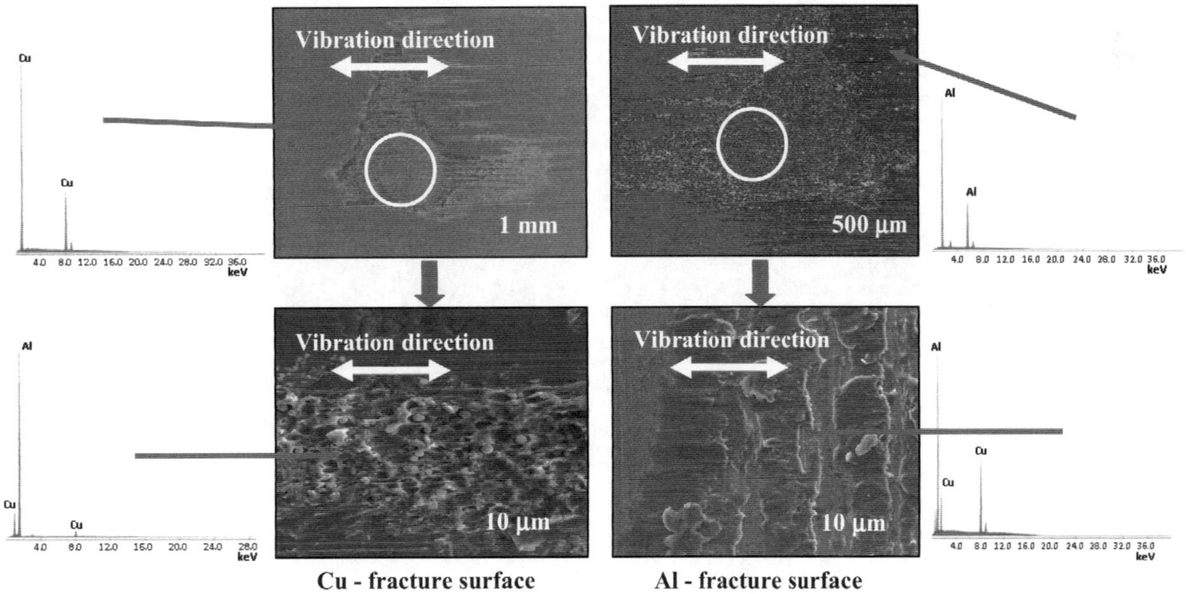

Figure 8. (SEM) fractograph and x-ray patterns from Al-Cu weld showing fracture surfaces for Cu and Al sides.

4. Conclusions

Design and fabrication of a lateral-drive ultrasonic spot welding system has been carried out to investigate the welding of thin metal strips. It has been observed that vibration amplitude, clamping force and material arrangement order, have a significant effect on weld strength. Al-Al welds are stronger than Cu-Cu welds and weld strength in both cases tends to increase with clamping force up to a critical value, where higher clamping forces restrict the lateral movements essential for welding. When dissimilar metals are considered, slightly stronger welds are obtained when the aluminium layer is placed on top and in direct contact with the ultrasonic horn and, in general, there is a decrease in weld strength when clamping forces above approximately 500 N are applied. Amplitude stepping (40-17 μm) allows better joining and slightly increases weld strength and lowers standard deviations in comparison with those tests that used a constant ultrasonic amplitude (17 μm) or (40 μm). However, tool/workpiece adhesion and part marking was worsened when higher clamping forces were used. Microscopic investigations show how observation of the fracture surfaces of tensile tested Al-Cu welds illustrate the differences on the Al side and on the Cu side. The fracture surfaces show clearer evidence of softer metal (Al) deformation in the weld zone than the harder metal (Cu). This relates to the material hardness, oxide layer hardness and surface roughness. SEM magnification determined the chemical compositions of the metal deposition on the fracture surfaces. A low percentage of Cu (11.29 wt %) confirms that the roughness and oxide layer of the relatively harder material significantly affects the bond strength and reduces weld quality. Higher weld quality is obtained through knowledge of the best combination of welding process parameters for the specific metal joining configuration.

5. References

[1] E. deVries, 2004, Mechanics and Mechanisms of Ultrasonic Metal Welding, PhD Thesis Ohio State University.

[2] K. Graff, 2005, Ultrasonic Metal Welding, in New Developmets in Advanced Welding Cambridge, 241-269.

[3] M. Bloss and K. Graff, 2009, Ultrasonic Metal Welding of Advanced Alloys: The Weldability of Stainless Steel, Titanium, and Nickel-Based Superalloys, *Trends in Welding Research Proceedings of the 8th International conference*, ASM International

[4] S. Kim, H. Sung, E. Kim and D. Park, 2010, Vibration Analysis of Ultrasonic Metal Welding Horn For Optimal Design, *Proceedings of the International Conference on Mechanical, Industrial, and Manufacturing Technologies*, Sanya China.

[5] P. Staff, 1997, *Handbook of Plastics Joining*, William Andrew Publishing/Plastics Design Library.

[6] ASTM International Codes, 2009, Standard Test Methods for Tension Testing of Metallic Materials, 1-24.

[7] British Standard Codes, 2009, Test Pieces and Test Methods for Metallic Materials for Aircraft, Metric units 1-7.

[9] M. Hiraishi and T. Watanabe, 2003, Improvement of Ultrasonic Weld Strength for Al-Mg Alloy by Adhesion of Alcohol - Ultrasonic Welding of Al-Mg Alloy, *Quarterly Journal of the Japan Welding Society*, **21**(2), 295-301.

Finite element analysis of drilling in carbon fiber reinforced polymer composites

V. A. Phadnis[1], A. Roy, V. V. Silberschmidt
Wolfson school of Mechanical and manufacturing Engineering
Loughborough University, Epinal way, LE11 3TU, UK

E-mail: V.A. Phadnis @lboro.ac.uk

Abstract. Carbon fiber reinforced polymer composite (CFRP) laminates are attractive for many applications in the aerospace industry especially as aircraft structural components due to their superior properties. Usually drilling is an important final machining process for components made of composite laminates. In drilling of CFRP, it is an imperative task to determine the maximum critical thrust forces that trigger inter-laminar and intra-laminar damage modes owing to highly anisotropic fibrous media; and negotiate integrity of composite structures. In this paper, a 3D finite element (FE) model of drilling in CFRP composite laminate is developed, which accurately takes into account the dynamic characteristics involved in the process along with the accurate geometrical considerations. A user defined material model is developed to account for accurate though thickness response of composite laminates. The average critical thrust forces and torques obtained using FE analysis, for a set of machining parameters are found to be in good agreement with the experimental results from literature.

1. Introduction

Driven by the advances in manufacturing technologies, carbon fiber reinforced plastic (CFRP) composites have gained widespread use as structural components in aerospace, naval, automotive and defence industries [1]. Drilling processes are extensively used to produce riveted and bolted joints for assembly operation of composite laminates with other components. For rivets and bolted joints, damage-free and precise holes must be drilled in the components to ensure high joint strength and precision. However, inherent characteristics of composite laminates such as non-homogeneous, anisotropic, and highly abrasive and hard reinforced fibres, result in difficulties during machining of composites[1, 2]. Several undesirable damages (such as delamination, and fiber pull-out) induced by drilling drastically reduce strength against fatigue, thus degrading the long-term performance of composite laminates [3].

To increase drilling efficiency of composite laminates with the least damage, it is essential to understand the effect of machining on CFRP. Several experimental studies have been carried out in the past [4-7]. Several analytical models [8-10] have also been developed to determine the critical thrust force and torque during drilling of composites, however these empirical formulae were solely based on

Vaibhav A. Phadnis[1], Loughborough University, UK
Phone no: 01509 227 634

Published under licence by IOP Publishing Ltd

the geometric configuration of tool- workpiece system, and do not account for the thick composite laminates and general process conditions.

Finite element (FE) models of complex machining processes can provide insight into the underlying mechanics of material behaviour especially at high strain and strain-rate regimes. An accurate and reliable FE simulation of drilling enables good predictions of strain and stress distributions; cutting forces and torque; and discrete damage modes in composites by taking into account the complex drill geometry and process parameters. With the advent of high power computing facilities, simulation of the drilling process using FE analysis provides an excellent tool for this challenging problem in applied mechanics.

In this study, a 3D finite element (FE) model of drilling in carbon fibre reinforced plastic composite (CFRP) - Al 2024 stack is developed. A user defined material model has been incorporated in general purpose FE solver ABAQUS/Explicit to accurately capture the through-thickness response of composite stack. Material anisotropy in CFRP has been modelled by taking into account discrete ply orientations with the help of local material co-ordinate systems. The model is primarily used to investigate the effects of cutting speed and feed rate on thrust force and torque in drilling of CFRP.

2. Finite element analysis of drilling in CFRP laminate

2.1. Methodology

A 3D Lagrangian finite-element (FE) model of drilling of CFRP composite was developed in a commercially available FE software ABAQUS/Explicit. The model accurately characterises the dynamic characteristics of the drilling process accounting for the complex contact interaction between a drill bit and CFRP laminate surface. A stress based failure criteria [18-19] were incorporated in user-defined material model for CFRP composite, to facilitate element deletion of mesh elements which have undergone severe deformations.

The FE model was used to study the effect of a set of machining parameters involving drill spindle speed and feeds on the average critical thrust force and torque during drilling. The details of this FE model are discussed in the following sections.

2.1.1. Geometry and meshing

A CFRP composite laminate with stack sequence of $[(90/\text{-}45/0/45)_2]_s$ was used in the FE analysis. A laminated CFRP plate with overall dimensions of 20 mm × 15 mm × 4.2 mm was modelled with an individual ply thickness of 0.26 mm. A plain carbide K-20 drill (118^0 point angle and 30^0 helix angle) of diameter Ø 8 mm was modelled as a discrete rigid body in order to reduce the computational efforts required to discretise the complex drill geometry. A lumped mass and rotary inertia was applied at a reference point located at the chisel edge of the drill bit, in order to model the kinematics of the drill bit accurately. The corresponding FE setup is shown in figure1.

The mesh used in this FE analysis was optimised through an extensive mesh convergence exercise. An optimum balance was achieved between mesh discretisation of highly complex drill geometry and stress distribution in CFRP laminate surface, for accurate prediction of thrust force and torque during drilling for the available computing resources. A refined mesh with a minimum size of 10 μm was used at and in the immediate vicinity of the workpiece volume to be drilled, while a coarse mesh was used to discretize the volume away from the process zone. The CFRP laminate was modelled using 8 node, 3D brick elements of type C3D8R with at least 3 elements in the z direction to accurately capture through-thickness stresses; while a twist drill was modelled with 4-node, 3D discrete rigid elements of type C3D4. The CFRP model consists of 1,021,000 elements while the drill was meshed with 15000 elements, with the smallest element size of 150 μm. The reference point of drill was assigned with a single node mass and rotary inertia element, where axial velocity and rotations were applied.

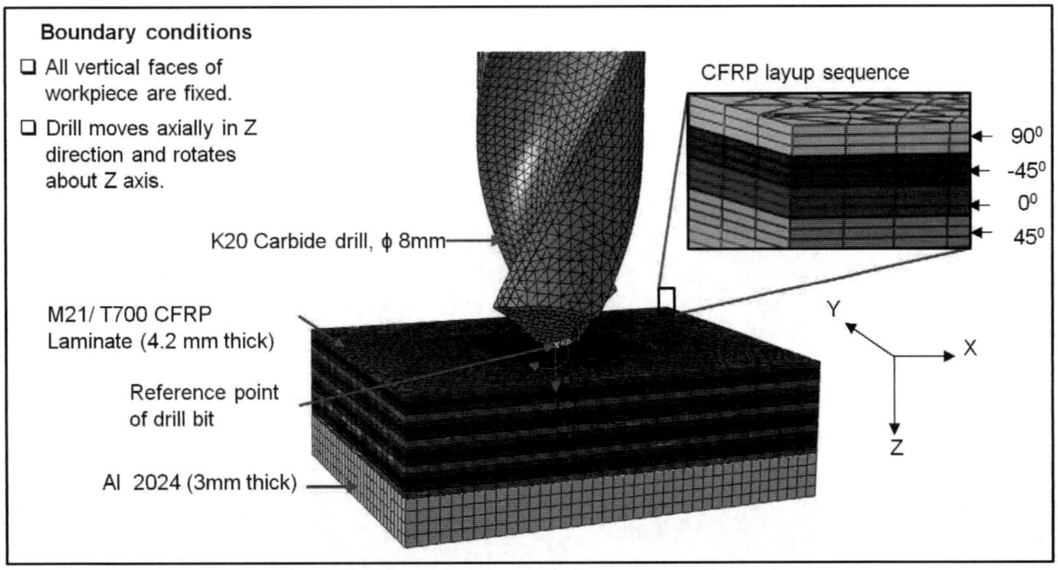

Figure 1. Setup showing FE model of drilling CFRP composite laminate

2.1.2. Boundary conditions and contact interactions

The drilling feed and spindle rotations were applied at a reference point using velocity boundary conditions in order to account for dynamic characteristics of drilling process. A reference point at the drill tip was constrained in X and Y directions while axial velocity corresponding to the imposed feed rate was applied in the Z direction. The drill was also made to rotate about Z axis while rotational degrees of freedom about X and Y directions were constrained. The laminated work piece is fixed at all four vertical faces. The combination of cutting parameters used in FE analysis, to study drilling induced thrust force and torque, is listed in table 1.

Table 1. Process parameters used in FE modelling of drilling [11]

Drill	K20 carbide , Ø 8 mm, point angle 118^0
Spindle speed (rpm)	1050, 2020, 2750
Feed (mm/rev)	0.05, 0.1, 0.15

In order to account for frictional effects during the drilling process, a Coulomb friction model is deemed appropriate and a coefficient of friction of 0.3[12] was used in the model. A general contact algorithm in ABAQUS/Explicit [13] was employed to model penalty friction based interaction between drill and workpiece. The drill was modelled as an element based master surface while CFRP workpiece was modelled as an element based slave surface. The FE model required on an average, 54 hours on a 24 Intel Quad-core processors with 48 GB RAM to complete. All simulations were performed at the High Performance Computing (HPC) facility available at Loughborough University.

2.1.3. Material modelling

The composite laminate was modelled as an orthotropic homogeneous material. The composite used in experimental study was Hexcel UD epoxy/carbon fiber laminate T700/M21 with 58% fiber content. The elastic properties of T700/M21 (Table 2) were obtained from the literature [11]. The fracture properties of similar composite were also taken from literature[14] to model composite failure. The material removal was controlled by using two distinct criteria which accounts for failure in fibres and matrix with the use of two distinct failure models. First, to account for fibre failure, : Hashin's damage model is implemented [15]. Element is removed from the mesh when damage parameter d (which equals d_{ft} or d_{fc} in tension and compression respectively, here d_{ft} and d_{fc} refers to empirical parameter accounting for fibre damage in tension and compression respectively which varies from 0 to 1, where 0 - no damage, 1- full damage) reaches 1 based on the corresponding failure strength in that direction, and it offers no subsequent resistance to the deformation. To account for failure in the brittle epoxy matrix, Puck's fracture plane model [18] was implemented. Similar stress based technique was applied to delete damaged elements. In this case, elements were removed when d_{mt} or d_{mc} = 1 (here d_{mt} and d_{mc} refers to empirical parameter accounting for matrix damage in tension and compression respectively which varies from 0 to 1, where 0 - no damage, 1- full damage). K-20 Carbide twist drill was modelled as an elastic rigid material. Al- grade 2024 material was modelled using classical Johnson-cook model; the details of the material properties were taken form material data base available from [19].

3. Drilling experiments and results

Drilling experiments were conducted on a CFRP/Al stack with 16 layers of T700/M21 with a thickness of 4.2 mm and a 3mm thick aluminium sheet of Grade 2024. The layup sequence of the CFRP is described in Section 2.1.1. Details of the drilling experiments are available elsewhere [11]. The FE model was used to simulate drilling in CFRP/Al stack. The results obtained from numerical experiments and FE model of the drilling process at a feed rate of 0.1 mm/rev and a spindle speed of 1050 rpm are shown in Fig 4. Discrete drilling regimes can be observed for CFRP and Al-2024 with different peak thrust and toque magnitudes. The computational results correspond closely with the experimental results.

The primary focus of this paper is to study drilling induced thrust force and torque in CFRP. Hence effect of drilling process parameters on thrust force and torque in drilling Al-2024 has not been discussed. The validated FE model was used to predict the effect of drilling feed rate and spindle speed on the thrust force and torque (section 4.2) for a range of drilling parameters (Table 1). The results were compared against experimental results as shown in figure 5 and 6.

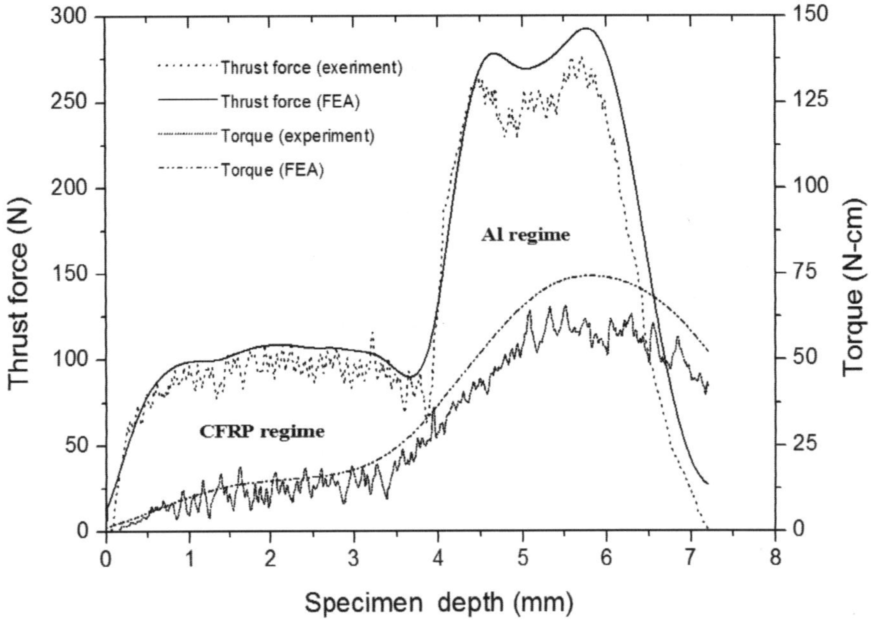

Figure 4. Thrust force analysis using FE modelling (feed 0.1 mm/rev, spindle speed = 1050 rpm)

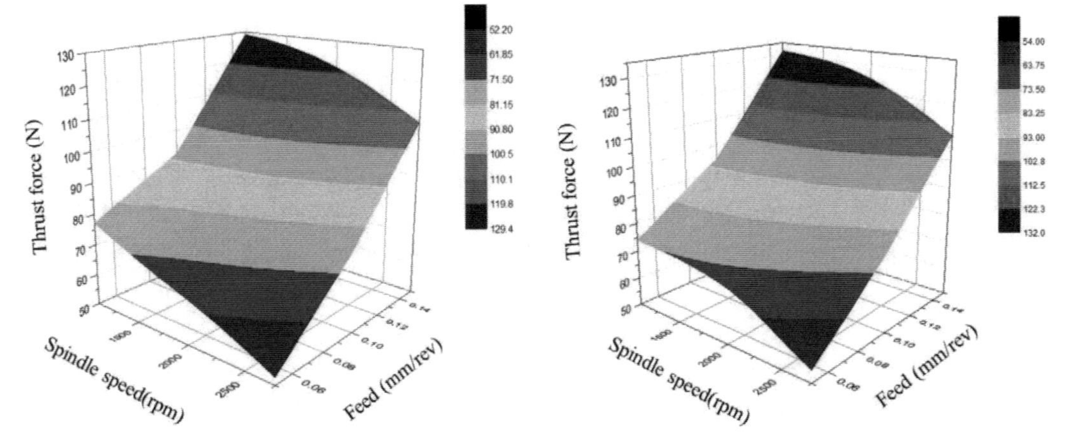

Figure 5. a) Effect of drill feed and spindle speed on thrust force using experiments [15] **b)** FE simulation

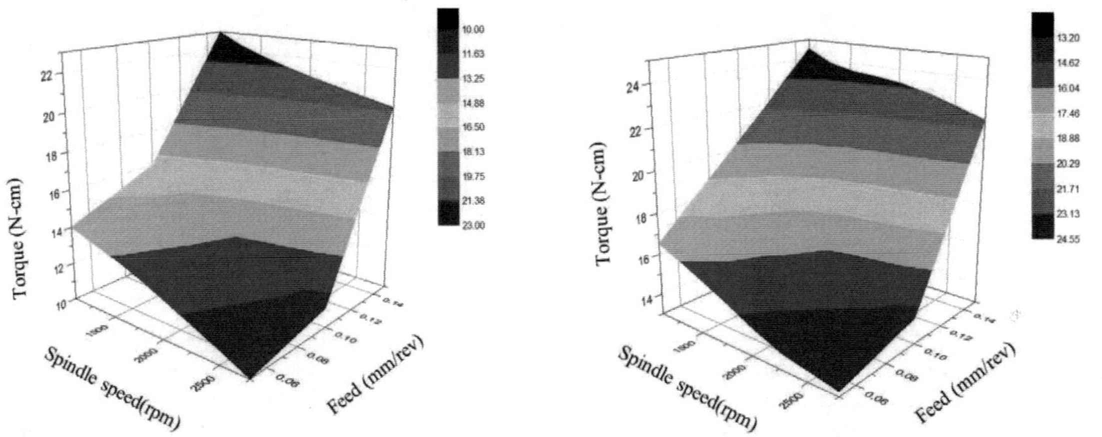

Figure 6. a) Effect of drill feed and spindle speed on torque using experiments [15] **b)** FE simulation

The overall results indicate that, thrust force decreases with the increase in spindle speed (Fig 5(a)). The thrust force was observed to highest at the cutting speed of 26.4 m/min (1050 rpm spindle speed), while the minimum cutting speed was observed at 69.1 m/min cutting speed (2750 rpm spindle speed). The results obtained from FE simulation showed that average thrust force (Fig 5(b) decreases from 20% to 25% when cutting speed was increased from 48% to 62%, while average thrust force increases from 29% to 40% when drill feed was increased form 50% to 67%. It was also noted that the percentage thrust force reduction was maximum for the combination of lowest feed and highest cutting speed, which was also observed in experimental results.

Figure 6(a) shows experimental analysis of drilling induced torque. The results indicate that drilling torque increases with an increase in feed rate. The results obtained from FE analysis (Fig 6(b)) revealed that torque was the highest at a feed rate of 0.15mm/rev and a spindle speed of 1050 rpm. The FE analysis also showed that, torque increases from 16% to 33% when feed was increased from 50% to 67%., while it decreases from 20% to 31% when cutting speed was increased form 48% to 62% .

Thus it is recommended to use low feed and high speed for drilling of CFRP composite material in order to reduce the thrust forces.

Journal of Physics: Conference Series **382** (2012) 012014 doi:10.1088/1742-6596/382/1/012014

4. Conclusion

A finite element model of drilling in CFRP composite has been developed. The effects of cutting speed and feed rate on the thrust force and torque was studied using numerical simulations. The FE model was validated using experimental results for a combination of drill feed and cutting speed; which was later used to predict thrust force and torque for varying process parameters. Our studies demonstrate the importance of feed rate and spindle speed with respect to the average drilling forces. We conclude that both thrust force and torque may be reduced by using a combination of low drill feed rate and high cutting speeds/spindle speed, while drilling CFRP composite materials.

5. References

[1] Abrão A M, Faria P E, Rubio J C C, Reis P and Davim J P 2007 Drilling of fiber reinforced plastics: A review *Journal of Materials Processing Technology* **186** 1-7

[2] Teti R 2002 Machining of Composite Materials *CIRP Annals - Manufacturing Technology* **51** 611-34

[3] Mishra R, Malik J, Singh I and Davim J P 2010 Neural network approach for estimating the residual tensile strength after drilling in uni-directional glass fiber reinforced plastic laminates *Materials & Design* **31** 2790-5

[4] Khashaba U A, El-Sonbaty I A, Selmy A I and Megahed A A 2010 Machinability analysis in drilling woven GFR/epoxy composites: Part I – Effect of machining parameters *Composites Part A: Applied Science and Manufacturing* **41** 391-400

[5] Tsao C C and Chiu Y C 2011 Evaluation of drilling parameters on thrust force in drilling carbon fiber reinforced plastic (CFRP) composite laminates using compound core-special drills *International Journal of Machine Tools and Manufacture* **51** 740-4

[6] Karnik S R, Gaitonde V N, Rubio J C, Correia A E, Abrão A M and Davim J P 2008 Delamination analysis in high speed drilling of carbon fiber reinforced plastics (CFRP) using artificial neural network model *Materials & Design* **29** 1768-76

[7] Lazar M-B and Xirouchakis P 2011 Experimental analysis of drilling fiber reinforced composites *International Journal of Machine Tools and Manufacture* **51** 937-46

[8] Tsao C C and Hocheng H 2008 Evaluation of thrust force and surface roughness in drilling composite material using Taguchi analysis and neural network *Journal of Materials Processing Technology* **203** 342-8

[9] Durão L M P, de Moura M F S F and Marques A T 2008 Numerical prediction of delamination onset in carbon/epoxy composites drilling *Engineering Fracture Mechanics* **75** 2767-78

[10] Strenkowski J S, Hsieh C C and Shih A J 2004 An analytical finite element technique for predicting thrust force and torque in drilling *International Journal of Machine Tools and Manufacture* **44** 1413-21

[11] Zitoune R, Krishnaraj V and Collombet F 2010 Study of drilling of composite material and aluminium stack *Composite Structures* **92** 1246-55

[12] Klinkova O, Rech J, Drapier S and Bergheau J-M 2011 Characterization of friction properties at the workmaterial/cutting tool interface during the machining of randomly structured carbon fibers reinforced polymer with carbide tools under dry conditions *Tribology International* **44** 2050-8

[13] 2010 Abaqus User's manual, version 6.11 RI: Hibbitt, Karlsson & Sorensen Inc.)

[14] Irisarri F-X, Bassir D H, Carrere N and Maire J-F 2009 Multiobjective stacking sequence optimization for laminated composite structures *Composites Science and Technology* **69** 983-90

[15] Hashin Z 1980 Failure Criteria for Unidirectional Fiber Composites *Journal of Applied Mechanics* **47** 329-34

[19] Johnson, G. R., and W. H. Cook 1985 Fracture Characteristics of Three Metals Subjected to Various Strains, Strain Rates Temperatures and Pressures *Engineering Fracture Mechanics* **21**. 31–48
[20] ASM Aerospace Specification Metals, Inc., Florida, US

Journal of Physics: Conference Series **382** (2012) 012015 doi:10.1088/1742-6596/382/1/012015

Damage in woven CFRP laminates subjected to low velocity impacts

H. ULLAH, A.A. ABDEL-WAHAB, A.R. HARLAND, V.V. SILBERSCHMIDT

Wolfson School of Mechanical and Manufacturing Engineering, Loughborough University, Leicestershire, LE11 3TU, UK

E-mail: H.Ullah@lboro.ac.uk

Abstract. Carbon fabric-reinforced polymer (CFRP) composites used in sports products can be exposed to different in-service conditions such as large dynamic bending deformations caused by impact loading. Composite materials subjected to such loads demonstrate various damage modes such as matrix cracking, delamination and, ultimately, fabric fracture. Damage evolution in these materials affects both their in-service properties and performance that can deteriorate with time. These processes need adequate means of analysis and investigation, the major approaches being experimental characterisation and non-destructive examination of internal damage in composite laminates. This research deals with a deformation behaviour and damage in woven composite laminates due to low-velocity dynamic out-of-plane bending. Experimental tests are carried out to characterise the behaviour of such laminates under large-deflection dynamic bending in un-notched specimens in Izod tests using a Resil Impactor. A series of low-velocity impact tests is carried out at various levels of impact energy to assess the energy absorbed and force-time response of CFRP laminates. X-ray micro computed tomography (micro-CT) is used to investigate material damage modes in the impacted specimens. X-ray tomographs revealed that through-thickness matrix cracking, inter-ply delamination and intra-ply delamination, such as tow debonding and fabric fracture, were the prominent damage modes.

1. Introduction

Woven-fabric reinforced composites offer good resistance to fracture and transverse rupture and high impact strength compared to their unidirectional and cross-ply tape counterparts [1]. These properties have attracted the sports industry to incorporate woven CFRP laminates in the design of sports products, which could be subjected to large-deflection bending and multiple impacts in service. Such types of dynamic loads can generate high local stresses, leading to complex damage modes, due to the heterogeneity and anisotropy of composite laminates. The damage mechanisms typically caused by out-of-plane impact loads are matrix cracking, fibre breakage and delamination at interfaces within the composite structure [2]. Impact damage and, in particular, delamination occurring at low velocity impact cause a significant decrease in the material's in-plane compressive strength and stiffness. Such internal damage mechanisms, which often cannot be detected by visual inspection, degrade the load bearing capacity of the structures. Therefore, it is important to study the damage suffered by the composites under impact loading conditions.

The low velocity impact response of woven-fabric composite laminates has been extensively studied in the literature [3-5]. However, the majority of the studies are dedicated to the impact behaviour of composites studied in drop-weight tests, which usually cause localised damage such as

Journal of Physics: Conference Series **382** (2012) 012015 doi:10.1088/1742-6596/382/1/012015

penetration and perforation. Large-deflection dynamic bending behaviour of laminate composites caused by a pendulum-type impactor is rarely investigated. Although such an instrumented impactor was used by Silberschmidt *et al.* [6] and Casas-Rodriguez *et al.* [7] to study damage in adhesively bonded CFRP joints under repeated impacts, the loading mode was tensile. The authors have previously studied large-deflection behaviour of woven laminates under quasi-static bending [8-11]. In the present work, a large-deflection bending behaviour of woven-fabric CFRP laminates subjected to impact loads is studied. Flexural impact tests were carried out using a pendulum type impact tester at various energy levels. The type and location of damage was investigated with X-ray micro-CT.

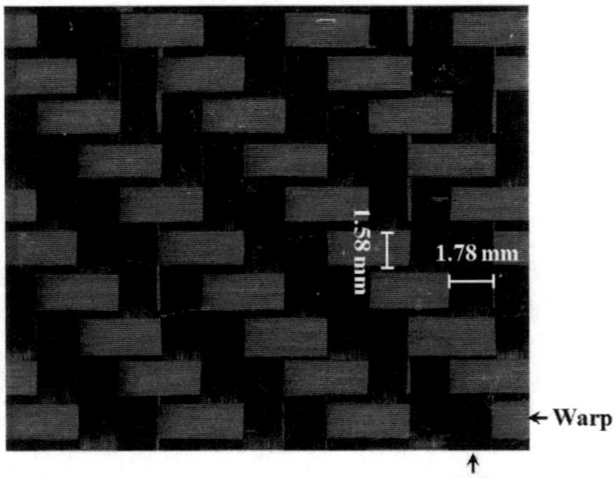

Figure 1. Two-dimensional image of CFRP 2/2 twill weave

Figure 2. Resil impact test set-up

2. Experimental

2.1 Material

The materials studied were laminates of woven fabric made of carbon fibres reinforcing a thermoplastic polyurethane (TPU) polymer matrix. The material was manufactured from 0°/90° prepregs in the form of four plies designated as $[0°,90°]_{2s}$, where 0° and 90° represent yarns in the warp and weft directions, respectively. The woven laminate had a 2/2 twill balanced weaving pattern shown in Figure 1; the fabric had the same number of yarns in the warp and weft directions. Un-notched rectangular specimens of 40 mm length and 25 mm width were prepared according to the ASTM D4812 standard [12]. The parameters of the 2/2 twill weave fabric composite are listed in Table 1.

Table 1. Parameters of CFRP twill 2/2 woven composite

Fibre type	Carbon T200
Yarn filament count	3000
End/pick count, yarns/cm	5.2/5.14
Fabric areal density (g/m^2)	200
Polymer	Thermoplastic polyurethane (TPU)
Laminate density (g/cm^3)	1.47
Fibre volume fraction (%)	45
Number of layers	4
Laminate thickness (mm)	1.0

2.2 Impact testing

Dynamic impact tests were carried out according to the ASTM D4812 standard on an instrumented pendulum type CEAST Resil impactor as shown in Figure 2. In the impact tests, the bottom of the specimen was fixed firmly in the machine vice as a cantilever beam. The upper 30 mm of the specimen was struck by the striking nose of the pendulum hammer with a controlled level of initial energy, resulting in dynamic large-deflection bending. The distance between the fixed support and the line of contact of the hammer's striking nose was 22 mm, according to the standard. In this work, a calibrated impact hammer with a mass of 0.6746 kg and a length of 0.3268 m was used. The hammer can generate maximum impact energy of 2 J at an impact velocity of 3.46 m/s, corresponding to an initial angle of 150° to the striking position. The magnitude of initial impact energy and velocity can be varied by changing the initial angle of the hammer. Increasing the initial angle of the hammer results in an increased impact energy and thus higher peak loads as can be seen in Figure 3 for 0.5 J and 0.6 J impacts corresponding to 58° and 64° angles. A piezoelectric force transducer was fixed rigidly to the hammer striking nose to capture the impact force signal. When the pendulum hammer is released from the pre-defined initial angle, the impact with the specimen generates a change in electrical resistance of the piezoelectric sensor that is captured by the data acquisition system (DAS 8000) connected to the Resil impactor. The signal is registered with a pre-defined sampling frequency of 833 kHz, with up to 8000 data points recorded per impact test. In order to decrease the data noise, a 1 kHz filter was used. CFRP specimens were tested at various energy levels to determine the energy inducing ultimate fracture of the specimen. Two samples per energy level were subjected to single impact loading. It was found that the specimen fractured at 0.6 J, corresponding to an initial angle of 64°. The rest of the tests were carried out at 0.5 J (corresponding to 58°) to study the behaviour of sub-critical damage modes such as matrix cracking and delamination. Typical records of force *vs*. time for un-fractured and fractured CFRP specimens are presented in Figure 3. Similarly, evolution of the specimen absorbed energy is shown in Figure 4.

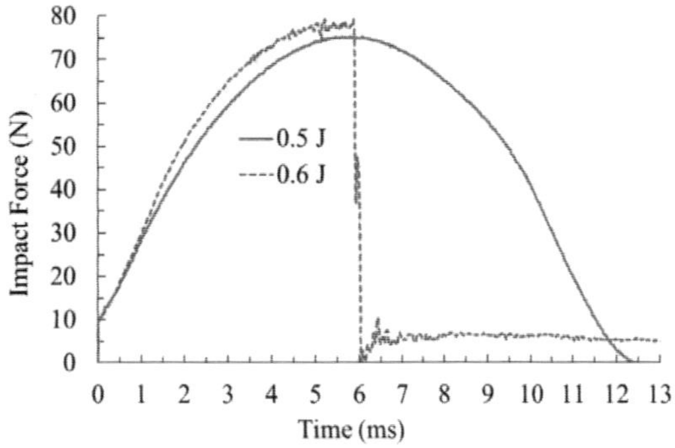

Figure 3. Typical force-time response of 2/2 twill woven CFRP laminate in impact tests

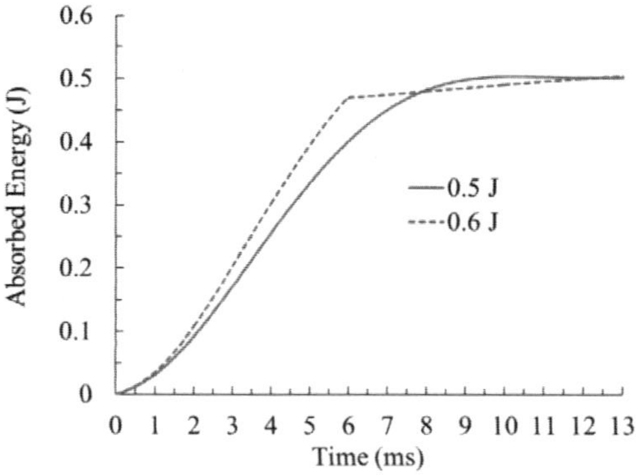

Figure 4. Typical absorbed energy-time response of 2/2 twill woven CFRP laminate in

3. X-ray micro computed tomography (micro-CT) for damage characterisation

Damage such as delamination and matrix cracking usually occurs inside composites laminates and is barely visible. The damage mechanisms need examination for reliable and accurate analysis of the behaviour and performance of composite laminates. A non-destructive evaluation method, micro computed tomography (micro-CT), was used to visualise the 3D microstructure and internal damage of the tested composite laminates. This technique has been used to investigate micro cracking and delamination in woven composite laminates at micron-range scale [13]. In this study, X-ray micro-CT measurements were performed using an XT H 225 X-ray scanner. The system consists of an X-ray detector and an electronic X-ray source, creating 2D cross-sections of the object. The source is a sealed X-ray tube operating at 25–225 kV with a 3 μm spot size. The specimen was positioned in a set up allowing two translations and one rotation, rotating and raising/lowering the sample to a specific region of interest for adjustment of the sample magnification and acquisition of tomographic data. Following acquisition, a software program builds a precise 3D map from 2D radiograph images by 'stacking' the individual slices one on top of the other; this process is known as reconstruction. As denser materials absorb more X-rays than voids and air, this attenuation contrast allows detection and characterisation of cracks and flaws in tomographic images. A high scan resolution is required to obtain the required internal details of damage in a composite laminate; as the resolution is increased, the field of view of the sample is reduced. However, samples must remain within the field of view to obtain the radiographs of the region of interest, thus there is a trade-off. Two samples, one from the fractured region and another from the impact region of the fractured CFRP laminate tested at 0.6 J were prepared to meet these requirements. Sizes of the fractured and impact samples were 16.4 mm length x 8.8 mm width x 1 mm thickness and 20.6 mm length x 10.2 mm width x 1 mm thickness, respectively. The data for those samples was collected at 75 kV and 80 μA. Transmission X-ray images were acquired from 3600 rotation views over 360° of rotation (0.1° rotation step) for 3D reconstruction. Those settings resulted in tomographs with a resolution of 12.7 μm and 14.7 μm for the fractured and impact specimens, respectively.

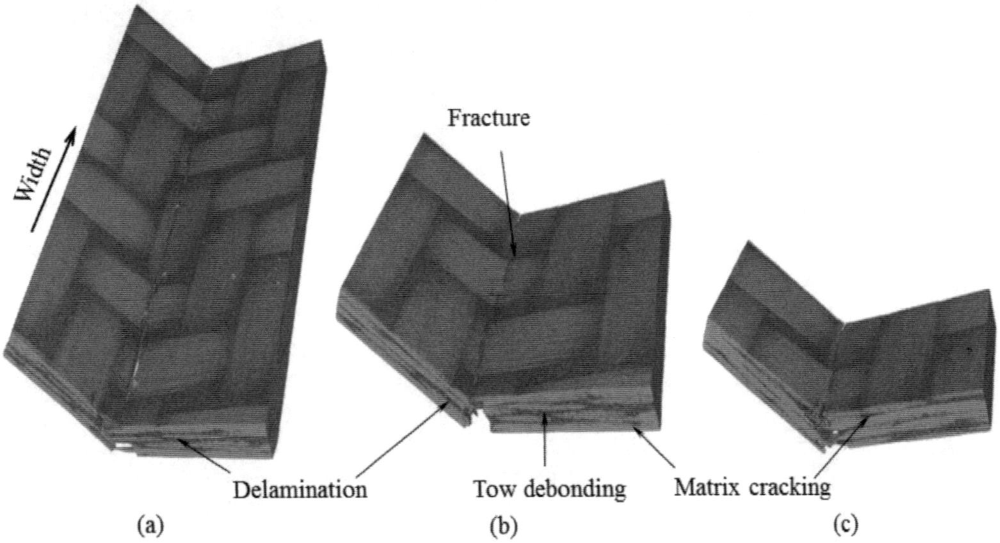

Figure 5. Reconstructed 3D images of a 2/2 twill CFRP specimen at the failure location across the width of the sample (resolution 12.7 μm): (a) edge; (b) 50% of width; and (c) 75% of width

4. Discussion of results

Results of experimental tests and micro-CT scanning of woven CFRP laminates subjected to low velocity impacts are presented in this section. Specimens were subjected to impact energies of 0.5 J and 0.6 J using an instrumented pendulum impact hammer. Figure 3 shows the variation of load with time for both energy levels. At the energy level of 0.5 J, the loading and unloading curve has a symmetrical parabolic shape suggesting that the respective stages during the contact duration are almost the same, and no significant damage occurred. Although the un-fractured samples underwent interlaminar damage before the structure lost its load-carrying capacity, the development of such inter-ply delamination was not reflected in the force-time plot for 0.5 J in Figure 3. The impact force at 0.6 J energy, causing the specimen's ultimate fracture, is higher. Here, a higher impact energy induces larger deformation and therefore, larger impact force. The load vs. time graph in Figure 3 shows oscillations due to significant damage in the specimen before its ultimate bending fracture. The ultimate fabric fracture is represented by a sudden drop in contact force, implying a momentary loss of contact between the impactor and specimen. Figure 4 shows the variation of the specimen's absorbed energy with time. The amount of energy transferred from the impactor to composite specimen at the end of impact events, i.e. absorbed energy, increases with the impact energy since a higher impact energy results in a more severe damage to a composite specimen. As shown in Figure 2, the absorbed energy increases with time during loading, reaches a maximum value and then decreases during unloading, and finally remains horizontal, i.e. reaches a constant value for 0.5 J impact energy. For 0.6 J impact energy, the absorbed energy reaches a value of 0.48 J at 6 ms and then gradually increases. The stabilized energy level indicates that the impactor has lost contact with the specimen. This constant value gives the total energy absorbed permanently by the composite specimens at the end of an impact event.

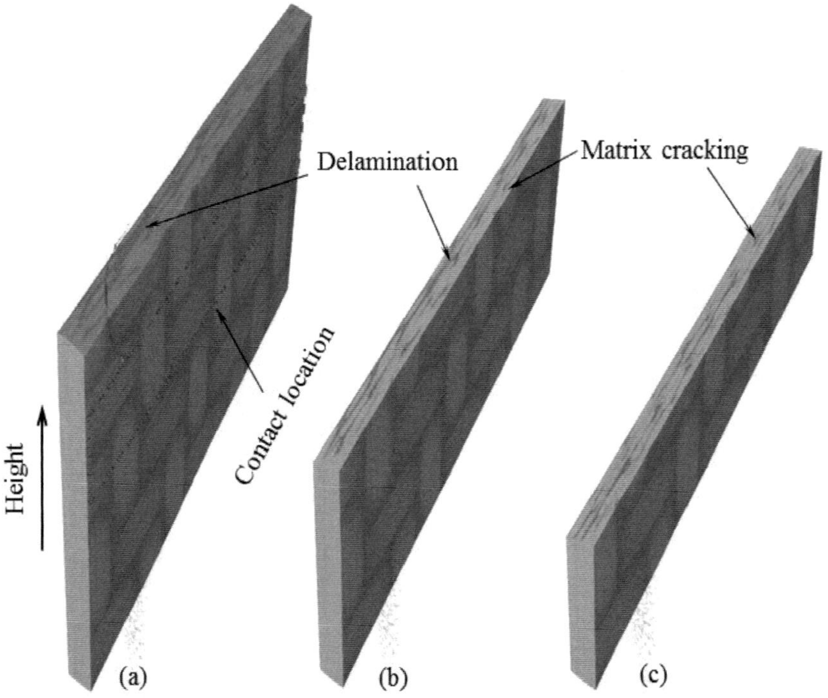

Figure 6. Reconstructed 3D images of twill 2/2 CFRP specimen at impact location across the height of sample (resolution 14.7 μm): (a) edge; (b) 50% of height; and (c) 75%of height

The reconstructed 3D images of the transversely fractured CFRP specimen at the bending (fractured) location are shown in Figure 5. Realisation of inter-ply and intra-ply damage mechanisms at the outer edge, 50% and 75% of the sample width is shown in Figure 5a, 5b, and 5c, respectively. Similarly, tomographs presented in Figure 6 shows matrix cracking and delamination at the hammer impact location along the specimen height. Dark grey regions in the images represent cracks and damage whereas light grey regions represent higher density material i.e. carbon-fibre yarns. The X-ray CT images have shown, before its ultimate fracture, the laminate exhibited matrix cracking and then delaminations and tow debondings. Matrix cracks developed in the weak resin-rich pockets around the tows. Inter-ply delamination and intra-ply delamination such as tow debonding can also be observed. Such delaminations normally appeared near the matrix cracks areas, which suggest that formation of the cracks initiated delamination. In the fibre-rich regions the damage was associated with debonding at the fibre/matrix interface. At the time of fabric fracture, as the analysis of the internal structure has showed, almost every ply was delaminated. All the tomographs showed that matrix cracking and inter-ply delamination are the prominent damage modes at the specimen's impact location, whereas at the bending (fractured) location, these modes are coupled with tow debonding and subsequently transverse ply fracture.

5. Conclusions

The large-deflection bending behaviour of woven CFRP laminates under low velocity impacts was studied using experimental tests and micro-CT scanning. Impact tests were carried out to characterise the material's dynamic behaviour and impact strength. Experimental results highlighted the energy absorbing capabilities of the CFRP composites. The micro-structural analysis was performed to observe different damage modes in the impacted specimens using micro-CT, providing a detailed picture of the damage modes and interaction among different failure mechanisms at various locations. It was observed that damage modes such as matrix cracking and delamination were prominent at the impact location, whereas these modes were coupled with tow debonding and fabric fracture at the bending location. The micro-structural examination has formed a basis for identification of interlaminar and intralaminar damage locations for future computational models. Further, the dynamic testing results will define the design load envelope for various sports products subjected to large-deflection dynamic bending.

References

[1] M. Aktaş, C. Atas, B.M. İçten, R. Karakuzu, Composite Structures, 87 (2009) 307-313.
[2] C. Menna, D. Asprone, G. Caprino, V. Lopresto, A. Prota, International Journal of Impact Engineering, 38 (2011) 677-685.
[3] N. Naik, Y. Chandra Sekher, S. Meduri, Composites Science and Technology, 60 (2000) 731-744.
[4] C. Evci, M. Gülgeç, International Journal of Impact Engineering, 43 (2012) 40-51.
[5] M.P.F. Sutcliffe, C. Monroy Aceves, W.J. Stronge, R.S. Choudhry, A.E. Scott, Composite Structures, (2012).
[6] V.V. Silberschmidt, J.P. Casas-Rodriguez, I.A. Ashcroft, Key Engineering Materials, 399 (2009) 71-78.
[7] J. Casas-Rodriguez, I.A. Ashcroft, V.V. Silberschmidt, Composites Science and Technology, 68 (2008) 2663-2670.
[8] H. Ullah, A.R. Harland, T. Lucas, D. Price, V.V. Silberschmidt, Computational Materials Science, 52 (2012) 147-156.
[9] H. Ullah, A.R. Harland, V.V. Silberschmidt, Applied Composite Materials, DOI: 10.1007/s10443-011-9242-7 (2011).
[10] H. Ullah, A.R. Harland, V.V. Silberschmidt, Computational Materials Science, 10.1016/j.commatsci.2012.05.036 (2012).
[11] H. Ullah, A.R. Harland, T. Lucas, D. Price, V.V. Silberschmidt, Journal of Physics: Conference Series 305 (2011) 12045.
[12] ASTM D4812 – 06: Standard Test Method for Unnotched Cantilever Beam Impact Resistance of Plastics (2006).
[13] P. Badel, E. Vidal-Sallé, E. Maire, P. Boisse, Composites Science and Technology, 68 (2008) 2433-2440.

Journal of Physics: Conference Series **382** (2012) 012016 doi:10.1088/1742-6596/382/1/012016

Ultrasonically assisted turning of Ti-6Al-2Sn-4Zr-6Mo

R Muhammad*, **A Maurotto, A Roy, V V Silberschmidt**
Wolfson School of Mechanical and Manufacturing Engineering, Loughborough University, Loughborough, Leicestershire, UK

*R.Muhammad@lboro.ac.uk

Abstract. Titanium alloys are widely used in the aerospace and offshore industries due to their high strength-to-weight ratio sustained at elevated temperatures, their fracture-resistance features and exceptionally good corrosion-resistance properties. However, poor thermal conductivity and high chemical affinity of these alloys to tool materials severely impair their machinability. As a result the machining processes of titanium alloys are typically characterized by low cutting feeds and speeds making production of components uneconomical.

Recently, a non-conventional hybrid machining technique, namely, ultrasonically assisted turning has been shown to significantly improve the machinability of intractable alloys with a concomitant improvement in material removal rates, thus improving machining economics.

In the current work, a 3D finite element model of turning of Ti-6Al-2Sn-4Zr-6Mo is developed in the commercial software, MSC Marc/Mentat. A constitutive behaviour of the workpiece material under large deformations and elevated temperatures is adequately represented by a Johnson-Cook material model. For validation of the developed numerical model, experimental tests were carried out. The numerical and experimental results were found to be in good agreement.

1. Introduction

Titanium alloys are extensively used in the aerospace, marine, medical and chemical processing industries owing to their exceptional corrosion-resistance, high–strength and light weight. However, their low thermal conductivity and high–strength severely impair the machinability of these alloys with conventional machining operations [1]. High cutting forces and high temperature at the tool-workpiece interface zone lead to premature tool failure, increasing manufacturing costs. In recent years, costs of machining involving the use of cutting fluids have increased substantially, primarily due to environmental concerns: the handling of cutting fluids as well as their disposal must obey strict rules of environmental protection. For manufacturing companies, the costs related to cutting fluids represent a large amount of the total machining costs. Consequently, elimination of cutting fluids, if possible, can be a significant economic incentive. Thus, dry machining attracts now great interest since it addresses current needs for environmentally friendly manufacturing,

Therefore, new hybrid-machining techniques are needed to improve the overall machinability of these high–strength alloys. Ultrasonically assisted turning (UAT) is a novel machining technique, in which high-frequency vibrations are superimposed on a cutting tool, preferably in a primary cutting direction, to improve the machinability of those high–strength alloys (Fig. 1). UAT has shown significant improvement in the processing of Ti-15333 [2] and Inconel 718 [3, 4].

Experimental observations indicate that various machining parameters, such as vibration frequency and amplitude, tool geometry and overall cutting speed, significantly affect the quality of machining in UAT. Numerical simulations based on finite-element analysis can be used to quantify the observed effects and to study various underlying physical characteristics of the turning of advanced materials with a vibrating tool.

The aim of the current study is to investigate the effect of ultrasonic vibration on machining cutting forces and temperature at the cutting region in turning of Ti-6Al-2Sn-4Zr-6Mo (a shorter notation – Ti–6246 – is used below). The current work elucidates the advantages of UAT in the processing of Ti–6246.

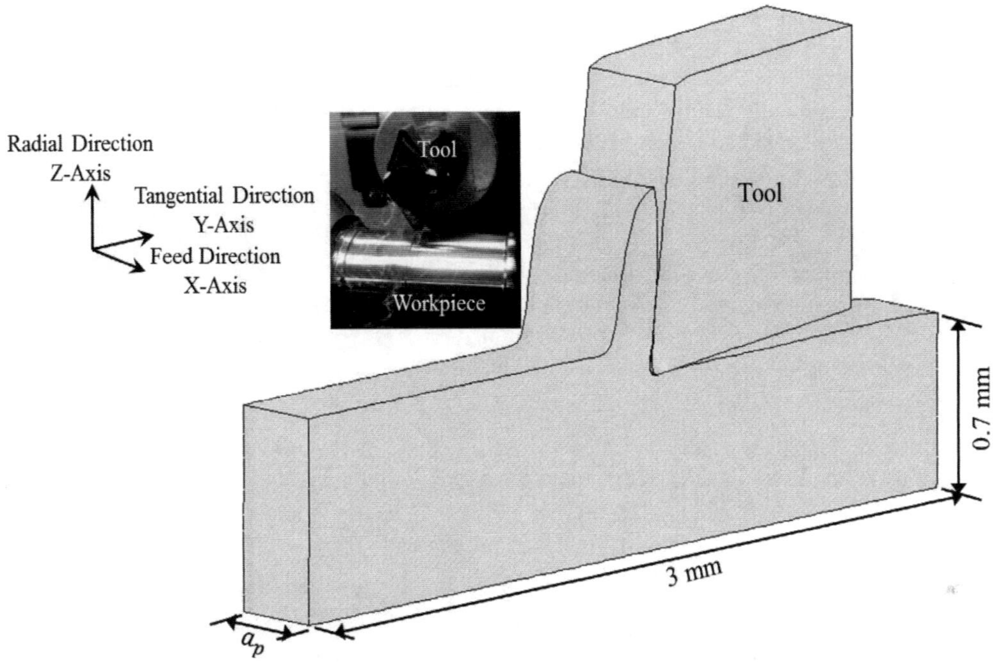

Figure 1: Schematic diagram of cutting model and principal direction of vibration

2. Experimental setup

A current UAT setup used in machining experiments consisted of a modified Harrison 300 lathe, with a custom-built ultrasonic transducer mounted on a wave-guide, a force-measurement and data acquisition systems, a charge amplifier and an A/D converter. The cutting-force components were measured with a Kistler piezoelectric dynamometer (KIAG SWISS Type 9257A), mounted on the cross slide of the lathe using a custom-built tool post. The ultrasonic transducer was mounted on the dynamometer with adequate insulation to protect it from electrical disturbances, generated during machining operations. Additionally, a micrometric dial gauge was used to ensure an accurate depth of cut.

An ingot of α+β-Ti-based alloy Ti–6246 was used as a workpiece material, and a commercial tungsten carbide insert, PVD coated with TiN (DNMG-150608 supplied by Seco) was employed as a cutting tool. All experiments were carried out in dry-cutting conditions without the use of any lubricants or coolants.

Several tests for conventional turning (CT) and UAT were carried out, and the primary cutting force in the tangential direction (see Fig. 1) was measured. Each experiment was repeated at least five times to get statistically consistent data. The cutting forces reduce significantly during ultrasonically assisted operations when compared to those obtained in CT. In addition to force reduction other substantial benefits were observed, in particular, elimination of the chatter noise during operations and better surface finish in the machined workpiece [4-6]. Furthermore, a FLIR ThermaCAM™ SC3000 camera was used to carry out thermal analysis in real time during the cutting operation. The system consists of a built in 20° lens, cables, connectors and range of optional hardware and software accessories. The new Stirling-cooled Quantum Well Infrared Photon (QWIP) photo-detector enables the system to capture images with a spectral response between 8 and 9 μm, temperature range of -20°C to 2000°C, 320 × 240 pixel resolution, low noise detection and high image stability and uniformity. In the current experiments, a temperature resolution of 30°C to 650°C was used. The ThermaCAM® QuickView™ software was used to analyse the data of the FLIR ThermaCAM™ SC3000 system.

3. Model and Simulations

A schematic of the workpiece and the cutting tool is shown in Fig. 1. The deformable workpiece had the following dimensions: 3.0 mm in length and 0.7 mm in height. Initially, the workpiece was meshed with 8-noded iso-parametric elements with a minimum element length of 0.1 mm; they were further converted to 5-noded tetrahedron elements as a result of re-meshing of the highly distorted areas [7]. The tool was meshed using 2500 four-noded iso-parametric quadrilateral elements with a refined mesh close to the cutting edge. In order to simulate UAT, harmonic oscillations in the cutting direction (tangential direction) with an amplitude (a) and frequency (f) were superimposed on the cutting tool's movement. The cutting conditions and parameters of tool geometry used in the implicit FE simulations are summarized in Table 1.

The chip-formation process involves substantial plastic deformation in the workpiece material, especially near the cutting tip [8]. As a result, the elements near the cutting tip distort substantially, leading to numerical convergence problems in simulations. The use of the re-meshing techniques helps to mitigate problems with convergence as highly distorted elements are replaced with geometrically consistent ones.

The ambient temperature was selected as 20°C for the numerical experiments. Additionally, the effect of friction on the tool-workpiece-interface was also accounted for [9, 10]. This was achieved by choosing a friction coefficient as shown in Table 1; $m_{fr} = 0.5$ corresponds to the dry cutting condition (without any lubricant). The number of simulation steps was kept at 5000 with a total simulated period of time of 0.0179 s.

Table 1: Cutting parameters used in FE simulations

Parameters	Value
Cutting speed, V (m/min)	10
Depth of cut, a_p (mm)	0.3
Cutting edge radius, r (μm)	25
Rake angle α, (degree)	15
Frequency, f (kHz)	5; 10; 15; 17.8; 20; 25
Amplitude, a (μm)	3; 6; 8; 10; 12
Coefficient of friction, m_{fr}	0.5 (dry cutting)

3.1. Material Model

Strain, strain rate, temperature and strain hardening have a major influence on the flow stress or instantaneous yield strength, at which the material starts to deform plastically or flow. Accurate and reliable flow stress models are necessary to represent a constitutive behaviour of the workpiece material under different cutting conditions. In the current study, the Johnson-Cook material model [11] was used to describe the mechanical behaviour of alloy Ti–6246 at high-strain, high strain-rate and elevated temperature. The effects of strain, strain-rate and strain-hardening are combined in a multiplicative manner:

$$\sigma_y = [A + B(\varepsilon_p)^n][1 + Cln(\dot{\varepsilon})][1 - (T_H)^m], \quad (1)$$

where σ_y is flow stress, ε_p is the effective plastic strain, $\dot{\varepsilon} = \frac{\dot{\varepsilon}_p}{\dot{\varepsilon}_o}$ is the strain rate, $\dot{\varepsilon}_p$ is the plastic strain rate, $\dot{\varepsilon}_o$ is the reference strain rate, $T_H = \frac{T - T_R}{T_M - T_R}$ is the homologous temperature, T is the absolute temperature, T_M is the melting temperature, T_R is the room temperature, A is the initial yield stress at the reference strain rate, B is the hardening modulus, C is the strain rate-sensitivity coefficient, n is the hardening coefficient and m is the thermal softening coefficient. All these parameters were determined experientially in a series of split-Hopkinson tests at Tampere University of Technology, Finland, and are given in Table 2.

Table 2: Johnson Cook Parameters of Ti-6246

Material	A (MPa)	B (MPa)	C	n	m	$\dot{\varepsilon}_o$ (s^{-1})	T_M (°C)	T_R (°C)
Ti-6246	1657	402	0.0054	0.218	0.8034	1	1625	20

4. Discussion and Analysis of Results

The effect of vibration parameters on the tangential cutting forces produced in the turning of Ti-6246 is discussed here. In conventional turning (CT), the tool is in contact with the workpiece throughout the turning operation, resulted a peak force of 119 N. However, in UAT the tool separates from the workpiece and fully engages in each vibratory cycle, which leads to a fluctuating cutting force, changing from zero to a peak value of 123 N, results an average cutting force of 38 N. The reduction in average cutting forces is attributed to a vibro-impact mechanism of the tool interaction with the workpiece material. Figure 2 demonstrates the tangential components of cutting force for both CT and UAT for a frequency of 20 kHz and an amplitude of 8 μm, both experimentally and numerically. A significant reduction of 67% in tangential force was observed when compared to CT. The FE simulations for various depths of cut and cutting speed for both CT and UAT were carried out and were in good agreement with experimentation [6].

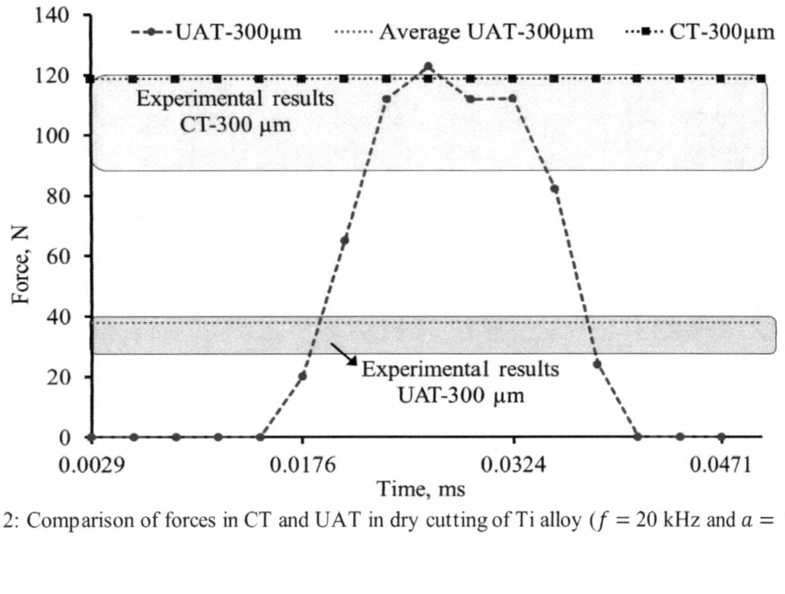

Figure 2: Comparison of forces in CT and UAT in dry cutting of Ti alloy (f = 20 kHz and a = 8 μm)

4.1. Effect of Vibration Amplitude

In UAT, the effects of amplitude on the average tangential cutting force were studied. The cutting depth was fixed at 300 μm, the feed rate at 100 μm and the amplitude changed from 3 μm to 12 μm. From Fig. 3, it is clear that the tangential component of force declines with an increase in vibration amplitude. It is observed that a higher vibration amplitude leads to an increase in the force reduction when compared to CT at the same machining parameters. However, high-power transducers are required to produce a stable high amplitude for UAT. A decrease of 41% in the average force level was recorded for an increase in the amplitude from 0 to 6 μm. A further 18% decrease was observed when the amplitude increased from 6 μm to 8 μm (Fig. 3). However, the reduction in cutting forces lessened above 8μm (approx.13% from 8 μm to 10 μm and ultimately 8% when increase from 10 μm to 12 μm). Therefore, in the current study, the amplitude of 8 μm was used in experimental analysis, resulting in a significant reduction in cutting forces.

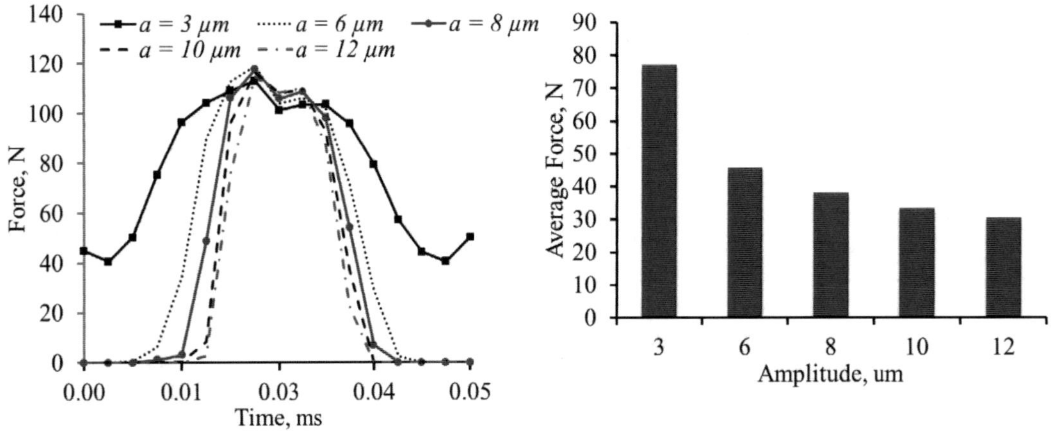

(a) (b)

Figure 3: (a) Tangential cutting forces in UAT of Ti alloy; (b) Average cutting forces (variable amplitude; frequency 20 kHz)

4.2. Effect of Vibration Frequency

Five levels of frequency (5 kHz, 10 kHz, 15 kHz, 20 kHz and 25 kHz) were analysed in FE simulations of UAT of Ti-6246. In all the cases, the depth of cut was fixed at 300 μm, cutting speed at 10 m/min and amplitude of vibration at 8 μm. The relative velocity of the cutting tool is proportional to the vibrational amplitude and frequency:

$$V_t = a\sin(2\pi f l), \quad (2)$$

where f is the frequency, V_t is the cutting speed and a is the amplitude.

Equation (2) was verified by using the above set of frequencies to demonstrate the predictability of the model. For all five cases, a reduction in cutting forces with an increase in frequency is summarised in Fig. 4. An average reduction of 11 N in the cutting forces was noted with application of 5 kHz frequency on the cutting tool. The decline increases to 42 N and 52 N with an increase in frequency to 10 kHz and 15 kHz, respectively, when compared to CT. A further increase in frequency from 20 kHz to 25 kHz, resulted in average reduction of 56 N and 58 N when compared to CT, respectively. Therefore, in the current study, a frequency of 20 kHz was used for UAT.

4.3. Temperature Distribution

Temperature in the process zone generated in CT and UAT was investigated with the developed FE model. In all the cases, the depth of cut was fixed to 300 μm and cutting speed at 10 m/min. The temperature in UAT in both simulation and experimentation is approximately 200°C higher when compared to CT in turning of Ti-6246, the possible cause for this is the effect of the additional factors linked to dissipation of the vibration energy [12]. Figure 5 shows the temperature distribution in the cutting region in CT and UAT. In the previous study of Ti-15333 and Inconel-718, the temperature in the process zone was 60-80°C higher when compared to CT [2, 4]. However, in the current study, a relatively higher temperature of the process zone was observed in UAT when compared to CT. Simulation results are in good qualitative agreement with the experimental ones.

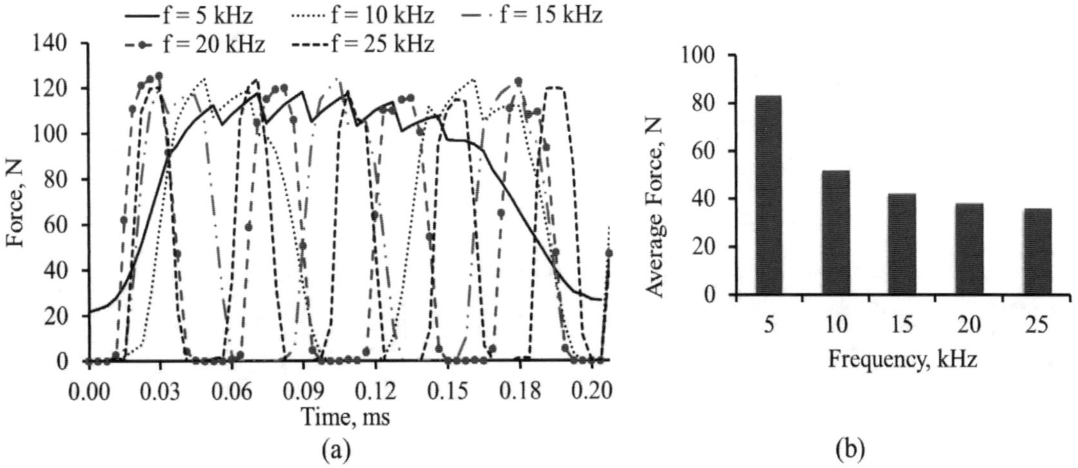

Figure 4: (a) Tangential cutting forces in UAT of Ti alloy; (b) Average cutting force (variable frequency; amplitude 8 μm)

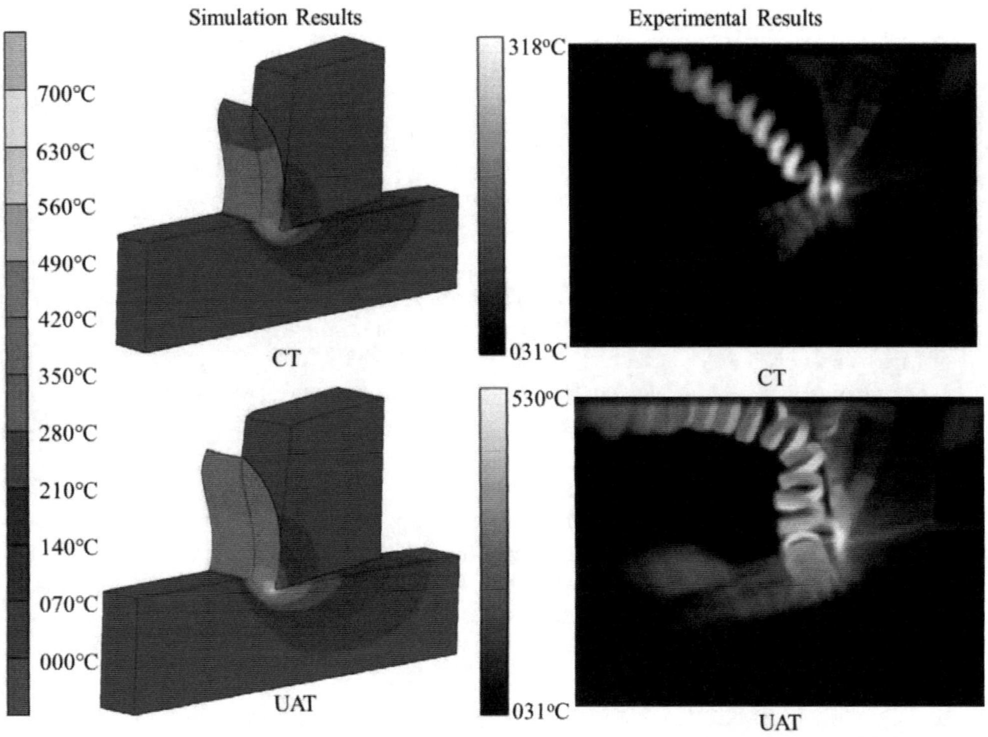

Figure 5: Temperature distribution of the cutting region in CT and UAT (length scale in images of simulation results is much larger than that in thermo-grams)

5. Conclusions

In this study, a 3D thermo-mechanically coupled finite element model for UAT and CT of Ti-6246 was developed. The model was used to undertake a parametric study of the UAT of Ti-6246. The use of this hybrid machining technique resulted in a substantial reduction in the cutting forces in the processing of Ti-alloy. The developed model was validated by experimental results that demonstrated a good agreement.

In future, the current model will be extended to 3D modelling of oblique machining in order to perform a parametric study of UAT using actual tool geometry and to suggest optimize cutting conditions for various type of materials.

Acknowledgement
The research leading to these results has received funding from the European Union Seventh Framework Programme (FP7/2007-2013) under grant agreement No. PITN-GA-2008-211536, project MaMiNa.

Authors would also like to acknowledge Tampere University of Technology, Finland, for providing material data for Ti-6246.

References

[1] Che-Haron, C.H., Tool life and surface integrity in turning titanium alloy. Journal of Materials Processing Technology, 2001. 118(1–3): p. 231-237.
[2] Muhammad, R., M. Demiral, N. Ahmed, A. Roy, and V.V. Silberschmidt, Computational study of ultrasonically-assisted turning of Ti alloys. Advanced Material Research, 2011. 223: p. 30-36.
[3] Ahmed, N., A.V. Mitrofanov, V.I. Babitsky, and V.V. Silberschmidt, 3D finite element analysis of ultrasonically assisted turning. Computational Materials Science, 2007. 39(1): p. 149-154.
[4] Muhammad, R., N. Ahmed, A. Roy, and V.V. Silberschmidt, Turning of advanced alloys with vibrating cutting tool. Solid State Phenomena, 2012. 188: p. 277-284.
[5] Muhammad, R., A. Maurotto, A. Roy, and V.V. Silberschmidt, Hot ultrasonically assisted turning of β-Ti alloy. Procedia CIRP, 2012. 1: p. 353-358.
[6] Muhammad, R., N. Ahmed, A. Roy, and V.V. Silberschmidt, Numerical modelling of vibration-assisted turning of Ti-15333. Procedia CIRP, 2012. 1: p. 364-369.
[7] MSC.Marc User's Guide Version 2011. MSC Software Corporation LA.
[8] A.J. Shih, Finite element analysis of orthogonal metal cutting mechanics International journal of Machine Tools and Manufacture, 1996. 36: p. 255-273.
[9] Muhammad, R., A. Maurotto, A. Roy, and V.V. Silberschmidt, Analysis of forces in vibro-impact and hot vibro-impact turning of advanced alloys. Applied Mechanics and Materials, 2011. 70: p. 315-320.
[10] R. Muhammad, N., Ahmed, Y.M. Shariff , V.V. Silberschmidt, Effect of cutting conditions on temperature generated in drilling process: A FEA approach. 2011. 223: p. 240-246.
[11] Johnson, G., Cook W, Fracture characteristics of three metals subjected to various strains, strain rates, temperatures and pressures. Eng Fract Mech, 1985. 2: p. 31-48.
[12] N. Ahmed, A.V., Mitrofanov, V.I. Babitsky, V.V. Silberschmidt, Analysis of forces in ultrasonically assisted turning. Journal of Sound and Vibration, 2007. 308: p. 845-854.

Journal of Physics: Conference Series **382** (2012) 012017 doi:10.1088/1742-6596/382/1/012017

Application of Smooth-Particle Hydrodynamics in Metal Machining

Abolfazl Zahedi, Simin Li, Anish Roy, Vladimir Babitsky, Vadim V. Silberschmidt

Wolfson School of Mechanical and Manufacturing Engineering, Loughborough University, Leicestershire, LE11 3TU, UK

S.Zahedi@lboro.ac.uk

Abstract. The finite element (FE) method has been extensively used to model complex cutting processes. However, due to large strains in a process zone, leading to increased element distortions, such simulations are confronted with numerical difficulties. Smooth-particle hydrodynamics (SPH) is a mesh-free computational method, which has been used to simulate multi-body problems. In this paper we present a 3D hybrid modelling approach for orthogonal micro-machining of a copper single crystal with the use of SPH and continuum FE. The model is implemented in a commercial FE software ABAQUS/Explicit. The study is used to gain insight into the effects of crystallographic anisotropy on the machining response of f.c.c. cubic metals.

1. Introduction

Machining of single-crystal metals is of interest due to a demand from various high-technology applications with single-crystal components due to their uniformity and reduced level of defects, important for manufacture of turbine blades and precision mirrors. Thus, there is a need for developing fundamental understanding of materials behaviour and material removal, since all crystalline materials are composed of grains and grain boundaries, which, to a large extent, define their mechanical properties. During mechanical micromachining of metals, a tool-workpiece interaction occurs entirely within either a single crystal or a few crystals of the workpiece material. Single-crystal copper is widely used in fabrication of precision instruments and civil communication devices due to its inherent favourable properties, such as excellent signal transportation and fatigue resistance [1, 2].

A number of finite element studies have confirmed that a machining response; including machining forces, chip shape and surface roughness, depends on the crystallographic orientation in f.c.c metals [3, 4]. These studies identified excessive element deformation as a fundamental problem for numerical modelling of the cutting processes. The problem with element distortion was overcome with the application of techniques such as adaptive meshing and an Arbitrary Lagrangian-Eulerian (ALE) scheme. Remeshing techniques have inherent limitations such as loss of data resolution and round-off errors from previous time-step increments. As a result, errors can accumulate in reported field variables [5, 6]. Hence, there is a need to develop numerical schemes to alleviate problems in numerical convergence in the modelling of high strain and strain-rate deformation regimes.

This has led to the development of element-free (or mesh-free) techniques in modelling high-deformation processes such as machining. Smooth particle hydrodynamics (SPH) is a mesh-free computational method, which has been used to simulate N-body problems, including, i. a., complex fluid flows and large-strain solid-mechanics problems.

In this paper we demonstrate a multi-scale modelling approach in the simulation of a micro-machining process. The philosophy of this approach was to attain close-to atomistic details in the regions of interest with a continuum formulation for the rest of the workpiece volume so as to reduce computational demand. To gain a physical comprehension of the cutting process at the micro level, a material model based on crystal plasticity was implemented

2. Multi-scale simulation model

A 3D workpiece with dimensions of 500 μm × 500 μm × 50 μm is selected as model. The workpiece was divided in two regions. A schematic of the tool and workpiece is shown in Fig. 1(a) and a 3D model is shown in Fig. 1(b). The right and bottom faces of the workpiece were fixed with regard to linear displacements and rotations. A tool with a cutting edge with a radius of 5 μm and $\theta = 45^{\circ}$ was used to cut a free surface of the workpiece. The workpiece was modelled as a deformable body and the tool as a rigid part. The region close to the cutting surface (150 μm × 150 μm × 50 μm) was simulated with resolution of 0.2 μm, while the remaining regions were modelled as a solid continuum. The geometry was meshed with eight-node brick elements (C3D8R) with reduced integration for a continuum FE region and PC3D for the SPH area, 668 particles were used for the SPH part. The Coulomb's friction law with a friction co-efficient of 0.1 was used to model the frictional interaction between the tool and the workpiece surface material.

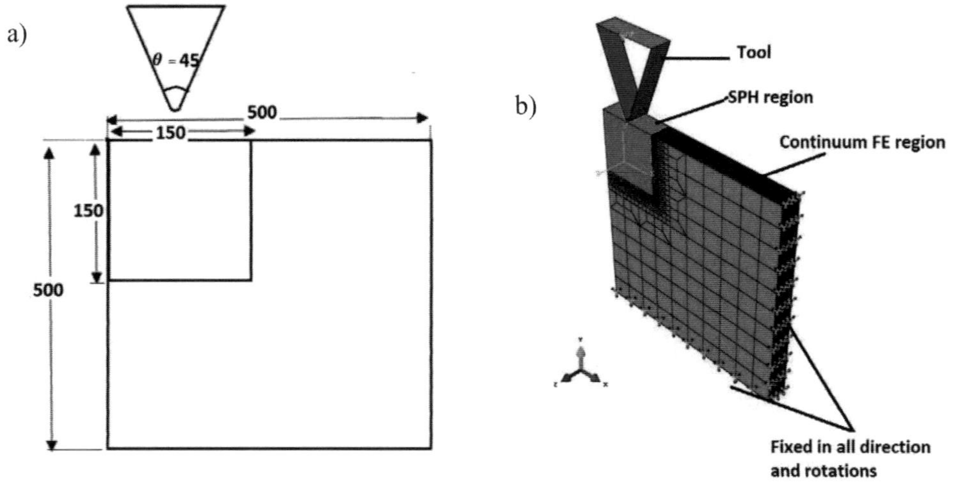

Fig.1 (a) Tool and workpiece (dimensions in μm) (b) 3D finite-element model of micro-machining

The crystal-plasticity model was implemented in a FE software package ABAQUS/Explicit using the user-defined material subroutine VUMAT [7]. Material parameters for single-crystal copper used in the FE simulations are listed in Table 1 [8].

3. Simulation results and discussion

All numerical experiments were carried out by indenting the workpiece to a depth of 50 μm. The indenter moved at a velocity of 50 mm/s. As expected, the workpiece material rearranged when the strain energy in the deformed lattice exceeded the binding energy of the workpiece particles. Consequently, that depth of penetration was observed to cause numerical convergence errors in a case of continuum finite element implementation. The stress distribution under different crystal orientation

Table 1. Material parameters of single-crystal copper [8]

Elastic parameters	Parameters of slipping rate	Parameters of hardening
$C_{11} = 168.0$ GPa		$h_0 = 180$ MPa
$C_{12} = 121.4$ GPa	$\dot{\gamma}_0 = 0.001\ \mathrm{s}^{-1}$ $m = 0.05$	$\tau_s = 148$ MPa
$C_{44} = 75.4$ GPa		$\tau_0 = 16$ MPa

with regard to axes x and y is shown in the Fig. 2.

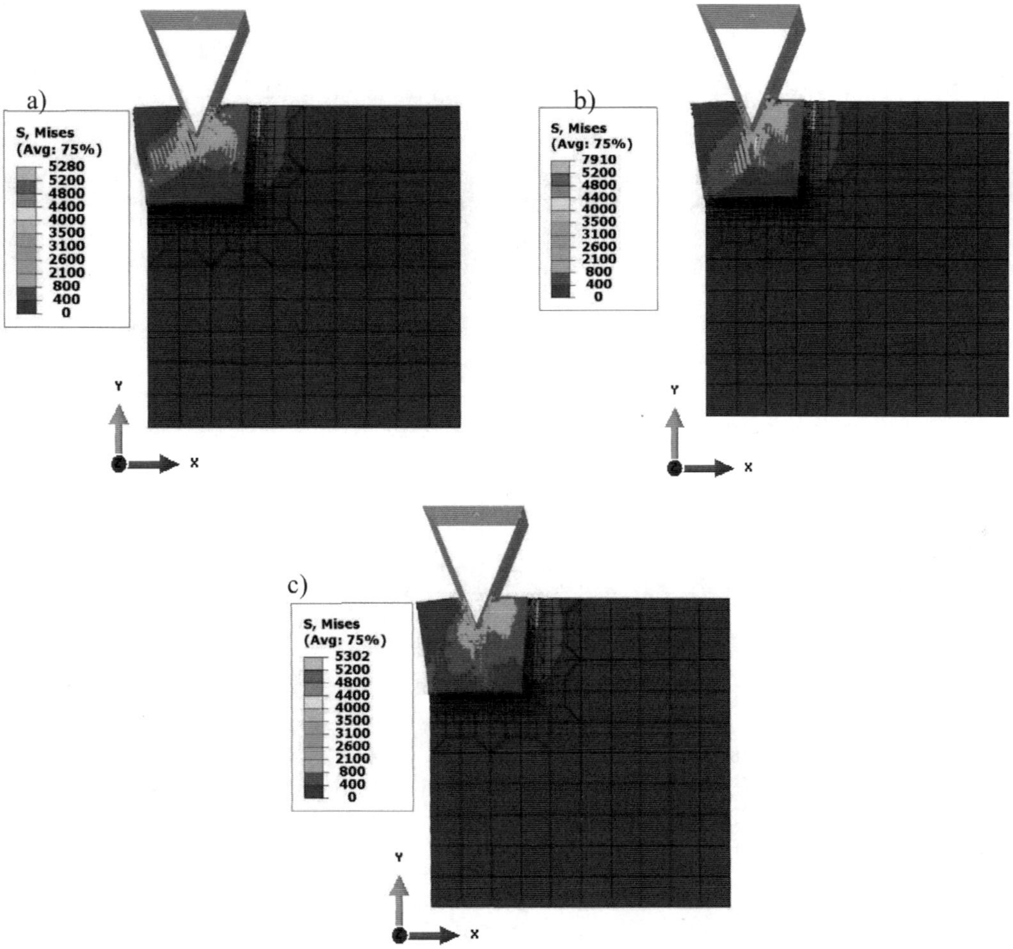

Fig. 2 Plastic stress distribution for cutting process for different orientation setups: (a) (111)[110]; (b) (111)[$\bar{1}$10]; (c) (111) [$\bar{2}$11]

The same plane (111) with three slip directions [$\bar{1}$10], [110] and [$\bar{2}$11] was selected for our numerical simulations. The results demonstrate that different surface profiles and stress distributions were obtained in the micro-machining for various orientations. For instance, von Mises stress is nearly 25% higher in the case of (111)[$\bar{1}$10] orientation setup compared to cases of (111)[110] and (111) [$\bar{2}$11].

Journal of Physics: Conference Series **382** (2012) 012017 doi:10.1088/1742-6596/382/1/012017

To find a compromise between the conflicting requirements of accuracy and computational cost, a mesh-sensitivity analysis of the model was initially performed. The vertical path (A-B-C) was chosen to present the distribution of von Mises stress. Selecting the vertical path was linked to location of nods in the workpiece to provide a shortest horizontal distance from the end of tool tip. There are only some changes in the stress along the path near point B (representing a boundary between SPH and FE domains). Hence, a mesh with a minimum element size of 0.2 μm at the region of machining was sufficient to characterize accurately the stress-distance behaviour during machining. The mesh-sensitivity analysis was performed for (111)[110] orientation setup of copper.

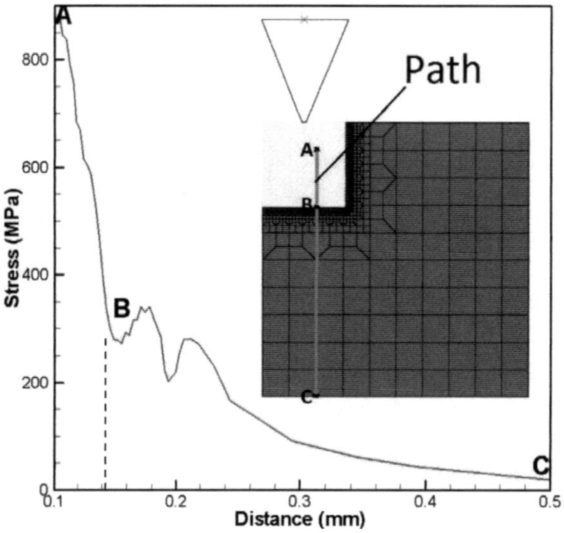

Fig. 3. Stress-distance distribution along path in workpiece material

4. Crystallographic orientation and cutting force

The primary indentation thrust forces along the direction of tool motion (F_y) at different crystal orientation are shown in Fig. 4. The value of F_y was found to be lowest in the (111)[110] orientation setup followed by (111)[$\bar{1}$10] and (111) [$\bar{2}$11] combination. This indicated that the F_y in the crystal orientation family (111) plane changed from [110] to [$\bar{1}$10] and [$\bar{2}$11] crystal orientation directions. This conclusion was consistent with the theory suggesting that a low tangential cutting force should be observed in the slip plane when oriented favourably to the deformation direction in the f.c.c. materials (i.e. (111) [110]).

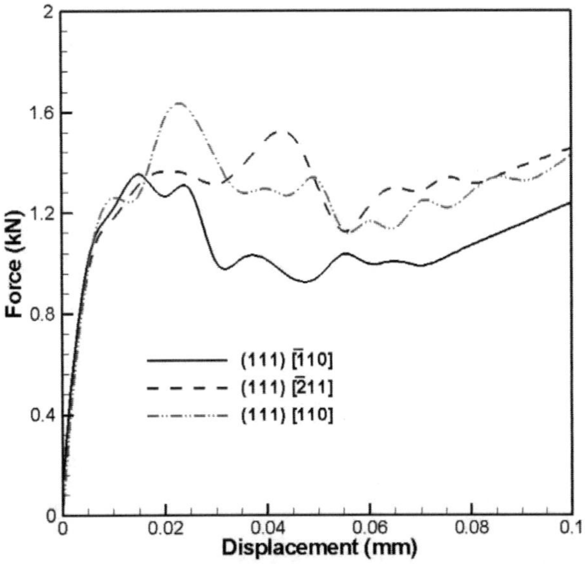

Fig 4. Effect of crystallographic orientation on load-displacement response during indentation

5. Conclusion

In the present work a 3D hybrid finite-element model of micro-machining was developed to study deformation of a single crystal of copper. The present model allows for high deformations in the underlying material. It was observed that the influence of cutting direction in micro machining of single copper crystal significantly affects evolution of cutting forces.

References

[1] N. Kota, O. B. Ozdoganlar, Orthogonal machining of single-crystal and coarsegrained aluminum, Journal of Manufacturing Processes 14 (2012) 126-134.

[2] H.M. Pen, Y.C. Liang, X.C. Luo, Q.S. Bai, S. Goel, J.M. Ritchie, Multiscale simulation of nanometric cutting of single crystal copper and its experimental validation , Computational Materials Science 50 (2011) 3431–3441.

[3] E. Demir, Taylor-based model for micro-machining of single crystal fcc materials including frictional effects Application to micro-milling process, International Journal of Machine Tools & Manufacture 48 (2008) 1592–1598

[4] S. Min, D. Dornfeld, I. Inasaki, H. Ohmori, D. Lee, M. Deichmueller, T. Yasuda, K. Niwa, Variation in Machinability of single crystal materials in micromachining, Annals of CIRP 55 (2006) 103-106.

[5] T. Özel, E. Zeren, Finite element modeling the influence of edge roundness on the stress and temperature fields induced by high-speed machining, International Journal of Advanced Manufacturing Technology 35 (2007) 255–267

[6] M. Movahhedy, M.S. Gadal, Y. Altintas, Simulation of the orthogonal metal cutting process using an arbitrary Lagrangian Eulerian finite-element method, Journal of Materials Processing Technology 103 (2000) 267-275

[7] A. Zahedi, M. Demiral, A. Roy , V. Babitsky, V. Silberschmidt, Indentation in f.c.c. single crystals, Solid State Phenomena 188 (2012) 219-225

[8] Y. Wang, D. Raabe, C. Kluber, Orientation dependence of nanoindentation pile-up patterns and of nanoindentation microtextures in copper single crystals, Acta Materialia 52 (2004) 2229–2238.

Strength of fibres in low-density thermally bonded nonwovens: An experimental investigation

***Farukh Farukh [(1)], Emrah Demirci [(1)], Memiş Acar [(1)], Behnam Pourdeyhimi [(2)], Vadim V. Silberschmidt [(1)]**

1- Wolfson School of Mechanical and Manufacturing Engineering
Loughborough University, UK

2- Nonwovens Cooperative Research Center
North Carolina State University, Raleigh, NC, USA

E-mail: F.Farukh@lboro.ac.uk

Abstract. Mechanical properties of nonwovens related to damage such as failure stress and strain at that stress depend on deformation and damage characteristics of their constituent fibres. Damage of polypropylene-fibre commercial low-density thermally bonded nonwovens in tension was analysed with tensile tests on single fibres, extracted from nonwovens bonded at optimal manufacturing parameters and attached to individual bond points at both ends. The same tests were performed on raw polypropylene fibres that were used in manufacturing of the analysed nonwovens to study quantitatively the effect of manufacturing parameters on tenacity of fibres. Those tests were performed with a wide range of strain rates. It was found that the fibres break at their weakest point, i.e. bond edge, in optimally bonded nonwovens. Additionally, failure stress and strain in tension of a fibre extracted from the fabric were significantly lower than those of virgin fibre. Since damage in nonwovens occurs by progressive failure of fibres, those experiments were used to establish criteria for damage initiation and propagation in thermally bonded nonwovens based on polypropylene fibres. Moreover, the results obtained from the experiments are useful to simulate the damage behaviour of nonwoven fabrics.

1. Introduction

A continuous growth in applications of nonwovens has led to a substantial increase in use of thermal bonding as compared to mechanical and chemical bonding processes for manufacturing of nonwovens since a higher production rate can be achieved by thermal bonding. Furthermore, it consumes the lowest amount of energy compared to other bonding processes with no hazardous substances involved as in chemical bonding. Therefore, thermal bonding is more suitable for the modern world with greater awareness of environmental impact [1, 2]. The growing demand for disposable and reliable products impels the research aimed at deformation and damage in low-density nonwovens. Thus, low-density

* Farukh Farukh, Loughborough University, UK
Phone no: 01509 227 634

thermally bonded nonwovens (planer density less than 50 g/m^2) are selected in this study to understand their damage behaviour.

In thermal bonding, fibres are passed through two hot calendar rolls under pressure; surface design of which can be smooth or have an embossed pattern. On embossed calendar rolls, bonding mainly occurs at raised areas resulting in bonded spots called "bond points". As higher temperature is involved in thermal bonding, fibre properties are characterized by the temperature level (and variation) during this manufacturing stage. Depending on the manufacturing parameters related to bonding performance, there are three types of thermally bonded nonwovens: under-bonded, well-bonded and over-bonded. Nonwovens manufactured at optimal temperature range (150°C to 160°C) are called well-bonded, whereas bonding at lower and higher than optimal temperature is called under-bonding and over-bonding, respectively [3]. Other manufacturing parameters such as calendaring pressure and manufacturing speed also affect the performance of bonding in nonwovens [2, 4]. All the manufacturing parameters influence each other, and only a careful selection of optimal temperature with adequate levels of calendaring pressure and processing speed can produce a well-bonded fabric.

Many researches were conducted to investigate the effect of bonding temperature on mechanical performance of nonwovens and their fibres during thermal bonding. According to literature, most of the fibres fail at bond periphery in well-bonded and over-bonded fabrics due to a rapid change in molecular orientation of fibres at bond edge [5-11], leading to reduction in modulus and yield strength of fibres at the bond edge. All these studies emphasize the origin and reasons of fibre failure in thermally-bonded nonwovens in tension rather than quantifying the loss of strength in a single fibre in terms of stress and strain. From a fundamental viewpoint, it is important to quantify the change in maximum stress and corresponding strain for fibres in a thermally bonded nonwoven as failure of multiple fibres occurs before final failure of fabric. This phenomenon was the main objective of this study.

This paper focuses on experimental analysis of mechanical characteristics of polypropylene fibres used in low-density thermally bonded nonwovens with different basis weight. A novel experimental approach was followed in this study in a way that single fibres attached to individual bond points at their both ends were tested to analyse the effect of manufacturing parameters on failure stress and strain of single fibre.

2. Material

The material type used in this study is low-density — planer density less than 50 g/m^2 — thermally-bonded nonwoven, composed of mono-component polypropylene (PP) fibres (manufactured by FiberVisions, USA). Three different basis weights of fabric (20 g/m^2, 30 g/m^2 and 40 g/m^2) manufactured at an optimal temperature of 156°C are used. Microstructure of the nonwoven material studied is shown in Fig. 1a. Apparently, these fabrics are composed of staple fibres having diameter of approximately 18 μm (Fig. 1b) with the length of 38.1 mm and linear density around 2.3 denier. All the fabrics have small rectangular-shaped bonds as shown in Fig. 1a.

3. Configuration of test samples

Test samples used in this study were single mono-component PP fibres utilised in manufacturing of low-density thermally bonded nonwovens. During manufacturing of this kind of nonwovens, web of fibres is passed through the hot calendars with engraved pattern on their surface. Fibres partially melt in respective regions and join together at a cooling stage. Molecular orientation of fibres changes from the bridging fibres into the bond in this manufacturing stage, ultimately leading to a decrease in the modulus and breaking strength of the fibres at bond edge. Thus, it is not possible to use the material properties of virgin fibres especially for damage analysis of the fabric. In this study, fibre samples were extracted from the fabric in a way that they are attached to individual bond points at their ends as shown in Fig. 2.

(a) (b)

Figure 1. SEM image of thermally bonded nonwoven fabric (30 g/m^2 PP) (a); and its PP mono-component fibres (b)

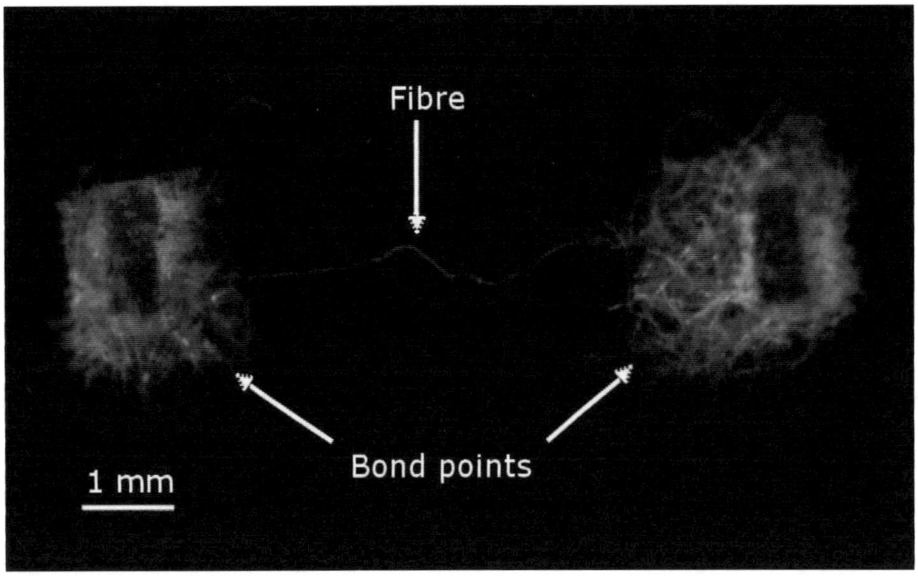

Figure 2. Single fibre attached to bond points at each end

Preparation of samples was tedious due to handling issues related to the size and softness of the specimens. In order to ensure removal of the fibres between the bond points so that only a single fibre is attached to two bond points at its ends, a magnifying glass was used. The single-fibre specimens were prepared without applying preload. To prevent slippage and dislocation of specimen during the tests, bond points at the edges of the fibre were attached to small pieces of cardboard with the help of a double-sided tape in such a way that the cardboard is as close to the edge of the bond towards the fibre

Journal of Physics: Conference Series **382** (2012) 012018 doi:10.1088/1742-6596/382/1/012018

as possible so that the accurate stress-strain behaviour of the processed fibre is obtained without amalgamating it with that of bonds. Additionally, specimens of unprocessed or virgin polypropylene fibres used to manufacture these nonwovens were also prepared by employing the same technique. The length of the fibre specimens varied between approx. 3 mm and 6 mm as it is not feasible to control their length due to the specimen-preparation technique and orientation of fibres. The samples were clamped with special grips mounted on Instron® Micro Tester 5848 having the Instron® ±5N high- precision load cell. A high-speed camera (Photron Fastcam SA3®) with advanced macro capabilities was used to make sure that fibre was not stretched during preparation and mounting of specimens. In addition to this, the camera was used to monitor the fibre to avoid slipping during the test. The complete arrangement of setup with the fibre mounted on tensile test rig is shown in Fig. 3. Fibre behaviour was determined by measuring the load continuously during the test along the cross-head displacement.

Figure 3. Test setup for fibre testing (Instron Micro Tester 5848; Photrom fastcam SA3)

4. Results and Discussions

Tensile tests were performed by stretching the fibre samples with three different levels of constant strain rates. Due to difficult control of the fibre's length, a constant strain rate $\dot{\varepsilon}$ was achieved by modifying the velocity of the cross-head with respect to the current length of the fibre. The relationship used for this purpose is shown in equation 1.

$$\frac{V}{l} = \dot{\varepsilon} = \text{const}, \tag{1}$$

Where V is the cross-head velocity and l is the length of fibre. Here, the constant strain rate corresponds to 0.01, 0.1 and 0.5 1/s. The tensile tests at each strain rate for processed and unprocessed fibres were repeated for at least ten times to assess the variability in results.

The fibre behaviour observed at different strain rates was similar to that of a typical polymer and is characterized by nonlinearity based on elastic and plastic stages as shown in Fig. 4. Variation of stress-strain curves with a change in the strain rate was due to viscous effects in the material. Those experiments did not result in a certain value of maximum stress and strain at maximum stress in fibres at any particular value of strain rate rather provided a range, within which these parameters fall. The mean values of these parameters with standard deviations are given in Table 1.

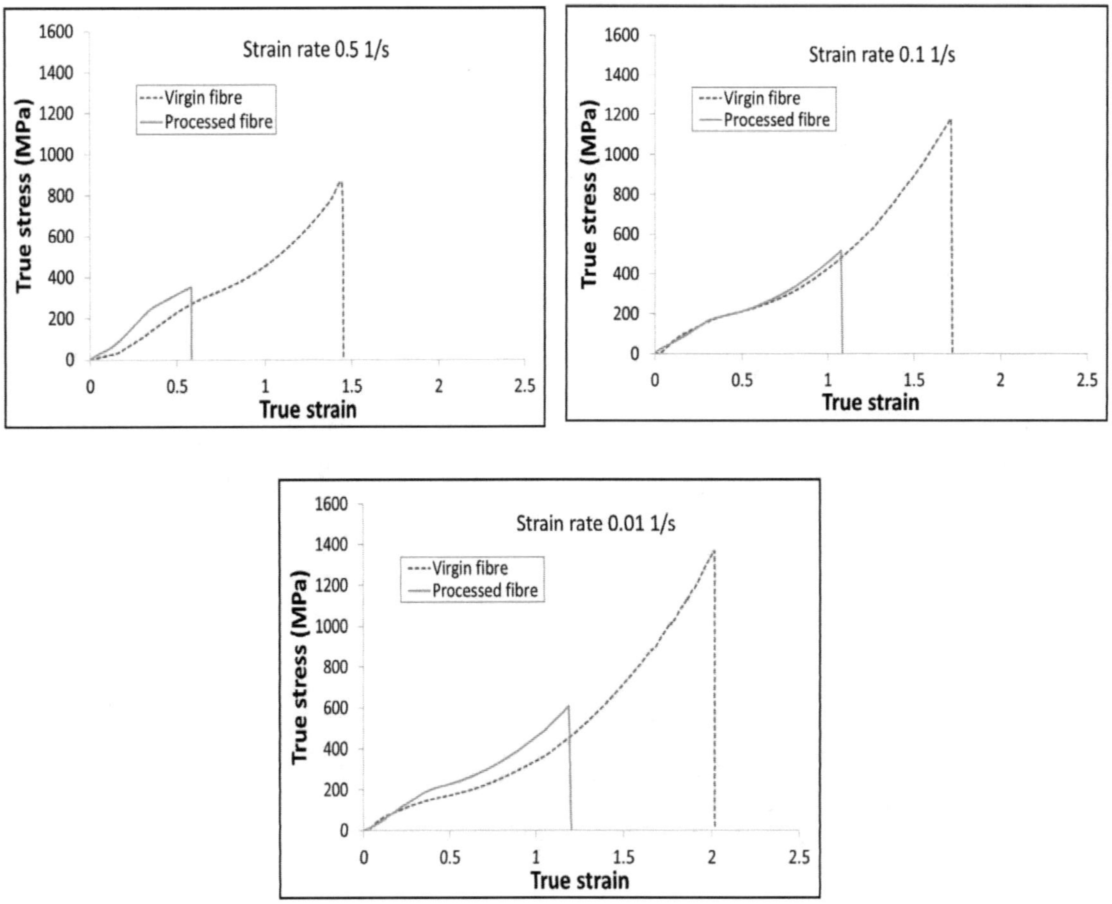

Figure 4. Mechanical behaviour of PP fibres at strain rate 0.05 1/s (a); 0.1 1/s (b); and 0.01 1/s (c)

A loss in tenacity of PP fibres due to manufacturing conditions of nonwovens can be analysed from the data given in Table 1. The data show that at optimum manufacturing parameters, the breaking stress of fibres decreased by some 64% and the corresponded strain decreased by 41%. This deterioration of performance of fibres in the fabric can be understood by examining the physics of bond formation. During thermal bonding, the fibres are pressed by the calendar rolls and thus weakened at the bond edge where the bond-point flattens the fibres due to calendar pressure. As this mechanical damage of fibres contributes almost negligibly to their loss of tenacity [10], therefore, the depth of bonding is not considered here. However, the temperature involved in thermal bonding causes the melting of fibres and diffusion of chain segment for bonding resulting in decrease of fibre strength.

Table 1. Mean values with standard deviation of failure stress and strain of virgin and processed fibres at various strain rates

	Strain rate (1/s)	Failure stress (MPa)	Strain at failure stress
0.01	Processed fibre	481 ± 48	1.29 ± 0.67
	Unprocessed fibre	1299 ± 95	2.93 ± 0.78
0.1	Processed fibre	315 ± 28	1.05 ± 0.11
	Unprocessed fibre	998 ± 57	1.77 ± 0.09
0.5	Processed fibre	240 ± 53	0.79 ± 0.1
	Unprocessed fibre	818 ± 60	1.49 ± 0.14

5. Conclusions

In this paper, a procedure for measuring directly the loss in tenacity of fibres in thermally bonded nonwovens was presented. Tensile tests at different strain rates were performed on carefully excised fibres from the nonwoven in a way that each fibre was attached to individual bond points at both ends. It was found that the breaking strength of polymeric fibres in the thermally bonded nonwoven at optimum processing conditions was less than that of virgin fibres. The failure stress of fibres and corresponding strain decreased by approx. 64% and 41%, respectively. The results reported here are in good agreement with the previous work [5, 6, 8, and 10]. The reason for the decrease in failure parameters of processed fibres is the rapid change in the orientation of their molecular chains at bond edge since the chains of polymers relax and diffuse due to high temperature and pressure during bonding process [5-11]. These molecular chain segments are cocrystallized to form the bond. The variation in failure parameters with strain rate is due to the viscous behaviour of PP fibres. The molecular chains in fibre are normally cross-linked which tend to stretch as result of uniaxial loading in tensile testing. The stretched conformation of the chains is not stable and chains tend to return their original upstretched form [14]. At higher strain rates, there is not enough time for the chain segments to regain their original form and fibres fail at lower level of stress and strain. During loading of nonwovens, rearrangement of fabric microstructure occurs due to reorientation of fibres in the direction of loading and more fibres start participating in load bearing. The fibres, with the applied stresses reaching their threshold, fail triggering damage in nonwoven. However, the complete failure does not follow immediately but occurs as a progressive failure of fibres [13]. Parameters of this continuous failure process realised in the fabric can be used to simulate the damage initiation and propagation behaviour of thermally bonded nonwovens.

Acknowledgement

We greatly acknowledge the support by the Nonwovens Cooperative Research Centre of North Carolina State University, Raleigh, USA. We also acknowledge the use of Photrom (Fastcam SA3), which was borrowed from the EPSRC Engineering Instrument Pool.

References

[1] Russell, S. J. Handbook of Nonwovens, Woodhead Publishing Limited, 2007.

[2] Albrecht, W.; Fuchs, H.; Kittelmann, W., W., Nonwoven Fabrics: Raw Materials, Manufacture, Applications, Characteristics, Testing Processes, Wiley-VCH, Weinheim, 2003.
[3] Kim, H. S.; Deshpande, A.; Pourdeyhimi, B.; Abhiraman, A.; Desai, P. Textile Res. J. 2001, 71, 157.
[4] Kim, H. S.; Pourdeyhimi, B.; Desai, P.; Abhiraman, A. Textile Res. J. 2001, 71, 965.
[5] Wang, X.; Michielsen, S. Textil Res. J. 2002, 72, 394.
[6] Michielsen, S.; Wang, X. Int. Nonwovens J. 2002, 11 (2), 35.
[7] Dharmadhikary, R. K.; Davis, H.; Gilmore, T. F.; Batra, S. K. Textile Res. J. 1999, 69, 725.
[8] Wang, X.; Michielsen, S. Textile Res. J. 2001, 71, 4251
[9] Hedge, R. R.; Bhat, G. S.; Campbell, R. A. J. Appl. Polymer Sci. 2008, 110 (5), 3047.
[10] Chidambaram, A.; Davis, H.; Batra, S. K. Inter Nonwovens J. 2000, 9 (3), 27.
[11] Bhat, G.S.; Jangala, P.K.; Spruiell, J.E., J. Appl. Polymer Sci. 2004, 92 (6), 3593.
[12] Michielsen, S.; Pourdeyhimi, B.; Desai, P., J. Appl. Polymer Sci., 2006, 99 (5), 2489.
[13] Farukh, F.; Demirci, E.; Sabuncuoglu, B.; Acar, M.; Pourdeyhimi, B.; Silberschmidt, V.V., Comp. Mat. Sci., in press
[14] Reusch, W. Virtual Textbook of Organic Chemistry, 1999, Available at http://www2.chemistry.msu.edu/faculty/reusch/VirtTxtJml/intro1.htm

Journal of Physics: Conference Series **382** (2012) 012019 doi:10.1088/1742-6596/382/1/012019

Cutting forces in ultrasonically assisted drilling of carbon fibre-reinforced plastics

Farrukh Makhdum, Luke T Jennings, Anish Roy and Vadim V Silberschmidt
Wolfson School of Mechanical and Manufacturing Engineering,
Loughborough University, Leicestershire, LE11 3TU, UK

E-mail: f.makhdum@lboro.ac.uk

Abstract. Ultrasonically assisted drilling (UAD) is a non-traditional hybrid machining process, which combines features of conventional drilling and vibratory machining techniques to obtain remarkable improvements in machinability of advanced materials. The experiments are conducted on commercially available samples of a carbon fibre-reinforced plastic (CFRP) at a feed rate of 16 mm/min. In this study, a thrust force reduction in excess of 60% is observed in UAD when compared to conventional drilling (CD). Lower delamination was observed when compared to CD techniques. Optical microscopy revealed that the material is removed as a continuous chip in UAD whereas in case of CD we observe powdered dust. Light and scanning electron microscopy of CFRP chips obtained in drilling elucidate fundamental differences in the underlying machining processes in UAD of CFRP.

1. Introduction

Carbon fibre-reinforced plastics (CFRPs) are widely used in aerospace, automobile and other structural application due to their superior mechanical and physical properties. CFRPs outperform conventional metals in durability, high strength to-weight-ratio, stiffness and density [1-4]. In modern aircraft the part of their total weight composed of composites ranges from 22% in the Airbus™ A380 to 50% in Boeing™ 787 [4, 5]. Typically, CFRP parts are manufactured to near net-shape, but machining cannot be avoided when it comes to component assembly [6]. Holes need to be drilled to facilitate riveting and bolting of components. It is well known that conventional drilling in CFRP induces different type of damages such as cracking, fibre pull-outs and delamination (the most critical type of damage) [7- 9]. Such barely visible damage in CFRPs decreases the load-carrying capacity of the composite parts and affects the structural strength adversely. Additionally, rapid tool wear is caused by abrasives of CFRPs resulting in increase of manufacturing cost.

With the development of new materials, novel machining techniques are continually being developed. Ultrasonically assisted drilling (UAD) is one such hybrid machining technique, which has been used to improve machining of conventional metals and advanced composites. In UAD high-frequency (typically in excess of 20 kHz) vibration is superimposed on a standard twist drill bit, preferably, in the axial direction. This vibration is generated by piezoelectric transducers and applied to enhance the cutting process [10]. There are several advantages of UAD over conventional drilling (CD) such as the reduction in thrust forces and torque, better surface finish, low tool wear and elimination/reduction in burr formation [11-14]. As drilling-thrust forces have a direct effect on machining-induced damage,

we consider the drilling forces as the primary factor affecting the quality of the drilled component. [15-16].
Several studies performed over the last few years [10, 13, 17-23] indicate that vibrating the drill bit in the axial direction yields the maximum reduction in drilling-induced damage. The paper presents the UAD and CD experiments on CFRPs conducted at Loughborough University together with damage analysis in the vicinity of the drilled hole carried out using advanced imaging technique.

2. Experimental setup

A standard M-300 Harrison lathe was adequately modified to incorporate an ultrasonic drilling head. A 6 mm TiN-coated Jobber carbide twist drill bit (Figure 1(b)) was mounted in the ultrasonic transducer and the transducer was held in a three-jaw universal chuck of the lathe. The workpiece was a 10 mm-thick industrial-grade CFRP clamped on a two-channel Kistler™ dynamometer (type 9271A). Finally, the dynamometer was mounted on to the cross slide of the lathe using an angle plate (see Figure 1(a)).

Our experiments were performed at a spindle speed of 40 rpm and feed rate of 16 mm/min. A resonance frequency of the ultrasonic transducer with the drill bit was achieved at 27.8 kHz and a free-vibration amplitude at that frequency was 12 μm. The thrust force and torque signals from the dynamometer were amplified using the charge amplifier. The signals from the drilling experiments were acquired with a Picoscope ™ 4424 PC oscilloscope and processed in Matlab™. Each test was repeated four times to consider reproducibility.

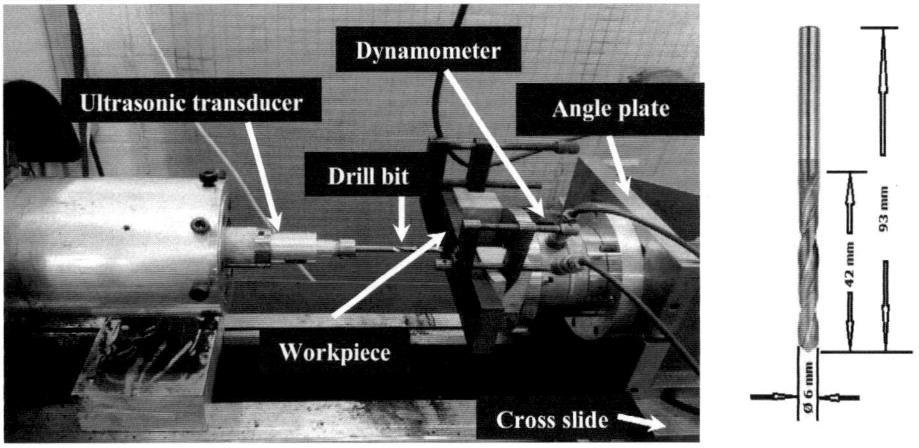

(a) (b)

Figure 1. (a) Experimental setup. (b) Drill bit

3. Results and discussions

3.1. Thrust force and torque

Remarkable reduction of thrust force and torque were obtained in UAD when compared to CD for the same drilling parameters. The average magnitudes of the peak thrust force in UAD were measured to be 151 N and while in CD it was 456 N (Table 1). The measured magnitude of torque in UAD vanished, with the torque fluctuating between ± 50 N-cm (Figure 2). The magnitudes for torque and thrust force were obtained based on several repetitions to assure the representativeness of results. The significant reduction of drilling force can be attributed to combinations of the following features of the UAD process:

1. The intermittent character of the drill engagement causes diminishment of the average drilling forces.
2. Dynamic frictional effects lead to a reduction in the effective co-efficient of friction at the drill-workpiece interface, and subsequently, a decline of the frictional component in cutting forces.
3. The ultrasonic softening effect which was observed in tensile tests of materials [27], can also contribute to force reductions.

The exact contribution of each of the above factors is difficult to quantify and needs to be investigated separately. Currently, an approach based on finite element modelling of UAD is being developed with the aim of elucidating the phenomena contributing to the reduction of drilling forces.

Table 1: Thrust force and torque

Feed rate (mm/min)	Spindle speed (rpm)	Average thrust force at full drill engagement (N)		Force reduction	Average torque at full drill engagement (N-cm)	
		CD	UAD		CD	UAD
16	40	456	151	66.8%	160	0

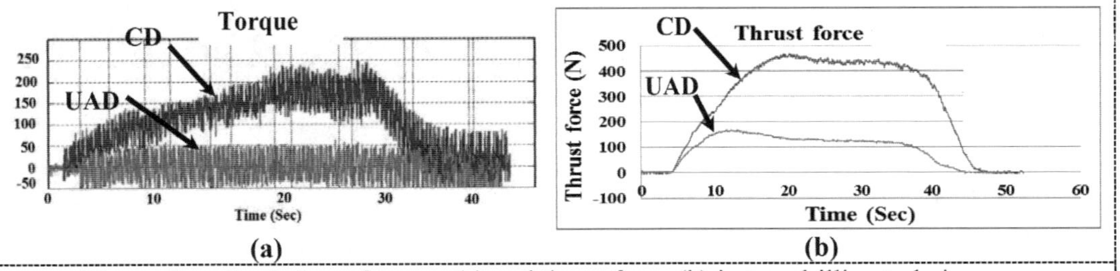

Figure 2. Evolution of torque (a) and thrust force (b) in two drilling techniques

3.2. Chip analysis

Light microscopy of chips obtained in our drilling experiments was carried out using a Nikon™ optical microscope model SM22T. The observed difference in the cutting behaviour in UAD manifests in completely different morphology of the chip. In case of CD the chip was highly fragmented, powder-like, as observed in traditional machining of CFRP, but in the case of UAD a continuous spiral chip was observed (Figure 3). This indicates that the underlying deformation mechanisms in UAD were significantly different from those in CD. Such a transition is similar to previous observations for ultrasonically assisted machining of brittle materials that demonstrated a brittle-to-ductile transition under the change from the UAD to CD regime. [26].

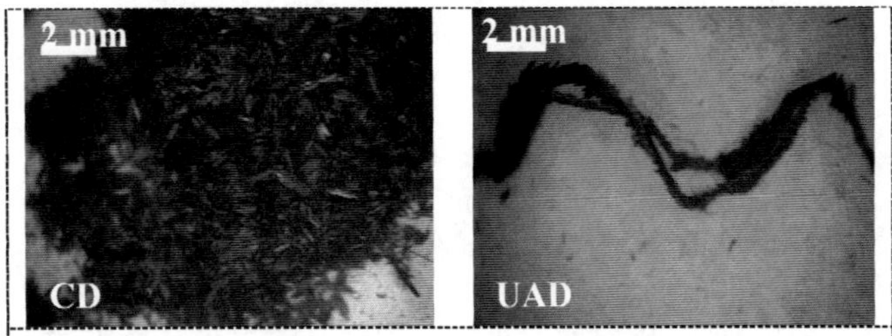

Figure 3. Optical microscopy

In addition to optical microscopy, scanning electron microscopy analysis was conducted using a Liks Analytical™ system (model 5431)

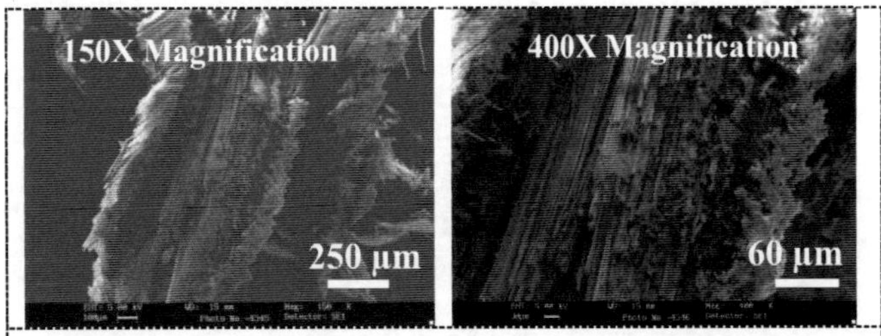

Figure 4. Scanning electron microscopy of UAD chip

machine on the chips obtained in the UAD experiments. The observations revealed fusing of carbon fibres into the matrix (Figure 4); this indicates elevated drilling temperatures in UAD. Further tests using a thermal camera are planned for the future to quantify the heat generated during this machining process.

3.3. Delamination study

Delamination in CFRP due to drilling is a major concern for successful application of fibre-reinforced composite materials in various industries. Determination of the size of the delamination zone after drilling is essential in assessing the effectiveness of UAD in comparison to CD. To this end, micro-computed tomography (μCT) was performed to quantify the damaged area of CFRP in the vicinity of the drilled hole. The measurements were carried out using an X-tech System™ XTH-160 machine. Each sample was exposed to X-ray radiation and rotated by 360° about a vertical axis to capture the images for 3-D re-construction. The X-ray voltage and current were set at 80 kV and 75 mA respectively [25]. The details of μCT for two drilling techniques are given in Figure 5. Analysis of composite plies near the drill entry and exit demonstrate significantly reduced delamination zones after UAD when compared to those in CD (for visualisation purposes the delaminated area is shaded in red in Figure 5).

(a) (b) (c) (d)

Figure 5. Delamination analysis: entry delamination after CD (a) and UAD (b); exit delamination after CD (c) and UAD (d)

Conclusions

A non-traditional hybrid machining process, UAD, was used to demonstrate improvements in the machinability of CFRP composites compared to CD. The following conclusions can be drawn from the experimental study carried out:

- UAD can reduce the thrust force by up to 60% when compared to a conventional drilling process. This was achieved at a drill spindle speed of 40 rpm and linear feed of 16 mm/min.
- The material removed in the form of a chip in case of UAD as compared to CD
- The chips obtained in UAD showed brittle-to-ductile transition of the composite material when subjected to ultrasonic vibration. A further study is planned to characterise this effect.
- Ply delamination was reduced significantly, both at entry and exit.

References

[1] Casas-Rodriguez JP, Ashcroft IA, Silberschmidt VV. 2008. Delamination in adhesively bonded CFRP joints: Standard fatigue, impact-fatigue and intermittent impact. *Composites Science and Technology* 68:2401-9

[2] Ullah H, Harland AR, Lucas T, Price D, Silberschmidt VV. 2011. Finite-element modelling of bending of CFRP laminates: Multiple delaminations. *Computational Materials Science* 52:147-156

[3] Xia ZH, Curtin WA. 2008. Damage detection via electrical resistance in CFRP composites under cyclic loading. *Composites Science and Technology* 68:2526

[4] Kolesnikov B, Herbeck L, Fink A. 2008. CFRP/titanium hybrid material for improving composite bolted joints. *Composite Structures* 83:368-80

[5] Lee CK. 2009. Structure, electrochemical and wear-corrosion properties of electroless nickel–phosphorus deposition on CFRP composites. *Materials Chemistry and Physics* 114:125-33

[6] Zitoune R, Krishnaraj V, Sofiane Almabouacif B, Collombet F, Sima M, Jolin A. 2012. Influence of machining parameters and new nano-coated tool on drilling performance of CFRP/Aluminium sandwich. *Composites Part B: Engineering* 43:1480-8

[7] Silberschmidt VV. 2005. Matrix cracking in cross-ply laminates: effect of randomness. *Composites Part A: Applied Science and Manufacturing* 36:129

[8] Iliescu D, Gehin D, Gutierrez ME, Girot F. 2010. Modeling and tool wear in drilling of CFRP. *International Journal of Machine Tools and Manufacture* 50:204-13

[9] Kyriazoglou C, Guild FJ. 2005. Quantifying the effect of homogeneous and localized damage mechanisms on the damping properties of damaged GFRP and CFRP continuous and woven composite laminates--an FEA approach. *Composites Part A: Applied Science and Manufacturing* 36:367

[10] Thomas PNH, Babitsky VI. 2007. Experiments and simulations on ultrasonically assisted drilling. *Journal of Sound and Vibration* 308:815

[11] Alam K, Mitrofanov AV, Silberschmidt VV. 2011. Experimental investigations of forces and torque in conventional and ultrasonically-assisted drilling of cortical bone. *Medical Engineering & Physics* 33:234

[12] Graff KF. 1975. Macrosonics in industry: 5. Ultrasonic machining. *Ultrasonics* 13:103

[13] Babitsky VI, Astashev VK, Meadows A. 2007. Vibration excitation and energy transfer during ultrasonically assisted drilling. *Journal of Sound and Vibration* 308:805-14

[14] Wiercigroch M, Wojewoda J, Krivtsov AM. 2005. Dynamics of ultrasonic percussive drilling of hard rocks. *Journal of Sound and Vibration* 280:739-57

[15] Pyo JJ, Woo KG, Yong LK. 2005. Critical thrust force at delamination propagation during drilling of angle-ply laminates. *Composite Structures* 68:391-7

[16] Tsao CC. 2012. Effect of induced bending moment (IBM) on critical thrust force for delamination in step drilling of composites. *International Journal of Machine Tools and Manufacture* 59:1-5

[17] Kim J-D, Lee E-S. 1994. A study of the ultrasonic-vibration cutting of carbon-fiber reinforced plastics. *Journal of Materials Processing Technology* 43:259-77

[18] Hocheng H, Hsu CC. 1995. Preliminary study of ultrasonic drilling of fiber-reinforced plastics. *Journal of Materials Processing Technology* 48:255

[19] Wang X, Wang LJ, Tao JP. 2004. Investigation on thrust in vibration drilling of fiber-reinforced plastics. *Journal of Materials Processing Technology* 148:239

[20] Takeyama H, Kato S. 1991. Burrless drilling by means of ultrasonic vibration. *CIRP Annals - Manufacturing Technology* 40:83

[21] Egashira K, Mizutani K, Nagao T. 2002. Ultrasonic vibration drilling of microholes in glass. *CIRP Annals - Manufacturing Technology* 51:339-42

[22] Devine J. 1985. *Ultrasonically Assisted Metal Removal*: American Society for Metals

[23] Makhdum F, Norddin DNP, Roy A, Silberschmidt VV. 2012. Ultrasonically assisted drilling of carbon fibre reinforced plastics. *Solid State Phenomenon* 188:170-5

[24] Mitrofanov AV, Babitsky VI, Silberschmidt VV. 2004. Finite element analysis of ultrasonically assisted turning of Inconel 718. *Journal of Materials Processing Technology* 153–154:233-9.

[25] Ullah H, Harland AR, Silberschmidt VV. 2012. Experimental and numerical analysis of damage in woven GFRP composites under large-deflection bending. *Applied Composite Materials* DOI:10.1007/S/0443-011-9242-7

[26] Astashev VK, Babitsky VI. 2007. Ultrasonic processes and machines: dynamics, control and applications. Berlin Heidelberg New York: Springer

[27] Langenecker B. 1961. Work-softening of metal crystals by alternating the rate of glide strain. Acta Metallurgica 9:937-40

Vibration Analysis of Frame Structures Using Wavelet Finite Elements

M Musuva[1] and C Mares[1]

[1]School of Engineering and Design, Brunel University, Kingston Lane, Uxbridge UB83PH, UK

Email: mutinda.musuva@brunel.ac.uk
cristinel.mares@brunel.ac.uk

Abstract. The recently developed Wavelet Finite Element Method (WFEM) involves combining the versatile wavelet analysis with conventional Finite Element Method (FEM) by utilizing the wavelet and scaling functions as interpolating functions, providing an alternative to the conventional polynomial interpolation functions used in FEM. In this paper, the B-Spline Wavelet on bounded Interval (BSWI) wavelets are utilized in the construction of the wavelet based finite elements for the dynamic analysis of a Vierendeel frame structure subjected to a moving load. The numerical results for these WFEM formulations are compared with the classical Euler Bernoulli elements.

1. Introduction

Moving load dynamic problems have been studied in the last decades with various forms of interactions and varying loads as outlined by Ouyang [1]. The majority of the simple moving load problems may be analytically solved as described in Fryba's monograph [2] and Yang et al. [3]. However, in the analysis of complicated moving load problems, particularly those with significant variations in the structural and loading properties as well as boundary conditions, conventional FEM is deemed to be the preferred solution method to obtain their approximate solutions. The use of this numerical analysis technique for problems with regions of the solution domain where the gradient of the field variables vary suddenly or rapidly, encounters difficulties. For this reason, research has been carried out in recent years by structural analysts to overcome these limitations and find more efficient and accurate tools to analyse these structural systems, particularly in moving load problems.

Initially used mainly by mathematicians as a decomposition tool for data functions and operators, the application of wavelet analysis has vastly grown in various disciplines at an exponential rate due to the varying properties of the increasingly different wavelet families. Daubechies [4] and Chui [5] monographs outline detailed descriptions and formulations of general wavelets. In recent years, work has been carried out exploring the fusion of conventional FEM and wavelet analysis to obtain an approximate solution for various engineering problems, widely referred nowadays as the wavelet finite element method (WFEM). These wavelet based elements are constructed using the scaling functions of the wavelet families as the interpolating functions, contrary to the polynomial interpolating functions used in conventional FEM. Minimum compact support of wavelets is a valuable wavelet property in the formulation of the WFEM that leads to fewer degrees of freedom (DOFs). Another key property, multiresolution analysis (MRA) which stems from the "two-scale" relation, provides the opportunity for the analysis of systems globally and most importantly, locally, at different scales without alterations to the original modelling. An individual element or group of elements can be analysed at finer scales to obtain better accuracy once the meshing of the system has been carried out. Furthermore, MRA helps ensure the convergence condition is met. However, the properties of different wavelet families vary, and therefore the decision on which family, order and scale is the most adequate, becomes paramount to its application.

Ko et al [6] were among the first to propose the WFEM when they applied the Daubechies wavelet family to construct a wavelet based finite element (WFE) and analyse one dimensional and two dimensional Neumann problems. The development of orthogonal wavelets such as the Daubechies wavelet family, led to the development of a new finite element strategy, via Galerkin formulations. Zhou et al [7] applied the method to the bending of beams and plates. Ma et al [8] and Chen et al [9,10] progressed this research by using the Daubechies WFE to prove the method's ability to solve static structural problems. Due to their minimum compact support, orthogonality and MRA properties, the Daubechies wavelets were initially preferred in the formulation of wavelet based finite elements (WFEs). However, their numerical implementation is challenging because of lacking a closed form, and because of the possibility of the orthogonality loss, when used in weak variational formulations for general boundary value problems. To overcome these problems, specialized numerical techniques are required [11], which increase the computational costs but maintain the approximation accuracy and orthogonality properties in the analysis. BSWI family of wavelets constructed by Chui and Quak [12], have also been used in some WFEM formulations [13-14]. This family of wavelets has, in addition to the multiresolution and compact support properties, explicit expressions and therefore there is no need to carry out additional calculations to obtain the integral of the products of the scaling function and/or their derivatives.

In this paper the WFEM is formulated using BSWI. The WFEs are used in the analysis of the dynamic response of a Vierendeel frame structure subjected to a moving load. The results are compared with analytical formulations and the conventional Euler Bernoulli beam elements in order to highlight the efficiency and general properties of such formulations. Section 2 of this paper highlights the general wavelet and MRA properties with a brief overview of BSWI formulations. Section 3 illustrates the wavelet Galerkin finite element formulations. Finally in section 4, simulated numerical examples for the dynamic response of a Vierendeel frame structure under a moving load are carried out to highlight the formulation efficiency as well as the difficulties encountered in the numerical implementation, followed by conclusions.

2. Wavelet Analysis

A brief overview of wavelet analysis and multiresolution analysis as well as their properties is carried out in this section. Multiresolution, representing an advantageous property of wavelets, refers to the simultaneous appearance of multiple scales in function decompositions in the Hilbert space $L^2(\mathbb{R})$ using a sequence of closed subspaces V_j which is represented mathematically as [5]:

$$\ldots \subset V_{-2} \subset V_{-1} \subset V_0 \subset \cdots \subset V_j \subset V_{j+1} \subset \cdots \quad (1)$$

These subspaces satisfy the following conditions:

1. $V_j \subset V_{j+1}$
2. $\cap V_j = \{0\}$ and $\overline{\cup} V_j = L^2$ (Completeness). (2)

By defining an orthogonal complement W_j to the subspace V_j, the next subspace V_{j+1} is completed

$$V_j \oplus W_j = V_{j+1} \quad (3)$$

and the space $L^2(\mathbb{R})$ can be represented as a direct sum:

$$L^2(\mathbb{R}) = \oplus\, W_j \quad (4)$$

2.1. B-Spline Wavelets on the Interval [0,1]

Given a knot vector, t^j comprising of a sequence of knots t_i^j on [0,1], via joining the piecewise polynomials between the knots, one is able to construct B-Splines at a scale $j \in \mathbb{N}_0$[12,15].

$$t_m^j := \{t_i^j\} \tag{5}$$

The *m*th order B-Splines are defined as:

$$B_{m,i}^j(x) := \left(t_{i+m}^j - t_i^j\right)\left[t_i^j, t_{i+1}^j, \ldots, t_{i+m}^j\right]_t (t-x)_+^{m-1} \tag{6}$$

Where $\left[t_i^j, t_{i+1}^j, \ldots, t_{i+m}^j\right]_t$, is the *m*th divided difference of the truncated power function $(t-x)_+^{m-1}$ with respect to variable *t*.

$$\psi_{m,i}^j(x) := \psi_m\left(2^j x - k\right) = \frac{1}{2^{2m-1}} \sum_{k=0}^{2m-2} (-1)^k N_{2m}(k+1)\, B_{2m,2i+k}^{j+1}(x) \tag{7}$$

$$supp\psi_{m,i}^j = \left[\frac{i}{2^j}, \frac{i+2m-1}{2^j}\right] \tag{8}$$

With values of $0 \leq i \leq 2^j - 2m + 1$, $supp\, \psi_{m,i}^j \subseteq [0,1]$. In order to have one inner wavelet on the interval [0,1], the following condition must be met [15].

$$2^{j_0} \geq 2m - 1 \tag{9}$$

Where j_0 is the smallest scale *j* of order *m* that will ensure that there is at least one inner wavelet on the interval [0,1]. Thus, the scaling functions, $\phi_{m,k}^j(x)$, and corresponding semi-orthogonal wavelet functions, $\psi_{m,k}^j(x)$ of BSWI of order *m* at scale $j \geq j_0$ (BSWI*mj*) are evaluated as follows:

$$\phi_{m,k}^j(x) = \begin{cases} B_{m,k}^{j_0}\left(2^{j-j_0}x\right) & -m+1 \leq k \leq -1 \\ B_{m,0}^{j_0}\left(2^{j-j_0}x - 2^{-j_0}k\right) & 0 \leq i \leq 2^j - m \\ B_{m,2^j-k-m}^{j_0}\left(1 - 2^{j-j_0}x\right) & 2^j \leq i \leq 2^j + m - 1 \end{cases} \tag{10}$$

$$\psi_{m,k}^j(x) = \begin{cases} \psi_{m,k}^{j_0}\left(2^{j-j_0}x\right) & -m+1 \leq k \leq -1 \\ \psi_{m,0}^{j_0}\left(2^{j-j_0}x - 2^{-j_0}k\right) & 0 \leq i \leq 2^j - m \\ \psi_{m,2^j-k-2m+1}^{j_0}\left(1 - 2^{j-j_0}x\right) & 2^j \leq i \leq 2^j + m - 1 \end{cases} \tag{11}$$

3. Wavelet Galerkin Finite Element Formulation

For a wavelet finite element formulation, the scaling functions specific for each wavelet formulation are used to construct the element displacement interpolation functions, instead of the conventional polynomial interpolating functions, with the variable field approximated using the Riesz basis of the fundamental space V_0:

$$\widetilde{w}(\xi) = \sum a_{m,k}^j{}_k\, \phi_{m,k}^j(\xi) \tag{12}$$

With $\{\alpha\}$ composed of projection coefficients in the wavelet space a_k and [*N*] the shape functions matrix composed of the wavelet basis functions. For the differential operator *L* associated with a specific analysis, it follows that:

$$[B] = L[N] \tag{13}$$

In the case of structural dynamics, using equation (12), the elemental stiffness, mass matrices and force vector are given as

$$\widetilde{K}_e = \int_0^1 [B]^T [B]\, d\zeta, \qquad \widetilde{M}_e = \int_0^1 [N]^T [N]\, d\zeta, \qquad \widetilde{R}_e = \int_0^1 q(x)\, [N]^T\, d\zeta \tag{14}$$

The matrices and the load vector are in wavelet space and thus α_k are the corresponding DOF in this wavelet space. It is necessary to transform the vector and matrices into physical space so as to satisfy the boundary conditions of the beam as well as the connectivity of the neighbouring elements that are linked by the respective nodes. Thus a transformation matrix T is calculated from the scaling functions from equations (10) and (12). The DOFs transformation expression is given as:

$$w = [T]\alpha \tag{15}$$

Where the vector $w = [u_1\, w_1 \theta_1 u_2 w_2 u_3 w_3 \cdots u_{nel+1} w_{n_{el}+1} \theta_{n_{el}+1}]^T$, consists of axial deformations and vertical displacements along the element and two rotations at the element ends to ensure effective connection between adjacent elements. Subsequent, the transformation from the wavelet space into the physical space for the stiffness and mass matrices as well as load vectors are given by:

$$K_e = a\,[T^{-1}]^T \tilde{K}_e [T^{-1}] \qquad M_e = \rho AL\,[T^{-1}]^T \tilde{M}_e [T^{-1}] \qquad R_e = [T^{-1}]^T \tilde{R}_e \tag{16}$$

For a rod element, $a = E\,A\,L^{-1}$ and for a beam element $a = E\,I\,L^{-3}$. E – Young's Modulus, I – moment of inertia, A – cross sectional area and L – Length. Equation (16) is formulated in local coordinates and therefore, transformation into global coordinates is carried out thereafter. Using equation (16), the system's dynamics for a moving load problem is then described by:

$$M\ddot{q}(t) + C\dot{q}(t) + Kq(t) = R\delta(x - ct) \tag{17}$$

where $\ddot{q}, \dot{q}$ and q are the accelerations velocities and displacements at the nodal degrees of freedom in the physical space, M, C, K are the mass, damping and stiffness matrices and R represents the load vector.

4. Numerical Analysis

The Vierendeel frame as illustrated in figure 1, is subjected to a moving point load of magnitude P = 2000N moving along elements 1 - 6. The normalised deflection at point A is presented in addition to the first 10 angular frequencies. The material properties of the beams are E = 207 Nm^{-2}, A = 8.06x10^{-5} m^2, ρ = 7.81x10^3 kg m^{-3}, I = 2.71x10^{-10} m^{-4} and L = 0.305 m. The load travels at a velocity c ms^{-1} corresponding to the speed parameter α of values 10^{-5}, 1 and 2 where $c = 6\alpha L\omega_1/\pi$ (ω_1 – First angular frequency). The absolute value deflection $\delta_0$at point A is obtained from static loading of the frame at the same point A with load P. The results using 8 BSWI WFEs at scale j = 3 to model the Vierendeel frame are compared with a similar analysis using 48 Euler Bernoulli elements.

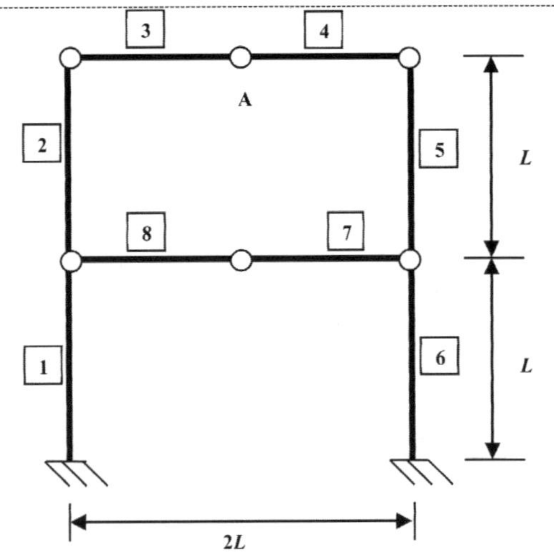

Figure 1. Layout of Vierendeel frame structure.

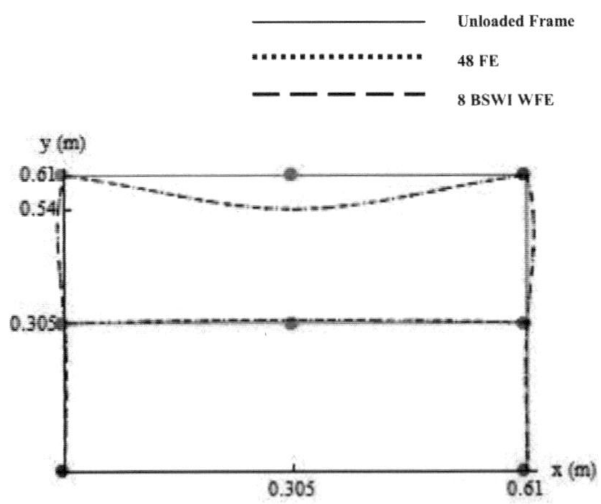

Figure 2. Deformation, δ_0 of Vierendeel frame subjected to static load P applied at Point A.

Results from figure 2 indicate that the deformation of the frame under a static load P using 8 BSWI WFEs gives accurate results, similar to 48 conventional finite elements. This is further validated in table 1, where the first 10 angular frequencies of the frame are compared with that from [16].

Table 1. First 10 angular frequencies of the Vierendeel frame structure (rads^{-1}).

Mode No.	Jara-Almonte's Exact Element[16]	48 Euler Bernoulli Elements	8 BSWI WFEs
1	107	107	107
2	377	337	377
3	397	397	397
4	476	475	475
5	1099	1098	1098
6	1316	1315	1315
7	1504	1503	1503
8	1912	1910	1910
9	2061	2060	2060
10	2248	2246	2246

Defining δ as the deflection at Point A as the load P moves across the beam at c ms^{-1}, the plots of the normalised dynamic deflection δ / δ_0 are plotted in figures 3, 4 and 5 for values of $\alpha = 10^{-5}$, 1 and 2 respectively. The plots show that using 8 BSWI elements gives similar results to 48 Euler Bernoulli finite elements. Using the WFEM requires significantly fewer elements in comparison to the conventional FEM with high levels of accuracy.

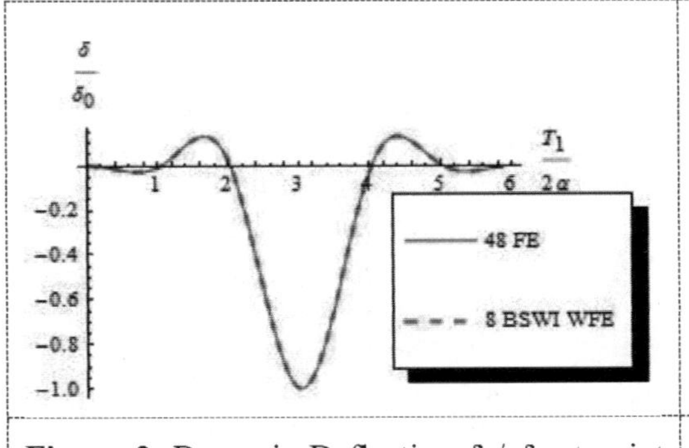

Figure 3. Dynamic Deflection δ / δ_0 at point A as load moves over the frame when $\alpha = 10^{-5}$

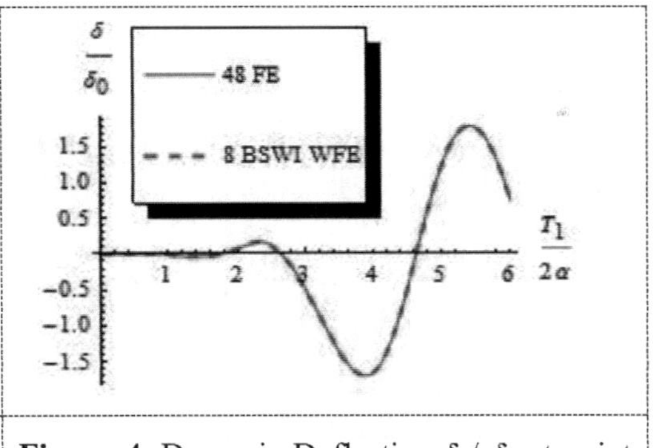

Figure 4. Dynamic Deflection δ / δ_0 at point A as load moves over the frame when $\alpha = 1$

Conclusion

The use of B-Spline Wavelets on bounded Interval (BSWI) in the formulation of wavelet finite elements (WFEs) and its application to vibration analysis of frame structures subjected to a moving load has been highlighted in this paper. The numerical example of the Vierendeel frame structure subjected to a moving load illustrates high levels of accuracy attained using WFEM with fewer elements in comparison to conventional FEM. The natural frequencies of the frame structure obtained using WFEM, particularly at higher modes, are also of better accuracy than the conventional FEM approach. However, the kind, order and scale of the wavelet family is dependent on the application and level of accuracy required, therefore an increase in either scale or order leads to an increase in computational complexity. Furthermore, with the property of multiresolution, WFEM allows for an

Journal of Physics: Conference Series **382** (2012) 012020 doi:10.1088/1742-6596/382/1/012020

individual element or group of elements to be analysed at finer scales to obtain better accuracy in regions of the solution domain where the gradient of the field variables vary suddenly or rapidly. Hence, the potential of WFEM for more complex moving load problems.

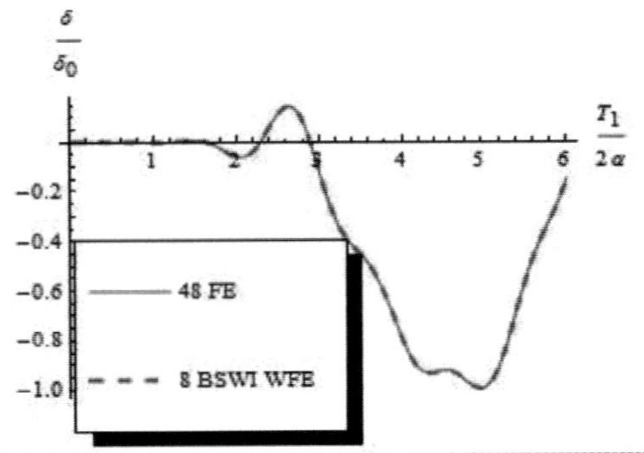

Figure 5. Dynamic Deflection δ / δ_0 at point A as load moves over the frame when $\alpha = 2$.

References

[1] Ouyang H 2011 *Mechanical Systems and Signal Processing* **25** 2039-60
[2] Fryba L 1972 *Vibration of Solids and Structures under Moving Loads* (Groningen: Noordhoff International Publishing)
[3] Yang B Y, Yau J D and Wu Y S 2004 *Vehicle-Bridge Interaction Dynamics* (World Scientific Publishing Co)
[4] Daubechies I 1992 *Ten Lectures on Wavelets* (Pennsylvania: Society for Industrial and Applied Mathematics)
[5] Chui C K 1992 *An Introduction to Wavelets* (Academic Press)
[6] Ko J, Kurdila A J and Pilant M S 1997 *Computer Methods in Applied Mechanics and Engineering* **146** 1-17
[7] Zhou Y, Wang J and Zheng X 1998 *Applied Mathematics and Mechanics* **1** 779-87
[8] Ma J, Xue J, Yang S and He Z 2003 *Finite Elements in Analysis and Design* **39** 965-75
[9] Chen X, Yang S, Ma J and He Z 2004 *Finite Elements in Analysis and Design* **40** 541-54
[10] Chen X, He Z, Xiang J and Li B 2006 *Computational and Applied Mathematics* **188** 228-45
[11] Latto A, Resnikoff H L and Tenenbaum E 1992 *The Evaluation of Connection Coefficients of Compactly Supported Wavelets* ed Y Maday, (Berlin: Springer)
[12] Chui C K and Quak E 1992 *Numerical Methods in Approximation Theory* **1** 53-7
[13] Xiang J W, Chen X F, He Z J and Dong H B 2007 *Computational Mechanics* **40** 325-38
[14] Xiang J, Chen C, He Y and He Z 2006 *Finite Elements in Analysis and Design* **42** 1269-80
[15] Quak E and Chui C K 1994 *Applied and Computational Harmonic Analysis* **1** 217-31
[16] Chen W H and Chen C W 1995 *Computational Mechanics* **16** 11-21

Rational Placement of a Macro Fibre Composite Actuator in Composite Rotating Beams

J Latalski, F Georgiades, J Warmiński
Lublin University of Technology
Applied Mechanics Department
Nadbystrzycka 36, 20-618 Lublin, Poland

E-mail: j.latalski@pollub.pl

Abstract. Abstract. In the presented research the dynamics of a thin rotating composite beam with surface bonded MFC actuator are considered. A parametric analysis aimed at finding the most efficient location of the actuator on the beam is presented. Gyroscopic effects resulting in the beam's initial strain and therefore non-zero voltage in PZT are taken into account. Within the frame of the study maximising the system's response observed in vibration modes for uncoupled and coupled motions is examined. The results are compared to the case of a nonrotating beam and also to the maximum response of the beam with the actuator placed at different positions. To perform the analysis an ABAQUS finite element model of an electro-mechanical system under consideration is developed. The multi-layer composite beam structure is modelled by shell elements according to a layup-ply technique; the MFC actuator is modelled by 3D coupled field piezoelectric elements. Both modal analysis and frequency response spectra are performed to obtain the structural modal parameters and response amplitude, respectively. The analysis is repeated for three different orientations of the beam's cross-section with respect to the plane of rotation (i.e. arbitrary assumed pitch angles); in all cases the condition constant angular speed is preserved. This work is fundamental for continuing the research for control of dynamics of rotating composite beams with active elements.

1. Introduction

In recent years the idea of possible use of smart materials (e.g. piezoelectric elements, shape memory alloys or magnetorheological fluids) in modern structures has been the area of extensive interest. The significant research is related to the active control of flexible systems in order to improve their operating performance and to increase their service lifetime [1, 2]. In active vibration control, apart from the standard structural and controller design, the rational placement of an actuator for the most effective structural response plays a significant role.

Among different structures to be controlled lightweight rotating composite beams are of particular interest because of their rich dynamics. Numerous studies yield qualitatively and quantitatively their unique properties, like the rotating speed-dependent natural frequencies and mode shapes, the large deformations under centrifugal force and couplings between displacements/ /deformations along different directions [3]. These effects observed for isotropic materials are even more complex and more pronounced in case of lightweight composite beams [4], and also in multi-physics systems.

Therefore, the scope of this paper is a numerical parametric analysis of the rotating composite beam with an embedded MFC actuator. To perform the analysis an ABAQUS finite element electromechanical model of the beam with MFC patch is developed. The model is used for identification of a rational position of the actuator considering the influence of gyroscopic effects.

The study is aimed at finding the rational location of the transducer on the host structure to achieve the maximum system's response in its uncoupled and coupled bending motions. Studies are performed for different rotating speeds ω and the results are compared to the case of a nonrotating beam. Moreover, different orientations θ of the beam's cross-section with respect to the plane of rotation (i.e. pitch angles) are analysed. Effects resulting in the beam's initial strain and therefore non-zero voltage in a d_{33} type PZT patch are taken into account.

2. Rotating beam model

The model of a structure under consideration is presented in Figure 1 – the composite beam is clamped at the hub rotating with constant angular velocity ω. Orientation of the beam's local coordinate system *xyz* with respect to the surface of revolution *XOY* is defined by θ angle. In the current analysis a Macro Fibre Composite actuator of d_{33} type is used (poling direction along the beam axis). This type of device, as opposed to traditional piezoceramic patch, is flexible and can therefore be applied to curved surfaces. In-plane poling is achieved by interdigitated electrode, which produces more induced actuating strain than monolithic piezoelectric ceramic patch and also more than d_{31} type elements. It is assumed that the MFC element is perfectly bonded to the surface of the beam.

Within the frame of analysis three different possible positions of the actuator are taken into account – i.e. at hub, at the middle and at free end (see the figure below) in order to compare the system's response to the harmonic excitation from the actuator.

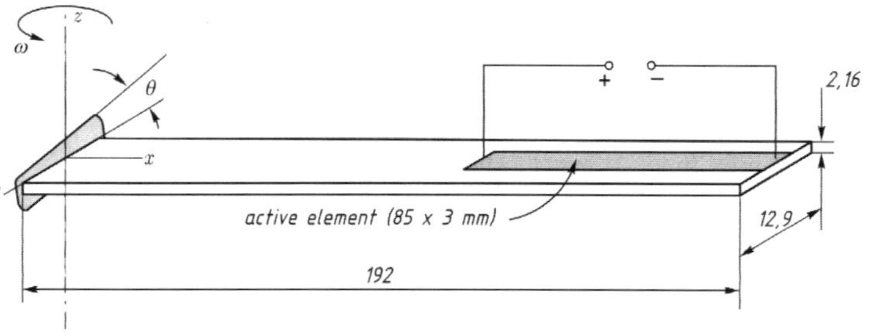

Figure 1. Rotating beam with single PZT element at free end

3. Numerical studies

The composite beam is made of unidirectional glass fibre tape and epoxy Prime 20; average composite density ρ = 1.85g/cm^3, Young's modulus along the fibres E_1 = 36 250 MPa, transversal Young's modulus E_2 = 6 662 MPa, shear moduli $G_{12} = G_{13} = G_{23}$ = 2 446 MPa and Poisson's ratio ν_{12} = 0.285. The layers of the composite are set in the following order: 0°/90°/+45°/–45°/+45°/90°/0° (according to the *Ox* axis pointing side edgewise). A Rayleigh damping model is assumed for the composite material with coefficients α = 15.351, β = 2.172e-5.

The MFC patch material data used in the numerical analysis corresponds to an M-8503-P1 element by Smart Material Corp. For the numerical calculations the following electric and elastic properties are used: isotropic ferroelectric material with constant permittivity $\xi = 8 \times 10^9$ F/m and mean value of piezoelectric coefficient $d_{33} = 400 \times 10^{-12}$ m/V; elastic orthotropic material having Young moduli $E_1 = E_2$ = 15 857 MPa, E_3 = 30 336 MPa, shear moduli $G_{12} = G_{13} = G_{23}$ = 5 515 MPa and Poisson ratios $\nu_{12} = \nu_{13} = \nu_{23}$ = 0.31; ρ = 5.40g/cm^3. Despite the modular structure of the transducer under consideration the piezo-elements have not been modelled in micro scale – a substitute body made of orthotropic, homogenous piezoelectric material is proposed.

The numerical model of a beam is developed in ABAQUS ver. 6.10 software. Specimen is defined as a lamina type according to the layup technique and meshed by linear shell elements with reduced integration. The actuator domain is modelled with second order continuum piezo-elements C3D20E. The model of the piezo-element is 'glued' to the host structure by TIE constraints, which results in joining appropriate DOF of both bodies.

The boundary conditions of the electrical part used for modal analysis are at one pole zero electrostatic potential and the other pole is free. Three different pitch angles ($\theta = 0°, 45°, 90°$) and at three rotating speeds ($\omega = 100$ rpm, 6000 rpm, 12000 rpm) are studied. Mode shapes and frequencies of natural vibrations are recorded.

Following these results a steady-state response of the system to a continuous harmonic excitation by piezo actuator is studied. For the current analysis a direct steady-state method is used which is a perturbation procedure, where the perturbed solution is obtained by linearization about the current base state. For the calculation of the base state the system may exhibit material and geometrical nonlinear behaviour (centrifugal force). Moreover a direct calculation option is used because the modal based procedures do not adequately transform the charge loads into modal loads [5].

4. Results

Results of the modal analysis of the beam with the MFC patch are given in Table 1 and in Figure 1. In the case of a 45° pitch angle strong coupling of bending modes is observed – see Figure 1. Stressing of the structure caused by centrifugal loading is confirmed by increased natural frequencies at higher rotating speeds. Moreover a 'load stiffness' effect caused by the change in direction of centrifugal loading if the vibration causes motion in the plane normal to the axis of rotation is observed – 0° pitch angle vs 90°.

Table 1. First two natural frequencies [Hz] and modes of the rotating beam with an MFC patch

Hub

0°			45°			90°		
100 rpm	6000 rpm	12000 rpm	100 rpm	6000 rpm	12000 rpm	100 rpm	6000 rpm	12000 rpm
39,28 xz	110,62 xz	178,31 xy	39,26 xz	79,87 xz^1	109,39 xz^1	39,24 xz	47,55 xz	41,84 xz
166,86 xy	170,37 xy	204,21 xz	166,87 xy	186,75 xy^2	247,97 xy^2	166,87 xy	197,50 xy	267,75 xy

Middle

0°			45°			90°		
100 rpm	6000 rpm	12000 rpm	100 rpm	6000 rpm	12000 rpm	100 rpm	6000 rpm	12000 rpm
35,99 xz	107,68 xz	171,72 xy	35,98 xz	75,65 xz^1	104,41 xz^1	35,96 xz	39,91 xz	30,44 xz
165,74 xy	167,75 xy	202,31 xz	165,74 xy	184,37 xy^2	243,82 xy^2	165,74 xy	195,21 xy	263,38 xy

Free end

0°			45°			90°		
100 rpm	6000 rpm	12000 rpm	100 rpm	6000 rpm	12000 rpm	100 rpm	6000 rpm	12000 rpm
34,16 xz	108,92 xz	171,17 xy	34,14 xz	77,08 xz^1	109,33 xz^1	34,12 xz	43,13 xz	48,32 xz
160,85 xy	163,84 xy	205,76 xz	160,85 xy	180,96 xy^2	244,16 xy^2	160,85 xy	191,87 xy	263,03 xy

xz – bending in xz plane, xy – bending in xy plane, xz^1 – coupled xz-xy bendings xz plane dominant, xy^2 – coupled xz-xy bendings xy plane dominant.

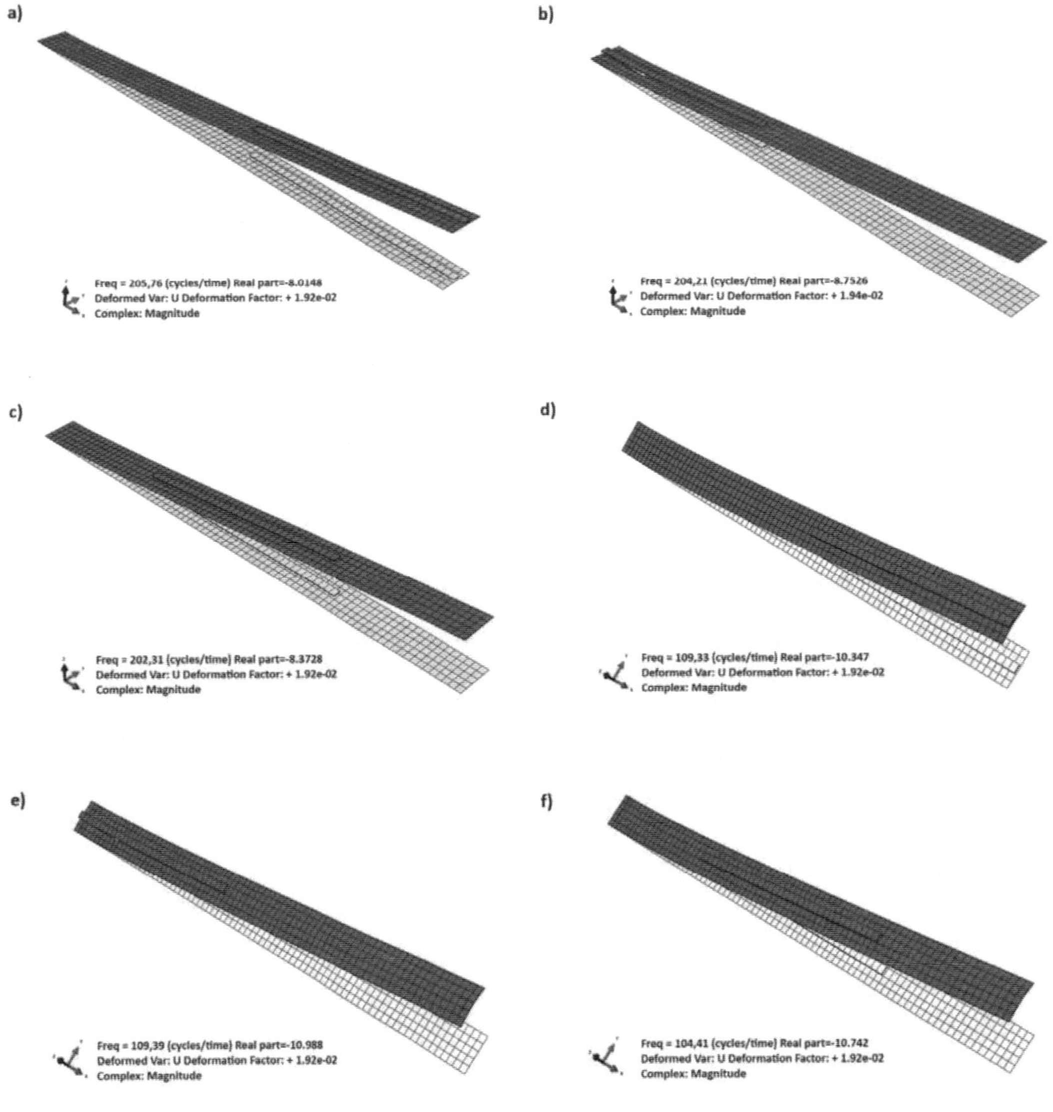

Figure 2. Exemplary mode shapes of the beam with an MFC actuator; pitch angle 0° (cases *a* to *c*); pitch angle 45° (cases *d* to *f*) – coupled *xz* and *xy* bending motions are clearly visible.

At the next stage a series of tests to examine the steady-state response of the system to the continuous excitation by piezo actuator are performed. The system is excited by a harmonic voltage signal $U(t) = U_0 \cos(\Omega t)$, $U_0 = 500$ V applied to the PZT poles – see Figure 1.

Frequency response spectra are depicted for all the cases in Figure 3. In Figures 3a and 3b the amplitude of the beam tip displacement in case of 0° and 90° pitch angle considering only *xz* motion is presented. This motion is excited directly by the active element. Careful study of these two cases indicates that the active element is more efficient when it is placed at the hub, and as long as the rotation speed is increasing the effectiveness of the active element decreases. The frequency response spectra in case of a 45° pitch angle for *xy* and *xz* bending are shown in Figure 3c and 3d respectively.

Although the active element excites xz bending motion directly – due to coupling – one can identify excitation also of xy motion (Figure 3c) upon rotation.

Therefore the rational placement of an actuator is at the hub where the MFC patch is most effective, and as long as the rotation speed is increasing in the case of direct excited motion the effectiveness of the active element is decreasing. In case of xy motion (Figure 3c) which is indirectly excited the phenomenon is opposite. This means the increase in rotation speed results in higher effectiveness of the active element due to stronger coupling in the equations of motion.

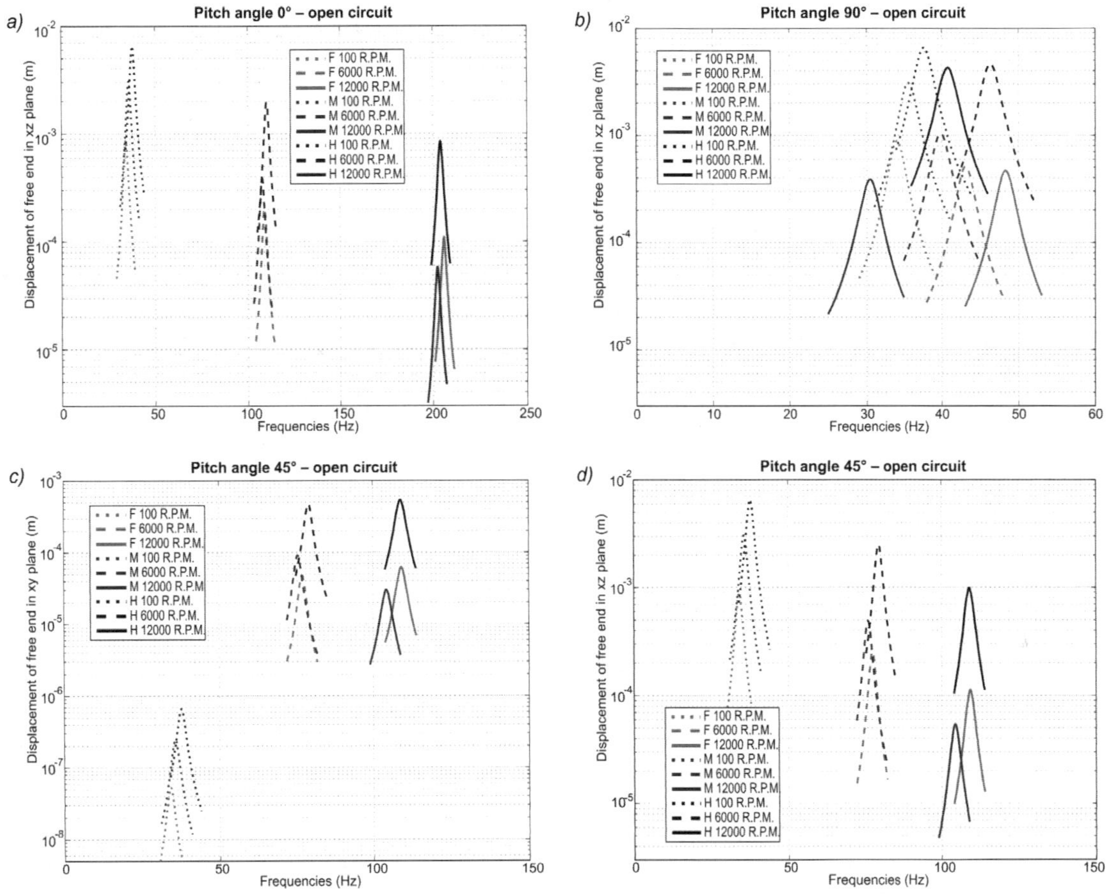

Figure 3. Steady state response of the system: a) pitch angle 0°, b) pitch angle 90°, c) pitch angle 45° – response in xy plane, d) pitch angle 45° – response in xz plane; notation: F – patch at the beam's free end, M – patch in the middle, H – patch positioned at the hub.

6. Conclusions

In the presented paper properties of a light, thin rotating composite beam with an active element are examined. Parametric studies aimed at finding the rational location of this patch with respect to maximum structural response to actuator excitation are summarized. Different rotating speeds and different orientations of the beam's cross-section with respect to the plane of rotation are considered.

The most important conclusions from the research undertaken are the following:

- In all the analysed cases locating the MFC actuator at the hub results in the highest system's response in first bending mode.

- While the rotating speed is increased the response from the system gets smaller because of an increase in the stiffness due to centrifugal force. Moreover, the softening effect of 'load stiffness' caused by the change in direction of centrifugal loading, if the vibration causes motion in the plane normal to the axis of rotation, is confirmed.
- Setting the orientations of the beam's cross-section with respect to the plane of rotation to 45° results in coupled bending motions – in-plane (*xz*) and out-of-plane (*xy*) directions.
- Since the actuator excites the beam in the *xz* plane the observed coupled response in the *xy* plane is significant only for the rotating system. Also, as opposed to directly excited motions the effectiveness of the active element in *xy* motion is increasing with an increase of the rotation.

Acknowledgements
The research leading to these results has received funding from European Union Seventh Framework Programme (FP7/2007-2013), FP7 – REGPOT – 2009 1, under grant agreement No: 245479. The support of Polish Ministry of Science and Higher Education – Grant No. 1471-1/7.PRUE/2010/7 – is also acknowledged by the first and third authors

References
[1] Liu JJ, Liaw BM 2004 Efficiency of Active Control of Beam Vibration Using PZT Patches. Proceedings of the SEM X International Congress & Exposition on Experimental and Applied Mechanics (Costa Mesa, USA (CA), June 7-10, 2004, Paper 385)
[2] Fahro F, Wang Y 1997 Optimal location of Piezoceramic Actuators for Vibration Suppression of a Flexible Structure. Proceedings of the 36th Conference on Decision & Control (San Diego, USA (CA), December 1997, pp. 1966-1971)
[3] Crespo Da Silva M 1998 A comprehensive analysis of the dynamics of a helicopter rotor blade. *International Journal of Solids and Structures* 35 (7-8) pp. 619–635
[4] Yoo HH, Kim SK, Inman DI 2002 Modal Analysis of rotating composite cantilever plates Journal of Sound and Vibration 258(2), 233–246
[5] –: ABAQUS User's Manual (6.10), Dassault Systèmes, 2010

Nonlinear stress analysis of plates under thermo-mechanical loads

P Khazaeinejad[1], A S Usmani[1] and O Laghrouche[2]

[1] School of Engineering, The University of Edinburgh, Edinburgh EH9 3JL, UK
[2] School of the Built Environment, Heriot-Watt University, Edinburgh EH14 4AS, UK

E-mail: p.khazaeinejad@ed.ac.uk

Abstract. The aim of the present work is to carry out a simplified mathematical modelling for nonlinear stress analysis of plates under temperature changes and mechanical transverse loads. The material properties of the plate are proposed to be temperature-dependent. The geometrically nonlinear plate theory is employed to understand the stress distribution due to thermo-mechanical loads. A set of coupled nonlinear partial differential equations are solved using harmonic series expansion to find the static responses. Two boundary conditions are considered for simply supported plates, namely, movable edges and immovable edges. The accuracy of the results is checked by comparing with the output of other solution methods.

1. Introduction

In plate theories the small deformation assumption is no longer acceptable when the deflection magnitude is of the same order as the plate thickness. Thus, large deformation theory is introduced which includes the effect of membrane forces and coupling between the axial reactions and the transverse deformation of the plate. In response to the need for new structural materials, materially nonlinear theory was also developed by introducing nonlinear constitutive relations for stresses and strains [1-3]. Later on, both geometrically and materially nonlinear theories were employed to analyse plates with large deflections, but the most common approach was using approximate solutions for plates with *movable* simply supported edges. For movable (stress-free) boundary condition the supported edges are free to move whereas the out-of-plane displacement is fixed. However, for *immovable* boundary condition, equivalent axial reaction loads could be defined to prevent both in-plane and out-of-plane displacements along the edges [4]. To the best authors knowledge few attempts were made to find a simplified mathematical procedure for large deflection analysis of rectangular plates under thermo-mechanical loads [5, 6].

In this paper an analytical method is developed to present a robust nonlinear analysis for plates under thermo-mechanical loads. Geometric nonlinearity theory is used to establish the nonlinear governing partial differential equations. The plate properties are assumed to vary with temperature according to the Eurocode [7]. The accuracy of the present methodology is checked by comparing the results with some exiting results for large deflections of plates.

2. Mathematical formulations

The inverse strain-stress relations for plane stress including the thermal strains are

$$E\epsilon_{xx} = (F_{,yy} - \nu F_{,xx}) + \alpha E \Delta T, \quad E\epsilon_{yy} = (F_{,xx} - \nu F_{,yy}) + \alpha E \Delta T, \quad E\gamma_{xy} = -2(1+\nu)F_{,xy} \quad (1)$$

where E, F, ν, α and ΔT are modulus of elasticity, stress function, Poisson's ratio, coefficient of thermal expansion and temperature change, respectively. The compatibility equation formed by strain components can be expressed by

$$\epsilon_{xx},_{yy} + \epsilon_{yy},_{xx} - \gamma_{xy},_{xy} = w,_{xy}^2 - w,_{xx} w,_{yy} \tag{2}$$

Substitution from relations (1) into (2) and then integration over the thickness leads to

$$h\,\nabla^4 F + N^T,_{xx} + N^T,_{yy} - Eh\left(w,_{xy}^2 - w,_{xx} w,_{yy}\right) = 0 \tag{3}$$

where h is the plate thickness and N^T is the thermal stress resultant defined by

$$N^T = E\alpha \int_{-h/2}^{h/2} \Delta T dz \tag{4}$$

The other nonlinear partial differential equation is given by

$$D\,\nabla^4 w - h\left(F,_{yy} w,_{xx} - 2F,_{xy} w,_{xy} + F,_{xx} w,_{yy}\right) + M^T,_{xx} + M^T,_{yy} - q(x,y) = 0 \tag{5}$$

where q is the applied load per unit area and M^T is the thermal moment resultants expressed by

$$M^T = \frac{E\alpha}{1-\nu} \int_{-h/2}^{h/2} \Delta T z dz \tag{6}$$

The following functions satisfy the two governing equations (3) and (5) for a plate with length a, width b and simply supported edges

$$(w, q, N^T, M^T) = \sum_{m=1}^{N} \sum_{n=1}^{N} (w_{mn}, q_{mn}, N^T_{mn}, M^T_{mn}) \sin(\alpha_m x) \sin(\gamma_n y) \tag{7}$$

$$F = -2P_x y^2 - 2P_y x^2 + \sum_{m=1}^{N} \sum_{n=1}^{N} F_{mn} \sin(\alpha_m x) \sin(\gamma_n y) \tag{8}$$

where

$$(N^T_{mn}, M^T_{mn}, q_{mn}) = \frac{4\left(-1+(-1)^m\right)\left(-1+(-1)^n\right)}{mn\pi^2}(N^T, M^T, q) \tag{9}$$

Here $\alpha_m = m\pi/a$ and $\gamma_n = n\pi/b$. Two different in-plane boundary conditions are considered for simply supported plate, SS1 where $w = w,_{xx} = 0$ and SS2 where $u = v = w = w,_{xx} = 0$. P_x and P_y are the tensile loads defined at $x = 0, a$ and $y = 0, b$, respectively, for SS2 case which can be obtained by the following relations. The purpose is that the plate edges should be restricted to move along the x and y directions. Then, the axial displacements can be expressed by

$$u = \int_0^a \left\{ \frac{1}{E}(F_{,yy} - \nu F_{,xx}) - \frac{1}{2} w_{,x}^2 + \alpha \Delta T \right\} dx \tag{10}$$

$$v = \int_0^b \left\{ \frac{1}{E}(F_{,xx} - \nu F_{,yy}) - \frac{1}{2} w_{,y}^2 + \alpha \Delta T \right\} dy \tag{11}$$

Substitution from relations (7) and (8) into the above relations gives

$$u = \frac{-4aP_x}{E} + \frac{4a\nu P_y}{E} - \frac{aF_{mn}[\nu\alpha_m^2 + \gamma_n^2]\left[-1+(-1)^m\right]\left[-1+(-1)^n\right]}{Emn\pi^2} + a\alpha\Delta T - \frac{a}{8}\alpha_m^2 w_{mn}^2 \tag{12}$$

$$v = \frac{-4bP_y}{E} + \frac{4b\nu P_x}{E} - \frac{bF_{mn}[\alpha_m^2 + \nu\gamma_n^2]\left[-1+(-1)^m\right]\left[-1+(-1)^n\right]}{Emn\pi^2} + b\alpha\Delta T - \frac{b}{8}\gamma_n^2 w_{mn}^2 \tag{13}$$

where for immovable edges $u = v = 0$ and the tensile loads will be expressed by

$$P_x = -\frac{\gamma_n^2 F_{mn}\left[-1+(-1)^m\right]\left[-1+(-1)^n\right]}{mn\pi^2} - \frac{Ew_{mn}^2}{8(1-\nu^2)}(\alpha_m^2+\nu\gamma_n^2) + \frac{N^T}{h(1-\nu)} \quad (14)$$

$$P_y = -\frac{\alpha_m^2 F_{mn}\left[-1+(-1)^m\right]\left[-1+(-1)^n\right]}{mn\pi^2} - \frac{Ew_{mn}^2}{8(1-\nu^2)}(\nu\alpha_m^2+\gamma_n^2) + \frac{N^T}{h(1-\nu)} \quad (15)$$

Substituting relations (7) and (8) into equations (3) and (5) leads to

$$\begin{aligned}&\left(\left(\alpha_m^2+\gamma_n^2\right)^2 F_{mn}h - \left(\alpha_m^2+\gamma_n^2\right)N_{mn}^T\right)\sin(\alpha_m x)\sin(\gamma_n y)\\ &\quad -Eh\left\{\left(\alpha_m\gamma_n w_{mn}\cos(\alpha_m x)\cos(\gamma_n y)\right)\left(\alpha_r\gamma_s w_{rs}\cos(\alpha_r x)\cos(\gamma_s y)\right)\right.\\ &\quad \left.-\left(\alpha_m^2 w_{mn}\sin(\alpha_m x)\sin(\gamma_n y)\right)\left(\gamma_s^2 w_{rs}\sin(\alpha_r x)\sin(\gamma_s y)\right)\right\} = 0 \end{aligned} \quad (16)$$

$$\begin{aligned}&\left(D\left(\alpha_m^2+\gamma_n^2\right)^2 - 4hP_x\alpha_m^2 - 4hP_y\gamma_n^2\right)w_{mn}\sin(\alpha_m x)\sin(\gamma_n y)\\ &\quad -h\left\{\left(\gamma_s^2 F_{rs}\sin(\alpha_r x)\sin(\gamma_s y)\right)\left(\alpha_m^2 w_{mn}\sin(\alpha_m x)\sin(\gamma_n y)\right)\right.\\ &\quad -2\left(\alpha_r\gamma_s F_{rs}\cos(\alpha_r x)\cos(\gamma_s y)\right)\left(\alpha_m\gamma_n w_{mn}\cos(\alpha_m x)\cos(\gamma_n y)\right)\\ &\quad \left.+\left(\alpha_r^2 F_{rs}\sin(\alpha_r x)\sin(\gamma_s y)\right)\left(\gamma_n^2 w_{mn}\sin(\alpha_m x)\sin(\gamma_n y)\right)\right\}\\ &\quad -\left(\left(\alpha_m^2+\gamma_n^2\right)M_{mn}^T - q_{mn}\right)\sin(\alpha_m x)\sin(\gamma_n y) = 0 \end{aligned} \quad (17)$$

Using the expansion theorem we obtain

$$\begin{aligned}&\left(\alpha_m^2+\gamma_n^2\right)^2 hF_{mn} - \left(\alpha_m^2+\gamma_n^2\right)N_{mn}^T\\ &\quad -4Eh\left(\alpha_m\gamma_n\alpha_r\gamma_s w_{mn}w_{rs}\zeta_{mnrs} - \alpha_m^2\gamma_s^2 w_{mn}w_{rs}\eta_{mnrs}\right) = 0 \end{aligned} \quad (18)$$

$$\begin{aligned}&D\left(\alpha_m^2+\gamma_n^2\right)^2 w_{mn} - 4h(\alpha_m^2 P_x+\gamma_n^2 P_y)w_{mn} - \left(\alpha_m^2+\gamma_n^2\right)M_{mn}^T - q_{mn}\\ &\quad -4hF_{rs}w_{mn}\left(\alpha_m^2\gamma_s^2\eta_{mnrs} - 2\alpha_m\gamma_n\alpha_r\gamma_s\zeta_{mnrs} + \gamma_n^2\alpha_r^2\eta_{mnrs}\right) = 0 \end{aligned} \quad (19)$$

where

$$\zeta_{mnrs} = \frac{(-m+2m(-1)^{2m+r}-m(-1)^r)(-n+2n(-1)^{2n+s}-n(-1)^s)}{\pi^2(4m^2-r^2)(4n^2-s^2)} \quad (20)$$

$$\begin{aligned}\eta_{mnrs} &= \frac{(-2m^2+2m^2(-1)^r-(-1)^r r^2+r^2(-1)^{2m+r})}{rs\pi^2(4m^2-r^2)(4n^2-s^2)}\\ &\quad + \frac{(-2n^2+2n^2(-1)^s-(-1)^s s^2+s^2(-1)^{2n+s})}{rs\pi^2(4m^2-r^2)(4n^2-s^2)} \end{aligned} \quad (21)$$

For uncoupled term approximation ($r = m$ and $s = n$), the above equations will be reduced to a cubic nonlinear equation as follows

$$\begin{aligned}&\frac{32Eh\alpha_m^4\gamma_n^4 H_{mn}^2}{(\alpha_m^2+\gamma_n^2)^2}w_{mn}^3 + \left(D\left(\alpha_m^2+\gamma_n^2\right)^2 - 4h(\alpha_m^2 P_x+\gamma_n^2 P_y) + \frac{8N_{mn}^T H_{mn}\alpha_m^2\gamma_n^2}{(\alpha_m^2+\gamma_n^2)}\right)w_{mn}\\ &-\left(\alpha_m^2+\gamma_n^2\right)M_{mn}^T - q_{mn} = 0 \end{aligned} \quad (22)$$

where

$$H_{mn} = \frac{-1+2(-1)^m+2(-1)^n-(-1)^{3m}-(-1)^{3n}-3(-1)^{m+n}+(-1)^{3m+n}+(-1)^{m+3n}}{3mn\pi^2} \tag{23}$$

The membrane stresses may be obtained by the stress function F as follows

$$\sigma_{xx} = F_{,yy} = \sum_{m=1}^{N}\sum_{n=1}^{N}\Big(-4P_x - \gamma_n^2 F_{mn}\sin(\alpha_m x)\sin(\gamma_n y)\Big) \tag{24}$$

$$\sigma_{yy} = F_{,xx} = \sum_{m=1}^{N}\sum_{n=1}^{N}\Big(-4P_y - \alpha_m^2 F_{mn}\sin(\alpha_m x)\sin(\gamma_n y)\Big) \tag{25}$$

$$\tau_{xy} = -F_{,xy} = -\sum_{m=1}^{N}\sum_{n=1}^{N}\alpha_m\gamma_n F_{mn}\cos(\alpha_m x)\cos(\gamma_n y) \tag{26}$$

The extreme-fiber bending stresses are expressed by

$$\acute{\sigma}_{xx} = \sum_{m=1}^{N}\sum_{n=1}^{N}\frac{Eh(\alpha_m^2+\nu\gamma_n^2)}{2(1-\nu^2)}w_{mn}\sin(\alpha_m x)\sin(\gamma_n y) \tag{27}$$

$$\acute{\sigma}_{yy} = \sum_{m=1}^{N}\sum_{n=1}^{N}\frac{Eh(\gamma_n^2+\nu\alpha_m^2)}{2(1-\nu^2)}w_{mn}\sin(\alpha_m x)\sin(\gamma_n y) \tag{28}$$

$$\acute{\tau}_{xy} = -\sum_{m=1}^{N}\sum_{n=1}^{N}\frac{Eh\alpha_m\gamma_n}{2(1+\nu)}w_{mn}\cos(\alpha_m x)\cos(\gamma_n y) \tag{29}$$

where the Einstein summation convention over repeated indices m, n, r, and s are used. For any assumed N, a set of nonlinear algebraic equations will be derived. Figure 1 describes this procedure for determining the transverse displacement of the plate.

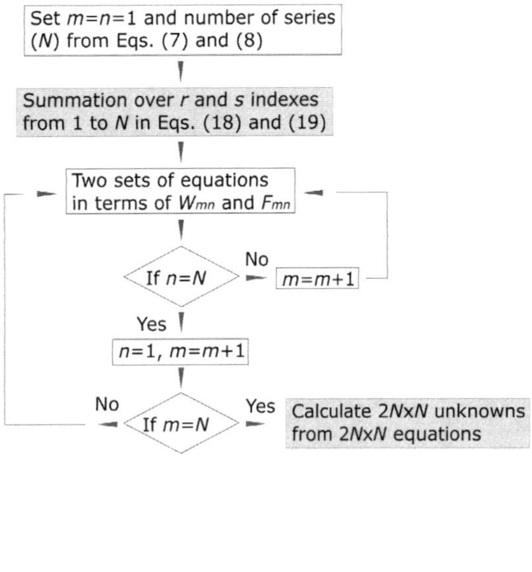

Figure 1. Flow chart of the solution programme.

3. Results and discussion

In this section, numerical examples for nonlinear analysis of plates under thermo-mechanical loads are presented to demonstrate the performance of the proposed method. The Poisson's ratio of the plate is assumed to be 0.3. Dimensionless parameters are considered as: centre deflection= w/h, load parameter= qa^4/Eh^4, stress= $(\sigma_{xx}, \acute{\sigma}_{xx}, \tau_{xy}')a^2/Eh^2$.

The centre deflection for an isotropic square plate ($E = 7.8 \times 10^6$ psi, $a = b = 1$) subjected to uniformly distributed load is firstly compared in Figure 2 with the results of a finite element (FE) solution [8]. Numerical results are presented for $N = 1$ and $N = 3$ and for two specific simply supported boundary conditions, SS1 and SS2. Solutions for one term approximation ($N = 1$) match closely with FE solutions. As mentioned before, the material properties of the plate are assumed to be temperature-dependent. For this purpose Eurocode [7] suggests a trend for reduction of elasticity modulus of carbon steel with temperature which is plotted in Figure 3. Figure 4 shows the variation of extreme-fibre bending and membrane stresses in a square plate under uniformly distributed load and linear temperature changes ($\Delta T = 0.5 + T_{,z}\, z$). The case associated with SS1 produces higher membrane and extreme-fibre bending stresses than SS2 case because of the existence of extension-bending coupling. It is observed that by increasing the value of load parameter, the plate is dominated by membrane stresses. For SS2 case, membrane stresses decrease due to the nature of the proposed stress function. The contour plots of extreme-fibre bending stresses for a square plate under thermo-mechanical loads are illustrated in Figure 5. The stress concentration for both patterns are similar to the finite element analyses.

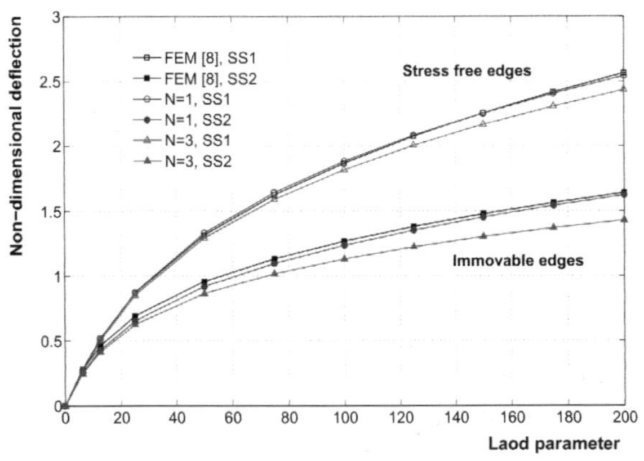

Figure 2. Convergence of the centre deflection for a square plate under uniformly distributed load.

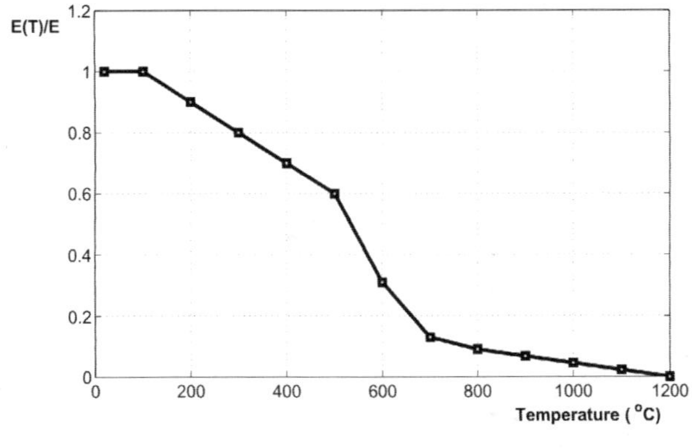

Figure 3. Reduction of elasticity modulus for carbon steel with temperature [7].

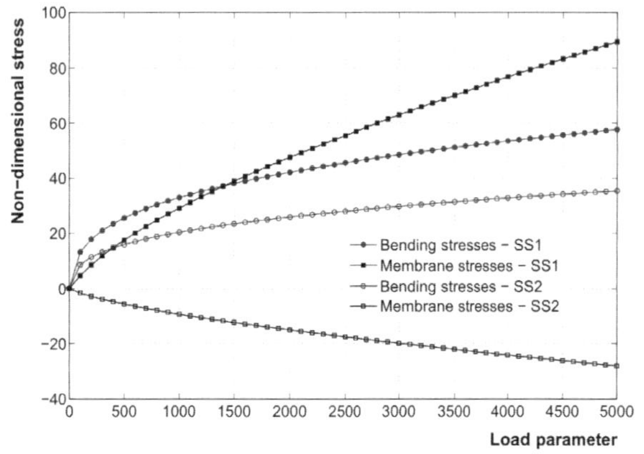

Figure 4. Bending and membrane stresses in a square plate under thermo-mechanical load ($N = 1, T_{,z} = 200$).

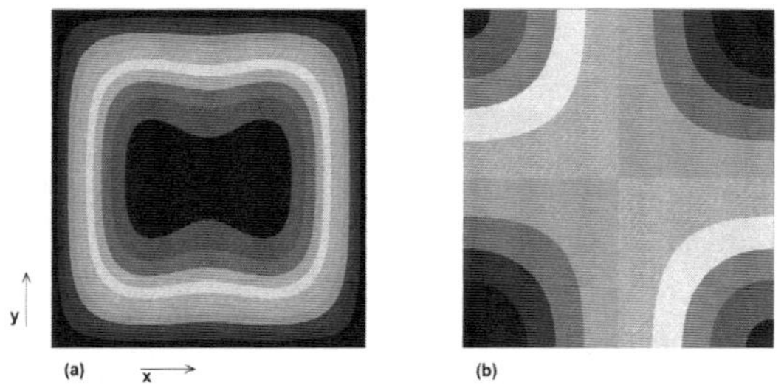

Figure 5. Contour plots of bending stresses for a square plate ($N = 3, T_{,z} = 800$, load parameter=100): (a) σ'_{xx}= 10.437; (b) τ'_{xy}= -8.517.

The present analysis has wide applications in structures under fire, particularly when elements of fire compartment boundaries such as wall panels are subjected to high temperatures and thermal gradients resulting in large displacements. Furthermore, evaluating new materials for aerospace applications could involve this kind of analysis.

4. Conclusions

A new mathematical formulation is developed for nonlinear stress analysis of plates with large displacements subjected to thermo-mechanical loads. The solution based on one term approximation was very close to those of other considered approaches whereas the solution of coupled terms provides more accurate results. The actual immovable edges can be simulated using an appropriate stress function. The results reveal that the effects of membrane action on large scale plates under thermo-mechanical loads are more than the effect of bending action.

References

[1] Wang C T 1948 *Nonlinear Large-Deflection Boundary-Value Problems of Rectangular Plates* (Washington, DC: NACA)
[2] Rushton K R and Hook P M 1974 *J. Strain Anal. Eng. Des.* **9** 178
[3] Reddy J N 2000 *Int. J. Numer. Meth. Engng.* **47** 663
[4] Timoshenko S and Woinowsky-Krieger S 1959 *Theory of Plates and Shells* 2nd ed (New York: McGraw-Hill)
[5] Wang D and El-Sheikh A L *J. Engng. Mech.* **131** 809
[6] Ohtaki S, Horiuchi T, Ohtsuka T 2008 *8th. World Cong. Compu. Mech. and 5th. Euro. Cong. Compu. Meth. Appl. Sci. Engng.* (Venice, Italy)
[7] EN 1993-1-2 2005 *Eurocode 3: Design of steel structures, Part 1-2: General rules - Structural fire design*
[8] Reddy J N 2004 *Mechanics of Laminated Composite Plates and Shells: Theory and Analysis* 2nd ed (Baca Raton: CRC Press)

Journal of Physics: Conference Series **382** (2012) 012023 doi:10.1088/1742-6596/382/1/012023

Energy harvesting from human motion: an evaluation of current nonlinear energy harvesting solutions

P L Green, E Papatheou, N D Sims

Department of Mechanical Engineering, University of Sheffield, Mappin Street, Sheffield S1 3JD, United Kingdom

E-mail: mep09plg@sheffield.ac.uk

Abstract. The concept of harvesting electrical energy from ambient vibration sources has been a popular topic of research in recent years. Recently, the realisation that the majority of ambient vibration sources are often stochastic in nature has led to a large body of work which has focused on the response of energy harvesters to random excitations - most of which approximate environmental excitations as being Gaussian white noise. Of particular interest here are recent findings which demonstrate the advantages that Duffing-type nonlinearities can introduce into energy harvesters. The aim of this paper is to identify how well these results can be applied to that of a real energy harvesting scenario. More specifically, the response of an energy harvester to excitation via human motion is studied using digital simulations in conjunction with acceleration data obtained from a human participant. As well as assessing whether Duffing-type nonlinearities can have a beneficial impact on device performance this paper aims to investigate whether Gaussian white noise can indeed be used as a good approximation for this particular ambient vibration source.

1. Introduction

Many recently published works on energy harvesting have been focused on gaining a better understanding of how energy harvesting devices respond to random excitations (often Gaussian white noise) which are thought to form a better approximation of ambient vibrations than mono-tone sinusoidal excitations. The main question which this paper will try to resolve is: how well can the results obtained from these works be applied to a real energy harvesting problem? More specifically, how well will the conclusions drawn about a device which is excited with white noise be transferred to a scenario where one is attempting to harvest electrical energy from human walking motion? The first part of this paper will focus on the response of a monostable energy harvester with nonlinearities similar to the Duffing oscillator. Of specific interest are the works of [1, 2] in which the Fokker-Planck-Kolmogorov (FPK) equation was used to show that, when excited by Gaussian white noise, Duffing type nonlinearities could be used to reduce the stroke of an energy harvesting device without effecting its power output. The second part of this paper will focus on the response of the bistable nonlinear device - specifically the idea that the useful bandwidth of such a device could be extended by having the system 'escape' from its potential well into a high energy solution [3, 4, 5] or by activating interwell dynamics [6, 7].

To this end, acceleration data obtained from the walking motion of an individual is used to excite digital simulations of different types of energy harvester, thus exploring the validity of

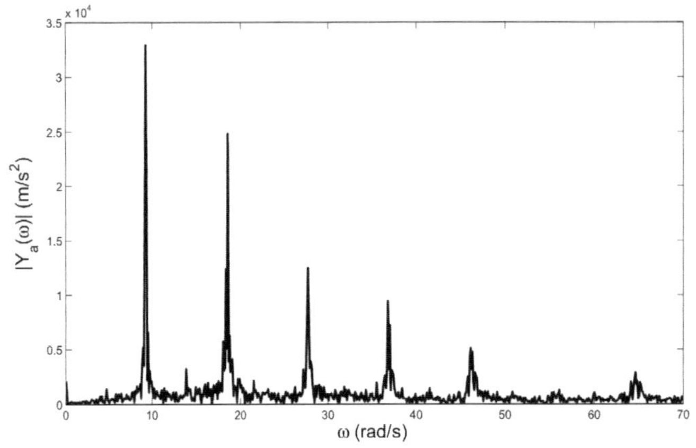

Figure 1. Fourier transform of walking excitation signal (Y_a).

extending the afore mentioned nonlinear design solutions to a real energy harvesting scenario.

2. Excitation

To acquire the desired excitation signal, a DC accelerometer was placed on a participant who was then asked to walk on a tread mill. This data was gathered as part of an investigation into the effect that harvesting energy from human motion can have on the gait of an individual [8, 9]. Consequently, for information about the test procedure the reader is directed towards [8]. This paper focuses on the motion of one of the participants, walking at a speed of 3.6 km/h.

The frequency content of the resulting excitation signal is shown in Figure 1 where it is clear that the power in the excitation is distributed over a relatively large bandwidth and that the dominant frequencies of excitation are relatively low. When one considers energy harvesters of the electromagnetic type delivering power to a load resistor, the device circuitry can be modelled as series combination of an inductor and resistor. Such a configuration is known to act as a low pass filter and as such would not have a significant impact on such a low frequency application. However, the circuitry of piezoelectric devices are often modelled as a capacitor and resistor in series - a combination which is known to create a high pass filter. Depending on the level of capacitance of the device it is worth noting that, when harvesting energy from such low frequencies, an electromagnetic device may be preferable to a piezoelectric device.

An approximation of the excitation probability density function was obtained by taking histograms of the acceleration signal and normalising such the area under the histogram was unity (Figure 2). It is immediately obvious that this type of excitation does not have a gaussian distribution. This emphasises the need to investigate specific excitation conditions rather than that of Gaussian white noise.

3. Monostable nonlinear energy harvesting

The nonlinear monostable device of interest here was first proposed by Mann and Sims [10]. The basic operating principle of this energy harvester is that a magnet is made to levitate by the opposing magnetic poles of two outer magnets which are affixed to the shell of the device. In [10] it was shown that the restoring force on the centre magnet created by such an arrangement could be accurately modelled as the force from a nonlinear spring, similar to a hardening-spring

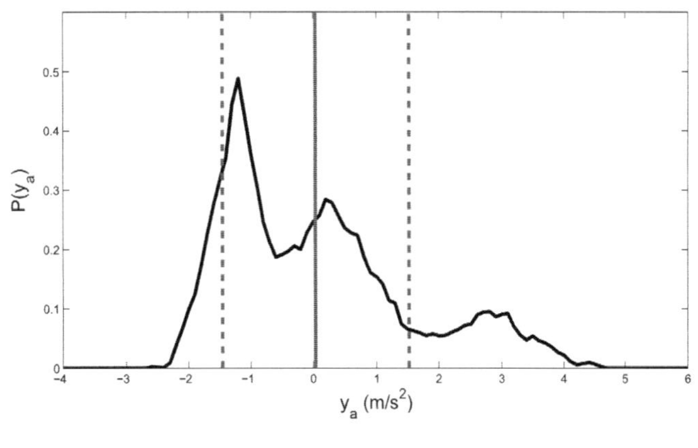

Figure 2. Excitation probability density where the mean and standard deviation is represented by the solid and dashed red lines respectively.

Duffing oscillator. Kinetic energy is converted into electrical using an electromagnetic coupling. The equation of motion of the device is:

$$m\ddot{z} + (c_m + c_e)\dot{z} + kz + k_3 z^3 = -m y_a(t) \tag{1}$$

where m is the mass of the centre magnet, c_m is the viscous damping due to mechanical losses, k is the linear stiffness, k_3 is the nonlinear stiffness and y_a represents the excitation of the base of the device due to the walking motion. The term c_e represents the damping introduced into the system as a consequence of the electromechanical coupling of the device (this is based on several assumptions which are detailed and validated experimentally in [2]).

The model was excited with the walking acceleration signal with the aim of finding whether one can indeed reduce the maximum displacement of the device in question without dramatically effecting its power output (a result that was proved for the case of Gaussian white noise excitation in [2]). It was immediately obvious that any increase in k_3 resulted in a drop in the optimum linear stiffness (a consequence of the 'skewing' effect that one can observe on the frequency response of a hardening spring Duffing oscillator). This shifting to higher frequencies made it difficult to harvest energy from the first harmonic of the excitation as, with an increase in k_3, the optimum level of k soon dropped below zero.

Figure 3 shows the variation of expected power and displacement variance with k for 2 different values of k_3. As the linear stiffness is increased (thus altering the natural frequency of the device) one can see peaks in the displacement and power when the device is tuned to the harmonics of the excitation. It can also be seen that, as stated in the previous paragraph, an increase in k_3 leads to a decrease in the optimum values of k. The key conclusion that can be drawn from Figure 3 is that, in general, any benefits that k_3 can have with regards to the displacement of the device has come at the expense of the power output. This is different to the white noise excitation case studied in [2] where it was shown that k_3 could reduce the displacement variance without effecting the power output.

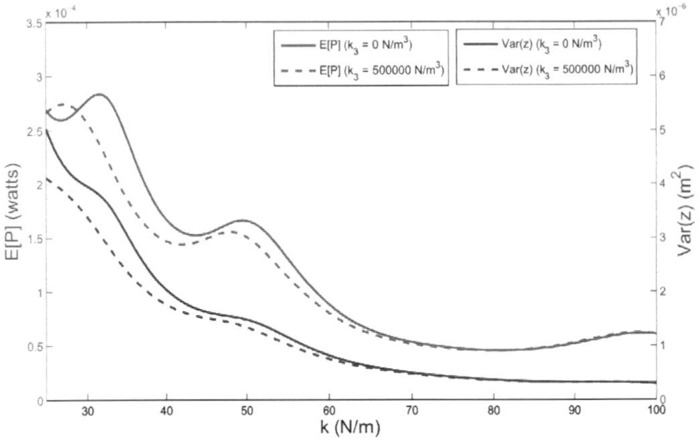

Figure 3. Variation of expected power (red) and displacement variance (blue) where $m = 0.02$, $c_m = 0.08$ and $c_e = 0.08$. Solid and dashed lines represent $k_3 = 0$ and 500000 N/m^3 respectively.

4. Bistable nonlinear energy harvesting

In this section the behaviour of bistable devices are considered. Such devices are designed such that they have two potential equilibrium points. As mentioned earlier, it is clear that power in the walking excitation signal is distributed over a relatively large range of frequencies. The purpose of this investigation then is to identify whether a bistable device can harvest energy over a larger bandwidth and therefore outperform a linear resonant device (as stated in [6, 7]). The equation of motion of such a device is very similar to that of the monostable energy harvester discussed in the previous section (equation (1)) except that k can be allowed to become negative, such that, upon the appropriate selection of k and k_3, a device with a bistable potential is created.

Figure 4 shows how the power delivered to the electrical domain varied for different values of k and k_3. Interestingly, the optimum amount of power is harvested when k is positive and k_3 approaches zero - in other words, the monstable linear device out performs the nonlinear bistable device.

The first point to note in Figure 4 is that the power output in region where $k \approx -10N/m$ is relatively small. In an effort to understand why, a phase portrait for the system with the afore mentioned value of linear stiffness was plotted for the case where $k_3 = 8000$ (Figure 5). It is clear that the system is stuck in one energy well and is unable to jump into the other. Figure 6 shows a similar plot for the case when $k = -3$. Clearly in this case, interwell dynamics have been activated although this has yielded little benefit with regards to power output. Additionally, there is also the school of thought that, even if a chaotic response could maximise the power delivered to the electrical domain, its unperiodic nature would make it difficult to collect and store the generated electrical energy [5]. Finally, Figure 7 shows the phase portrait for the linear monostable case. Clearly, periodic behavior is demonstrated. This not only outperforms that of the bistable nonlinear device but also provides an electrical signal which is preferable with regards to energy storage.

5. Discussion and conclusions

The aim of this paper was to investigate the applicability of previously developed nonlinear energy harvesting solutions to a very specific type of excitation condition. While it is clear that some of the proposed solutions were not effective in this case it is important to emphasise that

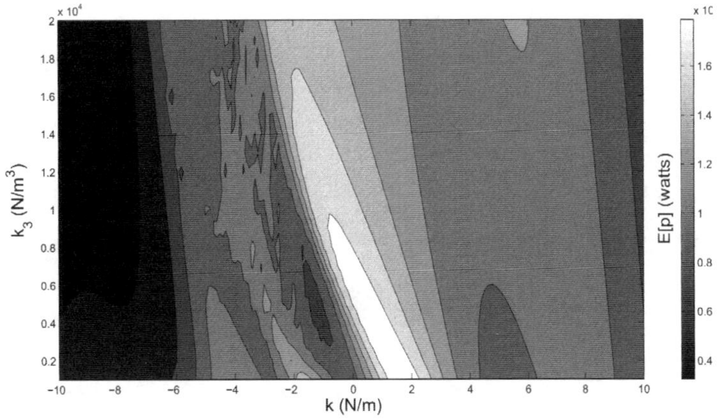

Figure 4. Variation of power delivered to the electrical domain with changes in k and k_3 ($k = 2.85$, $k_3 = 0$, $c_m = 0.08$, $c_e = 0.08$).

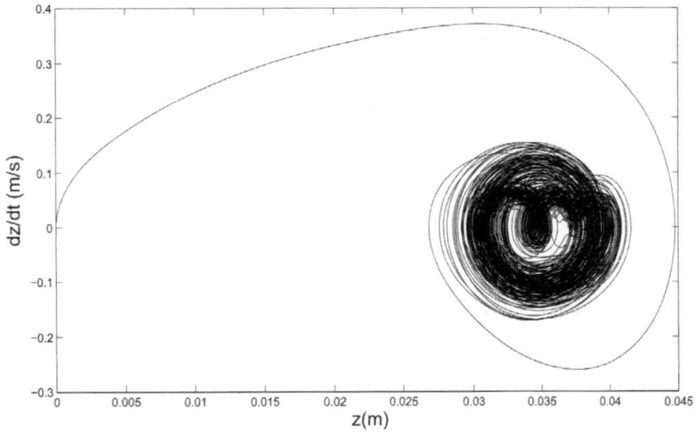

Figure 5. Phase portrait of the system with $c_m = 0.08$, $c_e = 0.08$, k = -10 and $k_3 = 8000$.

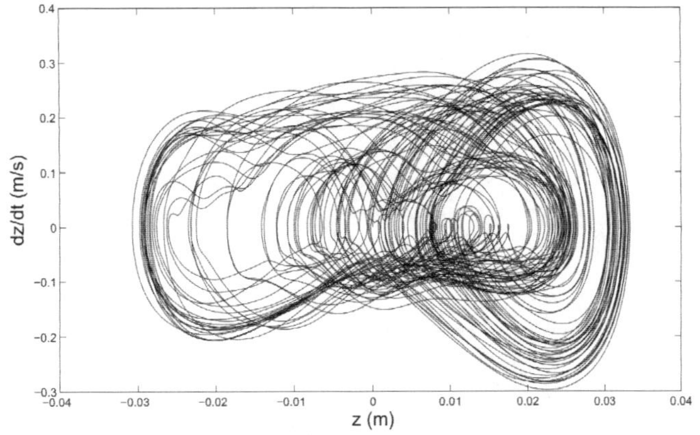

Figure 6. Phase portrait of the system with $c_m = 0.08$, $c_e = 0.08$, k = -3 and $k_3 = 8000$.

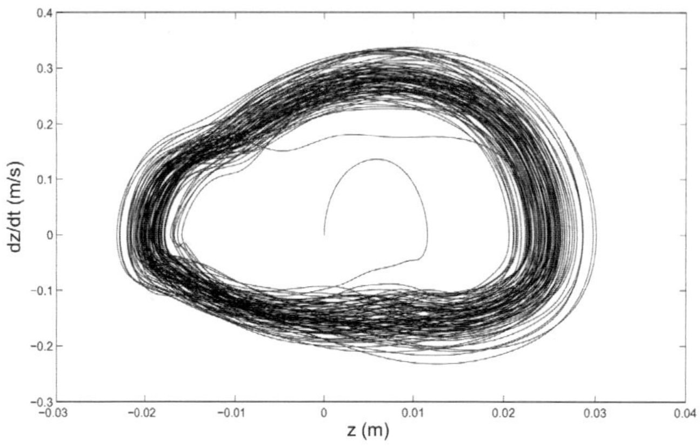

Figure 7. Phase portrait of the system with $c_m = 0.08$, $c_e = 0.08$, $k = 2.5$ and $k_3 = 0$.

the aim of this work was not to criticise the work of other authors but to highlight the difficulties that can arise when attempting to apply fairly generalised solutions to a very specific problem. Detailed here is only one of the potential ambient excitations that an energy harvesting device may be subjected to - the conclusions of the paper could have been completely different if, say, bridge excitation was considered. Essentially, this paper is intended to show that future work could be directed towards finding which types of ambient excitation will allow one to effectively use any of the afore mentioned design solutions.

This study was partly supported by the UK Engineering and Physical Sciences Research Council (EPSRC) and the Defence Science and Technology Laboratory (DSTL) through grant reference number EP/H020764/1.

[1] P L Green, K Worden, K Atallah, and N D Sims. The modelling of friction in a randomly excited energy harvester. *Proceedings of ISMA 2012, Conference on Noise and Vibration Engineering, Leuven, Belgium, 20-22 September 2010*, 2012.

[2] P L Green, K Worden, K Atallah, and N D Sims. The benefits of duffing-type nonlinearities and electrical optimisation of a mono-stable energy harvester under white gaussian excitations. *Journal of Sound and Vibration*, 0(0):0, 2012.

[3] B.P. Mann and B.A. Owens. Investigations of a nonlinear energy harvester with a bistable potential well. *Journal of Sound and Vibration*, 329(9):1215–1226, 2010. cited By (since 1996) 13.

[4] L. Gammaitoni, I. Neri, and H. Vocca. Nonlinear oscillators for vibration energy harvesting. *Applied Physics Letters*, 94(16), 2009. cited By (since 1996) 27.

[5] A. Erturk and D.J. Inman. Broadband piezoelectric power generation on high-energy orbits of the bistable duffing oscillator with electromechanical coupling. *Journal of Sound and Vibration*, 330(10):2339–2353, 2011. cited By (since 1996) 4.

[6] R. Masana and M.F. Daqaq. Relative performance of a vibratory energy harvester in mono- and bi-stable potentials. *Journal of Sound and Vibration*, 330(24):6036–6052, 2011. cited By (since 1996) 0.

[7] C.R. McInnes, D.G. Gorman, and M.P. Cartmell. Enhanced vibrational energy harvesting using nonlinear stochastic resonance. *Journal of Sound and Vibration*, 318(4-5):655–662, 2008. cited By (since 1996) 12.

[8] E. Papatheou and N.D. Sims. Developing a hardware in-the-loop simulator for a backpack energy harvester. *Journal of Intelligent Material Systems and Structures*, 23(7):827–835, 2012. cited By (since 1996) 0.

[9] E Papatheou, P L Green, V Racic, Brownjohn JMW, and N D Sims. A short investigation of the effect of an energy harvesting backpack on the human gait. *Proceedings of SPIE 2012, Smart Structures/NDE, San Diego, California*, 2012.

[10] B. P. Mann and N. D. Sims. Energy harvesting from the nonlinear oscillations of magnetic levitation. *Journal of Sound and Vibration*, 319(1-2):515–530, JAN 9 2009.

An experimental study of stochastic resonance in a bistable mechanical system

Honggang Hu[1], Kimihiko Nakano[2], Matthew P.Cartmell[3], Rencheng Zheng[4] and Masanori Ohori[5]
[1, 2, 4, 5] Institute of Industrial Science, University of Tokyo,
4-6-1 Komaba, Meguro-ku, Tokyo, 153-8505, Japan
[3] School of Engineering, University of Glasgow,
James Watt Building, Glasgow, G12 8QQ, UK

E-mail: huhg@iis.u-tokyo.ac.jp

Abstract. Potential applications for stochastic resonance have developed strongly in recent years. This paper presents a study of an application of stochastic resonance in a mechanical system. Since a linear system cannot normally exhibit stochastic resonance, a cantilever beam with an end magnet was used to constitute a bistable nonlinear oscillator. Excited by ambient random vibration, the elastic beam undergoes a modulation of the potential well by means of a periodic excitation and flips between bistable states as a result of this. By adjusting the distance between the end magnet and a fixed magnet it is possible to drive the system controllably between bistable states. An electromagnet was used to provide the periodical parametric excitation which can result in stochastic resonance. The conditions for the occurrence of stochastic resonance are also discussed in the paper. Furthermore, simulations and experimental studies have been implemented to illustrate the application. The experimental results prove that stochastic resonance can occur, and that it can be usefully applied in such a mechanical system under specific conditions.

1. Introduction

Stochastic resonance is a physical phenomenon that can occur in nonlinear systems, whereby generally feeble noise signals can be amplified and optimised by the assistance of a parametric excitation. The effect requires three basic ingredients: (a) an energetic activation barrier, (b) a weak coherent input such as a periodic signal, and (c) a source of noise that is inherent to the system [1]. Recently, stochastic resonance has been considered in several fields, such as signal processing [2], circuit experiments [3], and image visualisation [4]. However, experimental research on the application of stochastic resonance in real mechanical vibration systems is relatively rare. McInnes *et al* showed that stochastic resonance can be considered to be an effective method for enhancing vibration energy harvesting and confirmed that stochastic resonance can occur in certain mechanical vibration systems [5].

For a nonlinear mechanical system we can consider the ambient vibration as the source of noise which can excite the stochastically resonant system into a bistable nonlinear response. The presence of bistability makes the system capable of rapidly switching between the stable states under an external nonlinear force. This generates a double well potential. The probability of a transition between the two potential wells is determined by the Kramers rate, and will be small if the potential barrier between the two wells is very large. If we apply a weak parametric excitation to the system the double well potential periodically raises and lowers the potential barrier, and noise-induced hopping between the potential wells can become synchronised with the parametric excitation, leading to stochastic resonance [1]. On that basis an experimental study on stochastic resonance in a vibrating mechanical system is discussed in this paper.

2. Methodology

2.1. Modelling of a nonlinear bistable vibrating system

Based on the theory of stochastic resonance a bistable nonlinear system is conjectured, as in Figure 1 where there is a cantilever beam with an end magnet. This system is excited by ambient vibration, $Y(t)$, and there is an interaction between the elastic force of the beam and the magnetic force, generating a nonlinear response. By adjusting the distance d between the two magnets, the system can show bistability. A non-contact actuator is used to provide the source of the periodic parametric excitation.

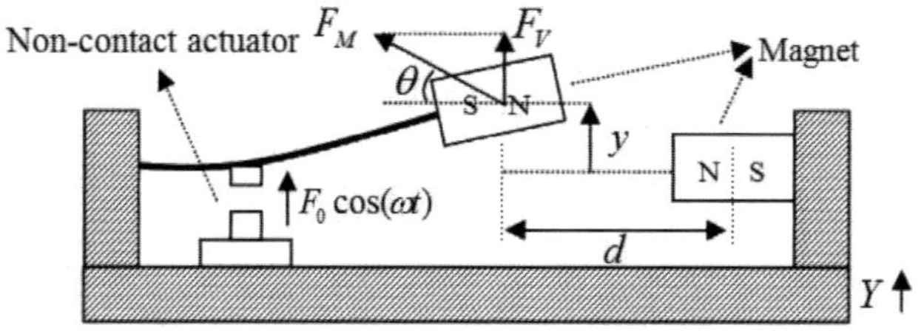

Figure 1. A bistable vibrating system

A repulsive force F_M acts between the two magnets. The vertical component, F_V, can be written in terms of F_M and expanded into a Taylor series, computed around $y = 0$ and truncated as follows,

$$F_V = F_M \sin\theta = \frac{F_M}{d} \frac{y}{\left[1 + (y^2/d^2)\right]^{1/2}} \approx \frac{F_M}{d} y - \frac{F_M}{2d^3} y^3 \tag{1}$$

It can be seen from this that F_V is a nonlinear force. For an elastic beam the stiffness can be approximated by a linear spring of stiffness k, defined by,

$$k = \frac{3EI}{l^3} \tag{2}$$

where E is Young's modulus, I is the area moment of inertia of the beam, and l is the length of the beam. The system in Figure 1 comprises an undamped nonlinear system containing a mass m and spring k, with the addition of the nonlinear source of force F_V, noting that in practice low inherent viscous damping will also be evident. The free movement of the mass through displacement y is given by the following differential equation,

$$m\ddot{y} + ky - F_V = -m\ddot{Y} \tag{3}$$

This can be re-written as,

$$m\ddot{y} + (k - \frac{F_M}{d})y + \frac{F_M}{2d^3} y^3 = -m\ddot{Y} \tag{4}$$

The potential energy $U(y)$ of the system can therefore be stated as,

$$U(y) = \frac{1}{2}(k - \frac{F_M}{d})y^2 + \frac{1}{4}\frac{F_M}{2d^3} y^4 \tag{5}$$

In equation (5) the distance d between the two magnets can be adjusted and therefore the value of $k - F_M/d$ can become negative. We symbolise $k - F_M/d$ as k - a, and $F_M/2d^3$ as b, and the potential energy is presented in Figure 2 for different $k - a$ and b values. It is clear from the Figure that the double potential wells appear when $k - a$ is negative, leading to the possibility of bistability.

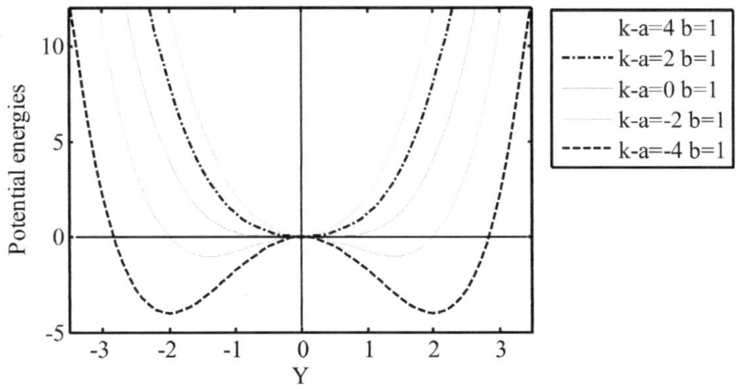

Figure 2. Potential energies $U(y)$ for different values of k-a and b

2.2. Forced vibration

In order to achieve stochastic resonance within the system of Figure 1 an electromagnet is used as a non-contact actuator. A small magnet is fitted to the beam and positioned above the electromagnet so that when a periodic voltage is applied to the electromagnet coil a periodic excitation excites the beam and this force correspondingly adjusts the height of the potential well to drive the system into stochastic resonance. We can consider h as the distance from the electromagnet to the static position of the beam, and y_0 as the vibration displacement of the magnet attachment point on the beam. Therefore, $(h+y_0)$ defines the distance between the electromagnet and the beam when the system vibrates, and we define F_E as the force between the electromagnet and the magnet, noting that this acts as the source of the parametric excitation.

When the distance between the electromagnet and magnet increases F_E will reduce, and it is clear that there is a nonlinear relationship between the force and the distance. In order to determine this relationship a calibration experiment for F_E was carried out. The result shows that F_E is in inverse proportion to the square of the distance, and so this relationship can be represented by $F_E = \lambda/(h+y_0)^2$, where $\lambda = 28$ for this physical system. For a different electromagnet and voltage the parameter λ will assume a different and unique value. Therefore, when a harmonically oscillating voltage at angular frequency ω is applied to the electromagnet the periodic force that is generated can be defined by the following,

$$\begin{aligned} F_E &= \frac{\lambda}{(h+y_0)^2}\cos(\omega t) = \frac{\lambda}{h^2}\frac{1}{(1+y_0/h)^2}\cos(\omega t) = \frac{\lambda}{h^2}\frac{(1-y_0/h)^2}{(1+y_0/h)^2(1-y_0/h)^2}\cos(\omega t) \\ &= \frac{\lambda}{h^2}\frac{1-2y_0/h+(y_0/h)^2}{\left[1-(y_0/h)^2\right]^2}\cos(\omega t) \end{aligned} \tag{6}$$

where y_0 is the peak vibration displacement of the magnet's attachment point on the beam. This has to be related to the displacement y at the location of the end mass. The static deflection formula for a cantilever can readily be used to define this relationship since this is physically similar in shape to the first bending mode. This is given by,

$$y_0(x,t) = \left(\frac{3x^2}{2l^2} - \frac{x^3}{2l^3}\right)y(l,t) \tag{7}$$

In equation (7) x is the horizontal location along the axis of the beam and l is length of the beam. For the practical experimental system this relationship is found to be $y_0 = 0.25y$. In addition, when the beam vibrates y_0 should really be a very small value compared with h which means that we can consider $y_0 << h$. So, if we neglect the nonlinear term $(y_0/h)^2$ and then apply the relationship $y_0 = 0.25y$, equation (6) can now be written as follows,

$$F_E = \frac{\lambda}{h^2}(1 - 2\frac{y_0}{h})\cos(\omega t) = (\frac{\lambda}{h^2} - \frac{\lambda y}{2h^3})\cos(\omega t) \tag{8}$$

The dynamics of the mechanism can now be defined by the following equation of motion,

$$m\ddot{y} + (k - \frac{F_M}{d})y + \frac{F_M}{2d^3}y^3 = -m\ddot{Y} + (\frac{\lambda}{h^2} - \frac{\lambda y}{2h^3})\cos(\omega t) \tag{9}$$

A nondimensional position coordinate $\delta = (F_M/2d^3)^{1/2}y$, and a free parameter $\mu = (k - F_M/d)$ can now be usefully defined. Equation (9) is therefore simplified and can be stated as,

$$\ddot{\delta} + \mu(1 + \eta\cos(\omega t))\delta + \delta^3 = N(t) \tag{10}$$

where $N(t) = -m(\ddot{Y} - \frac{\lambda}{h^2}\cos(\omega t))(\frac{F_M}{2d^3})^{1/2}$ and is considered to be the ambient excitation function, containing the noise term, and $\eta = (\frac{F_M}{8d^3})^{1/2}\frac{\lambda}{\mu h^3}$ represents the magnitude of the parametric excitation. From this the potential function can be defined as,

$$U(\delta) = \frac{1}{2}\mu(1 - \eta\cos(\omega t))\delta^2 + \frac{1}{4}\delta^4 \tag{11}$$

2.3. Analysis of stochastic resonance

It is seen from equation (11) that the periodic force can adjust the height of the potential well to push the system into stochastic resonance. The probability of this is determined by the Kramers rate, which is the characteristic escape rate from a stable state of a potential in the absence of a signal. For the system in this paper it can be defined by r as,

$$r = \frac{|\mu|}{\sqrt{2\pi}}\exp(\frac{-\mu^2}{4D}) \tag{12}$$

where D is the intensity of the ambient vibration. For example when the ambient vibration is white noise it can be defined as $N(t) = \sqrt{2D}g(t)$, where $g(t)$ is the Gauss function. It should be noted that by definition when the excitation frequency is half the maximum value of the Kramers rate, stochastic resonance will occur [6]. Contrary to this when the excitation frequency in Hz f exceeds that value then stochastic resonance cannot occur, where $f = \omega/2\pi$. Therefore the value of excitation frequency in Hz for stochastic resonance can be estimated by the following,

$$f = \frac{1}{2}r_{\lim} = \frac{1}{2}\frac{|\mu|}{\sqrt{2\pi}} = \frac{|\mu|}{\sqrt{8\pi}} \tag{13}$$

3. Experimental study

3.1. System design

The experimental system is shown in Figure 3 and mainly includes an aluminium alloy beam, two magnets, and an electromagnet to provide a periodic force. When a harmonic voltage is applied to the electromagnet coil a corresponding force is generated. The linear stiffness k of the beam is calculated to be 46.15 N/m using equation (2). Parameter μ from equation (10) changes while the beam vibrates because of the variation in F_M. A preliminary experiment showed that for $d = 4$ cm bistability ensues and the maximum peak displacement of the beam is then about 25 mm. The corresponding minimum value for F_M is then $F_{M(y=25mm)}$ = 2.83 N, therefore this gives a maximum numerical value for μ of $\mu_{\max} = (k - F_M/d) = -24.5$. For this value equation (13) predicts that the parametric excitation frequency is 2.76 Hz for stochastic resonance. When the system vibrates F_M will be the same as, or a little smaller than, $F_{M(y=25mm)}$, therefore the central value for the parametric excitation frequency should be around about, or a little bigger, than 2.76 Hz for stochastic resonance. Also, when the predominant frequency of the ambient vibration is near the resonance frequency of the beam, the vibration will also be enhanced. In order to distinguish between the effects of stochastic resonance and normal resonance it is necessary to ensure at the design stage that the parametric excitation frequency range for stochastic resonance is not coincident with the resonance frequency of the beam. The damped natural frequency of this beam was also measured by experiment and was found to be 5.87 Hz, which is well away from the value of 2.76 Hz that was calculated for stochastic resonance.

Figure 3. The experimental system

3.2. Experimental results

In the experiment, the ambient vibration was provided by a large-scale vibration generator. This generates a gaussian white noise defined by $n(t) = \sqrt{2D}g(t)$, with D = 6 mm and a bandwidth of 0 ~ 10 Hz. An amplified signal generator was used to provide a variable frequency voltage V of $3.5\cos(2\pi ft)$ to the electromagnet to generate the periodic force required for the parametric excitation, where f is the frequency in Hz. A laser vibrometer was used to measure the displacement of the beam.

Figure 4 (a) shows the displacement response of the end mass just under ambient vibration excitation and it can be seen that the beam is unable to jump between the two potential wells. Conversely, Figure 4 (b) presents the displacement response for the parametric excitation on its own, operating at 3.0 Hz. This excitation alone is unable to excite the beam into bistable vibration.

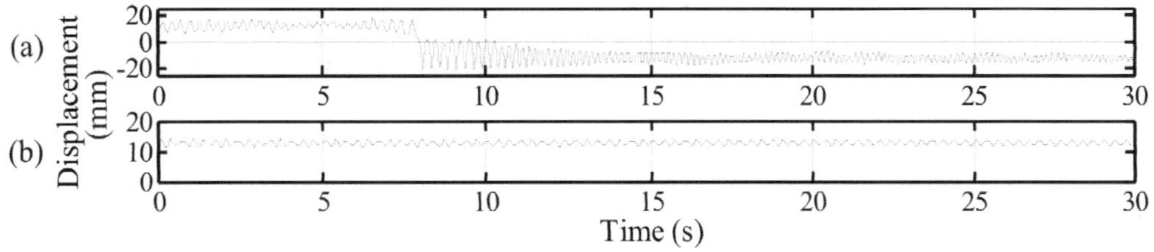

Figure 4. Displacement response under : (a) ambient vibration excitation, (b) parametric excitation

In order to observe stochastic resonance when the beam is being excited by ambient vibration, parametric excitations at different frequencies were tested, at 1.0 Hz, 2.0 Hz, 3.0 Hz, 4.0 Hz, 6.0 Hz, and 8.0 Hz, respectively. Clearly 3.0 Hz is near to the theoretically predicted parametric excitation frequency for stochastic resonance and 6.0 Hz is near to the measured damped natural frequency of the beam. Figure 5 shows the displacements of the beam response under different parametric excitation frequencies.

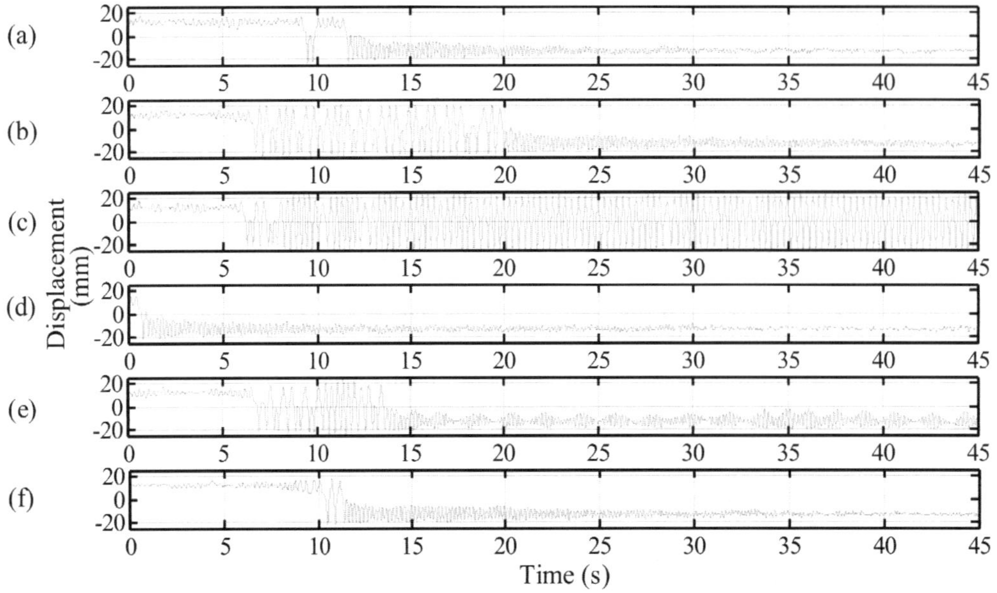

Figure 5. Beam responses under ambient vibration and parametric excitations at different frequencies: (a) 1.0 Hz, (b) 2.0 Hz, (c) 3.0 Hz, (d) 4.0 Hz, (e) 6.0 Hz, and (f) 8.0 Hz.

Figure 5 shows that when the parametric excitation frequency is 3.0 Hz the beam response is enhanced. Then, when the parametric excitation frequency is shifted to 4.0 Hz, the response amplitude decreases. A parametric excitation frequency of 6.0 Hz is near to the natural frequency of damped vibration of the beam so the response is more intense than at 4.0 Hz but weaker than at 3.0 Hz. It becomes weaker again when the parametric excitation frequency exceeds 6 Hz. Therefore it has been shown here that when the parametric excitation frequency is 3.0 Hz the response is considerably enhanced by stochastic resonance.

4. Conclusions

An experimental study was undertaken to investigate the possibility of generating stochastic resonance in real mechanical system and it has been shown that this is indeed possible under specific conditions. The frequency of the parametric excitation is confirmed as being one of the most important parameters for achieving this, and that when the parametric excitation frequency is half the maximum value of the predicted Kramers rate then stochastic resonance can be realised.

Reference

[1] L. Gammaitoni, P. Hanggi, P. Jung, F. Marchesoni, Stochastic resonance, *Reviews of Modern Physics* 70 (1) (1998) 223-287.
[2] B.E. Klamecki, Use of stochastic resonance for enhancement of low-level vibration signal componts, *Mechanical Systems and Signal Processing* 19 (2005) 223-237.
[3] W. Korneta, I. Gomes, C.R. Mirasso, R. Toral, Experimental study of stochastic resonance in a Chua's circuit operating in a chaotic regime, *Physica* D 219 (2006) 93–100.
[4] V.P. Rallabandi, Prasun Kumar Roy, Magnetic resonance image enhancement using stochastic resonance in Fourier domain, *Magnetic Resonance Imaging* 28 (2010) 1361–1373.
[5] C.R. McInnes, D.G. Gorman, M.P. Cartmell, Enhanced vibrational energy harvesting using nonlinear stochastic resonance, *Journal of Sound and Vibration* 318 (2008) 655-662.
[6] Y. Leng, Mechanism of parameter-adjusted stochastic resonance based on Kramer's rate, *Physica Sinica* 58 (8) (2009) 5916-5200.

Energy Harvesting for Aerospace Structural Health Monitoring Systems

M R Pearson, M J Eaton, R Pullin, C A Featherston and K M Holford

Cardiff School of Engineering, Cardiff University, Queen's Buildings, The Parade, Newport Road, Cardiff, CF24 3AA

pearsonmr@cardiff.ac.uk

Abstract. Recent research into damage detection methodologies, embedded sensors, wireless data transmission and energy harvesting in aerospace environments has meant that autonomous structural health monitoring (SHM) systems are becoming a real possibility. The most promising system would utilise wireless sensor nodes that are able to make decisions on damage and communicate this wirelessly to a central base station. Although such a system shows great potential and both passive and active monitoring techniques exist for detecting damage in structures, powering such wireless sensors nodes poses a problem. Two such energy sources that could be harvested in abundance on an aircraft are vibration and thermal gradients. Piezoelectric transducers mounted to the surface of a structure can be utilised to generate power from a dynamic strain whilst thermoelectric generators (TEG) can be used to generate power from thermal gradients. This paper reports on the viability of these two energy sources for powering a wireless SHM system from vibrations ranging from 20 to 400Hz and thermal gradients up to 50°C. Investigations showed that using a single vibrational energy harvester raw power levels of up to 1mW could be generated. Further numerical modelling demonstrated that by optimising the position and orientation of the vibrational harvester greater levels of power could be achieved. However using commercial TEGs average power levels over a flight period between 5 to 30mW could be generated. Both of these energy harvesting techniques show a great potential in powering current wireless SHM systems where depending on the complexity the power requirements range from 1 to 180mW.

1. Introduction

Structural health monitoring (SHM) for aerospace applications offers a real viable solution for full coverage continuous monitoring of aircraft structures. In essence it would allow for the optimal use of the structure, drastically alter maintenance regimes and minimises downtime, whilst also improving reliability and safety. Furthermore the implementation of an SHM system at the design stage could allow for more optimised structures in safety critical areas which would reduce weight [1]. This would lead to improved aircraft performance, lower fuel consumption and greater maximum range which would reduce the running costs of an aircraft.

However for these systems to become viable it would need to consist of wireless sensor nodes that can communicate information on damage wirelessly to a central base station. One option for powering a system is to harvest energy from sources that exist in abundance in the aircraft environment, two such sources are vibration and thermal energy. Research has shown that power levels in the μW range can be generated by piezoelectric devices and in the mW range by thermal electric generators (TEG) [2].

Piezoelectric materials can generate an AC voltage from an applied mechanical strain and hence harvest power from vibrations. Due to the high impedance nature of these devices they generate large voltage and a small current output. A cantilever piezoelectric beam with a proof mass generated 4mW for 1.53g at a resonant frequency of 5Hz for railway applications [3]. Microelectromechanical systems (MEMS) consisting of several piezoelectric cantilevers have been shown to generate 11nW at a resonant frequency of 35.8Hz and an acceleration level of 0.1g [4]. The above research has relied on a resonant devices, however in most applications the harvesters will be subjected to broad band vibrations. One way of producing broad band energy harvesters is to introduce non-linearities into the system [5]. Experimental studies of piezoceramic actuators for power generation found that MIDE Quickpak devices without interdigitated electrodes showed the greatest power output [6].

TEG generators utilise the Seebeck effect where a voltage can occur if a thermocouple consisting of two different semi-conductors is subjected to a temperature difference. Several of these thermocouples are joined together to form a TEG. Generally TEGs are lower impedance devices generate lower voltage and higher current levels. Experimental evaluation and modelling showed an average power of 50mW could be generated for temperature gradients that exist between the fuel tank and outer wing skin on an aircraft [7]. A TEG energy harvester was developed for elevated temperatures generated 40mW from a hot side temperature of 200 °C [8] whilst a novel TEG design for aircraft applications was able to generate 20mW for a temperature gradient of 20°C [9].

This papers reports on investigation of assessing the feasibility of both vibration and thermal energy harvesting for generating power of aerospace SHM applications.

2. Experimental Procedure and Simulation

2.1. Vibration Energy Harvesting

An investigation into the feasibility of using a piezoelectric device to provide power for a SHM system from typical aircraft vibrations was undertaken. Depending on whether the aircraft is taking off, landing or at cruising altitude vibrations occur at frequencies between 0-300Hz. The MIDE Quickpack QP10n was utilised as the harvesting device as shown in Figure 1. It consists of a rectangular sheet of piezoelectric material housed in a flexible copper clad polyimide laminate. The device was adhered using M-Bond AE-10 strain gauge adhesive manufactured by Vishay Precision group to the centre of a 300mm diameter circular disc made from aerospace grade BS1470 6082-T6 aluminium with a thickness of 0.7mm. Figure 1 shows the bespoke test rig that was developed in order to promote curvature in one direction. An LDS V201 electromagnetic shaker driven by a PA25E amplifier was placed underneath the centre of the panel. A brass connecting rod bonded to a bar was attached to the shaker in order to vibrate the panel.

(a) (b)

Figure 1. Experimental set-up for the energy harvesting device (a) and novel test rig (b)

The panel was subject to a range of vibrations between 20 and 400Hz using a sinusoidal driving signal, at each particular frequency the load resistance was altered from 1MΩ to 5kΩ in order to gain the maximum power transfer. The resulting voltage traces were recorded on an oscilloscope which recorded the peak and RMS voltage.

2.2. Thermal Electric Energy Harvesting

Certain technological advances have been made in TEG development since previous work conducted at Cardiff University [7]. Micropelt have developed a series of high-tech thin film TEGs including the MPG-D751 which was selected for the investigation into the feasibility of powering an SHM system from temperature gradients that exist on an aircraft. Micropelt has also developed a simulation tool 'mypelt' which enables the evaluation of their devices, this generates a three dimensional function of the output voltage from the device in terms of the average temperature, T_m, and the temperature gradient, ΔT.

$$V = 0.0783\ \Delta T + 3.6097 \times 10^{-8} .\ T_m \tag{1}$$

Utilising this three dimensional function and temperature data taken from thermocouples placed at various positions on an aircraft (as seen in Figure 2) it is possible to simulate output voltage of an individual TEG. In addition by knowing the internal resistance of the TEG it is then possible to derive the power output for a particular temperature scenario, integrating the instantaneous power output and dividing by the time of the simulation gives the average power levels.

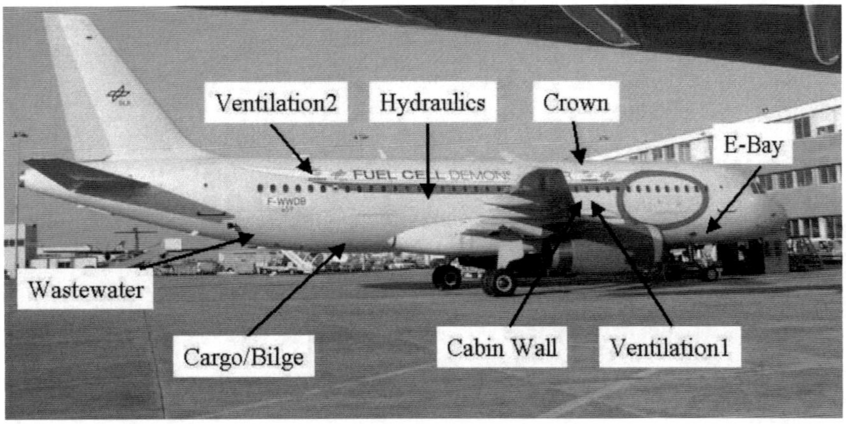

Figure 2. Locations of thermocouples used for the TEG power output simulation

3. Results and Discussion

3.1. Vibration Energy Harvesting Results and Discussion

Figure 3 shows the peak power output for a variety of input frequencies of vibrations in terms of the matched load resistance. The figure shows that for each frequency there is a peak in the power output for the device which corresponds to a particular load resistance. Also as the frequency of vibration is altered the load resistance at which the peak power is observes changes. This becomes increasingly prevalent when considering broadband nature of the frequency of vibration for aerospace applications.

Figure 4 shows the peak, RMS power and matched load resistance in order to achieve maximum power transfer and hence maximum power output. The figure shows that two narrow band peaks exist where significantly more power can be generated from the harvester. These peaks occurs at 40Hz and 300Hz where a peak power of 420 and 1100 μW and a RMS power of 220 and 440μW can be generated respectively. The peaks in power outputs are most likely due to the resonance of the device, plate and test rig itself. The figure also shows the resistance at which the maximum power generated plateaus for increasingly frequency bands this was due to not having the resolution in terms of resistance values to which the power levels were measured. However the figure does show the decrease in load resistance from 200kΩ to 5kΩ. As previous stated in order to maximise power transfer the load resistance needs to match the internal resistance of the device. If the internal resistance changes for increasing frequencies and the load is constant this can lead to inefficient transfer.

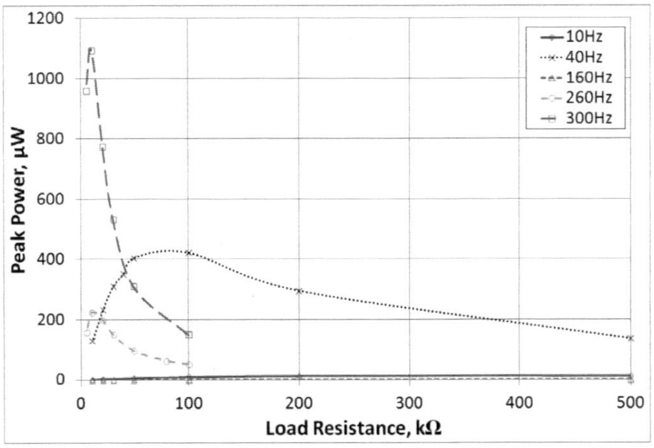

Figure 3. Peak power output from QP10n device for a variety of frequencies and load resistance

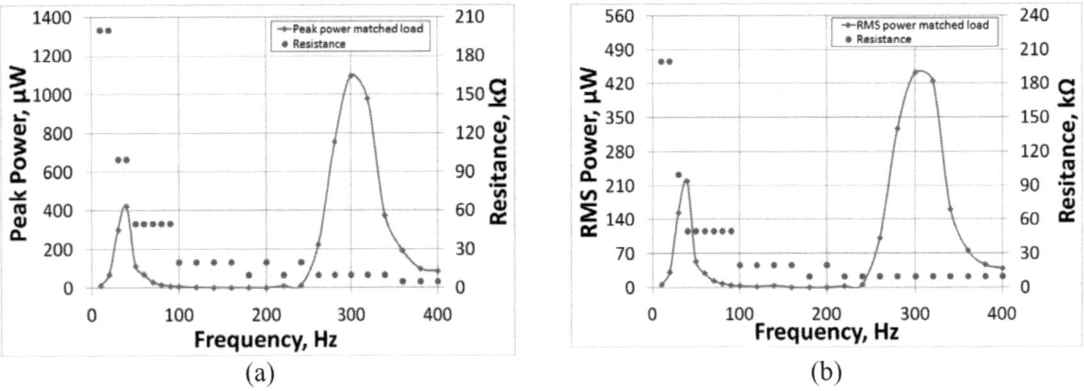

Figure 4. Peak Power (a) and RMS power (b) for the QP10n device in terms of the matched load resistance for different frequency of vibrations

3.2. Thermoelectric Energy Harvesting Results and Discussion

A comparison of the power output for an individual MPG-D751 module utilising the 'mypelt' simulation tool for various different locations on an aircraft was undertaken. Figure 5 shows example results for the temperature profiles for two hydraulics pipelines on an aircraft. The figure shows that there is a fairly constant temperature differential of around 10°C with a maximum of 20°C. This is advantageous because the temperature profiles remain relatively constant throughout the flight therefore not reaching equilibrium between the two where no further heat transfer would occur. Also due to the close proximity of these two lines to one another it would be relatively easy to mount the TEG between the two pipelines. Figure 6 shows the corresponding voltage and power levels of the TEG for that particular temperature gradient and shows a peak power of 8mW is generated at a voltage of 1.75V. Table 1 shows the resulting investigation for various different locations on an aircraft. It shows the peak temperature differential, peak power and average power over the duration of each particular flight. It shows a peak power 34.15mW can be generate by a single TEG for a temperature gradient of 40°C resulting with an average power 22.58mW. These TEGs can also be arranged in series or parallel or a combination of the both in order to the increase the power output, however careful consideration of the weight and cost of the energy harvesting system is necessary.

When comparing the power outputs from both vibrational and thermoelectric energy harvesting it has been showed that the QP10n generates power in the μW range while a TEG can generate power in the mW range. This investigation has shown that both types of harvesting are feasible for powering an SHM system where depending on the complexity the power levels can vary between 1-180mW.

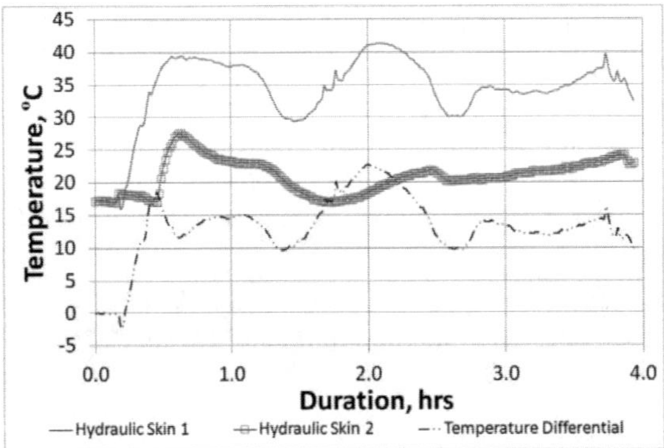

Figure 5. Temperature profiles over a particular flight for two hydraulic pipelines

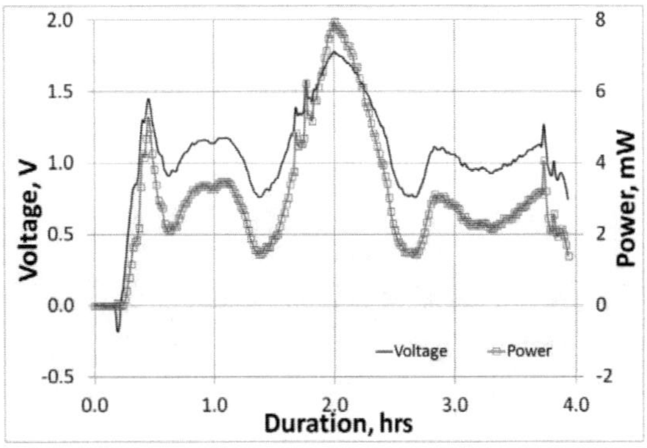

Figure 6. Simulated voltage and power profiles for the hydraulic temperature profiles

Table 1. Simulated power levels for temperature gradients on an aircraft.

	Peak Temperature Differential, °C	Peak Power, mW	Average Power, mW
Cargo skin Cargo Primary Insulation	40	34.15	22.58
Hydraulic Pipeline 1 Hydraulic Pipeline 2	20	7.97	3.07
Waste water tank Waste water ambient	15	5.46	2.99
E-bay fuselage skin E-bay primary insulation	35	18.72	6.42
Cabin wall fuselage skin Cabin wall primary insulation	30	13.36	3.97
Cabin wall fuselage skin Cabin wall secondary insulation	40	30.06	11.70

4. Conclusions

A novel test rig was developed in order to evaluate MIDE Quickpack device for the use of harvesting power from ambient vibrations on an aircraft. Experimental results showed that peak power levels from 1-1000μW and RMS power level of 0.5-400 μW could be generated. The power levels were highly dependent on the frequency and internal resistance of the device.

A numerical analysis was undertaken utilising the 'mypelt' simulation tool to determine the power output for an individual Micropelt MPG-D751 TEG for various different locations on an aircraft. Peak power levels ranging from 5.46-30.06mW and average power levels between 2.99-11.70mW were simulated from the temperature data. However careful considerations of the assumptions of the simulation are necessary firstly the analysis assumed a matched load resistance in order to achieve maximum power transfer, in reality it may not be possible to physically harvest from some of these locations and it might also not be possible to achieve the specified temperature gradient across the TEG module.

Both energy harvestings techniques shows a real potential for powering an SHM system, however in reality these devices would not be able to directly power a SHM system due to the varying nature of the sources. Therefore a further power management system would be necessary in order to provide a stable power source from the dynamic nature of the energy sources in order to power an SHM system.

References

[1] Chang, F.-K. and J.F.C. Markmiller, A New Look in Design of Intelligent Structures with SHM. Structural Health Monitoring 2006: *Proceedings of the Third European Workshop on Structural Health Monitoring*, 2006: p. 5 - 20.

[2] Vullers, R.J.M., et al., Micro Power Energy Harvesting. *Solid-State Electronics*, 2009. 53(7).

[3] Pasquale, G.De, Soma A, Fraccarollo F. Piezoelectric energy harvesting for autonomous sensor network on safety-improved railway vehicles *Proceedings of the Institution of Mechanical Engineers, Part C: Journal of Mechanical Engineering Science* **226** (4), 2012 1107-1117

[4] Liu H, Quan C, Tay C.J., Kobayashi T, Lee C. A MEMS-based piezoelectric cantilier patterned with PZT thin film array for harvesting energy from low frequency vibrations. *International Conference on Optics in Precision Engineering and Nanotechnology*. **19**. 2011,129-133

[5] Ferrari M, Ferrari V, Guizzetti M, Ando B, Baglio S, Trigona, C. Improved Energy Harvesting from Wideband Vibrations by nonlinear piezoelectric converters. *Sensors and Actuators A* **162**, 2, 2010 425-431.

[6] Sodano, H.A., J. Lloyd, and D.J. Inman, An experimental comparison between several active composite actuators for power generation. *Smart Materials and Structures*, **15**(5), 2006, 1211 - 1216.

[7] Featherston C A , Holford K M, Waring G. Thermoelectric Energy Harvesting For Wireless Sensor Systems in Aircraft. *Key Engineering Materials* **413-414** (2009) pp 487-494.

[8] Sodano H.A, Simmers G.E, Dereux R, Inman D.J.. Recharging Batteries using Energy Harvested from Thermal Gradients. *Journal of Intelligent Material systems and structures*. **18**(1) 2007 3-10

[9] Samson D, Kluge M, Becker TH, Schmid U. Wireless sensor node powered by aircraft specific thermoelectric energy harvesting. *Sensors and Actuators* A **172** (2011) 240-244.

Amplitude dependent damping from granular viscoelastics

B Darabi* and J A Rongong[1]

Department of Mechanical Engineering, The University of Sheffield, Sheffield, S1 3JD, UK
*Corresponding author
E-mail: b.darabi@sheffield.ac.uk
[1]E-mail: j.rongong@sheffield.ac.uk

Abstract. The ability of a granular medium to dissipate vibrational energy is studied at different frequencies and amplitudes. The filler comprises relatively large particles with significant viscoelasticity and is placed in a rectangular box-shaped container and vibrated perpendicular to the direction of gravity. The performance of a model based on wave behaviour that is suitable for very low amplitude vibrations is compared with discrete elements and experimental results. Frequency dependent behaviour for the viscoelastic material is taken into account. The effects of vibration amplitude on performance are considered carefully – especially at the point where particles begin to move relative to each other. One interesting finding is that internal and interface loss mechanisms are closely interrelated – reduction in internal loss increases the mobility of individual particles and therefore more energy dissipation via friction. As a result, the overall effectiveness of the granular medium is less sensitive to material and configurationally parameters than might be expected.

1. Introduction

Levels of vibration damping in hollow structures can be increased significantly by placing granular materials within the cavities. For low amplitude vibrations, the most effective fillers tend to be made from low modulus materials with high loss factor [1]. However, as the vibration amplitude increases, causing the particles to lose contact with one another temporarily, it has been found that the damping level can drop significantly [2]. At these higher amplitudes, interface friction becomes an important loss mechanism [3] allowing the use of harder materials with low internal loss such as metals. It has also been shown that while interface friction and the internal loss both contribute to the overall damping level, increases in one tend to reduce the other and vice versa [4]. This paper considers the effects of amplitude on the damping performance of a granular medium made from viscoelastic polymer particles whose properties change significantly over the frequency range considered. Experimental results are compared with those from two different numerical approaches. The first numerical approach involves Finite Element (FE) analysis to calculate the effect of internal waves in the granular medium while the second approach relies on the Discrete Element Method (DEM) to represent the behaviour of individual particles [5].

2. Granular system

The study was conducted for an open topped rectangular box with rigid walls and cavity area 190×120 mm that was filled to a nominal depth of 35 mm. The granular medium consisted of 15.1 mm diameter spheres made from a synthetic elastomer. The effectiveness of this system in dissipating vibration energy was studied for dynamic forcing applied perpendicular to gravity.

Published under licence by IOP Publishing Ltd

For the physical experiments the container was constructed using blocks of Perspex 30 mm in thickness. This provided high rigidity and good visibility as shown in Figure 1. The container was suspended using nylon line and light metal springs to simulate free boundary conditions. Excitation was provided via an electrodynamic exciter.

Figure 1: Experimental set-up showing container and particles

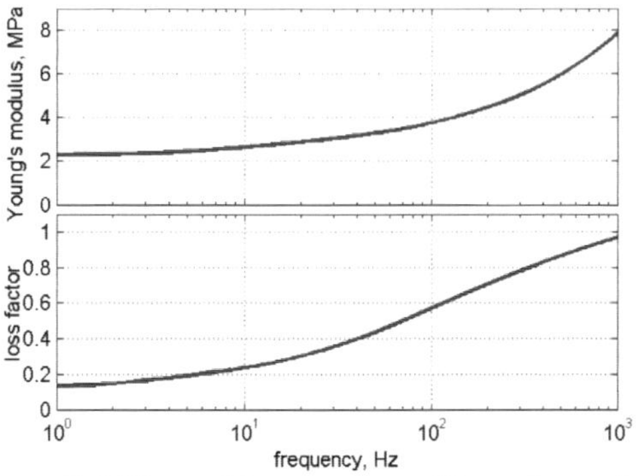

Figure 2: Effect of frequency on Young's modulus and loss factor of elastomer at 20 °C

Initial tests showed that flexible modes of the container were above 500 Hz while the rigid body modes were below 5 Hz. The container was then filled with 260 randomly placed particles and the power dissipation under sinusoidal loading measured at various amplitudes and frequencies. Following the approach used by Yang [6] and Wong [3], the average power transmitted per cycle is,

$$P_{av} = \frac{1}{2}\sum f_n v_n \cos(\alpha_n - \beta_n) \tag{1}$$

where f_n and α_n are the magnitude and phase of the force while v_n and β_n are the magnitude and phase of the velocity. The experiment was repeated with the container empty to estimate the power dissipated by boundary conditions and electronics.

3. Material properties

The material displayed significant viscoelasticity in the temperature and frequency ranges considered. The material was characterised using Dynamic Mechanical Thermal Analysis (DMTA) equipment to generate the master curves that define properties at different temperatures and frequencies. Complex modulus curves used in this work are shown in Figure 2.

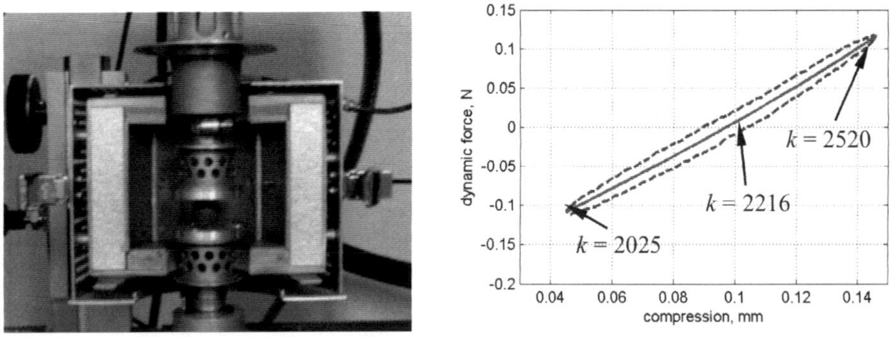

Figure 3: Hysteresis loop measurement for particle at 2.5 Hz and 20°C with 0.1 mm pre-compression

Theoretically, the axial load-deflection behaviour of an elastic sphere follows Hertz's law. The hysteresis loop of a single sphere was measured at low frequency – the test rig and result are presented in Figure 3. Note that only the dynamic component of the force is shown in the hysteresis loop. The underlying elastic curve (solid green line in Figure 3) is clearly nonlinear. The loss factor obtained by dividing the area of the hysteresis loop with the peak stored dynamic strain energy was 0.142 which is close to the value of 0.155 obtained directly from the master curve (Figure 2). The density of the elastomer was 1170 kg/m^3 and the Poisson's ratio was assumed to be approximately 0.45.

4. Model for low amplitude vibration of a granular medium

It has been shown that the damping of structural resonances can be increased significantly if standing waves are induced in an attached medium with high energy dissipation capacity [7]. The approach taken involves representing the granular medium as an equivalent solid using FE analysis.

3.1 *Equivalent homogenous solid*

A granular medium subjected to a confining pressure and vibrating at low amplitude, where particles remain in contact, can be approximated as a homogeneous solid. For uniform spheres with random packing and rough contacts Walton [8] showed that the bulk modulus could be estimated as,

$$K = \frac{1}{6}\left(\frac{3E^2\phi^2 n^2 p}{\pi^2(1-\nu_{eff}{}^2)^2}\right)^{1/3} \qquad \text{with} \qquad \nu_{eff} = \frac{\nu}{2(5-3\nu)} \tag{2}$$

where E and ν are the dynamic Young's modulus and the Poisson's ratio of the sphere material respectively, ϕ is the packing fraction, n is the number of contact points and p is the confining pressure. The effective Young's modulus and density of the equivalent homogeneous medium are,

$$E_{eff} = 3(1-2\nu)K \qquad \rho_{eff} = \rho\phi \tag{3}$$

with ρ the density of the sphere material. The confining pressure in granular medium is related to the container geometry and the distance from the free surface y. For a rectangular box filled with randomly arranged rough spheres Janssen's model [9, 2] gives,

$$p = \frac{\rho_{eff}\, g}{0.75S}\; 1 - \mathrm{e}^{-0.75Sy} \qquad \text{with } S = \frac{2(b+w)}{bw} \tag{4}$$

For a shallow box Janssen's model gives the same results as standard hydrostatic pressure. In this work, it was assumed that the structure had an amorphous arrangement with $\phi = 0.64$ and $n = 6$ [10].

3.2 *Standing waves*

Solid elements with 20 nodes were used to model the fill medium and the material was assumed to be fixed rigidly at all five contacting walls. Standing wave information (i.e. natural frequency, mode shape and effective mass for horizontal motion) was obtained for the internal cavity using a standard elastic eigenvalue extraction routine assuming an effective Young's modulus based on properties at 100 Hz. Natural frequencies were adjusted to account for frequency dependence using,

$$\omega = \omega_{100Hz}\sqrt{\frac{E_{actual}}{E_{100Hz}}} \tag{5}$$

while mode shape and effective mass were assumed not to change. Properties of typical modes are shown in Figure 4. This approach decoupled the effect of the granular system to that of many SDOF units where k_{ieff} and m_{ieff} are the effective mass and stiffness of each mode while,

$$c_{ieff} = \eta\sqrt{k_{ieff} m_{ieff}} \tag{6}$$

where η is the loss factor of the viscoelastic material at that particular frequency. As the experiment described in Section 2 represents base motion of the system, the relative displacement amplitude (where $z=x-y$ and ω and ω_n are the exciting and natural frequencies) is given by,

$$z = \frac{m_{ieff}\, y\omega^2}{k_{ieff} - \omega^2 m_{ieff} + j\omega c_{ieff}} \tag{7}$$

The total power dissipated is obtained from the real part of the total complex power [6,3] from all SDOF units and is calculated using [11],

$$\text{Power dissipated} = \frac{1}{2} c\omega^2 \Re(z)^2 \tag{8}$$

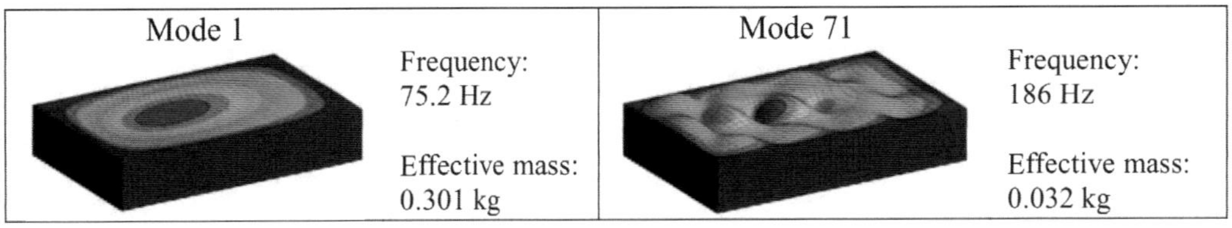

Figure 4: Typical mode shapes with significant horizontal contribution

5. Discrete element model

The calculations performed in the Discrete Element Method (DEM) alternate between the application of Newton's Second Law to the particles and a force-displacement law at the contacts. In this work DEM, is based on PFC3D (Particle Flow Code in 3 Dimensions) software [12]. In Figure 5, the contact parameters considered are shown. These include mass, normal and shear stiffness (k_n and k_s), normal and shear damping (c_n and c_s) and the friction coefficient. A description of the approach to finding these parameters can be found in the literature [4] while actual values used are presented in Figure 6. Note that stiffness and damping values were modified to reflect the excitation frequency using the same approach as for the modal properties – see Equation 5.

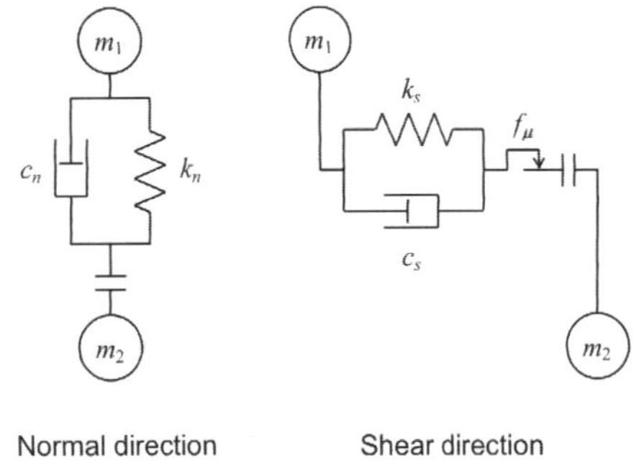

Item	Units	Value
m	Grams	2.1
k_n	N/m	3146
k_t	N/m	2233
c_n	Ns/m	1.04
c_t	Ns/m	0.732
μ		1

Figure 5 Contact model of particles, data at 100Hz

6. Comparison of theoretical models and experiment

The DEM model has been shown to be accurate at higher amplitude but over a limited frequency range [4]. In this section, a comparison is made between the damping predicted by the two calculation methods and experiments at very low amplitudes. As a large frequency range was of interest, results are compared for displacement amplitudes of 10^{-7}, 10^{-6} and 10^{-5} metres depending on the frequency. These values were selected to ensure maximum accelerations remained below 0.3g. Results are presented in Figure 6. The results show that both prediction methods perform well in the frequency range 80 to 400 Hz. At higher frequencies increased damping in the experiment is thought to arise from the presence of container wall resonances. Below 80 Hz, the wave approach does not match the other methods as it only considers standing waves – the lowest of which is near this figure.

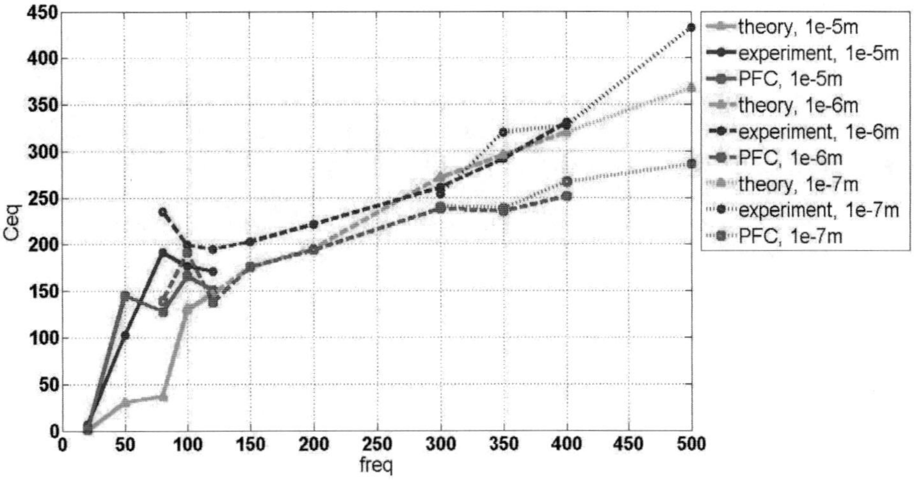

Figure 6: Comparison of equivalent damping for three different displacement amplitude levels

7. Parametric studies

The sensitivity of damping to excitation amplitude was studied using DEM – see Figure 7.

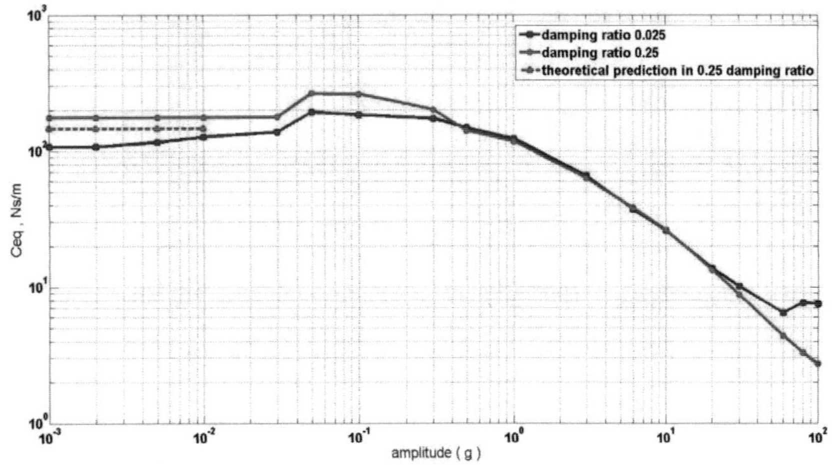

Figure 7: Equivalent damping versus excitation amplitude, at 100Hz excitation

It can be seen that in the very low amplitude (solid) region where the particles are in contact permanently performance is close to that predicted using the low amplitude theory. When the amplitude increases (accelerations of 0.03g to 0.3g) the particles start to slide over each other and damping increases: the fraction of energy dissipated by friction increases from 1% at 0.03g to 32% at 0.05g. At higher amplitudes still, the particles slide and roll more freely over each other (convection). In this case there is a trade off between damping due to friction and viscous effects [4] resulting in similar damping levels irrespective of the material loss factor. Increasing the amplitude to extremely high levels, the particles separate from each other (gas region). It can be seen the particles with lower damping reach the gas region earlier because they are less sticky than the other type and more collision can happen so although the damping for each individual particle is less but the total damping increases.

8. Conclusions

This paper considered the damping from polymeric granular materials. The simulation approaches are in good agreement with experiment. In a very low amplitude excitation the theory shows fairly good agreement with simulation and experiment particularly higher than the first natural frequency. The equivalent damping is independent of low vibration amplitudes (solid region) and is higher than the convection and gas regions where the particles collide and move with each other. It can be seen that in the convection region the comparison between two different materials - very high and very low loss factor –shows that the overall damping levels are close to each other and less sensitive to material.

9. References

[1] House J R 1990 *Vibration Damping Materials* World Intellectual Property Organization International Patent Classification F16F7/00,G10k11/16 , International publication Number WO90/01645
[2] Rongong J A and Tomlinson G R 2002 Vibration damping using granular viscoelastic materials *proc. Of ISMA 27 Noise and Viration Engineering Conference*(Leuven: Belgium) 431-40
[3] Wong C X, Daniel M C and Rongong J A 2009 Energy dissipation prediction of particle damperes *J. Sound and Vibration* **319** 91-118
[4] Darabi B and Rongong J A 2012 Polymeric particle dampers under steady-state vibrations *J. Sound and Vibration* **313** 3304-16
[5] Cundall P and Strack O 1979 A Distinct Element Model for Granular Assemblies , *Geotechnique* **29** 47–65
[6] Yang M Y 2003 *Development of master design curves for particle impact dampers* Doctoral Thesis (The Pennsylvania State University)
[7] Ungar E E and Kerwin E M 1964 Plate damping due to thickness deformations in attached viscoelastic layers *J. the Acoustical Society of America* **36** (2) 384-92
[8] Walton K 1987 The effective elastic moduli of a random packing of spheres, *J.Mech.Phys.Solids* **35** 213-26
[9] Janssen H A 1895 Experiments on grain pressure in silos *Transactions of the Association of German Engineers* **39** 1045-49
[10] Zamponi F 2008 Packing close and loose *Natue* **453** 606-7
[11] Rao S S and Yap F F 2010 *Mechanical vibration* (Pearson Education)
[12] Itasca Consulting Group Inc. 2008 *PFC 4D V4.0 manual* (Minneapolis/Minnesota : USA)

Strength knock-down assessment of porosity in composites: modelling, characterising and specimen manufacture

Alison J McMillan[1], Edward Archer[2], Alistair McIlhagger[2], Guillaume Lelong[3]
[1] Faculty Advanced Technology, University of Glamorgan, Pontypridd, CF37 1DL
[2] Faculty of Engineering, University of Ulster
[3] CESI School of Engineering, Rouen, France.
E-mail: alison.mcmillan@physics.org

Abstract. Porosity and inclusion of foreign material is known to reduce the strength of materials, and this paper addresses the particular problem of strength knock-down assessment due to porosity in composite materials. Porosity is often measured in terms of percentage of voids per unit volume of a component, because this can be related directly to ultra-sound absorption. Nevertheless, this is a poor indicator of actual strength knock-down, as it provides little information about void size, shape, orientation and whether they are evenly distributed or are clustered. Characterisation of void clustering enables a link between a cluster characteristic and the strength knock-down. Laboratory based testing achieves controlled porosity in specimens by introducing pin-holes into the RTM in-flow pipework, which entrains voids into the body of the preform within mould tooling. Specimens are manufactured to create resin regions bounded by a fibre reinforced picture frame, to allow for easy load application. Strength knock-downs from test are related to the theoretical expectations.

1. Introduction

Porosity and inclusion of foreign material is known to reduce the strength of materials, and this paper addresses the particular problem of strength knock-down assessment due to porosity in composite materials. Porosity is often measured in terms of percentage of voids per unit volume of a component, because this can be related directly to ultra-sound absorption, which is a straight-forward Non-Destructive Evaluation (NDE) procedure. Nevertheless, this is a poor indicator of actual strength knock-down, as it provides little information about void size, shape, orientation and whether they are evenly distributed or are clustered [1]. It is well known that manufacture process parameters influence porosity, and the effect of particular process parameters on particular porosity characteristics is a topic of on-going research [2].

Computational models of material with randomly distributed voids can be undertaken, but the results have shown that the stress-raiser effect on the surface of clustered voids depends critically on cluster characteristics. Schemes for characterising void clustering are proposed, and put into practise using synthetic model geometry, thus enabling a link between a cluster characteristic and the strength knock-down.

Laboratory based testing first requires a method of introducing controlled porosity into a specimen. This is achieved by introducing pin-holes into the RTM in-flow pipework (see figure 2), which

[1] To whom any correspondence should be addressed.

entrains voids into the body of the preform within mould tooling. In order to allow for convenient load application in the test machine, the specimen preform is modified to create resin regions bounded by a fibre reinforced picture frame. Strength test results are related to the theoretical expectations.

2. Porosity modelling challenges

The main challenge in modelling porosity is to create geometry that is representative of reality. On the one hand, it is desirable to keep things simple, for example to make each pore or hole in the material the same size and shape. On the other, it is necessary to allow sufficient variation to capture the more important emergent details. For example, it is well known to punch a row of slots into paper cheques and tickets to assist in tearing off counterfoils. A more scattered pattern of slots, or slots with different orientations would not be suitable for a tear-off counterfoil, so in this case, the emergent detail is the alignment of the slots.

Another intuitively obvious emergent detail would be clustering of holes. In this case, a number of holes might be said to be "near" to each other, and that this relative proximity might have some impact on the material properties. While such a geometrical configuration might be easy described in human language, to give it a precise mathematical description requires significantly more thought.

It is not simply a matter of choice of words; creating an appropriate mathematical description is the first step to being able to create a geometrical model with the appropriate features.

2.1 Placement of holes

The most obvious approach to placement of holes is to select numbers at random to represent the coordinates of the centroid of the hole within the material domain. While a very simple model would have circular holes, a somewhat more complex model might use ellipses, with the orientation of the ellipse defined by another random number [3]. Choosing numbers at random does not result in a well distributed pattern of holes, and so a system is required to define exclusion zones around currently placed holes and features, and to reject subsequent random number combinations which would otherwise have placed a hole into an exclusion zone [4].

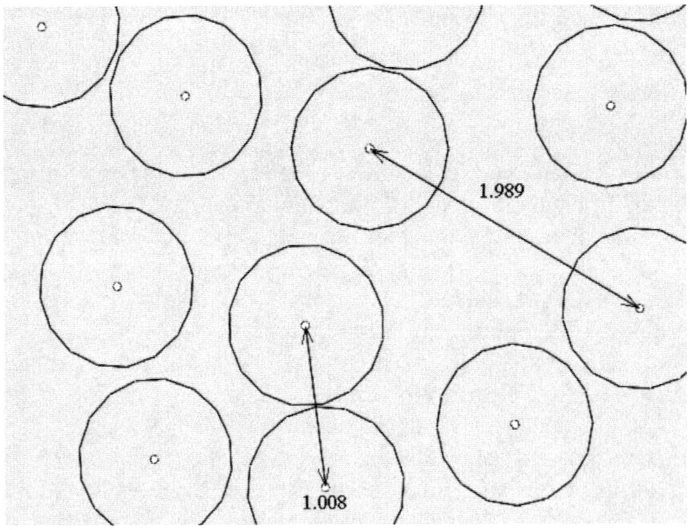

Figure 1. Placement of circular holes; larger circles describe the semi-exclusion zone.

Figure 1 shows a number of holes (the small circles) placed at random, each with an exclusion zone of 1 unit. The large circles have a radius equal to half the exclusion zone distance, thus to obey the exclusion rule, the large circles must not overlap. Holes are placed until it becomes impossible to place any further without compromising the exclusion rule. This technique is known as a fully dense

Poisson Disc fill [5]. A partially dense fill will contain fewer holes than a fully dense fill, and will (almost always) be different to a fully dense fill with a larger exclusion zone between holes. As such, some groups of holes might be said to be clustered.

This describes a convenient method for creating material models, but it is not the only way. Variations on this method might allow for the exclusion zone to shrink as more holes are placed. Such a method might show rather different characteristics to the simpler approach.

Another approach might be to deflect the location of existing holes in order to make space for the new hole to be added. While this might be complicated to implement, it would seem physically reasonable, particularly for modelling of foamed material.

In each case discussed so far, it has been assumed that each hole is of a constant size and shape. A more complex example might start with holes of unit volume, and where two holes are placed sufficiently closely, they merge to form a hole of twice the volume. Rules for surface tension and attraction/repulsion might control the formation of larger holes or deflection of holes that come close but not close enough.

3. Characterising porosity

The description given above relates to the creation of model geometry by means of rules. The method chosen to create the model does not necessarily characterise the model. Further methods for characterising as-created model geometries will be required [4].

Such methods should also be applied to as-measured real material geometries, to confirm that model geometries are truly representative of real materials, or to provide an indication as to how to improve the geometry creation methodology.

4. Manufacture of specimens

Composite plaques were manufactured using the VaRTM method with 8 layers of 8-harness satin weave HS grade carbon fibre (200gsm) cut to 350mm x 350mm using a scalpel blade on a glass cutting table. An initial preforming stage was undertaken to stabilise the preform under vacuum bag and then a caul plate was used in conjunction with a flexible membrane to consolidate the fabric reinforcements under full vacuum, prior to the injection of resin by peripheral gating at a tool temperature of 75°C (see figure 2). Araldite LY564 and Hardener HY2954 (based on bisphenol A epoxy and a cycloaliphatic amine hardener) were mixed and degassed at 2.86:1 by weight before transfer of the resin. After the addition of the resin and hardener, vigorous mixing for approximately five minutes at a temperature of 30°C was undertaken. After injection, a ramp was applied to 100°C. The temperature was held isothermally for 60 minutes; the composite plaque was then de-moulded and post cured for a further 180 minutes at 140°C.

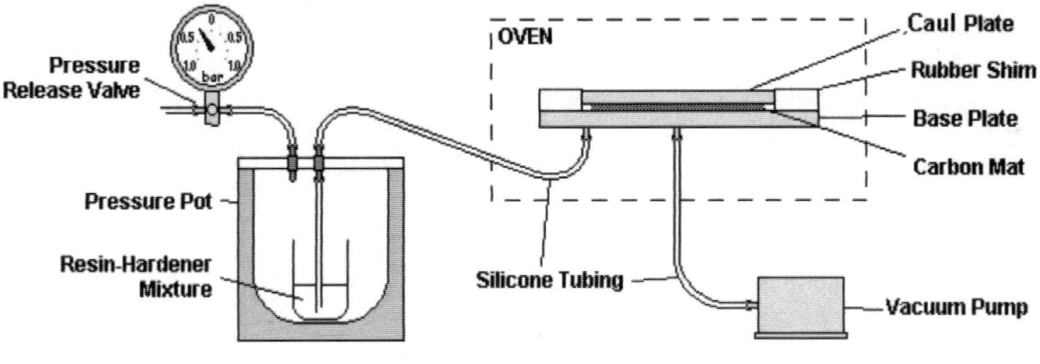

Figure 2. VaRTM apparatus.

For validation of analytical results, a baseline plaque with minimum achievable percentage of defects was manufactured under full vacuum after a vacuum drop test was conducted on the system and a loss of less than 10mBar in 5 minutes was detected. The degassing process described above removes air in the resin/hardener mix that is introduced during mixing. It is essential to remove these bubbles as they can increase the occurrence of gas within processed composite plaques. Degassing was carried out within a water-cooled/oil heated steel chamber with a ToolTemp TT-156E maintaining a degassing temperature of 30°C. The lid of the pressure pot is connected to a vacuum pump to reduce the pressure in the vacuum chamber to 0 mbar absolute. The reduction in pressure caused the trapped air bubbles to rise from within the resin toward the upper rim of the resin container. Once the resin reached the rim of the container the vacuum pipe was sealed, thereby stopping the movement of the resin over the container sides. This process was repeated until no bubbling was observed under the vacuum pressure. This process takes one hour at 30°C.

Two further plaques were manufactured with a range of degassing methods to introduce extra porosity into the composite. These were:

- No degassing
- Degassing for 20 minutes only

Furthermore a final plaque was manufactured with the introduction of deliberate through thickness flaws of various diameters ranging from 1mm to 5mm; during the VaRTM process these holes would be filled with resin and could replicate neat resin regions with picture frames of reinforcement to enable the loading like in the theoretical studies. This was done at the fabric preform stage where the 8 layers of 8-harness were arranged on a cutting table, vacuum consolidated, and holes were carefully drilled in the preform at predetermined locations using a high-speed drill of the required diameter. The locations were recorded to ensure the flaw could be placed at the centre of the subsequent test coupon. A clamping fixture was used to minimise yarn damage in the vicinity to the drilled holes and to ensure that the layers would remain in the preformed position.

5. NDE and testing of specimens

After VaRTM and post cure test coupons were extracted using an OMAX 2626 WaterJet Machining Centre. Fibre volume fraction was measured using the density buoyancy method in which the sample's mass in air is recorded before weighing the sample again in distilled water according to ASTM D792-91 [6]. Therefore, having obtained the specific gravity and knowing the densities of the constituent parts i.e. the fibre and resin, the percentage content of the fibres was calculated. Prior to mechanical testing the specimens were NDT tested using a USN 58 Krautkramer portable A-Scan.

Optical microscopy was carried out on the composite samples in order to ascertain the geometric characteristics of the fabric and the real extent of porosity. A Struers DAP-7 automatic polishing machine was utilised to prepare the samples for microscopy by removing any surface defects left from the cutting process and to also reveal the desired cross section for analysis. Prior to polishing, samples were mounted in Struers Epofix® polymer resin in either 25mm diameter or 50mm diameter circular moulds and rectangular moulds of dimension 75mm x 25mm, depending on the dimensions of the sample. Mounting in Epofix ® ensured that the sample can be firmly gripped in the DAP-7 sample holder and thereby ensuring that the sample is held flat against the polishing discs and so polishing occurs in the same axis as the direction of the intended subsequent micrograph. Samples were polished progressively with silicone carbide paper from 500 grit to 2400 grit. The final finishing polish was carried out using an MD-DAC fabric disc lubricated in DiaPro® diamond suspension polishing solution.

Tensile specimens were tested in the warp and weft direction on a Zwick Z100 universal tester. The testing was performed to CRAG standard 300 [7]. An MTS 632.85F biaxial extensometer was used to determine tensile modulus and Poisson's ratio and a series of 1mm gauge length, 120 Ohms strain gauges were bonded to the specimen to determine changes in strain near the deliberate through thickness flaws.

6. Discussion

Whether the material is real or modelled, random numbers are involved, and the results obtained will all be somewhat different. The key issue is to link particular aspects of the characterisation to particular aspects of the model or physical test results.

In the case of a computational model of a material containing porosity, a block of material is defined, a finite element mesh created, and then boundary conditions and some loading applied. Results of the finite element analysis are then viewed: in previous analysis [3, 4] the maximum and minimum principal stresses were examined, and the peak values in the material domain were recorded. This is approximate, because the values obtained depend on the mesh refinement, nevertheless maintaining a consistent approach means that results can be compared from one model to another. A more adventurous approach is to include the effects of plasticity and/or crack growth as the load is increased so that local stresses exceed the material limits [8].

In the case of real material, an NDT inspection can be performed, to capture actual porosity geometry before and after testing. Differences in geometry will point to localised failure or plastic yield, and the extent to which that has occurred. Comparing these results with computational models will provide insight into the relative importance of porosity hole size, shape, orientation or clustering in the sequence of failure.

Acknowledgments

The authors wish to acknowledge assistance of the ERASMUS programme, which enabled G Lelong to spend four months internship at University of Ulster.

References

[1] Paciornik S and d'Almeida J R M 2009 *J. Composite Materials* **43** 101-12.
[2] Olivier P, Cottu J P and Ferret B 1995 *Composites* **26** 509–15.
[3] McMillan A J 2012 *Journal of Reinforced Plastics & Composites* **31**(1) 13-28.
[4] McMillan A J 2012 Defect indentification and characterisation algorithms for assessing effect on component strength *Proceedings of 15th European Conference on Composite Materials,* (Venice, Italy, 24-28 June 2012).
[5] Lagae A and Dutré P 2008 *Computer Graphics forum* **27**(1) 114-129.
[6] ASTM D792-91, Standard Test Methods for Density and Specific Gravity (Relative Density) of Plastics by Displacement.
[7] CRAG Standard Method 302 "Method of test for the Tensile Strength and modulus of multidirectional Fibre Reinforced Plastics".
[8] McMillan A J 2011 Stresses and crack propagation in porosity clusters *Proceedings of The 16th Composites Durability Workshop,* (Seoul National University, South Korea, 28-30 August 2011).

Efficient Parametric Optimisation of Support Loss in MEMS beam resonators via an enhanced Rayleigh-Ritz method

H.T.D.Grigg[1], B.J. Gallacher[2]

[1]Ph.D. Candidate, School of Mechanical and Systems Engineering, Newcastle University-Stephenson Building, Claremont Road, Newcastle upon Tyne NE1 7RU, UK

E-mail: harry.grigg@newcastle.ac.uk

[2]Senior Lecturer, School of Mechanical and Systems Engineering, Newcastle University-Stephenson Building, Claremont Road, Newcastle upon Tyne NE1 7RU, UK

E-mail: barry.gallacher@newcastle.ac.uk

Abstract: MEMS resonators offer attractive prospects in several application areas, including high-performance, low cost sensors, among several others. The performance of many resonant MEMS depends critically on the Q factor, and an important, poorly quantified contribution to the overall Q is the support loss. Additionally, the parameter space for the geometry can be of moderately high dimension, making FEA based parametric optimisation computationally inefficient. Thus motivated, a numerical method based on the Rayleigh-Ritz substructure synthesis using quasicomparison functions is developed, applicable to a wide and important class of beam resonators. It is shown to be highly efficient by comparison with classical FEA methods, facilitating a detailed examination of the support Q as a function of position in parameter space. Selected results are presented and briefly discussed, with particular attention given to convergence, computational efficiency and design optimisation. General design principles for multiply-supported framelike beam resonators are considered in the light of the results, and extensions to the modelling are briefly covered.

KEYWORDS: MEMS resonators, Rayleigh-Ritz, numerical vibration modelling, optimisation, Xylophone Bar Resonator, Support Loss, Q factor.

1. Introduction.

The last decade has seen considerable interest devoted to MEMS devices based on resonators, including RF MEMS[1], resonant accelerometers and gyroscopes[2], magnetometers[3], energy harvesters, biosensors, and other applications. The performance and behaviour of many of these devices depends critically on the quality (Q) factor of the underlying resonator.

A common model for the Q factor of a MEMS resonator can be stated as:

$$Q^{-1} = Q_{TED}^{-1} + Q_{INTRINSIC}^{-1} + Q_{GAS}^{-1} + Q_{SUPPORT}^{-1} + Q_{SURFACE}^{-1} + Q_{OTHER}^{-1} \quad (1)$$

Little study has been dedicated to the subject of support loss([4], [5], etc). The phenomenon consists of energy transport from the resonator to the environment via elastic waves excited in the substrate by the stresses at the interface between the resonator supports and the substrate.

1.1. Xylophone Bar Resonator

The method was developed in the context of performance optimisation of a Xylophone Bar Resonantor (XBR) based magnetometer. The resonator consists of a main sense beam element supported transversely by ancillary support beams, of comparatively small flexural stiffness, forming an H-shaped frame.

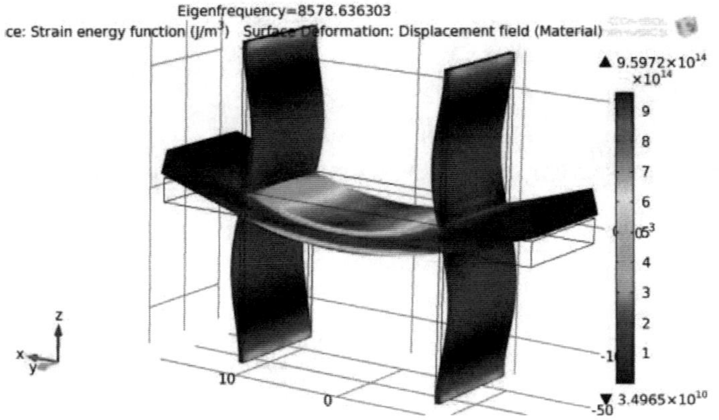

Figure 1. Finite Element based visualisation of the operational mode of a XBR. Hue denotes Von Mises stress. The substrate, omitted here, is attached to the free ends of the transverse beams.

A periodic voltage is applied to the device such that any out-of-plane component of magnetic flux excites the mode of interest at its natural frequency, with a steady-state amplitude proportional to the flux density and to the device Q factor.

2. Resonator Model.

To model the resonator dynamics, a Rayleigh-Ritz substructuring model[6] based on the decomposition of the frame structure into 1D elements intersecting at nodes and imposing geometric conditions of compatibility there and at the substrate interfaces. The field Lagrangian was chosen as

$$L = T^* - V = \left\{ \int_0^L \rho A \left[\left(\frac{\partial U}{\partial t} \right)^2 + \left(\frac{\partial V}{\partial t} \right)^2 \right] dx \right\} - \left\{ \int_0^L EI \left(\frac{\partial^2 U}{\partial x^2} \right)^2 dx + \int_0^L EA \left(\frac{\partial V}{\partial x} \right)^2 dx \right\} \quad (2)$$

Where U is a transverse displacement and V an axial displacement. Polynomial trial functions were employed for the axial field, while quasicomparison functions were developed and used based on beam modes under appropriate constraints for the transverse field.

3. Substrate Model.

The problem of elastic wave radiation in a semi-infinite half space has been studied previously. Miller and Pursey's 1953 paper[7] arrives at closed-form expressions for the energy radiated by localised stress sources on the surface of an infinite linear elastic half-space.

It can be shown that[4]:

$$W = \int_0^{2\pi} Re(\overline{F_T v_T} + \overline{F_N v_N})\, dt = \frac{\pi b K}{\rho S^2}\{|\sigma_N|^2 Im(\Upsilon_N) + |\sigma_T|^2 Im(\Upsilon_T)\} \quad (3)$$

Where

$$Im(\Upsilon_N) = 0.2215, \qquad Im(\Upsilon_S) = 0.3350$$

This formula is an explicit relation between the distal forces of constraint acting at the interface between the resonator and the substrate, and the corresponding support loss, in a compact form suited to evaluating the Q factor.

4. Results and discussion.

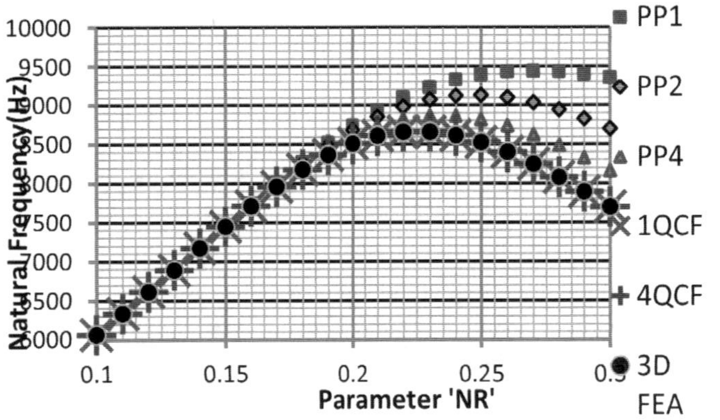

Figure 2. Natural frequencies obtained by simulation using the commercial FEA package COMSOL;Rayleigh-Ritz models using 1, 2, and 4 Pinned-Pinned mode shapes per substructure as trial functions; and Rayleigh-Ritz-Meirovitch models using 1 and 4 quasi-comparison trial functions. Agreement between the 3D FEA and the RRM models is superb.

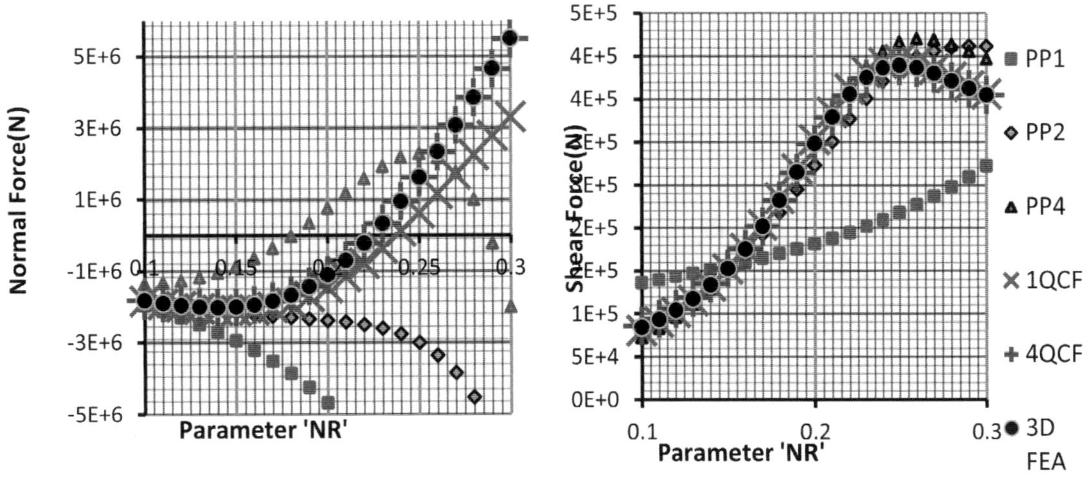

Figure 3. Shear and normal forces at the support-substrate interface, obtained by simulation. Models are as in Fig.I. Note that the PP models do not converge to the QCF/FEA solutions, even in the regime where the natural frequencies have converged.

Excellent convergence and agreement are observed. It is worth noting that over 10000 DOF were needed for 3D FEA and 2000 DOF for 2D FEA to converge to this result, while the 4QCF model contained 71 DOF. From an optimisation point of view, then, the RRM approach is at least two orders of magnitude more efficient under the conditions considered.

Figure 3, taken together with the form of Equation 1, gives an insight into the relationship between resonator geometry and support Q factor. The zero crossing of the normal force that occurs between NR=0.22 and 0.23 for the converged results. This matches to the third significant figure the result for the node points of a free-free beam. This corresponds to the nulling of the component of support Q due to axial forcing of the support, and is a corollary of the impedance tuning of the resonator.

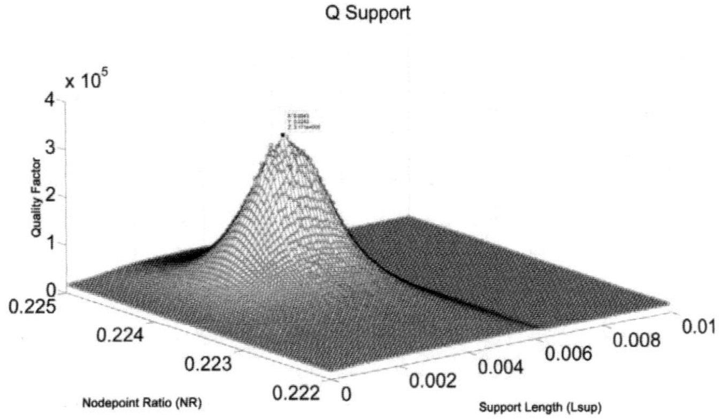

Figure 4. Support quality factor vs. parameters NR and $Lsup$. The parameters at the peak are $Lsup = 4.3mm$; $NR = 0.2242$. The support Q is 7.541×10^5.

Figure 3 illustrates a local maximum of the Q factor in the *NR* - *Lsup* parameter plane as a surface plot. There are 10^4 parametric combinations on display in a 100x100 grid, and the solve time was of the order of 1 minute on a desktop system. This compares favourably with the commercial benchmarks. The half-maximum dimensons of the local Q maximum are 0.094% in NR and 0.0026 m in Lsup, quantifying the sensitivity of the support Q to mistuning in these parameters.

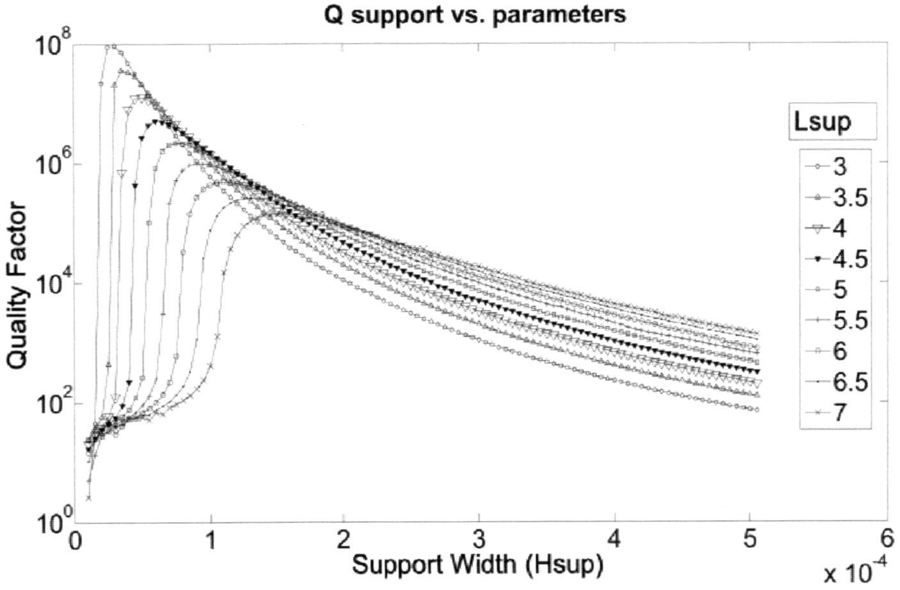

Figure 5. Support quality factor of the device as a function of *Hsup*, plotted for several values of *Lsup*.

The unconstrained global support Q maximum for the XBR system is unphysical, and cannot be approached closely , requiring an infinitely thin, infinitely short support. Of more interest for practical problems is the constrained maximum obtained by applying manufacturing considerations to the geometry. Figure 5 is of interest in this context. If an arbitrary 100 micron limit is imposed from below on the parameter Hsup, then the constrained optimal value for Lsup is not a limiting value, but is close to 500 microns. Using the RRM approach allows for detailed examination of the response in the parameter space and optimisation of the dynamics for support loss.

5. Conclusions and further work.

Dynamical modelling of XBR geometry using an RRM approach was shown to be a valid and efficient route to examine and optimise the resonator for support loss. The results imply the efficacy of impedance tuning the resonator-support geometry as a means to mitigate support loss.

Further work would include extending the RRM analysis to include nonlinear axial effects, and developing and integrating more complete and representative substrate models. Integration of a constrained optimisation method such as simulated annealing, PSO or genetic algorithms as appropriate should be a straightforward extension of utility for more complex frame topologies, for which a qualitative understanding of the impedance tuning conditions is not accessible or satisfactory.

6. References.

[1] J. Basu and T. K. Bhattacharyya, "Microelectromechanical resonators for radio frequency communication applications," *Microsystem Technologies*, vol. 17, no. 10–11, pp. 1557-1580, Aug. 2011.

[2] K. M. Harish, B. J. Gallacher, J. S. Burdess, and J. a Neasham, "Experimental investigation of parametric and externally forced motion in resonant MEMS sensors," *Journal of Micromechanics and Microengineering*, vol. 19, no. 1, p. 015021, Jan. 2009.

[3] A. L. Herrera-May et al., "Mechanical design and characterization of a resonant magnetic field microsensor with linear response and high resolution," *Sensors and Actuators A: Physical*, vol. 165, no. 2, pp. 399-409, Feb. 2011.

[4] Z. Hao, "An analytical model for support loss in micromachined beam resonators with in-plane flexural vibrations," *Sensors and Actuators A: Physical*, vol. 109, no. 1–2, pp. 156-164, Dec. 2003.

[5] B. Chouvion and I. Lyon, "Vibration Transmission and Support Loss in MEMS Sensors," University of Nottingham, 2010.

[6] A. Hale and L. Meirovitch, "A general substructure synthesis method for the dynamic simulation of complex structures," *Journal of Sound and Vibration*, vol. 69, no. 2, pp. 309-326, Mar. 1980.

[7] G. F. Miller and H. Pursey, "The Field and Radiation Impedance of Mechanical Radiators on the Free Surface of a Semi-Infinite Isotropic Solid," *Proceedings of the Royal Society A: Mathematical, Physical and Engineering Sciences*, vol. 223, no. 1155, pp. 521-541, May 1954.

Experimental validation of the orthogonalised reverse path method using a nonlinear beam

[1]P Muhamed, [2]K Worden and [2]ND Sims

[1]*I*-Kohza of Conditioning Monitoring, Malaysia-Japan International Institute of Technology, Universiti Teknologi Malaysia International Campus, 54100 Kuala Lumpur, Malaysia

[2]Dynamics Research Group, Department of Mechanical Engineering, University of Sheffield, Mappin Street, Sheffield S1 3JD, England.

E-mail: k.worden@sheffield.ac.uk

Abstract. The Orthogonalised Reverse Path (ORP) method is a new algorithm of the 'reverse path' class but developed in the time-domain. Like the Conditioned Reverse Path (CRP) method, the ORP approach is capable of identifying the underlying linear FRF of a system or structure in the presence of nonlinearities and may well also lead to simplifications in the estimation of coefficients of nonlinear terms. The method has shown itself to be numerically robust not only for simple simulated SDOF systems but also for simulated MDOF systems. The aim of this paper is to discuss an application of the ORP method to an experimental test set-up based on a nonlinear beam rig.

1. Introduction

The Orthogonalised Reverse Path (ORP) method is a new algorithm of the Reverse Path (RP) class for nonlinear system identification [1]. The RP methods are based on the idea of removing the effects of any nonlinearity from system input and output data in such a way that one can estimate the Frequency Response Function (FRF) of the underlying linear system. Unlike the previous state-of-the-art - the Conditioned Reverse Path (CRP) method [2] - which removed nonlinear components from spectra, the ORP method removes the nonlinear components from the initial time data, thus allowing completely standard FRF estimation algorithms to be used to compute the 'linear' FRFs. When the ORP was introduced in [1], it was validated only on simulated data generated from a number of lumped mass Multi-Degree-of-Freedom (MDOF) systems. The objective of the current paper is to demonstrate how the ORP method fares when faced with experimental data from a continuous system. The system of interest here consists of a cantilever beam with a geometrical nonlinearity. Data from the experiments were provided by Professor Gaetan Kerschen of the University of Liege, Belgium. The Liege researchers provided an analysis of their data based on the CRP method in [3], and this will allow a comparison with the ORP results obtained in the current paper.

2. Experimental set-up and problem formulation

The description of the experimental setup is taken from [3]; it involves a clamped beam with a thin beam part at the end of the main beam as shown in Figure 1. Both beams were made of steel. The dimensions of the thick (main) beam were: length 0.7 m, width 1.4 cm, thickness 1.4cm. The

Published under licence by IOP Publishing Ltd

dimensions of the thin beam were: length 4 cm, width 1.4 cm, thickness 0.5 mm. Seven accelerometers were placed at regular intervals on the main beam to measure responses. In addition, there was a displacement sensor located at the end of the main beam at position 7 (Figure 1). Force was applied at the location of the third sensor. The excitation was a white-noise sequence, band limited into the 0-1000 Hz range. The shaker at position 3 excited the structure in the horizontal plane in order to minimise the effects of gravity. Different excitation levels were considered in the1.4 to 22 N RMS range.

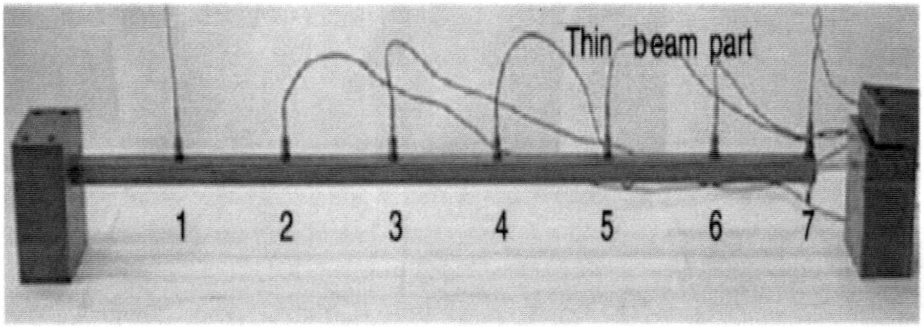

Figure 1. Experimental setup taken from [3].

The stiffening effect of the thin beam part at the end of main beam was modelled in [1] with a grounded symmetrical nonlinear term of the form $z_1(y) = a_1|y|^{\alpha} sign\,(y)$, where y is the displacement at the end of the beam and a_1 is the nonlinear coefficient term. The exponent of α was determined by seeking the maximum value for the spectral mean of the averaged cumulative coherence of all the seven sensors as explained in detail in [1]. Because one might argue that there are good physical grounds for believing the nonlinearity to be cubic [4], the nonlinear term used in the main analysis here took the simpler form $z_1(y) = a_1 y^3$.

3. The ORP Algorithm

The CRP algorithm has been described in numerous publications including [2] and [3] cited here, so it will not be presented here. In contrast, the ORP algorithm is only recently developed, so a basic and brief summary will be given for completeness. As previously stressed, the ORP method carries out the decorrelation procedure that removes the effects of nonlinearity directly in the time-domain. Then, in obtaining the underlying linear FRF, one uses only conventional FRF estimation approaches. For a single nonlinear term, the conditioned output of the individual responses which are rid of the nonlinearity is given as follows.

$$y_{1(-1)}(t) = y_1(t) - \int_{-\infty}^{\infty} l_{y1}(t-\tau) z_1(t) dt$$

$$\vdots$$

$$y_{7(-1)}(t) = y_7(t) - \int_{-\infty}^{\infty} l_{y7}(t-\tau) z_1(t) dt \qquad (1)$$

The subscripts refer to the seven measured responses from the accelerometers; where it is unlikely to cause confusion the sensor subscripts will be omitted. In general one has,

$$y_{(-1)}(t) = y(t) - \int_{-\infty}^{\infty} l_y(t-\tau)z(t)dt \tag{2}$$

where the integrand, $l_y(t-\tau)$ is the inverse Fourier transform of the 'FRF',

$$L_y(\omega) = \frac{S_{zy}(\omega)}{S_{zz}(\omega)} \tag{3}$$

where the S denote standard cross- and autospectra. Equation (2) is then approximated by a discrete sum in order to accommodate samples $y_i, i = -M, \cdots, M$, where $y_i = y(i\Delta t)$ and Δt is the sampling interval;

$$y_{i(-1)} = y_i - \sum_{j=-M}^{M} l_y((i-j)\Delta t)z(j\Delta t)\Delta t = y_i - \sum_{j=-M}^{M} l_{(i-j)y} z_j \Delta t$$
$$= y_i - \sum_{j=-M}^{M} C_{ij} z_j \tag{4}$$

The response data are assumed to consist of N sample points, but the number of lags (M) to be removed, or the correlating terms of $z_1(t)$ (as only one nonlinearity exists), is assumed much smaller ($N \gg M$). The details can be found in [1], but essentially ORP analysis proceeds by applying a Gram-Schmidt orthogonalisation algorithm (here using the built-in Matlab function *orth* [5]) to compute the orthogonal complement of the nonlinear terms within the measured response. This is an iterative process which finally yields,

$$\{y_{(-1)}\} = \{y\} - \sum_{j=-M}^{M} \alpha_j \{Z_j\} = \{y\} - \sum_{j=-M}^{M} \beta_j \{W_j\} \tag{5}$$

Where the $\{W_j\}$ are a set of orthonormal vectors spanning the space spanned by the (nonorthogonal) $\{Z_j\}$ generated by the nonlinear terms. To condition the input force x_3 in the time domain a similar recursion can be used,

$$\{x_{3(-1)}\} = \{x_3\} + \sum_{j=-M}^{M} \gamma_j \{W_j\} \tag{6}$$

The conditioned (underlying linear) FRF can now be computed using standard (unconditioned) spectral estimation methods on the conditioned input and output time-series. Note that all suspected nonlinearities must be included in the equivalent of equation (1) in order that the sequence of removals of orthogonalised nonlinear terms will work in general. The linear FRFs from the input and output time-series are estimated with the H_1 and H_2 estimators,

$$[H_1(\omega)] = [S_{xx}(\omega)]^{-1}[S_{xy}(\omega)] \tag{7}$$
$$[H_2(\omega)] = [S_{yx}(\omega)]^{-1}[S_{yy}(\omega)] \tag{8}$$

Once the conditioned FRF (or FRF matrix for MDOF) has been obtained, the coefficients of the nonlinear terms can be estimated as in the standard CRP method [2]. There are two ways of using the method for finding the nonlinear coefficients as used for the ORP method in this paper, the details of the algebra can be found in [6]. The first approach (denoted ORP_1) yields the coefficient estimate (note that this is actually a spectrum),

$$a_1(\omega) = \frac{S_{z1x3}(\omega)H_{33}^{c} - S_{z1y3}(\omega)}{H_{37}^{c}(\omega)S_{z1z1}(\omega)} \tag{9}$$

where H_{ij}^c denotes the conditioned FRF between excitation point i and response point j. The second approach (ORP_2) yields the superficially similar,

$$a_1(\omega) = \frac{H_{33}^c(\omega)S_{x3x3}(\omega) - S_{x3y3}(\omega)}{H_{37}^c(\omega)S_{x3z1}(\omega)} \tag{10}$$

4. Discussion of experimental results

There are two sets of experimental results presented here corresponding to a very low excitation, and a high level of excitation. Using different excitation levels allows one to get some idea of the influence of nonlinearity in the system. It was shown in [1] that the H_2 FRF estimator [6] proved most effective in the nonlinear system analysis, so this estimator was used throughout here, although both H_1 and H_2 estimators were initially used on the data from the lowest level of excitation. Figure 2 shows the magnitude of $H_{73}(\omega)$ for the lowest excitation of the beam (1.4 N RMS) using both classical estimators. The FRFs clearly identified four peaks from the system; however closer inspection showed the magnitude of the FRFs dropping when using the H_1 estimator. Figure 2 shows that influence of the nonlinearity is absent as the FRFs plot are smooth and clear. The outcomes of the results are based on the following simulation and signal processing parameters used: the sampling frequency was 2560 Hz, spectra were calculated on the basis of 20 ensemble averages of 2^{13} (8192) FFT points, a Hanning window was used, total number of samples =163840.

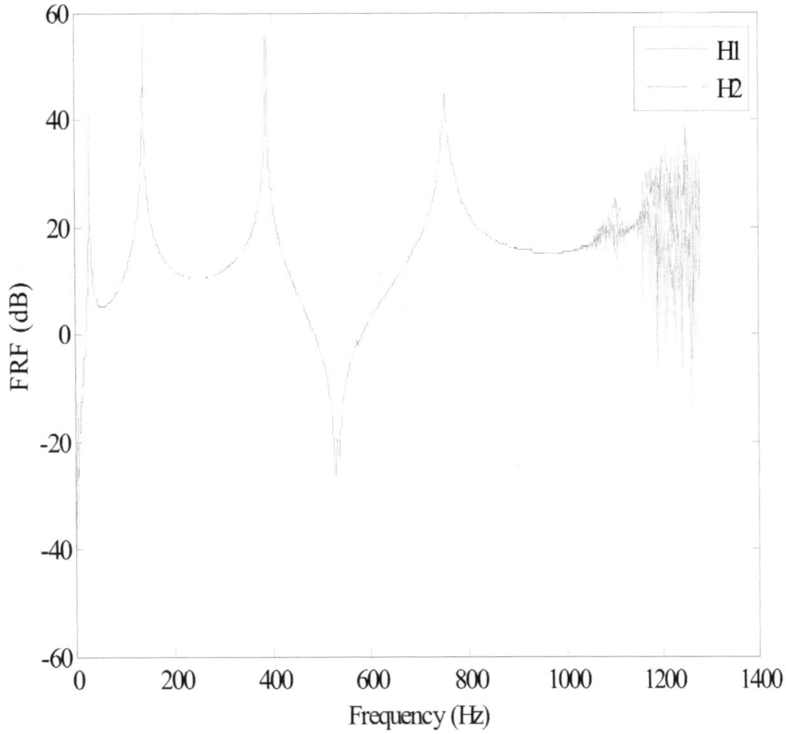

Figure 2. Classical (unconditioned) H_1 and H_2 estimates for the magnitude of $H_{73}(\omega)$ from the lowest excitation data (1.4 N RMS).

In order to see the influence of the nonlinearity on the structure, the highest excitation level (22 N RMS) data were then used to compute the magnitude of $H_{73}(\omega)$ which is shown in Figure 3. It was found that distortions appeared in the FRF when the excitation level was increased. The third and fourth modes were not much affected by the nonlinearity, so further analysis was concentrated on the range of 0-500 Hz.

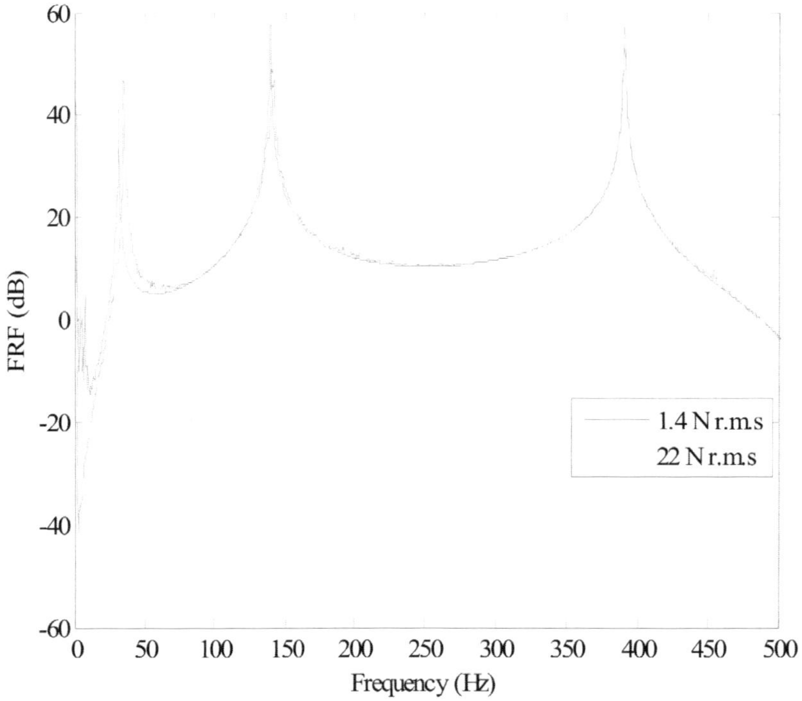

Figure 3. Classical (unconditioned) H_2 estimate for the magnitude of $H_{73}(\omega)$ from the highest excitation data (22 N RMS).

It now remains to illustrate the capability of the ORP method in removing the FRF distortion and also to obtain the nonlinear coefficient. The CRP results provided in [3] will be used as a benchmark. Also, the FRF for the 1.4 N RMS level estimated using the H_2 (known as H_{2lin}) will be used as a proxy for the linear FRF for comparisons with the CRP and ORP estimators later. Note that the nonlinearity considered here is the simple cubic stiffness.

Figure 4 shows the first two modes of the FRFs for the highest level of excitation of 22 N RMS. It can be observed that the first two natural frequencies from the unconditioned estimators are shifted towards higher frequencies, and this is due to the stiffening effect of the thin beam part. It is very clear that both the CRP and ORP methods purge the effects of the nonlinearity and give very close agreement with the FRF of the 'linear' system.

Figure 5 displays the real and imaginary parts of the estimate of the nonlinear coefficient a_1 found from CRP and equations (9) and (10) for the ORP methods. As stated earlier, the coefficients are frequency dependent and a spectral mean has to be performed to obtain a scalar value. Table 1 presents the spectral means of the estimated nonlinear coefficients from ORP for the 22 N RMS level of excitation and also for a level intermediate between the lowest and highest. The spectral means were taken between 20 to 250 Hz to avoid the noise at low frequencies in both methods. It is worth pointing out that the imaginary part of the coefficient, which is without any physical meaning, is always at least one order of magnitude below the real part.

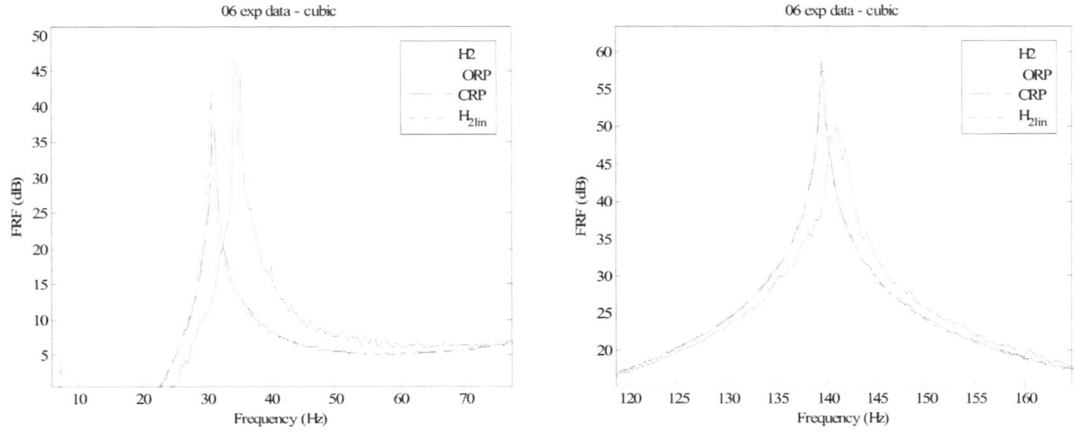

(a) (b)

Figure 4. Details of first (a) and second (b) resonance peaks for FRFs computed from data at highest level of excitation (22 N RMS).

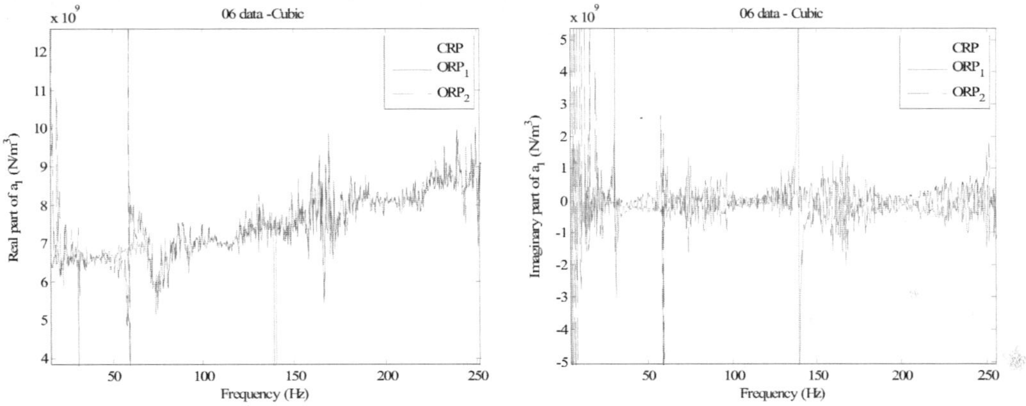

(a) (b)

Figure 5. Real (a) and imaginary (b) parts of nonlinear coefficient spectra from data at highest level of excitation (22 N RMS).

Excitation level (N RMS)	Nonlinear coefficient, a_1 (N/m^3)		
	ORP$_1$	ORP$_2$	CRP
11	$7.77\text{x}10^9 - i6.73\text{x}10^7$	$7.86\text{x}10^9 - i2.02\text{x}10^8$	$7.72\text{x}10^9 - i1.61\text{x}10^8$
22	$7.49\text{x}10^9 - i5.76\text{x}10^6$	$7.69\text{x}10^9 - i1.59\text{x}10^7$	$7.43\text{x}10^9 - i1.34\text{x}10^8$

Table 1. Spectral mean estimates of cubic nonlinear term coefficients at two higher levels of excitation.

In order to make a comparison with reference [3], the ORP analysis was repeated with a nonlinear term of the form $z_1(y) = a_1|y|^{\alpha} sign\ (y)$, the results of this analysis are presented in Table 2. The agreement with the results from [3] is excellent.

Excitation level (N RMS)	Nonlinear coefficient, a_1 (N/m$^{2.8}$)		
	ORP_1	ORP_2	CRP (from [3])
11	$1.94 \times 10^9 - i1.61 \times 10^7$	$1.96 \times 10^9 - i1.93 \times 10^7$	$1.94 \times 10^9 + i\,1.09 \times 10^6$
22	$1.98 \times 10^9 + i3.26 \times 10^6$	$2.04 \times 10^9 + i4.76 \times 10^7$	$1.96 \times 10^9 + i1.55 \times 10^7$

Table 2. Spectral mean estimates of fractional power nonlinear term coefficients at two higher levels of excitation.

Summary

Nonlinearities introduce distortions in system FRFs which can be seen with the H_2 estimate at higher excitation levels. Both ORP and CRP estimates show excellent results in purging the influences of nonlinearity and yielding the FRFs of the underlying linear system. Previous work with the ORP concentrated on simulated data; the aim of this paper has been to test the ORP method by applying it in an experimental test scenario where the CRP had already been found to be successful. The results were excellent and the ORP approach has showed its ability to deal with real experimental MDOF nonlinear systems under realistic conditions. A stable value for the nonlinear coefficient was identified that agreed well with a previous CRP estimate and the imaginary part of the coefficient was several orders of magnitude below the real part.

Acknowledgements

The authors would like to thank Professor Gaetan Kerschen of the University of Liege, Belgium, for providing the nonlinear beam data and also for numerous helpful discussions on reverse path analysis.

References

[1] Richards CM and Singh R 1998 Identification of multi-degree-of-freedom non-linear systems under random excitation by the "reverse path" spectral method. *J. Sound Vib.* **213** pp.673-708.
[2] Muhammad P, Sims ND and Worden K 2012 On the orthogonalised reverse path method for nonlinear system identification. *J. Sound Vib.* In Press.
[3] Kerschen G, Lenaerts V and Golinval JC 2003 Identification of a continuous structure with a geometrical non-linearity, part 1: conditioned reverse path method. *J. Sound Vib.* **262** pp.889-906.
[4] Gifford SJ 1989 *Volterra Series Analysis of Nonlinear Structure*. PhD Thesis, Department of Mechanical Engineering, Heriot-Watt University.
[5] The Mathworks 2004 *Matlab v7*.
[6] Muhammad P 2011 *On the Conditioned and Orthogonalised Reverse Path Methods for the Identification of Nonlinear Systems*. PhD Thesis, Department of Mechanical Engineering, University of Sheffield.

Journal of Physics: Conference Series **382** (2012) 012030 doi:10.1088/1742-6596/382/1/012030

An inverse vibration-based approach towards modelling and damage identification in nonlinearly vibrating structures. Application for delamination detection in a composite beam.

Irina Trendafilova
Reader, *Department of Mechanical Engineering the University of Strathclyde,* 75 Montrose street, Glasgow G1 1XJ, UK

E mail: Irina.Trendafilova@strath.ac.uk

Abstract This study explores the possibilities for inverse analysis and modelling from data of a nonlinearly vibrating structure. We are suggesting a statistical approach based on singular spectrum analysis (SSA). The method is based on a free decay response, when the structure is given an initial disturbance and is left to vibrate on its own. The measured vibration response is decomposed into new variables, the principal components, which are used to uncover oscillatory patterns in the structural response. In this study an application of the methodology for the purposes of delamination detection in a composite beam is explored.

1. Introduction.

The free decay response of a linearly vibrating structure is a decaying signal which in the frequency domain is a function of the natural frequencies of the structure and will have peaks at these natural frequencies. Most real structures have nonlinearities induced by their material, the geometry of the structure, as well as by its supports and joints. In some occasions it is possible to approximate a nonlinear vibratory behaviour with a linear one. But on a number of cases the presence of nonlinearities should be taken into account. Structures made of composite materials are one such example. It is known that composite materials and structures made of composites demonstrate quite complex vibratory behaviour which on most occasions cannot be approximated as linear. The free decay response of composite structures is quite difficult or on some occasions even impossible to analyse in the frequency domain especially because of the nonlinearities present in the structure which result in nonlinear vibration response signal. An example of such a free decay signal is given in Figure 1 where the signal is given in the time-domain and the frequency domain. It can be seen that the signal is nonlinear and it is difficult to draw conclusions for the structure's behaviour from these signal representations. The presence of noise and other uncertainties coming from the environmental conditions should not be forgotten because they also contribute to the nonlinear and non-stationary nature of the obtained signal.

There are different attempts to treat the problem for nonlinear structural vibration analysis and identification ranging from completely deterministic approaches to approaches based on probability and statistical methods. Deterministic modelling methods are based on deterministic analytical or numerical models for the nonlinear and non-stationary behaviour of certain structures, *vis.* composite shells, beams and or plates [1,2]. It should be noted that such methods are on most occasions oriented and developed for a certain structural and material type and in general they cannot be easily or automatically applied to other structures or materials. Inverse identification methods for such vibratory problems can be deterministic but in the majority of cases they are based on statistical time series models like e.g. ARMA and NARMA ones. A more detailed review and classification of such methods can be found in e.g. [3].

Another big class of modelling approaches attempt to model the structural response time series using nonlinear time series analysis methods. Although these methods are generally oriented towards modelling deterministic systems most of the methods operate on a statistical basis making statistical assumptions. Probabilistically based methods are generally based on certain hypotheses and when applied for identification purposes treat the parameters being identified as random variables.

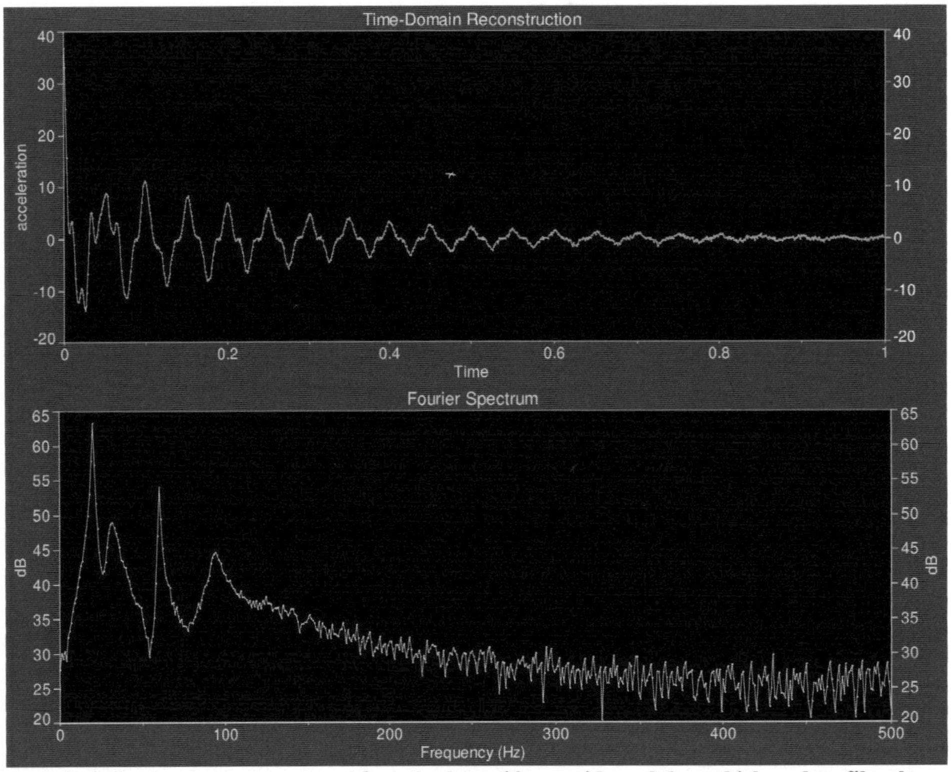

Figure 1. Free decay response measured for a 1m long, 20mm wide and 6mm thick carbon fibre beam

The method suggested here uses a time series approach where the time lagged vibration response measurements are considered as measured variables and are analysed using principal components analysis. This is used in order to reduce the number of time domain components of the measured signal and also to uncover oscillation patterns in it. This is certainly not the first paper to explore a statistical approach and not the first paper to use PCA for the purposes of damage assessment. But there are rather few applications of statistical methods and nearly no papers for the application of PCA for delamination detection in composite structures. In this study the time domain measured vibration response is treated as a random time dependent multidimensional variable which is analysed using singular spectrum analysis (SSA) which is an alternative of PCA. The main idea of PCA is to reduce the number of the measured time or frequency domain components in the measured discrete vibration response, retaining most of their variance. It has been previously applied in the time domain as well as over frequency domain data by e.g. [4] and the author of the present study [5]. As a result a vibrating system is represented by a new but much smaller set of variables *vis.* the several principal components. The new principal components can be associated to certain important properties of the structure including the presence of damage as well as its location in the structure and its size/extent. SSA is an alternative of PCA which is developed for time dependent variables [6]. It has the same properties as PCA but in addition it is applied in order to uncover oscillation patterns in the measured time series. This study suggests to use SSA rather than conventional modal analysis in order to analyse the oscillation patterns in the vibration response of a nonlinearly vibrating structure. The oscillations in the vibration response signal of a linearly vibrating structure can be associated with its modal behaviour and they correspond to its natural frequencies and are characterised by its mode shapes. Modal analysis cannot be applied to structures with nonlinearities. Also experimentally oscillations and harmonics are quite difficult or even impossible to uncover form the measured time or frequency domain response. For structures made of composites it is known that they usually demonstrate complex vibratory beahaviour which might exhibit double or very close resonances [7]. Our suggestion is to use SSA in place of modal analysis to uncover the oscillations/periodicity in the vibration response of nonlinearly vibrating structures. To our knowledge this is the first study to suggest this idea. The advantage of SSA is that it can uncover oscillations at any frequencies rather than at a certain set of frequencies.

The rest of the paper is organised as follows. The next paragraph introduces SSA as and the way it is applied to vibrating structures. &3 considers the case study of a vibrating composite beam and the problem considered. &4 is dedicated to the analysis of the obtained results.

2. The suggested SSA method and its application to vibrating structures

The vibration response of structure is measured by measuring its acceleration, velocity or displacement. These quantities are time series. In this investigation the measured accelerations are used. So one will be in possession of time series which on most occasions consists of equidistant time measurements, i.e.

$$\mathbf{x}' = (x_i) = x(t_i) = x(t_0 + (i-1)\Delta t) \quad \textbf{(1)}$$

where Δt is the sampling interval.

The idea of SSA is simple: a principal components analysis is applied to variables analysed being lagged versions of a single time series variable. More specifically our p variables are given by equation (2) where $i=1,2,..,p$. It is also assumed that we have m realisations $\mathbf{x}'_j = (x_{ij}) = (x_{1j}, x_{2j},.., x_{pj})$ of the p-dimensional variable x. So one is in possession of a data matrix **X**, which elements are X_{ij}, $i=1,2,..,p$ and $j=1,2,..,m$ and it has the vectors $\mathbf{x}_i$ as its rows. The new variables, the principle components, z_i are linear functions of x_i found using the covariance matrix of $\mathbf{X}$, $\mathbf{\Sigma}$ according following the next relations

$$\begin{gathered}\mathbf{z} = \mathbf{A}'.\mathbf{x} \\ (\Sigma - \lambda\mathbf{I}).\mathbf{A}' = 0\end{gathered} \quad \textbf{(2)}$$

where **I** is the identity matrix , λ are the eigen-values of $\mathbf{\Sigma}$ and **A** is the matrix of its eigen-vectors. Usually the first several PC's, corresponding to the first largest eigen values, contain most of the variance of the original variables and thus the variables $x_j, j=1,2,..,p$ can be represented by a much smaller number of new variables $z_k, k=1,2,..,q$, $q<<p$ which are independent.

If a time series has an oscillatory component, SSA will have an associated pair of eigenvectors with identical eigen-values. We are suggesting to use SSA in order to uncover periodicity in the measured vibration response. These periodic components will correspond to the natural frequencies of the structure under investigation. The corresponding eigen-vectors can be also regarded as characteristics of the vibrating system as the new principal components. Another important issue about PCA and SSA is that for the case of two or more categories of variables the distance between the new PC's from the same category tends to decrease while in the same time the distance between PC's from different categories increases.

3. Damage assessment using SSA

Damage assessment using vibration measurements is based on the fact that changes introduced in the structure result in change in its vibration response. It is generally argued that damage causes stiffness reduction in the structure. It is known that some damage scenarios change the natural frequencies of the structure under investigation. In this study we consider free decay response and thus the oscillatory patterns are associated with the natural frequencies of the structure. As a result of the SSA analysis one obtains the corresponding loading vectors for the PC's (the eigen vectors) and also the eigen values. Thus if one does an analogy to modal analysis SSA gives the possibility to easily estimate the eigen vectors as the eigen values from equation (2) which characterise the time series in question. Our argument is that these characteristics will change as a result of damage since the structural properties are changed and the response signal will be changed. And we prove this in the following example.

4. The structure and the experiment performed

The structure in consideration is a composite beam made of carbon fibre with the following dimensions: length 500 mm, width 50 mm and thickness 6mm. They are made of ten layers. The beam is clamped at both ends. The experiment performed is the so-called 'bump test' where the beam is given a bump by hammer and is then left to vibrate on its own. The acceleration in the middle of the beam is measured. The time domain signals are used to perform the SSA analysis while the frequency domain signals are used to estimate the natural frequencies of the structure. The tests are repeated twenty times and thus twenty different signal realisations $x_i, i = 1,2,...,n', n' = 20$ are recorded. Each signal is measured for one second and is sampled at 500 Hz so that we can uncover the first several natural frequencies of the beam.

For the SSA one is in possession of $n' = 20$ signals $\boldsymbol{x}_i$ each one made of *p=1000* points. Thus our data matrix $\mathbf{S}(n', p)$ is made of 20 rows and 1000 columns.

We introduce delamination across the whole width of the beam between two layers. Different delamination sizes are introduced by changing the length of the delamination. We introduce three different delamination sizes, namely small delamination-*10mm*, medium dlamination-*20mm* and large delamination *-30mm*. Delamination is also introduced in three different locations with respect to the beam length, namely 10 mm from the left end of the beam, in the middle of the beam and 10 mm from the right end of the beam. These are referred to as *left, centre* and *right* delamination, respectively. We also introduce three different positions of delamination across the beam thickness, between the 1st and 2nd layers, between the 5th and 6th layers and between the 9th and the 10th layers, which are referred to as *upper, middle* and *lower* delaminations respectively.

5. Delamination detection and estimation-Analysis of results.

5.1. Delamination detection using the natural frequencies.

We first look at the frequency domain vibration response signals for the intact beam and try to estimate its natural frequencies of vibration. One such signal is given on Figure 1. We were able to get the first three natural frequencies from all the recorded signals. The fourth and the fifth natural frequencies were obtained from 15 tests and in five tests we were not able to obtain them. Secondly it should be noted that the obtained estimates of the natural frequencies possess quite some variance which is between 5% and nearly 8%. Furthermore the changes of the natural frequencies due to delamination turned out to be in the range of the errors of the estimates. These excluded the natural frequencies as possible features for delamination detection.

5.2. Delamination detection using singular spectrum analysis

Next we consider the results from the SSA. Let us first look for oscillation patterns in the free decay behaviour. Table 1 gives the first twelve PC's with the corresponding individual and cumulative variances and the eigen-values/frequencies for the undamaged beam. It can be seen that there are five pairs of eigen-vectors with equal or very close eigen-values in the first twelve PC's. The first two PC's have equal variances of 18.2% and altogether they are responsible for 36.4% of the total variance. A similar pattern can be observed for the third and the fourth PC's and thus can be assumed to define an oscillation pattern in the beam behaviour.The sixth and the seventh PC's also define oscillation and the same is true for the 9-th and the 10-th and the 11-th and the 12-th PC's which suggest two more oscillation patterns in the beam behaviour. The fifth and the eighth PC's do not seem to define oscillation patterns.

PC No	Individual variance	Cumulative variance	Eigen value/frequency	Oscillation pattern
1	18.2	18.2	39.5829(6.3 Hz)	+
2	18.2	36.4	39.5829(6.3 Hz)	+
3	12	48.4	194.773(31 Hz)	+
4	12.2	60.6	197.9145(31.5 Hz)	+
5	10	70.6		-
6	4.1	74.7	571.753(91 Hz)	+
7	3.9	78.6	571.753(91 Hz)	+
8	3	81.6		-
9	2.6	84.2	753.96(120 Hz)	+
10	2.5	86.7	757.1015(120.5 Hz)	+
11	2	88.7	980.148(156 Hz)	+
12	2	90.7	977.0065(155.5 Hz)	+

Table 1. Variances in *per cent* and eigen-values for the first twelve PC's

We now look at the corresponding eigen-values which define the frequencies of oscillation for the PC's with close/equal variances (see Table 1). It should be noted that they are quite close to the natural frequencies of the beam. So as was expected the oscillation patterns found by SSA correspond to the natural frequencies of vibration of the beam. Thus obviously SSA was able to find these natural frequencies while in this particular case it was difficult to obtain them experimentally. SSA got these oscillation patterns quite easily and readily.

Table 2 gives the percentage changes for the first PC with delamination for the different delamination locations and sizes. It seem to contain information for the presence of delamination and its size. But it does not seem to change a lot as a result of the position of delamination. The second PC has very similar behaviour: again it changes with delamination but is not sensitive to its location.This behaviour is to be expected and it can be explained by the fact that the first natural frequencies are global structural characteristics. The percentage changes for the third PC are given in Table 3. This PC seems to contrast between central and side (left/right) dealamination. It also changes as a result of the size of the delamination but these changes are smaller than the ones experienced by the first two PC's. The fourth PC has similar behaviour as the behaviour of the third PC- it also contrasts between central and side delamination. The sixth and the seventh PC's seem to contrast between the two positions of delamination according to the beam thickness-namely between upper/lower and middle delamination. Table 4 gives the percentage changes of the sixth PC with delamination. It can be perceived that the changes for delamination close to the surface of the beam –upper and/or lower delamination- are much smaller as compared to the changes that this PC experiences for middle delamination. The seventh PC has similar properties.

6. Some comments and conclusions

This work considers the use of singular spectrum analysis, which is an alternative of principal components analysis, for the purposes of description and analysis of the vibratory behaviour of a freely vibrating structure. In this paper the application of SSA is demonstrated for the purposes of delamination assessment in a composite beam. It is shown that the PC's contain a lot of information regarding the presence and the location of delamination in the beam. So it can be argued that the PC's obtained can be used with much more success than the natural frequencies of the structure for the purposes of delamination assessment.

	small	**medium**	**large**
Left/right(side)			
Upper/lower	6.0	8.6	12.0
Middle	6.6	8.9	12.0
Centre			
Upper/lower	6.1	10.0	14.0
Middle	6.7	10.8	14.9

Table 2. PC1 percentage change with delamination.

	small	**medium**	**large**
Left/right(side)			
Upper/lower	4.1	6.0	10.0
Middle	4.7	6.8	11.3
Centre			
Upper/lower	7.2	9.1	13.0
Middle	7.8	9.8	14.1

Table 3. PC3 percentage change with delamination

	small	**medium**	**large**
Left/right(side)			
Upper/lower	5.1	6.0	9.0
Middle	8.0	10.6	15.6
Centre			
Upper/lower	5.6	6.8	11.0
Middle	8.9	11.0	16.1

Table 4. PC6 percentage change with delamination

7. References

[1] Askari, D., & Ghasemi-Nejhad, M. 2008 *Proc.ASME International Mechanical Engineering Congress and Exposition* **10**(part A) 289.
[2] Chiang, C., Mei, C., & Gray Jr., C. 1991 *Journal of vibration, acoustics, stress, and reliability in design* **113** 3 309.
[3] Poulimenos A.G., Fassois S.D., 2006, *MSSP* **20** 763
[4] De Boe P, Golinval J.-C.,2003 *Structural Health Monitoring* **2** 2 137.
[5] Trendafilova I, Cartmell M, Ostachowicz W., 2010 Journal of Sound and Vibration 3 3-5 560
[6] Jolliffe I. *Principal components analysis* 2002 Springer
[7] Minak G,Palazzetti R.,Trendafilova I., Zucchelli 2010 *Mechanics of composite materials* **46** 4 387

Analysis of Reciprocity Breakdown in Nonlinear Systems

Graeme Manson, Keith Worden and Michael Wood

Department of Mechanical Engineering, University of Sheffield, Mappin Street, Sheffield, S1 3JD, UK

E-mail: graeme.manson@sheffield.ac.uk

Abstract. The breakdown of the principle of reciprocity is a well-known phenomenon of nonlinear systems. A structure or system is said to exhibit reciprocity when the response at some point j to an input at some point i is identical to the response at point i when the same input is applied at point j. This paper seeks to explain this phenomenon by adopting a functional series representation which describes the input-output relationship. The frequency-domain Volterra Series representation utilises Higher-Order Frequency Response Functions (HFRFs) (generalisations of the linear FRF) to explain the behaviour of nonlinear systems. This breakdown in reciprocity may be observed through a breakdown in symmetry of HFRFs.

1. Introduction

There exists a number of techniques for identifying whether a system is linear or nonlinear. The most well-known of these techniques are the principle of superposition, and its more restricted forms of harmonic distortion and homogeneity, and the coherence function [1]. The subject of the current paper is the property of reciprocity which, for a linear system, states that the response at some point j to an input at some point i is identical to the response at point i when the same input is applied at point j. An alternative way of expressing this behaviour is that the frequency response function measured at response point j to an input at point i, $H^{(j:i)}(\omega)$, is equal to $H^{(i:j)}(\omega)$. Although the breakdown of reciprocity is often used to detect the presence of nonlinearity, such as in the study of Farrar et al. [2] where the reciprocity breakdown in a reinforced-concrete bridge column was observed due to the presence of multiple fatigue cracks, there has been little investigation to explain this behaviour or attempt to exploit it.

The current work adopts a Volterra series approach to attempt to better understand the breakdown of reciprocity in nonlinear systems and is the latest in a series of papers [3, 4, 5] which extend the fsunderstanding of structural dynamics observables using the Volterra functional series approximation of nonlinear system behaviour.

It is assumed throughout this work that the Volterra series representation exists for the systems. Sufficient conditions for this requirement are known [6] [7]. Essentially, it is required that the system nonlinearity be analytic. Another restriction is that the nonlinear system must have finite memory. Essentially this means that the system will reach a steady-state condition not governed by the initial conditions. This requirement precludes the use of the Volterra series for hysteretic systems and systems with limit cycles. Of greater concern is the issue of convergence. The Volterra series being a functional form of the Taylor series suffers similar limitations [8] but, in the current work, the issue of series convergence is not relevant.

2. Volterra Series Representation of Nonlinear System Response

The Volterra series representation [9] [10] of the output response of a nonlinear system, $y(t)$, to an input, $x(t)$, is a generalisation of the input-output relation for linear systems and takes the form of an infinite series:

$$y(t) = y_1(t) + y_2(t) + y_3(t) + \dots \tag{1}$$

where

$$y_n(t) = \int_{-\infty}^{+\infty} \dots \int_{-\infty}^{+\infty} d\tau_1 \dots d\tau_n \, h_n(\tau_1, \dots, \tau_n) x(t-\tau_1) \dots x(t-\tau_n) \tag{2}$$

The functions $h_n(\tau_1, \dots, \tau_n)$ are known as the *Volterra kernels* of the system.

As in the linear case, there exists a dual frequency-domain representation for nonlinear systems. The *higher order FRFs (HFRFs)* or *Volterra kernel transforms* $H_n(\omega_1, \dots, \omega_n)$, $n = 1, \dots, \infty$ are defined as the multi-dimensional Fourier transforms of the kernels, i.e.

$$H_n(\omega_1, \dots, \omega_n) = \int_{-\infty}^{+\infty} \dots \int_{-\infty}^{+\infty} d\tau_1 \dots d\tau_n \, h_n(\tau_1, \dots, \tau_n) e^{-i(\omega_1\tau_1 + \dots + \omega_n\tau_n)} \tag{3}$$

which leads to the frequency-domain dual of equation (1):

$$Y(\omega) = Y_1(\omega) + Y_2(\omega) + Y_3(\omega) + \dots \tag{4}$$

where

$$Y_n(\omega) = \frac{1}{(2\pi)^{n-1}} \int_{-\infty}^{+\infty} \dots \int_{-\infty}^{+\infty} d\omega_1 \dots d\omega_{n-1} H_n(\omega_1, \dots, \omega - \omega_1 - \dots - \omega_{n-1})$$

$$\times X(\omega_1) \dots X(\omega - \omega_1 - \dots - \omega_{n-1}) \tag{5}$$

3. Volterra Series Representation of Reciprocity Breakdown

In this section the Volterra series will be used to demonstrate the breakdown of reciprocity in nonlinear systems under sinusoidal forcing and investigate the conditions under which reciprocity does not breakdown despite the presence of nonlinearity.

3.1. Volterra Series Representation of Response to Sinusoidal Input

For the purposes of the current work, a sinusoidal input force will be employed in order to illustrate the breakdown of reciprocity in nonlinear systems but, it should be made clear that this breakdown can occur regardless of the nature of the forcing input.

If an input is applied at some point i, the Volterra series representation of the response at some point j will be:

$$y^{(j:i)}(t) = y_1^{(j:i)}(t) + y_2^{(j:i)}(t) + y_3^{(j:i)}(t) + \dots \tag{6}$$

Note that this is simply equation (1) with the addition of superscripts to reflect that this response is at some point j to an input at some point i. If a cosinusoidal input, $x^i(t) = X\cos(\Omega t)$, is applied at point i, it may be shown [1] that the individual response components will be:

$$y_1^{(j:i)}(t) = X \left| H_1^{(j:i)}(\Omega) \right| \cos\left(\Omega t + \angle H_1^{(j:i)}(\Omega)\right)$$

$$y_2^{(j:i)}(t) = \frac{X^2}{2} \left\{ \left| H_2^{(j:i)}(\Omega, \Omega) \right| \cos\left(2\Omega t + \angle H_2^{(j:i)}(\Omega, \Omega)\right) + H_2^{(j:i)}(\Omega, -\Omega) \right\}$$

$$y_3^{(j:i)}(t) = \frac{X^3}{4}\left\{\left|H_3^{(j:i)}(\Omega,\Omega,\Omega)\right|\cos\left(3\Omega t + \angle H_3^{(j:i)}(\Omega,\Omega,\Omega)\right)\right.$$
$$\left. + 3\left|H_3^{(j:i)}(\Omega,\Omega,-\Omega)\right|\cos\left(\Omega t + \angle H_3^{(j:i)}(\Omega,\Omega,-\Omega)\right)\right\} \tag{7}$$

etc. Note that the addition of the superscripts on the HFRFs also represent the response point j and the input point i.

Examination of the previous components shows that each odd-order HFRF contributes a component to each odd harmonic of that order or less and each even-order HFRF contributes a component to each even harmonic of that order or less: this includes a DC component.

The FRF of a linear system can be thought of as the magnitude and phase relationship between an input sinusoid and the output sinusoid of the same frequency and it is possible to adopt this representation for nonlinear systems to construct describing functions which the authors have termed composite frequency response functions [1]. The first sinusoidal composite FRF for response point j to an input at point i, $\Lambda_{s1}^{(j:i)}(\Omega)$, may be defined as:

$$\Lambda_{s1}^{(j:i)}(\Omega) = \frac{Y^{(j:i)}(\Omega)}{X^{(i)}(\Omega)} = H_1^{(j:i)}(\Omega) + \frac{3X^2}{4}H_3^{(j:i)}(\Omega,\Omega,-\Omega) + \frac{5X^4}{8}H_5^{(j:i)}(\Omega,\Omega,\Omega,-\Omega,-\Omega) + \dots \tag{8}$$

The s subscript is to emphasise that this composite FRF is for a nonlinear system under sinusoidal forcing as opposed to other forcing, such as random in which case an alternative composite FRF can be defined [3]. By examining the response at other harmonics it is possible to obtain the higher-order composite FRFs. e.g.

$$\Lambda_{s2}^{(j:i)}(\Omega,\Omega) = \frac{2Y^{(j:i)}(2\Omega)}{X^{(i)}(\Omega)^2} = H_2^{(j:i)}(\Omega,\Omega) + X^2 H_4^{(j:i)}(\Omega,\Omega,\Omega,-\Omega)$$
$$+ \frac{15X^4}{16}H_6^{(j:i)}(\Omega,\Omega,\Omega,\Omega,-\Omega,-\Omega) + \dots \tag{9}$$

and

$$\Lambda_{s3}^{(j:i)}(\Omega,\Omega,\Omega) = \frac{4Y^{(j:i)}(3\Omega)}{X^{(i)}(\Omega)^3} = H_3^{(j:i)}(\Omega,\Omega,\Omega) + \frac{5X^2}{4}H_5^{(j:i)}(\Omega,\Omega,\Omega,\Omega-\Omega)$$
$$+ \frac{21X^4}{16}H_7^{(j:i)}(\Omega,\Omega,\Omega,\Omega,\Omega,-\Omega,-\Omega) + \dots \tag{10}$$

3.2. Conditions for Reciprocity in a Nonlinear System

In order for reciprocity to hold between two points i and j in a nonlinear system it is necessary that the response at point j to an input at point i be identical to the response at point i to the same input applied at point j, i.e.

$$y^{(j:i)}(t) = y^{(i:j)}(t) \tag{11}$$

Examination of equations 6 to 10 shows that, for this to be the case, it is necessary that each of the composite FRFs must display reciprocity, i.e.

$$\Lambda_{s1}^{(j:i)}(\Omega) = \Lambda_{s1}^{(i:j)}(\Omega) \quad ; \quad \Lambda_{s2}^{(j:i)}(\Omega,\Omega) = \Lambda_{s2}^{(i:j)}(\Omega,\Omega) \quad ; \quad \Lambda_{s3}^{(j:i)}(\Omega,\Omega,\Omega) = \Lambda_{s3}^{(i:j)}(\Omega,\Omega,\Omega) \quad \dots \tag{12}$$

For this to be the case for all levels of forcing, X, (validity of the Volterra series permitting), it will be necessary that all of the HFRFs (or at least their diagonal terms in the case of sinusoidal forcing) display symmetry, i.e.

$$H_n^{(j:i)}(\Omega,\Omega,\dots,-\Omega,-\Omega) = H_n^{(i:j)}(\Omega,\Omega,\dots,-\Omega,-\Omega) \tag{13}$$

The form of these HFRFs for a particular multi-degree-of-freedom system will now be considered.

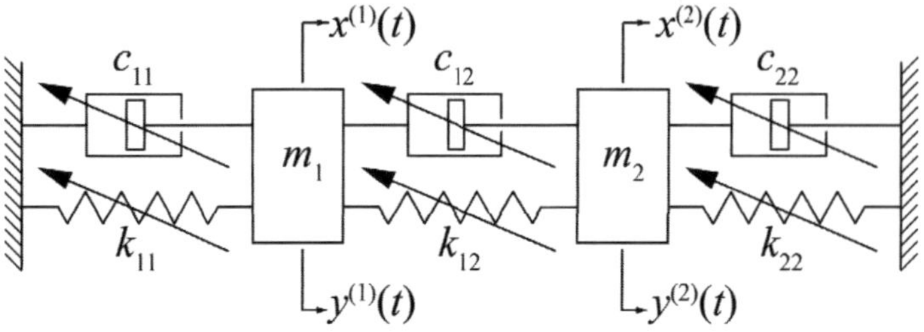

Figure 1. 2 degree-of-freedom nonlinear system

3.3. HFRF Calculation via Harmonic Probing

In order to determine the analytical form of the HFRFs, the method of harmonic probing was introduced by Bedrosian and Rice in [11] specifically for systems with continuous-time equations of motion. The procedure essentially consists of setting the input to be one or more pure harmonics (a single harmonic is used to calculate the linear FRF expression and multiple harmonics are used to calculate the higher-order FRFs), calculating the relevant terms of the Volterra response to this harmonic input then substituting the input and response expressions into the equations of motion and equating terms at frequencies of interest. This is best illustrated by considering a particular system.

A 2 degree-of-freedom system with the possibility of quadratic and cubic stiffness and damping elements at each location is shown in Figure 1. The order of the nonlinear elements is given by the subscript to the left of the c or k and the connectivity of the elements are indicated by the subscripts to the right of the c or k (where 11, 22 and 12 indicate a grounded element connected to mass m_1, a grounded element connected to mass m_2 and an internal element between masses m_1 and m_2 respectively).

This system has the following equations of motion:

$$\begin{aligned} m_1\ddot{y}^{(1)}(t) + \sum_{p=1}^{p=3} \Big\{ {}_p c_{11}\dot{y}^{(1)}(t)^p + {}_p c_{12}\left(\dot{y}^{(1)} - \dot{y}^{(2)}(t)\right)^p & \\ + {}_p k_{11} y^{(1)}(t)^p + {}_p k_{12}\left(y^{(1)}(t) - y^{(2)}(t)\right)^p \Big\} &= x^{(1)}(t) \\ m_2\ddot{y}^{(2)}(t) + \sum_{p=1}^{p=3} \Big\{ {}_p c_{22}\dot{y}^{(2)}(t)^p - {}_p c_{12}\left(\dot{y}^{(1)}(t) - \dot{y}^{(2)}(t)\right)^p & \\ + {}_p k_{22} y^{(2)}(t)^p - {}_p k_{12}\left(y^{(1)}(t) - y^{(2)}(t))\right)^p \Big\} &= x^{(2)}(t) \end{aligned} \tag{14}$$

In order to calculate an expression for the linear HFRF matrix, a periodic excitation composed of a single harmonic is applied at mass 1 (the subscript $p1$ denotes that this is the first probing input):

$$x_{p1}^{(1)}(t) = e^{i\Omega t} \tag{15}$$

Substituting this expression into equation (2) and forming the total response as in equation (1) gives an expression for the responses at mass j where $j = 1, 2$ due to the input at mass 1:

$$y^{(j:1)}(t) = H_1^{(j:1)}(\Omega)e^{i\Omega t} + H_2^{(j:1)}(\Omega, \Omega)e^{i2\Omega t} + H_3^{(j:1)}(\Omega, \Omega, \Omega)e^{i3\Omega t} + \dots \tag{16}$$

It transpires that only the first term in the above expression is required for calculating the H_1 matrix and therefore the probing expression for the output responses will be:

$$y_{p1}^{(j:1)}(t) = H_1^{(j:1)}(\Omega)e^{i\Omega t} \tag{17}$$

Substituting the expressions for the first probing input and first probing responses of equations (15) and (17) respectively into the equations of motion (14) and equation coefficients of $e^{i\Omega t}$ leads to:

$$\left\{ \begin{array}{c} H_1^{(1:1)}(\Omega) \\ H_1^{(2:1)}(\Omega) \end{array} \right\} \left[\begin{array}{cc} -\Omega^2 m_1 + (i\Omega)({}_1c_{11} + {}_1c_{12}) & -(i\Omega){}_1c_{12} \\ +({}_1k_{11} + {}_1k_{12}) & -{}_1k_{12} \\ & \\ -(i\Omega){}_1c_{12} & -\Omega^2 m_2 + (i\Omega)({}_1c_{12} + {}_1c_{22}) \\ -{}_1k_{12} & +({}_1k_{12} + {}_1k_{22}) \end{array} \right] = \left\{ \begin{array}{c} 1 \\ 0 \end{array} \right\} \tag{18}$$

Repeating this procedure but applying the single harmonic at mass 2 and combining with the above result leads to the familiar expression for the H_1 matrix for a linear 2 degree-of-freedom system:

$$\begin{bmatrix} H_1^{(1:1)}(\Omega) & H_1^{(1:2)}(\Omega) \\ H_1^{(2:1)}(\Omega) & H_1^{(2:2)}(\Omega) \end{bmatrix} = \left[\begin{array}{cc} -\Omega^2 m_1 + (i\Omega)({}_1c_{11} + {}_1c_{12}) & -(i\Omega){}_1c_{12} \\ +({}_1k_{11} + {}_1k_{12}) & -{}_1k_{12} \\ & \\ -(i\Omega){}_1c_{12} & -\Omega^2 m_2 + (i\Omega)({}_1c_{12} + {}_1c_{22}) \\ -{}_1k_{12} & +({}_1k_{12} + {}_1k_{22}) \end{array} \right]^{-1} \tag{19}$$

Reciprocity up to this order is confirmed by the fact that $H_1^{(1:2)}(\Omega) = H_1^{(2:1)}(\Omega)$ for all Ω.

In order to obtain an expression for the H_2 matrix, the input probing expression takes the form of a two harmonic input, applied first at mass 1 and then at mass 2:

$$x_{p2}^{(i)}(t) = e^{i\Omega_1 t} + e^{i\Omega_2 t} \tag{20}$$

and the resulting probing response expressions are of the form

$$y_{p2}^{(j:i)}(t) = H_1^{(j:i)}(\Omega_2)e^{i\Omega_1 t} + H_1^{(j:i)}(\Omega_2)e^{i\Omega_2 t} + 2H_2^{(j:i)}(\Omega_1, \Omega_2)e^{i(\Omega_1+\Omega_2)t} \tag{21}$$

Substituting the above two probing expressions into the equations of motion and equating coefficients of $e^{i(\Omega_1+\Omega_2)t}$ leads to the following expression for the H_2 matrix:

$$\begin{bmatrix} H_2^{(1:1)}(\Omega_1,\Omega_2) & H_2^{(1:2)}(\Omega_1,\Omega_2) \\ H_2^{(2:1)}(\Omega_1,\Omega_2) & H_2^{(2:2)}(\Omega_1,\Omega_2) \end{bmatrix} = \begin{bmatrix} H_1^{(1:1)}(\Omega_1+\Omega_2) & H_1^{(1:2)}(\Omega_1+\Omega_2) \\ H_1^{(2:1)}(\Omega_1+\Omega_2) & H_1^{(2:2)}(\Omega_1+\Omega_2) \end{bmatrix} \begin{bmatrix} N_2^{(1:1)} & N_2^{(1:2)} \\ N_2^{(2:1)} & N_2^{(2:2)} \end{bmatrix} \tag{22}$$

where N_2 is the matrix which depends upon the quadratic nonlinearities present in the system under consideration. The rows of N_2 are given by:

$$\begin{aligned} N_2^{(1:i)} &= -\{(i\Omega_1)(i\Omega_2){}_2c_{11} + {}_2k_{11}\} H_1^{(1:i)}(\Omega_1)H_1^{(1:i)}(\Omega_2) \\ &\quad - \{(i\Omega_1)(i\Omega_2){}_2c_{12} + {}_2k_{12}\} \left[H_1^{(1:i)}(\Omega_1) - H_1^{(2:i)}(\Omega_1)\right] \left[H_1^{(1:i)}(\Omega_2) - H_1^{(2:i)}(\Omega_2)\right] \\ N_2^{(2:i)} &= -\{(i\Omega_1)(i\Omega_2){}_2c_{22} + {}_2k_{22}\} H_1^{(2:i)}(\Omega_1)H_1^{(2:i)}(\Omega_2) \\ &\quad + \{(i\Omega_1)(i\Omega_2){}_2c_{12} + {}_2k_{12}\} \left[H_1^{(1:i)}(\Omega_1) - H_1^{(2:i)}(\Omega_1)\right] \left[H_1^{(1:i)}(\Omega_2) - H_1^{(2:i)}(\Omega_2)\right] \end{aligned} \tag{23}$$

for $i = 1, 2$. Whilst it is not relevant for the current work, it is also possible to calculate the expression for the H_2 cross-kernel transform, which describes the interaction due to inputs at multiple points, using an extension to the harmonic probing procedure [12].

A similar procedure to that above may be followed to obtain the H_3 matrix but, due to space considerations (each element of the N_3 matrix is composed of eight components), it will not be included here.

3.4. Conditions for HFRF Symmetry and therefore Reciprocity

Expansion of equation (22) gives the following expressions for $H_2^{(1:2)}(\Omega_1,\Omega_2)$ and $H_2^{(2:1)}(\Omega_1,\Omega_2)$:

$$\begin{aligned} H_2^{(1:2)}(\Omega_1,\Omega_2) &= H_1^{(1:1)}(\Omega_1+\Omega_2)N_2^{(1:2)} + H_1^{(1:2)}(\Omega_1+\Omega_2)N_2^{(2:2)} \\ H_2^{(2:1)}(\Omega_1,\Omega_2) &= H_1^{(2:1)}(\Omega_1+\Omega_2)N_2^{(1:1)} + H_1^{(2:2)}(\Omega_1+\Omega_2)N_2^{(2:1)} \end{aligned} \tag{24}$$

Examination of the above equations show that each of the $H_2(\Omega_1,\Omega_2)$ expressions are made up of a summation of two complex products of a linear H_1 term and a nonlinear N_2 term which is some function of the nonlinear parameters and H_1 (in an m-dof system there will be m complex products for each HFRF expression). The only two terms which are guaranteed to be equal are the $H_1^{(1:2)}(\Omega_1+\Omega_2)$ and $H_1^{(2:1)}(\Omega_1+\Omega_2)$ terms, due to linear reciprocity as shown in equation (19). It appears that the only situation which will result in the above two expressions being equal (for all Ω_1 and Ω_2) is for the underlying linear system to be symmetrical in its parameters i.e. $m_1 = m_2$, ${}_1c_{11} = {}_1c_{22}$ and ${}_1k_{11} = {}_1k_{22}$ and also for the grounded quadratic parameters to also be equal i.e. ${}_2c_{11} = {}_2c_{22}$ and ${}_2k_{11} = {}_2k_{22}$. It should also be noted that these expressions will not, in general, be equal if either of the quadratic nonlinearities between the two masses are non-zero. So, the final condition for HFRF symmetry for this 2 dof system is ${}_2c_{12} = {}_2k_{12} = 0$. The reason for this final condition is the manner in which the quadratic stiffness and damping forces from the elements between two masses are defined. Under the current definition they always act to push the masses apart for positive values of ${}_2c_{12}$ and ${}_2k_{12}$.

Although the construction of the H_3 matrix was not included above, it may still be examined to see under what conditions the HFRFs $H_3^{(1:2)}(\Omega_1,\Omega_2,\Omega_3)$ and $H_3^{(2:1)}(\Omega_1,\Omega_2,\Omega_3)$ will be equal. It transpires that this is simply a extension of the conditions for H_2: in addition to those conditions, it is also necessary that ${}_3c_{11} = {}_3c_{22}$ and ${}_3k_{11} = {}_3k_{22}$, i.e. continuing symmetry of the parameters. An interesting departure is that the two H_3 terms will be equal, assuming that the previous conditions are met, even if the cubic nonlinearities between the two masses, ${}_3c_{12}$ and ${}_3k_{12}$, are non-zero.

4. Illustrative Examples

In the previous section it was shown that the HFRFs $H_2^{(1:2)}(\Omega_1,\Omega_2)$ and $H_2^{(2:1)}(\Omega_1,\Omega_2)$ will be equal and $H_3^{(1:2)}(\Omega_1,\Omega_2,\Omega_3)$ and $H_3^{(2:1)}(\Omega_1,\Omega_2,\Omega_3)$ will be equal if the underlying linear system were symmetric and the nonlinear parameters were symmetric as long as the internal nonlinearity were of odd order (i.e. has no quadratic component). This will be shown to be the case by considering a numerical example. Consider the system shown in Figure 1 and let the linear parameters be $m_1 = m_2 = 1$ kg, ${}_1c_{11} = {}_1c_{12} = {}_1c_{22} = 20$ N/(m/s) and ${}_1k_{11} = {}_1k_{12} = {}_1k_{22} = 1\times10^4$ N/m. The nonlinear parameters are set to ${}_2c_{11} = {}_2c_{22} = 500$ N/(m/s)^2, ${}_2k_{11} = {}_2k_{22} = 1\times10^7$ N/m^2 and ${}_3k_{11} = {}_3k_{12} = {}_3k_{22} = 5\times10^9$ N/m^3 thereby satisfying the conditions outlined above.

The system was constructed in Simulink and simulated using 4th-order Runge-Kutta. First, an input was applied at mass m_1 given by $x^{(1)}(t) = 10\sin(100t)$ and the response $y^{(2:1)}(t)$ was plotted then the same input was applied at mass m_2 ($x^{(2)}(t) = 10\sin(100t)$) was applied and the response $y^{(1:2)}(t)$ was plotted. The resulting responses for the first 0.5 seconds are shown in Figure 2 and it may be seen that they are identical, as expected. Figure 3 shows the two responses when a quadratic stiffness of ${}_2k_{12} = 1\times10^7$ N/m^2 is added to the previous nonlinear system. It may be seen that, as predicted by the calculations in the previous subsection, this addition has resulted in a breakdown in reciprocity with the two responses no longer being the same.

It should be stated that, whilst a Volterra series approach was used to identify which nonlinear systems displayed reciprocity, there is no necessity that the Volterra series representation be

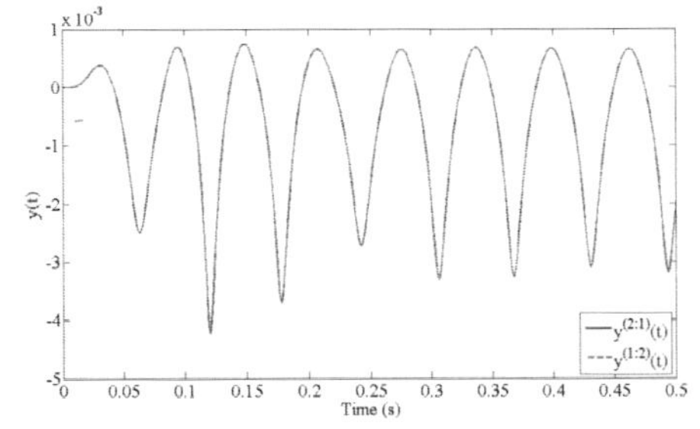

Figure 2. Displacement responses from symmetric nonlinear system with no quadratic nonlinearity between masses 1 and 2.

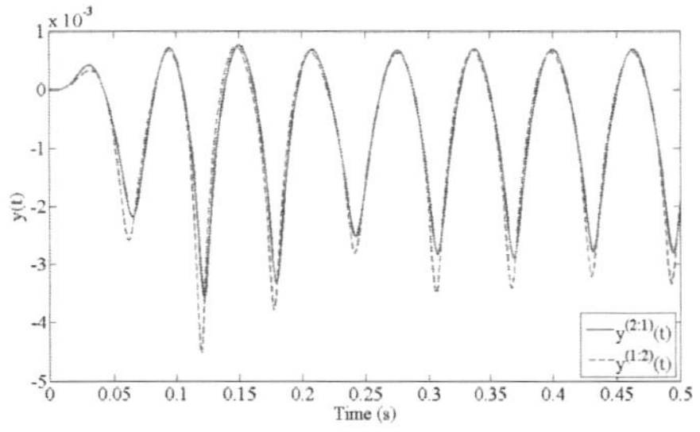

Figure 3. Displacement responses from same nonlinear system as Figure 2 with addition of quadratic stiffness between masses 1 and 2.

convergent for the forcing levels being applied in the example: this would only be necessary if the representation were being used to construct the responses.

5. Discussion and Conclusions

This paper has attempted to illuminate the subject of reciprocity breakdown in nonlinear systems by adopting a Volterra series approach. It should be emphaized that the conclusions are somewhat limited to low-order nonlinearities and that going higher-order may be computationally prohibitive. To a certain extent, the finding that nonlinear systems which are symmetrical in nature are capable of maintaining reciprocity may seem somewhat facile with the possible exception of the demonstration that even-ordered nonlinear elements between masses causes reciprocity breakdown (irrespective of symmetry). That said, it is the authors' opinion that a functional series approach is capable of providing valuable insight into the behaviour of nonlinear systems which may be exploited in such areas as system identification: this will be the subject of forthcoming work.

6. References

[1] Worden K and Tomlinson G R 2001 *Nonlinearity in Structural Dynamics: Detection, Identification and Modelling* (Bristol: Institute of Physics Publishing)
[2] Farrar C R and Worden K 2012 *Structural Health Monitoring: A Machine Learning Perspective* (John Wiley & Sons)
[3] Worden K and Manson G 1998 Random vibrations of a Duffing oscillator using the Volterra series *J. Sound Vibrat.* **217** 781–89
[4] Worden K and Manson G 1999 Random vibrations of a a multi-degree-of-freedom nonlinear system using the Volterra series *J. Sound Vibrat.* **226** 397–405
[5] Worden K and Manson G 2005 A Volterra series approximation to the coherence of the Duffing oscillator *J. Sound Vibrat.* **286** 529–47
[6] Palm G and Poggio T 1977 The Volterra representation and the Wiener expansion: validity and pitfalls *SIAM J. App. Math.* **33** (2) 195–216
[7] Rugh W J 1981 *Nonlinear System Theory - The Volterra/Wiener Approach* (John Hopkins University Press)
[8] Barrett J F 1965 The use of Volterra series to find region of stability of a non-linear differential equation *Int. J. Control* **1** (3) 209–16
[9] Volterra V 1959 *Theory of Functionals and Integral equations* (New York: Dover Publications)
[10] Schetzen M 1980 *The Volterra and Wiener Theories of Nonlinear Systems* (New York: John Wiley Interscience Publication)
[11] Bedrosian E and Rice S O 1971 The output properties of Volterra systems driven by harmonic and Gaussian inputs *P. IEEE* **59** 1688–707
[12] Worden K, Manson G and Tomlinson G R 1997 A harmonic probing algorithm for the multi-input Volterra series *J. Sound Vibrat.* **201** 67–84

A study of the use of artificial neural networks to estimate dynamic displacements due to dynamic loads in bridges

S Ok[1], W Son[2] and Y M Lim[2]

[1] Hyundai Engineering & Construction, 75 Yulgok-ro, Jongno-gu, Seoul 110-920, Korea
[2] Department of Civil & Environment Engieering, Yonsei university, 50 Yonsei-ro, Seodaemun-gu, Seoul 120-749, Korea

E-mail: yunmook@yonsei.ac.kr

Abstract. Measurement of dynamic displacement is one of the most essential aspects of a structural behavior because it portrays history of the global behavior of structure. In general, structural engineers are accepted these response as reliable physical quantities to evaluate the conditions of a structure. The reason is that these physical quantities can easily generate strain as well as stress, velocity and acceleration at the measuring points. However, it is difficult to directly measure the displacement of the bridge due to problems such as test conditions and the limitations of equipment. Therefore, in this study, an artificial neural network (ANN) demonstrates how it could overcome such limitations and utilize the random dynamic load to obtain the reliable estimations. Numerical analysis is conducted to obtain learning data about the axial strain as well as vertical displacement with time frame at multi-points and then applied to the ANN. The scenario centered on a variety of dynamic loads from the analysis of an urban bridge that was selected based on its general volume of traffic. The analysis was performed to estimate its displacement, which corresponds to the strain on the bridge caused by arbitrary loads of leaning results from the ANN. Then, it is confirmed that the estimated displacements of ANN show well agreements with that of an independent set of traffic scenario.

1. Introduction

The displacement of dynamic responses of structures is closely related to the magnitude of damage experienced under ultimate loads. It is a long-term history expressing global behavior, and it is thus important to secure the data (Powell and Allahabadi 1988; Jang and Kim 2008). By measuring the responses of structures to vehicle loads, not only can the characteristics of loads on a bridge be predicted (Moses 1979), but the state of the structure can also be intuitionally judged. Also, measuring the deflection changes of a bridge would be the basis for the criteria of structural degradation (Cho and Kim, 2008). There are various ways and means to measure the physical quantity, yet it is hard to do in a real construction site. To overcome difficulties at the site, in this study, an ANN was used. A way to write more realistic dynamic load scenarios that will be applied to ANN is also suggested.

The ANN has an advantage in that the relationship between strain and displacement can be numerically decided without complicated assumptions and limits. Through the dynamic scenarios and the ANN, we suggest a way to fairly precisely estimate the real-time vertical displacements of a bridge, which are generated by arbitrary dynamic vehicle loads.

Journal of Physics: Conference Series **382** (2012) 012032 doi:10.1088/1742-6596/382/1/012032

2. Numerical Analysis

This study does not offer particular results from a specific bridge, but proposes a way to estimate vertical displacements with time through strain values that are applicable to typical bridges we can see anywhere. Moreover, there are many problems with measuring strain on-site under all of the various load cases. Accordingly, testified numerical analysis needs to be conducted to provide the ANN with data for learning.

2.1. Vehicle Model

A bridge is affected by many factors, such as its weight, length, suspension system, and the natural frequencies of vehicles. There are sophisticated interactions among them, which makes it tough to practically consider every component when making a model. Thus, it was assumed there were only three kinds of vehicles: trucks, cars, and buses; these vehicles were driven at speeds of 40km/h, 60km/h, and 80km/h. As shown by Figure 1, the vehicle model used in the analysis was a three-axle truck consisting of two rigid bodies. To be applied to this numerical analysis, Zeng and Bert's (2003) truck model was modified, and car and bus models were made by eliminating trailers from the truck. Each model's specifications were from Li's (2005) truck, Zuo and Nayfeh's (2003) car, and Ahmed et al.'s (1997) bus.

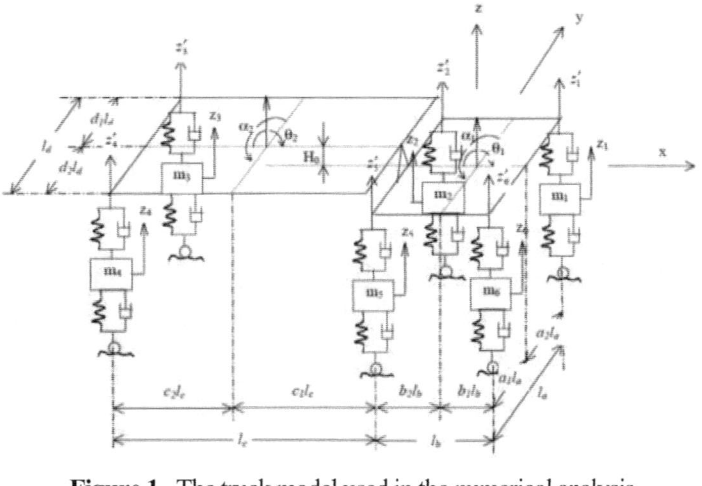

Figure 1. The truck model used in the numerical analysis

2.2. Bridge model

A simply supported slab-girder bridge was selected for the analysis. Each slab had a 36-m span and was 15 m wide (four lanes). A type I steel girder was used. Verification of the model was carried out by experiment results. In modeling, the moment and design loads, as well as the spacing of girders and diaphragms, were calculated in accordance with AASHTO LRFD (AASHTO 2002; AASHTO 2007). Since we assumed that there were small effects by slippage in anchorage, relaxation, shrinkage, and so on, they were negligible in the analysis and only vehicle movement was considered as a load on a bridge. A 3D solid element (linear hexahedral element) was used as an element of the slab, and a 3D shell element (linear quadrilateral element) was used for the girder for the finite element method. The shell element was constrained from rotation around a vertical direction by a tie-constant at the boundary between solid and shell elements. An approach slab was deployed at both ends of the bridge in order to reduce measuring errors, which can occur if vehicles enter and exit.

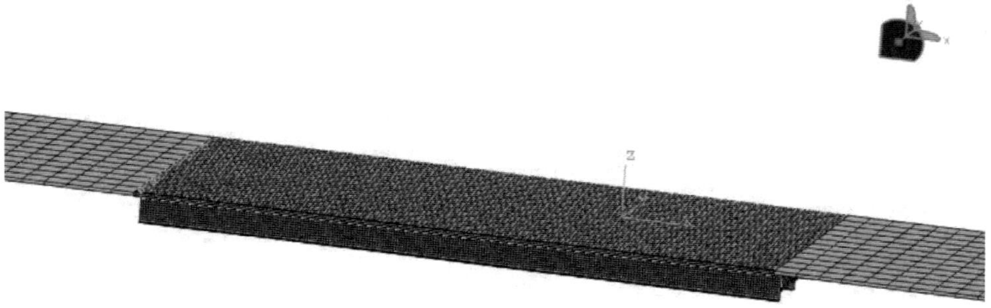

Figure 2. The bridge model in finite element method analysis

2.3. Verifying numerical model
A comparison of the results of the numerical analysis and those of the experiment regarding the behavior of the bridge, the model of the static-state conditions, and the dynamic state was verified. Statically, the experimental values of the Woodruff Bridge in Michigan were used. Dynamically, those of high-speed highway bridges in Florida were used.

3. Dynamic Moving Load Scenario
Information about the kinds of vehicles and distances between them was gained at one bridge in Seoul, Korea. All of the vehicles using the bridge were classified according to three velocities for writing scenarios: 40km/h, 60km/h, and 80km/h. To apply vehicle distances to the scenarios, we introduced a distribution function of the Pearson Type III distribution of traffic theories (May 1990).

$$\mathrm{f(t)} = \frac{\lambda}{\Gamma(K)}[\lambda(t-\alpha)]^{K-1}\, e^{-\lambda(t-\alpha)}$$

In the equation above, **f(t)** represents probability density functions. Influencing a form of the distribution, **K**—ranging from 0 to ∞—is a coefficient determined by the user. If K is 1, the distribution becomes the negative exponential function. $\boldsymbol{\alpha}$ is nonnegative and affects the distribution location. If α is zero, there is no movement. If α becomes 0.5, it means that a time headway under 0.5 s is considered to be zero. $\boldsymbol{\lambda}$ is a coefficient of the function of vehicle distance as determined by K and α. **t** is a time (in seconds) difference of the distance. **Γ(K)** is a gamma function. The observed data being applied to the distribution, a theoretical distribution curve of vehicle movements on the bridge can be obtained, which can describe real traffic situations.

We computed time headway by taking advantage of the above distribution and the data measured at the bridge. Figure 3 shows the measured percentage of vehicles with time headway and the calculated percentage of vehicles with time headway by the distribution in the first and second lanes. Acquiring this theoretical probability of headway, we could portray the simulated traffic situations as being closer to reality. Applying the headway information based on the Pearson Type III distribution, we composed 25 scenarios covering vehicle ratios of headway and arbitrary vehicle distances.

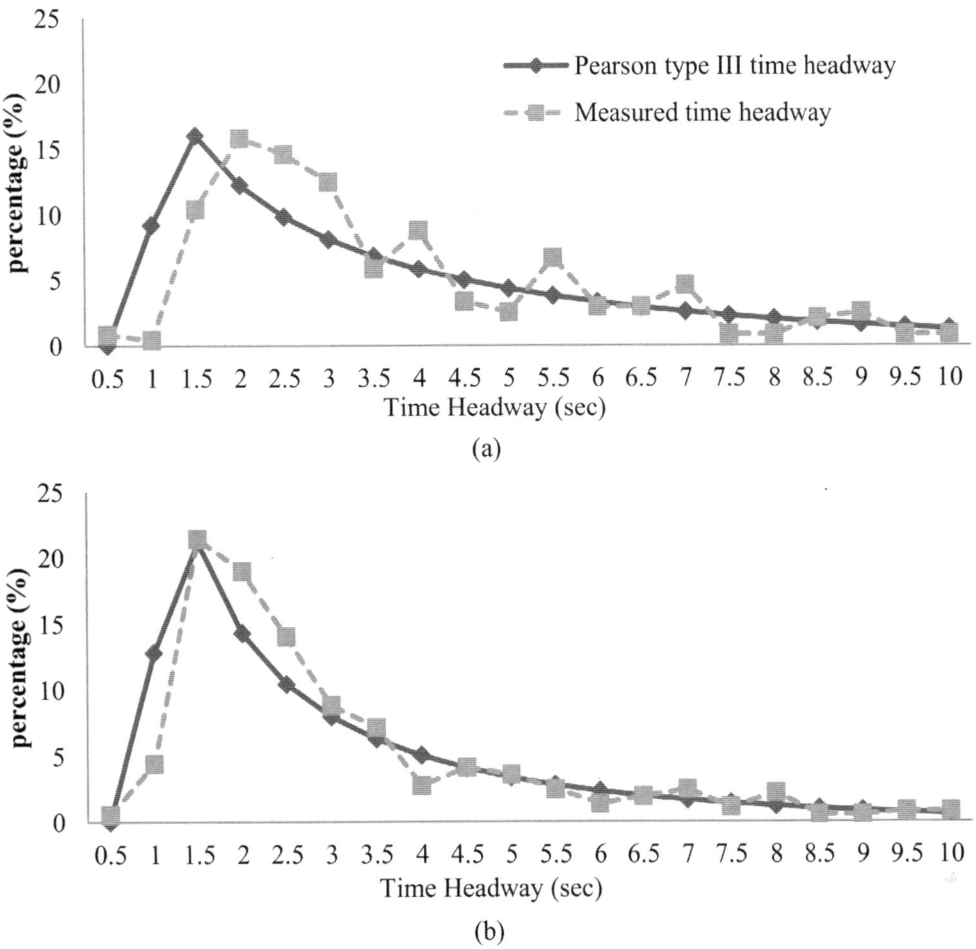

Figure 3. The percentage of vehicles with time headway (a) in the first lane and (b) in the second lane

4. The Artificial Neural Network

There are two steps—*Feedforward* and *Feedback*—in an ANN error back-propagation learning algorithm. In the Feedback step, the output adjusts the synaptic strength between a hidden layer and an output layer to minimize errors related to target values. It repeatedly arranges the strengths between an input layer and the hidden layer through output from the hidden layer. This is called a *generalized delta rule*. In the Feedforward step, errors between the computed values of each unit are found; the errors are reduced by feedback calculation. The generalized delta rule can obtain the synaptic strength, which minimizes the error through the least mean square (LMS) learning rule. This is a way to yield the change volume of the synaptic strength proportional to the value resulting from differentiating the synaptic strength. The process repeats until the sum of the generated errors converges on a threshold of the error being previously determined until the system is stable.

The dynamic strain and dynamic displacement data of a bridge were obtained at 11 points on the bottom of each flange of each girder. The spacing between each point was 3.6m. There was a total of 55 points on five girders. Through the numerical analysis, the strain and displacement by dynamic vehicle loads from the scenarios were measured in 0.01-second process. There were 25 cases depending on each velocity that were arbitrarily made following the traffic theory. Twenty-four cases were used for learning data, and the remainder was used for test data to verify the ANN. The test data were randomly extracted. Fifty-five input-layer elements, 40 hidden layer1elements, 40 hidden layer2 elements, and 55 output-layer elements constituted an ANN structure. The learning ratio was set to 0.1 and the momentum term to 0.9. Non-linear functions, such as sigmoid or tan-sigmoid functions, were suitable for transfer functions since the ranges of the input and output data had upper load limits through a combination of dynamic vehicle loads.

5. Analyses of the Dynamic Load Scenarios and the ANN

The learning by the ANN through the 24 cases converged after 4,962 iterations. The correlation of the ANN after learning was 99.98%, so we can confidently state that the learning worked well. The result of inputting the extracted test data into the learned ANN was 0.0018 mm of the mean square error (MSE). Figure 4 shows the output from the ANN and the test values at a point in the middle of the *second*, *third*, and *fifth* girders. It represents the vertical displacements with time by dynamic vehicle loads. A blue line represents results from the ANN, and a red line represents results from the FEM.

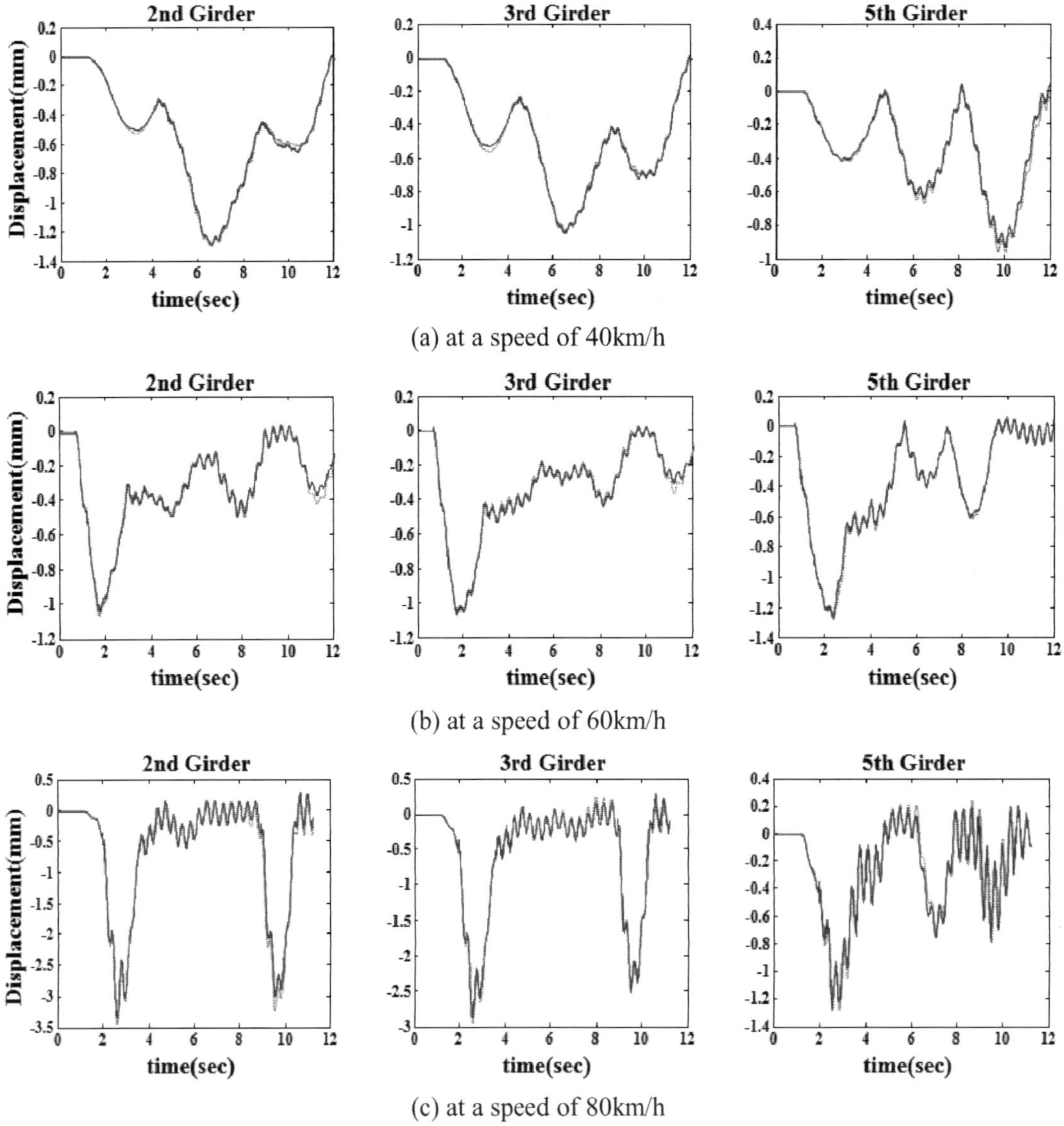

(a) at a speed of 40km/h

(b) at a speed of 60km/h

(c) at a speed of 80km/h

Figure 4. The Dynamic Displacements due to the Dynamic Loading

Both lines mostly match up. It can be confirmed that a time-displacement curve from the numerical analysis using a finite element method fairly corresponds with that from the ANN through the MSE and the figures.

6. Conclusions

The ANN was used to estimate the dynamic vertical displacements of a bridge. There were plenty of constraints to directly measure its strain value. This means that only limitary information about the behavior of the bridge can be gained in the field.

This study proposed how to understand the global behavior of bridges through the limited information on specific points of girders. A bridge model was made to obtain the bridge's data on strain and displacement through the numerical analysis, and dynamic vehicle load scenarios by the Pearson Type III distribution of traffic theory were written to reflect real traffic situations. Finally, we could fairly accurately estimate the vertical dynamic displacement through the ANN, which had been leaned upon by FEM results from loads based on real situations.

As we are going to widen the study ranges from bridges to other structures with difficulties in estimating overall displacement with time due to various load cases and complicated structures, only the measuring strain and displacement at a few points will make it possible to get precise displacement data and maintain sophisticated structures such as LNG tanks and more complex bridges than one we previously considered.

Acknowledgements

This research was supported by a grant from the Korea Gas Corporation.

References

[1] Powell G.H and Allahabadi R 1988 Seismic Damage Prediction By Deterministic Methods : Concept And Procedure *Earthquake Engineering And Structural Dynamics* **16** 719-734

[2] Cho N and Kim N 2008 Prediction of the Static Deflection Profiles on Suspension Bridge by Using FBG strain Sensors *Ksce Journal of Civil Engineering* **28 5A** 699-707

[3] Moses F 1979 Weigh-in-Motion System Using instrumented Bridges *Transportation Engineering Journal* **105 3** 233-249

[4] Zeng H and Bert C.W 2003 Dynamic Amplification Of Bridge/Vehicle Interaction: A Parametric Study For A Skewed Bridge *Journal of Structural Stability and Dynamics* **3 1** 71-90

[5] Li H 2005 Dynamic response of highway bridges subjected to heavy vehicles (Ph.D. diss) The Florida State University

[6] Zuo L and Nayfeh S.A 2003 Structured H2 Optimization of Vehicle Suspensions Based on Multi-Wheel Models *Vehicle system dynamics* **40 5** 351-371

[7] AASHTO 2002 AASHTO Standard Specifications for Highway Bridges 17th Edition Washington, D.C.

[8] AASHTO 2007 AASHTO LRFD Bridge Design Specifications 4th Edition Washington, D.C.

[9] Chun P 2010 Skewed bridge behaviors: Experimental, analytical, and numerical analysis (Ph.D. diss.) Wayne state University

Modern Practice in Stress and Vibration Analysis 2012 (MPSVA 2012) IOP Publishing
Journal of Physics: Conference Series **382** (2012) 012033 doi:10.1088/1742-6596/382/1/012033

A simple frequency-based delamination detection and localization method without baseline model

A Guechaichia[1] **and I Trendafilova**
Department of Mechanical Engineering University of Strathclyde, James Weir Building, 75 Montrose street, Glasgow, G1 IXJ, United Kingdom

E-mail: abdelhamid.guechaichia@strath.ac.uk

Abstract. This study suggests a novel non-model-based method for structural vibration-based health monitoring for composite laminated beams which utilises only the first natural frequency of the beam in order to detect and localise delamination. The method is based on the application of a static force in different positions along the beam. It is shown that the application of a static force on a damaged beam induces forces that push the delaminated layers together resulting in an increase of stiffness to a maximum when the static force is applied on the top and the middle of the delamination area. This stiffness increase in turn causes changes in the structural natural frequencies. The method does not require the frequency of the beam in its baseline condition. A very simple procedure for damage detection is suggested which uses a static force applied at only three points along the beam to detect and localise delamination. The method is numerically validated for a simply supported beam, using a finite element model of the beam. Our results show that the frequency variation with the change of the force application point can be used to detect, localize and in the same time quantify very precisely single delamination.

1. Introduction.

Composite materials, are attractive for structural applications and especially in mechanical and aerospace engineering where high stiffness/weight ratio is required .One of the benefits of composite is that they can be used to design structures with required strength and stiffness by changing stacking sequence, fibre orientation and type of fibre. The complexity of composite materials makes it difficult to fully understand their failure mechanism and to detect damage as most of them occur below the

[1] To whom all correspondence should be addressed

surface .This resulted in an over-dimensioning by using a safety factor by designer and not taking the full advantages of the cost benefits of composite materials.
The most common and critical defect in composites is the delamination between adjacent layers. The delaminations can occur at the manufacturing phase of the composite material, as a result of an impact by a foreign object during installation or in service , effect of temperature , fatigue loading and machining. Delamination is well known to cause a reduction in stiffness and the strength capabilities of the composite, this reduction varies with composite element size, the delamination size and location. All this changes will an affect the vibration behaviour of the composite element.
Delamination cannot be detected by visual inspection, to locate delamination non-destructive inspection methods are required. [1] presented a review on vibration-based model dependent delamination detection methods and health monitoring techniques for composite structures. the sensitivity of the measurement technique used and the parameters chosen for the identification methodology play a key role to damage identification.[2] studied the effect of delamination on the natural frequencies of a laminated composite beam, results showed that an increase in delamination length increases the reduction in the natural frequency.[3] presented the effect length of delamination on the free vibration of composite pre-twisted shells with different stacking sequences. A finite element model has been presented by [4] to analyse the natural frequencies of delaminated composite beams. The effect of the length of the delamination and its position on the bending natural frequencies of a cantiler beam was investigated. An experimental modal analyses was presented by [5] on a simply supported beam to study the effect of delamination on the natural frequencies of composite laminate. In this paper the results from the experiments are checked against a results from a finite element model using the classical beam theory.

Frequency response method in SHM are cheap to implement and can be used to detect damage in a structural component as a defect causing a stiffness reduction will affect the natural frequencies. They can indicate information about the global condition of the system but lot of sensors have to be used to extract some information about the local damaged area.

Most papers that use the natural frequencies for delamination identification are only used for detection and need a baseline frequency.The method we are suggesting is capable to do detect the delamination,localize it and estimate its length. The method that we suggest is based on the application of a force on the beam in different positions. It does not require any baseline frequency. It only needs one application point of the force for detection, two application points are needed for localization and an extra two points are required for length estimation. Nearly all of The mentioned research related to delamination identification is based on the use of several natural frequencies. The method suggested here only uses the first natural frequency of the beam.

2. Methodology.

The method is based on the work of the author [7] where a transverse static force is applied at different position along the damaged beam made of an isotropic material. The damage is detected and located using just the first natural frequency of the beam. No model is assumed and no model is needed for the suggested procedure. The method uses a static force which is applied in different positions along the beam length. The natural frequency of the beam for each position of the static force is measured.

In the case of a composite laminate beam with a delamination, the static force is applied in different location along the delaminated beam which will results in different bending moments values at the centre of the delamination area causing the delaminated layers to be pressed together which will result in a stiffness increase (as compared to the case when no static force is applied) causing the the first natural frequency to increase. As for the case of a simply supported beam the maximum bending moment occurs at the point of application of the transverse force, the stiffness of the delaminated beam will go up as the static force position is moved towards the delamination, reaching a maximum value when the force is applied at the centre of the delamination. At this position the deflection of the beam is maximum and the rotation angle of the deflection is close to zero. So the delaminated layers are pressed closely together which ensures a maximum load transfer.

It has been found in e.g. [6] that tension in a structure increases the natural frequencies, while compression reduces them. In the case of an un-delaminated beam under a transverse static force, as the beam is under bending the part above the neutral line will be under compression while the part beneath the neutral line is under tension. Both forces are equal in amplitude but with opposite signs so their effect on the natural frequencies is cancelled and the natural frequency of the undamaged beam with the static force is the same as the one for the same beam with no force applied. From this we can conclude that the frequency of the delaminated beam with a static force applied on centre of the delamination will increase as compared to the frequency measured without any force applied. to a maximum value equal of that of the undelaminated beam and will never go beyond that.

In this study we use a percentage frequency shift which is introduced as follows:

$$\Delta f = \frac{(f_f - f_{nf})}{f_{nf}}.100 \tag{1}$$

Where f_{nf} is the frequency with no force applied and f_f is the frequency with a static force applied for the delaminated beam correspondingly.

2.1. *Detection*

The detection is done by applyinng a static force in one position along the beam and compring the first natural frequency of the beam to the one with no static force applied. We suggest that the best position for the static force is at midspan because in this case the bending moment at any point of the beam is quite substantial and will have an effect on any delamination along the length of the beam. As a result if there is delamination in the beam there will be a frequency shift when the force is applied in the middle. If there is no frequency shift this will mean there is no delamination or the delamination is rather small or close to the supports of the beam. In such a case other points of application to the left and the right of the the centre are used to look for smaller delamination away from the centre of the beam.

2.2. *Localization*

From the finite element analysis results it has been found that the frequency shift variation with the static force position is almost linear, but as the static force position position reaches the edge of the delamination and moves toward its centre the line changes its gradient and become steeper. Fig(3) shows the frequency shifts for a number of application points of the static force for one such a case.

For the localization, the static force is applied at two symmetric points about the middle of the beam and preferably close to the supports, e.g. at about L/5 to L/4 from the ends of the beam .Let's denote by A an B these two points (see Fig 1). If the frequency shift when the static force is at A is bigger than the one when the force is at B, this means that the centre of the delamination is to the left of the middle of the beam and *vice versa*.The intersection of the segments Δf_A - Δf_O and Δf_B - Δf_C gives point P whose projection on the beam indicates the centre of the delamination (see Fig 1). The distance from the left hand support and the centre of the delamination is given by expression (2) where a and b are the distances of points A and B from the left hand support respectively .

$$x_d = \frac{\dfrac{\Delta f_B}{(L-b)}}{\dfrac{\Delta f_B}{(L-b)} + \dfrac{\Delta f_A}{a}} \tag{2}$$

2.3. Quantification.

To estimate the length of the delamination the static force is applied at two additional points M and N to the the left of the centre of the delamination (see Fig 1). The corresponding frequency shifts, denoted by Δf_M and Δf_N ,define the line of the frequency changes over the delamination. The intersection of the lines $\Delta f_A - \Delta f_O$ and Δf_M - Δf_N gives the point D which indicates the edge of the delamination. Once we know the centre and the end of the delamination its length can be determined. The distance between the left hand support and the edge of the delamination is given by Eq(3).

$$x_e = \frac{(n.\Delta f_M - m.\Delta f_N).a}{(\Delta f_M - \Delta f_N).a - \Delta f_A(m-n)} \tag{3}$$

Where a, m, n are the distances from the left hand support and the points A, M, N respectively. The length of the delamination d is given by the following expression.

$$d = 2(x_d - x_e) \tag{4}$$

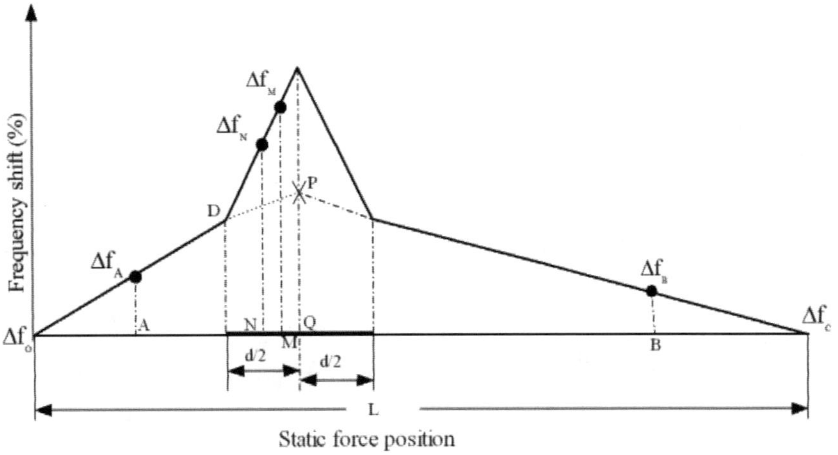

Figure 1: The beam with the delamination and the frequency shift lines.

3. The composite beam model.

The material used is a unidirection carbon / epoxy AS4D/9310 composite with E1 = 133.86 GPa, E2 = E3 = 7.706 GPa, G12 = G13 = 4.306 GPa, G23 = 2.76 GPa,$\nu_{12} = \nu_{13} = 0.301$, $\nu_{23} = 0.396$ and ρ = 1520 kg/m^3. The geometrical parameters of the specimen are length L = 0.35 m, thickness H = 0.002 m, width b = 0.03m.

The 10-layered structure has [0_/90_/0_/90_/0_]s as stacking sequence. All the cases are modeled as a through width delamination and the delaminated layers are free to move and are not constrained. Due to the fact that modal analysis in Ansys cannot deal with nonlinearities ,gaps or contacts, The undamaged beam was analysed using modal and transient analysis to make sure that both methods give the same results . On the delaminated beam a transient analysis was conducted and the results are compared to the ones from the prestressed transient analysis when an 8 N static force is applied.

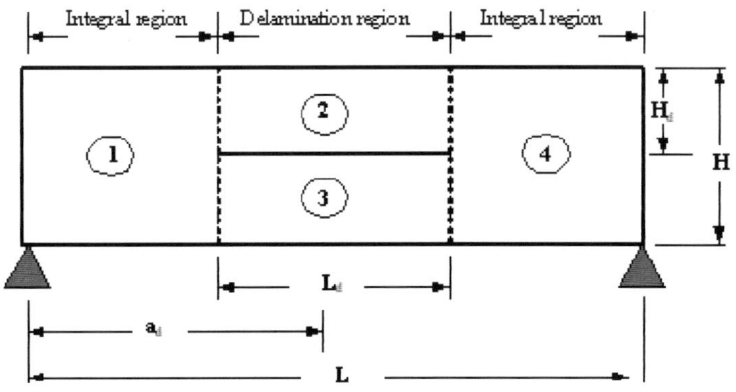

Fig 2: Schematic of the composite Beam

4. Results and discussion.

All the cases considered and the results are summarized in Table 1. The points A,B, are chosen to be at 0.1L and 0.9L from the left hand support of the beam while N and M are located at .025L and 0.05L from the centre of delamination respectively see Fig 1. The length of delamination that can be estimated is dependant of the choice of points M and N because they should be on the delamination itself. The smallest delamination length that can be estimated is equal to twice the disance between the centre of the delamination Q and point N, so if point M is chosen to be on Q and point N to be as close to Q as posssible so the length of a shorter delamination can be estimated. The estimated delamination lengths and the associated errors when compared to the real length value are presented in Table 1. It can be seen that for all cases that the frequency shift Δf_Q are quite high which indicates the existance of a delamination. This can be considered as an indication of the reliability of the detection methodology by applying a static force at mid span of the beam. The position of the centre of the delamination was precisely localised for all cases using equation 2, the delamination was quantified by estimating its length using equation 3 and 4 with a maximum error of 7.66 % when compared to the real value.

Table 1: Estimated delamination location and length for cases 1-12.

cases	a_d (L)	L_d (L)	H_d (L)	ΔFa (%)	ΔF_B(%)	Δf_M (%)	Δf_N (%)	ΔF_Q (%)	Est x_d (L)	Est x_e (L)	Est d (L)	Error (%)
1	0.50	0.10	0.50	1.19	1.19	5.83	5.30	6.39	0.50	0.45	0.10	2.13
2	0.50	0.10	0.30	1.11	1.11	5.71	5.09	6.28	0.50	0.45	0.11	7.35
3	0.30	0.10	0.50	1.65	0.73	4.96	4.17	3.86	0.30	0.25	0.11	7.66
4	0.30	0.10	0.30	1.74	0.75	5.22	4.25	3.82	0.30	0.25	0.09	6.94
5	0.50	0.20	0.50	1.13	1.13	6.09	5.53	6.76	0.50	0.40	0.19	4.04
6	0.50	0.20	0.30	1.16	1.16	6.39	5.81	6.98	0.50	0.40	0.20	0.86
7	0.30	0.20	0.50	1.64	0.69	5.75	4.93	4.11	0.30	0.20	0.19	3.25
8	0.30	0.20	0.30	1.70	0.72	5.93	5.09	4.24	0.30	0.20	0.20	1.88
9	0.50	0.30	0.50	1.32	1.32	7.91	7.25	8.57	0.50	0.35	0.30	0.51
10	0.50	0.30	0.30	1.36	1.36	8.14	7.46	8.82	0.50	0.35	0.30	0.98
11	0.30	0.30	0.50	1.81	0.78	7.23	6.33	4.78	0.30	0.15	0.30	1.33
12	0.30	0.30	0.30	1.87	0.80	8.67	7.49	4.96	0.30	0.15	0.30	1.07

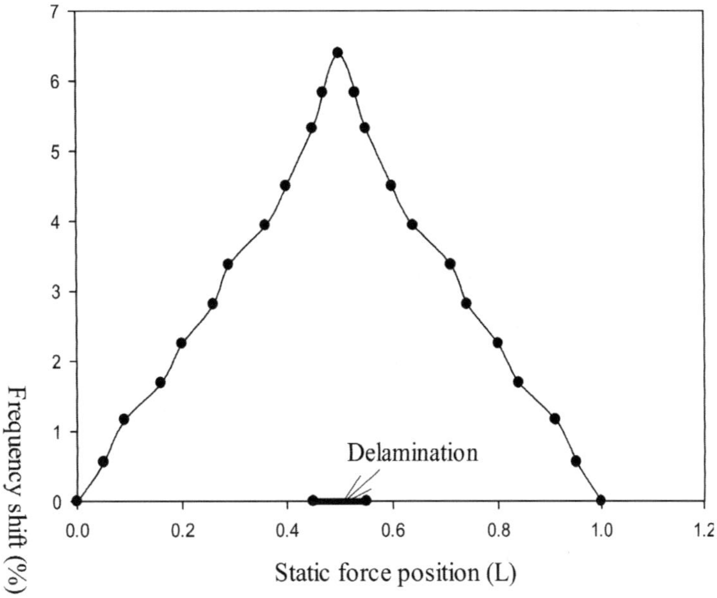

Figure 3: Frequency shift vs Static force position for case 1

5. Conclusion.

This paper suggests a very simple method for complete delamination identification in composite beams which is based on the first natural frequency of the beam and no base model is needed. The procedure involves the application of a static force which is applied in different positions along the composite beam. The result show that the delamination locations obtained using the FE model are the same as the real ones. The estimated delamination lengths are very close to the real values with a maximum error of 7.66 %. The method is being developed for the case of multiple delaminations.

6. References.

[1] Zou Y, Tong L, Steven GP. Vibration-based model-dependent damage (delamination) identification and health monitoring for composites structures – a review. J Sound Vib 2000; 230(2):357–78.
[2] R.M.Gadelrab. The effect of delamination on the natural frequencies of a laminated composite beam. J Sound Vib 1996; 197(3),283–292.
[3] A.Karmakar,H.Roy,K.Kishimoto,Free vibration analysis of delaminated composite pretwisted shells, Aircraft Engineeringand Aerospace Technology 77 (6)(2005)486–490.
[4]M.Krawczuk,W.Ostachowicz,A.Zak,Analysis of natural frequencies of delaminated composite beams based on finite element method, Journal of Engineering Mechanics 4 (3)(1996)243–255.
[5] J.J.Tracy,G.C.Pardoen,Effectofdelaminationonthenaturalfrequenciesofcompositelaminates, Journal ofCompositeMaterials 23 (1989)1200–1215.
[6] Bedri R. and Al-Nais M.O , 2005, *Lecture Notes in Computer Science* 3401pp 171-178
[7] Guechaichia, Abdelhamid and Trendafilova, Irina (2011) A simple method for enhanced vibration-based structural health monitoring. Journal of Physics Conference Series, 305 (1). ISSN 1742-6588

Journal of Physics: Conference Series **382** (2012) 012034 doi:10.1088/1742-6596/382/1/012034

3D Digital Image Correlation (DIC) analysis in relation to damage development in third generation AA2050-T851 aluminium alloy

M Lopez-Pedrosa[1], M Zanganeh, Y H Tai and C Pinna
Dept of Mechanical Engineering, The University of Sheffield, Mappin street, Sheffield, S1 3JD, UK

E-mail : m.pedrosa@sheffield.ac.uk

Abstract. Damage evolution and mechanisms of deformation have been investigated in the 3rd generation AA2050 aluminium alloy.This Al-Li-Cu alloy is characterised by the presence of second-phase particles which are preferred sites for damage nucleation. Digital Image Correlation (DIC) and high resolution microscopy have been used to establish a correlation between strain distribution and particles/voids volume fraction along the gauge length of tensile specimens. Results show that damage nucleates at about 11% applied strain with a strain to fracture of about 35%.

1. Introduction

The 3rd generation AA2050 aluminium alloy is one of the most attractive materials in the aerospace industry and for military applications. This aluminium alloy also called Al-Li-Cu alloy is attractive in terms of weight savings for aircraft structures compared to 2xxx and 7xxx aluminium alloys due to its low density and high modulus of elasticity [1]. This alloy is characterised by the presence of second-phases particles and Al_2CuLi precipitates obtained through heat treatment to strengthen the material [2]. The fracture properties of this relatively new alloy are not well characterised with very few studies reported in the literature. In general, it is believed that initial damage occurs by debonding or cracking of second-phase particles nucleating a void followed by growth and coalescence of several voids triggering final fracture [3]. This work therefore aims at characterising the fracture process of this alloy by focusing on the effect of second phase particles on damage development. Particular emphasis is put on finding a correlation between applied strain levels measured using 3D Digital Image Correlation and damage evolution. Results can then be used for the development of predictive physically-based models of damage and fracture for AA2050.

[1] Author to whom any correspondence should be addressed

2. Experimental Work

The 3rd generation AA2050 aluminium alloy with chemical composition 3.51%Cu, 0.91%Li, 0.40%Mg, 0.38%Mn, 0.36%Ag, 0.09%Zr, 0.03%Si and 0.02%Zn and wt%Al-balance was supplied in a 15mm thick plate.

A tensile test was performed on standard cylindrical smooth specimens shown in Figure1.a. A 3D DIC system was used during the test to obtain 3D displacement fields at the surface of the specimen up to fracture. The commercial VIC-3D 2009 software package from Correlated solutions was used to acquire images using two digital 5MPixels Pike cameras. Both cameras were set up at 600mm distance and 30° angle to the specimen. Previous to the test, a good calibration of the 3D DIC system is required [4]. A calibration target containing target points with a known distance between them is imaged simultaneously in both cameras at different angles. The synchronised images are used to fully calibrate the system establishing the camera orientation and calculating intrinsic parameters. After acquiring 10 to 12 calibration images to the project, the calibration is computed and a final score of the calibration is displayed. The smaller the number the better the calibration is. After all, a cross-correlation function was used and displacement vectors were calculated using VIC-3D 2009 software by correlating square subsets of 17x17pixels and 5 pixels steps spacing to carry out the correlation process for the undeformed and deformed images. The cross-correlation algorithms are based on tracking variations in intensity in the grey level of each pixel [5].

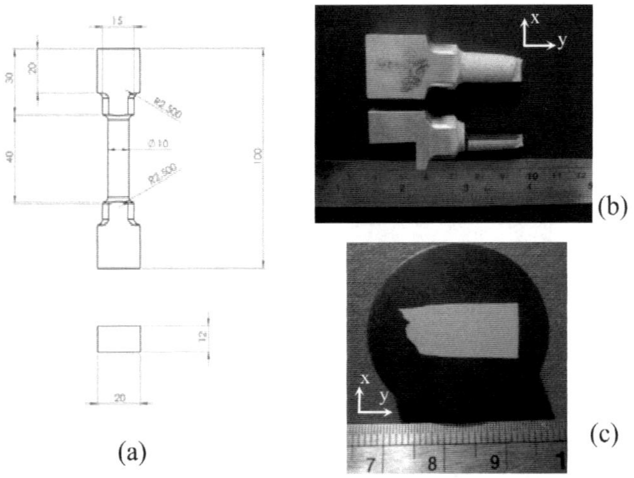

Figure 1. (a) Schematic diagram of the smooth cylindrical specimen, dimensions in mm, (b) fractured specimen after the tensile test, the undeformed part was used for metallographic analysis and (c) surface analysed in the SEM corresponding to the middle plane of the specimen shown in (b). x and y are macroscopic axes and correspond to the transverse and tensile axis respectively

Microstructural analysis was carried out by observing the middle plane of the broken specimen (Figure1.b). This plane where the damage process can be more clearly seen due to a higher stress triaxiality state during tensile loading has been revealed through a careful cutting and polishing process. Figure 1.c shows the specimen mounted in a synthetic conductive resin with the surface of interest for microstructure analysis. The surface was carefully ground with silicon carbide abrasive papers from 240 to 1200-grit size and polished with 6- and 1-μm diamond paste. Polishing was completed in a 0.04μm-grit size polishing cloth with a chemical suspension of silicon dioxide (SiO_2) in distilled water [6].

In order to develop a methodology to quantify voids and particles volume fractions (f_v and f_p respectively) in the material Backscatter Electron Images (BEIs) were obtained in the Inspect F high resolution Field Emission Gun Scanning Electron Microscope (FEGSEM). BEIs were acquired at 600x magnification frame size, each image covered an area of 427x368μm^2 with spatial resolution of 0.416μ/pixel. The stage of the microscope was moved consequently by steps of a magnitude corresponding to one image frame along the x axis from the fracture surface of the specimen to a total distance of 5978μm from the fracture surface. f_v and f_p have been quantified using the Image J imaging processing program and Corel PHOTO-PAINT X4.

3. Results and discussion

Figure 2.a shows the ε_{yy} logarithmic strain field map obtained under uniaxial tensile test in the 3rd generation AA2050 aluminium alloy before fracture. High strain gradients are clearly visible in the necking region with 35% local strain values before fracture. The logarithmic strain components and displacements values were exported as ASCII files from the commercial DIC software and processed using a MATLAB script to interpolate data over a grid whose pitch size matches the size of BEIs obtained usingthe FEG SEM. An example of the extracted ε_{yy} profile for a line along the gauge length of the specimen, centred at the middle of the necking region and with a step size of 427 μm is shown in Fig. 2b.

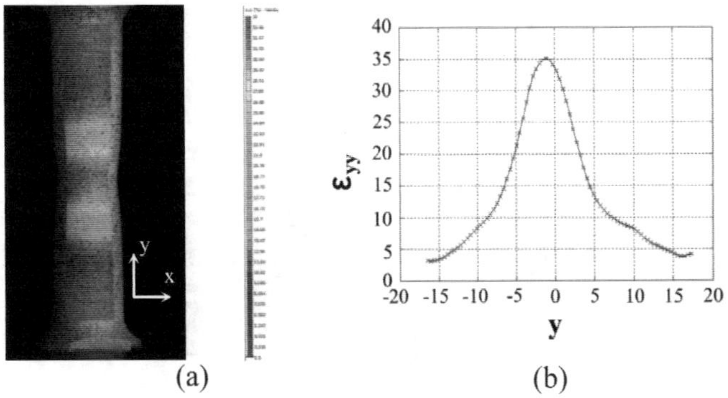

Figure.2. (a) Henky or logarithmic strain map before fracture and (b) ε_{yy} strain values profile along the y axis (the 0 value represents the centre position of the necking region).

Figure 3.a shows a BEI obtained close to the fracture surface. It can be observed that particles (white) are orientated parallel to the straining direction (y) which corresponds to the rolling direction. with some of them broken and displaying clear voids (black).. Every single BEI acquired in the FEG SEM is then processed and analysed using the Image J imaging software. Voids (Figure 3.b) and particles (Figure 3.c) are highlighted and automatically counted. f_v and f_p are then calculated for each individual BEI by dividing the cumulative area of detected voids/particles by the total frame area. Hence, f_v and f_p for the BEI close to the fracture surface (Figure 3) are 0.38% and 0.31% respectively considering particles >1μm^2.

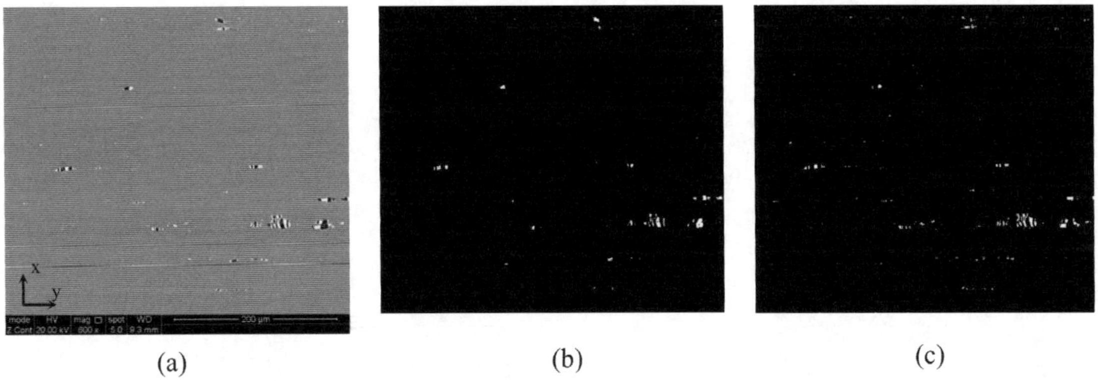

Figure 3. (a) BEI obtained close to the fracture surface. Images with only voids (b) and particles (c) were extracted independently from the BEI using Image J. x and y are macroscopic axes and correspond to transverse and tensile axes respectively. Scale in (a) is applied to (b) and (c)

As mentioned above, the 3D DIC dataset was used in this work to correlate local strain values (Figure 2) with f_v calculated for each BEI along the gauge length and Figure 4.a shows the results of the correlation. It can be observed that the maximum value of f_v corresponds to the location of the fracture surface with a local logarithmic strain value of 35.14% with a progressive decrease of f_v as strain values decrease. Figure 4.b shows the variation of f_p from the fracture surface to about 8mm distance from the fracture surface. It can be seen that the volume fraction of particles is homogeneous throughout the gauge length of the specimen which shows that the selected image size to carry out the analysis is large enough to cover a representative area of the microstructure and to provide statistically meaningful results.

A few more BEIs were acquired further away from the fracture surface in order to identify a local strain value to nucleate a void. BEIs were compared to a BEI of the as-received material and a local strain value of 11.28% was found to nucleate a void in a second-phase particle.

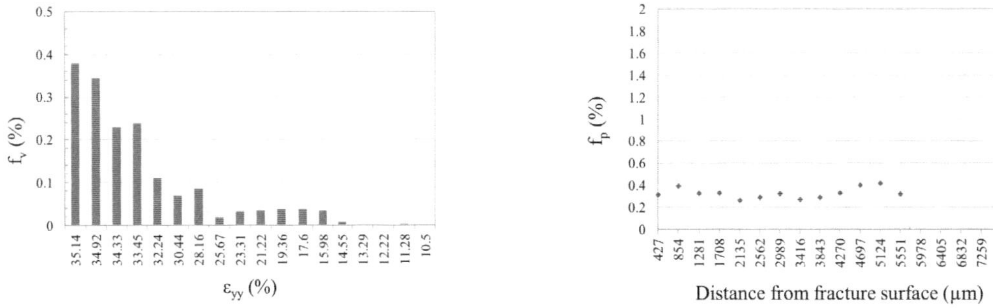

(a) (b)

Figure 4. (a) Void volume fraction (f_v) as a function of local logarithmic strain values (ε_{yy}) and (b) particle volume fraction (f_p) as a function of distance from fracture surface.

4. Conclusion

Tensile testing, 3D DIC technique and FEGSEM have been used to investigate the mechanism of deformation and damage evolution in the 3rd generation AA2050 aluminium alloy. The methodology used in this work has found a correlation between strain gradients along the gauge length of the specimen and void development in second-phase particles.

A nucleation strain for the damage process of about 11% has been found while fracture occurred at a local strain value of about 35% with a large amount of coalescence.

5. References

[1] Lequeu P h, Smith K P and Danielou A 2010 *Aluminium-Copper-Lithium alloy 2050 developed for medium to thick plate*, J. Mater. Eng. Perf. 19(6), 841-847

[2] Polmear I 2006 *Light alloys: From traditional alloys to Manocrystals*, Fourth edition, Elsevier

[3] Thomason P F 1990 *Ductile fracture of metals*, Pergamon Press Inc., Oxford

[4] *Vic-3D 2010 Reference manual*, Correlated Solutions

[5] Roux S, Réthoré J and Hild F 2009 Digital Image Correlation and fracture: an advance technique for estimating stress intensity factors of 2D and 3D cracks, J. Phys. D: Appl. Phys.42, 214004

[6] *ASM specialty Handbook: Aluminium and Aluminium alloys* 1993 edited by Davies, J.R. and associates, The materials information society, ASM international

Journal of Physics: Conference Series **382** (2012) 012035 doi:10.1088/1742-6596/382/1/012035

Calibration of a 3-D Digital Image Correlation system for large deformation contact problems

XH Tan[1,2], YL Kang[1] and EA Patterson[2]

[1]Tianjin Key Laboratory of Modern Engineering Mechanics, School of Mechanical Engineering, Tianjin University, Tianjin 300072, People's Republic of China, 300072
[2]School of Engineering, University of Liverpool, United Kingdom

Email: tju_ylkang@yahoo.com.cn

Abstract. In this work, the non-contact full-field Digital Image Correlation (DIC) technique has been utilized for measuring and analyzing the contact behaviour of soft materials undergoing large deformation. A vulcanized silicone rubber in contact with a wedge-shaped rigid indenter was investigated. In order to provide confidence in the measured results from the DIC system, an in-plane strain calibration procedure was conducted. Further, a procedure of out-of-plane displacement calibration was also deduced basing on the law of propagation of uncertainty.

1. Introduction

Over the past century, contact problems have occupied an important place in Solid Mechanics[1,2]. A practical application is a rigid indenter pressed onto an elastic body. It is difficult to get accurate analytical solutions for this kind of contact problem, so asymptotic methods were commonly implemented in theoretical analysis [3]. Since an indenter with blunted apex is more applicable to real problems, Korsunsky illustrated the influence of small variations in punch shape on the contact behaviour for a blunted Hertzian indenter and a rounded cone [4]. Then Hills and his co-workers have addressed a series of contact problem with various profiles [5]. It can be seen that the previous methods are useful to resolve some engineering problems, however, it should be noted that many of the following assumptions were employed in majority of the solutions for contact problems: the strains are small and within the elastic limit; each body were considered as elastic half-plane; there are no shearing tractions present, which requires the surfaces to be frictionless; the contacting bodies are elastically similar; the external angle of the wedge or cone must be very small.

The focus of this study is to remove the idealization of previous researchers, and investigate contact properties of a soft material indented by a rigid wedge in real conditions. Contact mechanics of soft materials is rather complex due to the geometrical and material nonlinearities and changing boundary conditions. Relatively little experimental work has been carried out so far because it is difficult to measure contact behavior by traditional techniques such as strain gauges or photo-elasticity [6,7]. Therefore, in this paper, the non-contact full-field Digital Image Correlation (DIC) technique has been utilized for measuring contact behavior of soft materials undergoing large deformation. Specifically, a vulcanized silicone rubber in contact with a rounded-tip rigid wedge was investigated.

2. Experimental setup and procedure

The specimen used in the experiment was cured from a Room Temperature Vulcanization (RTV) Rubber. It was cast into a rectangular block with dimensions 60mm×60mm×30mm. A very thin coat of quick-drying white paint (Matt Super White 1107, Plasti-kote, UK) was sprayed onto the surface using an aerosol can and then a fine coating of black speckle (Matt Super Black 1102, Plasti-kote, UK)was sprayed on top of the white coating. During the DIC experiment, the surface deformation of specimen can be deduced from the randomly distributed small paint dots.

The wedge indenter was manufactured using 2024 aluminum with a tip radius R=1.68mm and external angle θ=73.45°. It had the same thickness as the rubber specimen (30mm). The specimen was not pre-stressed before loading and placed in the centre of an aluminum disc which was mounted on the bottom grip of an Instron "Electropuls" machine, while the indenter was mounted in the top grip of the machine. Afterwards, the loading system and DIC system (Q-400, Dantec Dynamics GmbH, Germany) were triggered simultaneously. The test machine control software (Bluehill, Instron, UK) was set to apply the load at a constant rate of 1 mm/min up to the maximum load of 34.5N, and the DIC software (ISTRA 4D) was programmed to record specimen images every 30 seconds.

After the experimental measurements, data processing was performed using the DIC software. In order to save data-processing time and get better results near the contact interface, a small rectangular area around the contact interface (mask area) was analyzed. It was found that the best results could be obtained with a facet size of 25 pixels and grid spacing of 3 pixels. A close-up of the evaluated information near contact area is shown in Figure 1.

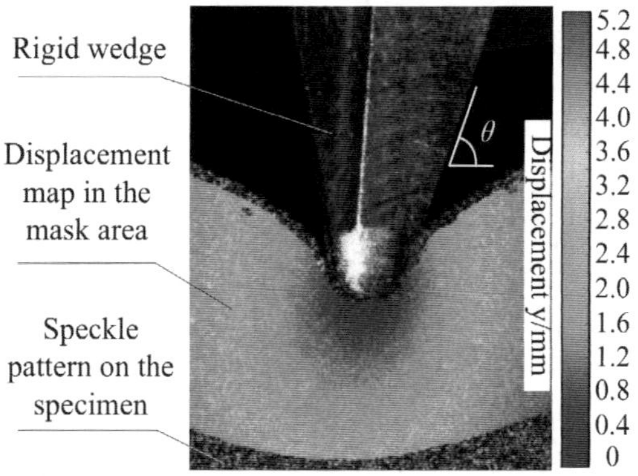

Figure 1: Displacement field in the loading direction evaluated by DIC system, overlapping with experimental image, which was captured by the left camera. For interpretation of the references in colour in this and all other figures, the reader is referred to the electronic version of this paper.

3. Calibration of DIC system

3.1. In-plane strain measurement calibration

According to the procedure described in the SPOTS[3] guideline, a successful calibration of a strain measurement system requires a known strain field to compare with the results generated by the measurement system. This can be achieved by using a Reference Material (RM). For an in-plane strain measurement calibration, the Reference Material is designed as a central horizontal beam loaded in symmetric four-point bending via an outer loading frame (shown in Figure 2). The RM used in this paper was manufactured from a single piece of 5 mm thick 2024-T351 Aluminum. The beam depth (*W*) in the loading direction was 15mm; the distance between the loading points (*a*) was 45mm; and the distance between inner loading point and beam centre (*c*) was 15mm. The specifications and methodologies for its use in calibration can be found in the SPOTS guideline [8].

According to SPOTS guideline [8], there are three sources of total measurement uncertainty of a system. The first uncertainty primarily comes from the manufacture of Reference Material (RM), including material property uncertainty and geometric uncertainty. These values can be found by repeating the measurement enough times to get a good estimate of the standard deviation of the values.

[3] SPOTS: Standardization Project for Optical Techniques of Strain measurement, www.opticalstrain.org

The evaluated poisson's ratio of RM was υ=0.33 and its corresponding uncertainty was u(υ)=0.005 in this study, while the geometric uncertainties were u(W)=0.017, u(a) =0.017 and u(c) =0.017 respectively.

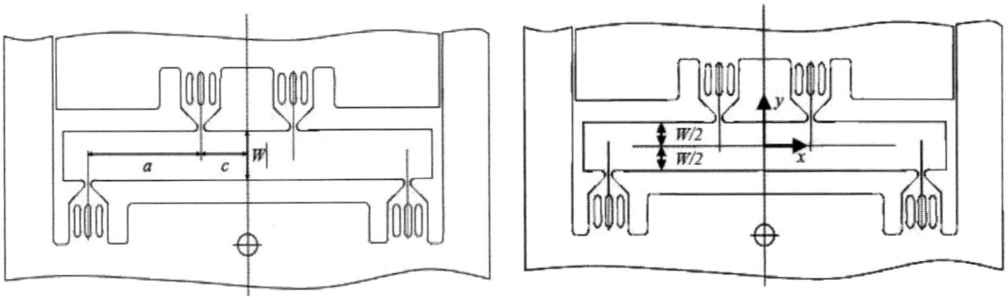

Figure 2: Schematics showing the central horizontal beam of RM. The axis system is located at the beam's geometric centre.

The second uncertainty primarily comes from the optical instrument, which means the DIC system in this study. For all data points (i, j) within the gauge area of Reference Material, this value can be found from the evaluation of mean square residual deviation after the linear fit of $d_k(i, j)$, which is the map of the difference between the predicted values and the measured values of strain. For each increment of load step k , the measurement uncertainty is:

$$u^2(d_k) = \frac{1}{N}\sum_{i,j}[d_k(i,j)]^2 - \alpha_k^2 - \beta_k^2\frac{1}{N}\sum_{i,j} y_j^2 \qquad (1)$$

where α_k and β_k are the fit parameters for a linear least-squares fit of the deviation $d_k(i, j)$.

The third source of uncertainty is the displacement measurement and the constraint imposed by the attachment to the monolithic frame. In standard practice, the applied displacement should include at least 3 increments over a range between 10% and 90% of the maximum allowable load. In this study, the average of the measurement values in three displacement steps were $\bar{v}_1$=112 µm, $\bar{v}_2$=152 µm and $\bar{v}_3$=214.5 µm respectively. Their associated uncertainty was provided by the calibration certificates purchased with digital displacement indicators (Mitutoyo 543-392, Kawasaki-shi, Japan), which was $u(\bar{v})$=0.89 µm. The strain expressions for an ideal beam should be modified due to the constraint imposed by whiffle-trees which support the beam, so the more accurate expression is:

$$\varepsilon_{xx} = \frac{\bar{v}}{6W^2}(\kappa y+\eta),\ \varepsilon_{yy} = -\frac{\upsilon\bar{v}}{6W^2}(\kappa y+\eta) \text{ and } \varepsilon_{xy} = 0 \qquad (2)$$

The correction factors κ and η and the corresponding uncertainties were determined using a pair of strain gauges on the top and bottom surfaces of the beam. The values obtained in this study were κ=0.986, η=-0.068 mm, and their associated uncertainty values were u(κ)=0.006 mm, $u(\eta)$=0.006 mm.

After all of the uncertainty values mentioned above were determined, the combined uncertainty of the RM for each displacement step were calculated using the following equation:

$$u_{RM}^2(\varepsilon_{xx}) = (\varepsilon_{xx})^2\left(\frac{u^2(\bar{v})}{\bar{v}^2} + \frac{u^2(\kappa)}{\kappa^2} + \frac{16u^2(W)+u^2(a)+u^2(c)}{4W^2}\right) + \left(\frac{\bar{v}}{6W^2}\right)^2 u^2(\eta) \qquad (3)$$

These values were used to obtain the expanded uncertainty in the RM, $U_{RM} = 2u_{RM}$. Finally, for each of the three displacement load steps, the mean residual deviations $(\alpha_k + \beta_k y) - 2u(d_k)$ were plotted along with the expanded uncertainty of the calibration specimen, $\pm U_{RM}$ in Figure 3.

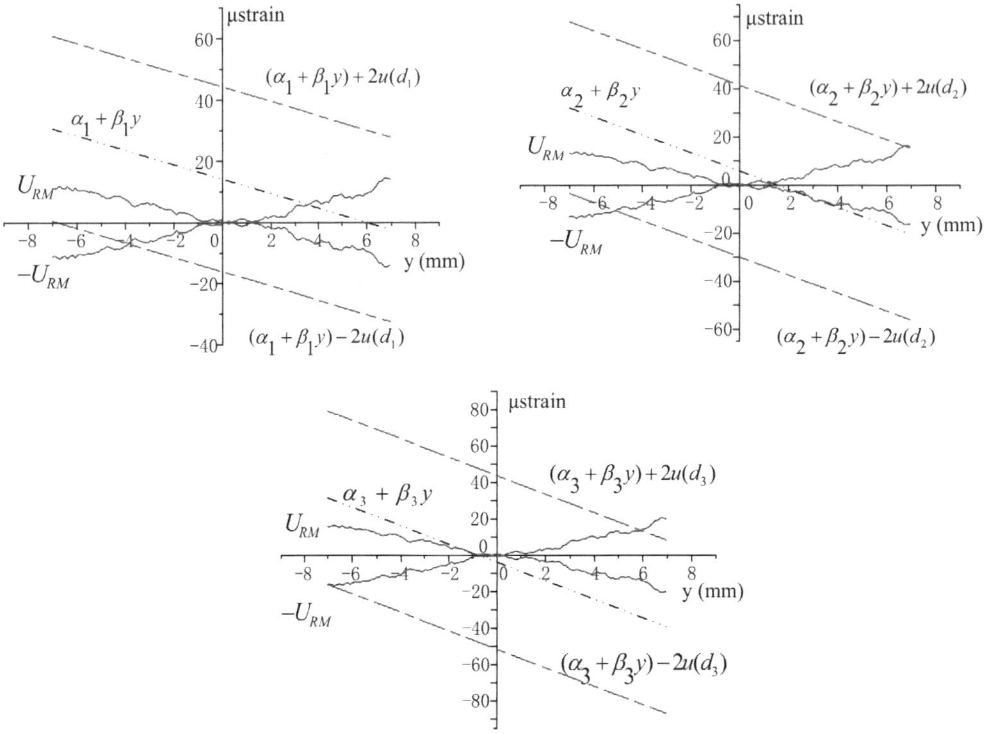

Figure 3: Plots of expanded uncertainty of Reference Material, $\pm U_{RM}$ (red solid lines) as a function of y, the distance from the geometric centre of the beam in the direction of applied load, together with the mean residual deviations $(\alpha_k+\beta_k y)-2u(d_k)$ (blue long dash lines) for each load step: 112 μm (top left), 152 μm (top right), 214.5 μm (bottom). $\pm 2u(d_k)$ indicates the width of the scatter band around the long dash double dot line $(\alpha_k+\beta_k y)$.

In an ideal system, the measured strain value should be the same as the predicted one, that is, the deviation $d_k(i,j)=\alpha_k+\beta_k y$ should be zero and no adjustments would be necessary. It can be observed from figure 3 that, for every load step, the scatter band $\pm 2u(d_k)$ on $\alpha_k+\beta_k y$ and the area bounded by $\pm U_{RM}$ are completely overlapped, which means the measurement values coincide well with the predicted values, and $d_k(i,j)$ contains random noise only. In this case, no correction is needed for the DIC system.

3.2. Out-of-plane displacement calibration

A homogeneous isotropic steel cantilever beam was utilized as a Reference Material to calibrate the out-of-plane displacement. Although the simple cantilever was designed by ADVISE[4] to calibrate full-field optical systems for transient and non-linear deformation cases, it can also be used for three-dimensional deformation fields induced by static loading. Specifications of the simple cantilever can be found in *ADVISE deliverable D3.5* [9]. The design is fully parametric based on the cantilever thickness T, which is 8mm in this study. Other dimensions can be expressed as length 20T and width 5T, the enlarged portion at one end was thickness 3T and length 10T, giving the cantilever an overall length of 30T. During calibration, the enlarged portion of the cantilever was clamped on a rigid,

[4] ADVISE: Advanced Dynamic Validation by Integrated Simulation and Experimentation, www.dynamicvalidation.org

immovable table and five displacement increments were applied to the tip of cantilever. One of the experimental images captured by DIC system is shown in Figure 4.

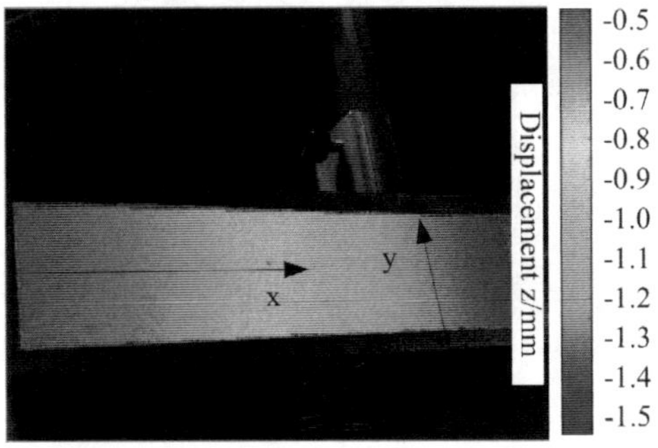

Figure 4: Out-of-plane displacement field evaluated by the DIC system for the 2760 μm displacement load, overlapping with experimental images of Reference Material.

The tip deflection $\bar{v}$ was measured using a calibrated displacement transducer (the same one used in in-plane strain calibration). The deflection δ in the cantilever can be related to the tip deflection $\bar{v}$ using theory of elasticity:

$$\delta = \frac{\bar{v}x^2}{2a^3}(3a - x) \qquad (0 \le x \le a) \tag{4}$$

where a is the length from the fixed end to loading point. By repeating the measurement twenty times, the average value of $a = 147.05\text{mm}$ was determined and the associated uncertainty $u^2(a) = 0.483mm$ was estimated from the standard deviation of the measured values. The expression for the combined uncertainty of the Reference Material deflection values, u_{RM} can be deduced based on the law of propagation of uncertainty using the root-sum-of-squares (RSS) method [10]. It is given in the following form:

$$u_{RM}^2 = \delta^2 \left(\frac{u^2(\bar{v})}{\bar{v}^2} + (\frac{a}{3a - x} - 1)^2 \frac{u^2(a)}{a^2} \right) \tag{5}$$

Three displacements steps were chosen for analysis, which were 428, 1280, and 2760 μm. Ideally the deflection at any section at a distance, x from the cantilever end should be constant for a given applied load. The differences $d_k(i,j)$ between the predicted values and the measured values of deflection were calculated over the gauge area. Similar to in-plane strain measurement calibration, a linear least-squares fit was applied to the field of deviations $d_k(i,j)$, which yields fit parameters α_k and β_k for each load step.

The uncertainty values obtained from the calibration experiment are summarized in Table 1. The resulting calibration uncertainty of the system u_{cal} was found using the maximum displacement value from each of the load steps and the results from the combination of the uncertainties of the Reference Material and of the displacement measurements. For each of the three displacement load steps, the mean residual deviations $(\alpha_k + \beta_k y) - 2u(d_k)$ were plotted along with the expanded uncertainty of the calibration specimen, $\pm U_{RM}$ in Figure 5.

Table 1. Summary of the uncertainty values for each load step applied to the cantilever.

$\bar{v}_k$ (μm)	Measurement Uncertainty	$u(d_k)$	u_{RM}	u_{cal}	$w_{y=\max} - w_{y=\min}$
428	0.89μm	0.24	0.65	0.69	4.38
1280	0.89μm	0.39	0.64	0.75	1.82
2760	0.89μm	0.28	0.65	0.70	-2.16

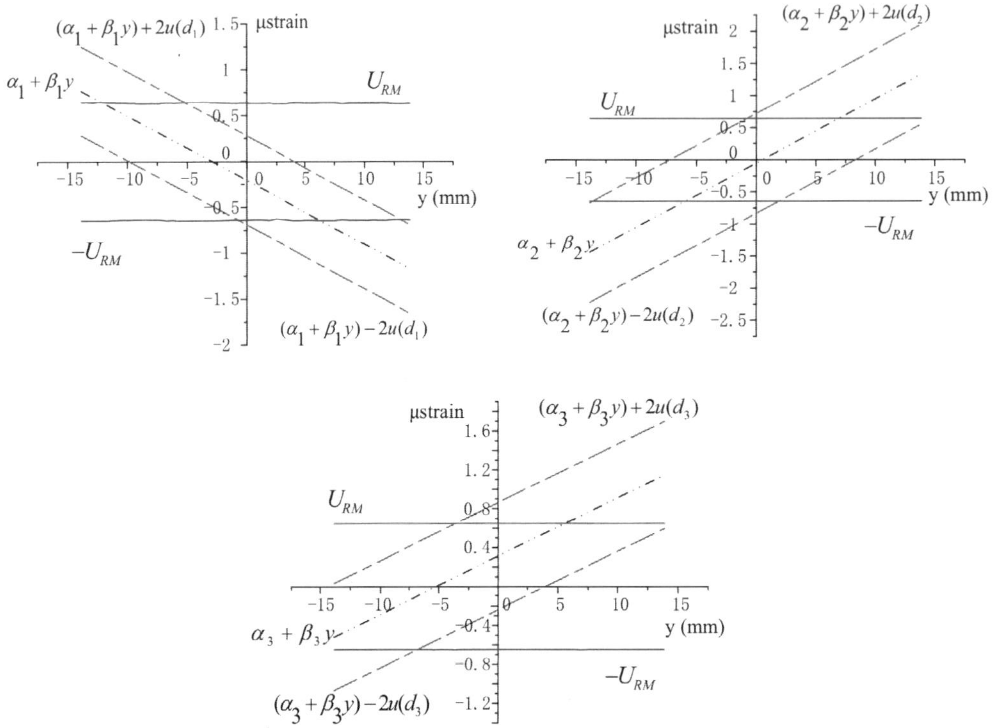

Figure 5: Plots of expanded uncertainty of Reference Material, $\pm U_{RM}$ (red solid lines) as a function of y, the distance from the geometric centre of the cantilever in the direction of width together with the mean residual deviations $(\alpha_k+\beta_k y)-2u(d_k)$ (blue long dash lines) for each load step: 428 µm (top left), 1280 µm (top right), and 2760 µm (bottom). $\pm 2u(d_k)$ indicates the width of the scatter band around the long dash double dot line $(\alpha_k+\beta_k y)$.

For every load step, a complete overlap between the scatter band $\pm 2u(d_k)$ on $\alpha_k+\beta_k y$ and the area bounded by $\pm U_{RM}$ can be observed in Figure 5, which indicates the measurement values coincide well with the predicted values, and $d_k(i,j)$ contains random noise only. In this case, the quality of the measured out-of-plane displacement made by the DIC system is satisfactory and no correction is needed.

4. Conclusion

The 3-D Digital Image Correlation technique has been applied to an investigation of contact problem in soft materials. Two calibration experiments demonstrate that the DIC system is appropriately calibrated, and the results obtained from the calibrated system were accurate and reliable. The experimental data will be used to compare with theoretical solutions in our future work.

Acknowledgement

This Project was financially supported by The University of Liverpool, National Basic Research Program of China (No. 2012CB937500), and National Natural Science Foundation of China (No. 11127202); EAP is the recipient of a Royal Society Wolfson Research Merit Award and XT is the recipient of and China Scholarship Council (No.2011625016).

References

[1] Johnson K L 1985 *Contact Mechanics* (Cambridge: Cambridge University Press)
[2] Barber J R and Ciavarella M 2000 Contact mechanics *Int. J. Solids Struct.* **37**(1-2) 29-43
[3] Dini D, Barber J R, Churchman C M, Sackfield A and Hills D A 2008 The application of asymptotic solutions to contact problems characterised by logarithmic singularities *Eur. J. Mech. A-Solids* **27**(5) 847-858
[4] Korsunsky A M 2001 The influence of punch blunting on the elastic indentation response *J. Strain Anal. Eng. Des.* **36**(4) 391-400
[5] Ciavarella M, Hills D A and Monno G 1998 Contact problems for a wedge with rounded apex *Int. J. Mech. Sci.* **40**(10) 977-988
[6] Xiao X, Kang Y L, Hou Z D, Qiu W, Li X L and Li X Q 2010 Displacement and Strain Measurement by Circular and Radial Gratings Moiré Method *Exp. Mech.* **50**(2) 239-244
[7] Burguete R and EA Patterson 1997 A photoelastic study of contact between a cylinder and a half-space *Exp. Mech.* **37**(3) 314-323
[8] EA Patterson, Hack E, Brailly P, Burguete RL, Saleem Q, Siebert T, Tomlinson R A and Whelan M P 2007 Calibration and evaluation of optical systems for full-field strain measurement *Opt. Lasers Eng.* **45**(5) 550-564
[9] Hack E and Whelan M 2011 Specification of 'transient' reference material published on website *ADVISE Deliverable D.35*www.dynamicvalidation.org
[10] Taylor B N and Kuyatt C E 1994 Guidelines for evaluating and expressing the uncertainty of NIST measurements results NIST Technical Note 1297 (Washington, D.C.: Government Printing Office)

Modern Practice in Stress and Vibration Analysis 2012 (MPSVA 2012) IOP Publishing
Journal of Physics: Conference Series **382** (2012) 012036 doi:10.1088/1742-6596/382/1/012036

Optimal placement of sensors and actuators for active vibration reduction of a flexible structure using a genetic algorithm based on modified $H_{\_infinity}$

J M Hale and A H Daraji
School of mechanical systems and engineering, Newcastle University, NE1 7RU, UK

Email: jack.hale@ncl.ac.uk a.h.a.daraji@ncl.ac.uk

Abstract. This paper is concerned with active vibration reduction of a square isotropic plate, mounted rigidly along one edge to form a cantilever. Optimal placement of ten piezoelectric sensor/actuator pairs is investigated using a genetic algorithm to suppress the first six modes of vibration. A new objective function is developed based on modified $H_{infinity}$ to locate the sensor/actuator pairs. The plate, with piezoelectric sensor/actuator pairs bonded to its surfaces, is modelled using the finite element method and Hamilton's principle based on first order shear deformation theory including bending, membrane, and shear deformation effects. The effects of piezoelectric mass, stiffness and electromechanical coupling are taken into account. The first six natural frequencies are validated by comparison with the finite element ANSYS package using two dimensional SHELL63 and three dimensional SOLID45 elements and also experimentally. Vibration reduction for the cantilever plate with piezoelectric patches bonded in the optimal location was investigated to attenuate the first six modes of vibration using a linear optimal control scheme. The new fitness function has reduced the computational cost and given greater vibration reduction than other previously published results.

1. Introduction

Vibration reduction of a flexible structure can be optimised by locating piezoelectric sensors and actuators in efficient locations using optimisation methods such as genetic algorithm. The complexity of optimisation process is directly proportional to the fitness function type and search space number of possible solutions which equal to the statistical combination of finite element discretisation and sensors or actuators number to be optimised. The reduction of optimisation problems for large scale structure has attracted researchers to reduce search space, computer time calculation, exploring global optimal sensors and actuators location to achieve controller optimality.

The genetic algorithm has been utilized for the optimal location of piezoelectric sensors and actuators by many researchers for flat plates and shells. The Rayleigh-Ritz method was implemented by Sadri et al to model a thin plate, with modal and grammian controllability taken as an objective function to locate two actuators [1]. Two actuators and piezofilm sensors were optimized using controllability, observability and spillover as an objective function to suppress the first three modes of vibration by Han and Lee [2] . Quek, et al optimized two piezoelectric sensor/actuator pairs based on modal controllability to suppress the first two modes of vibration [3]. Optimal placement of ten sensor/actuator pairs studied by Kumar and Narayanan using minimization of linear quadratic index as an objective function to suppress the first six modes of vibration [4]. The half and quarter chromosome technique has been developed by Daraji and Hale to reduce the genetic algorithm search space by 99.99% to locate ten and eight sensors/actuators pairs for a flat plate using linear quadratic index minimization as an objective function [5]. A new placement strategy including conditional filter is proposed by Daraji and Hale to reduce genetic algorithm search space and explore the global optimal configuration of ten and four piezoelectric pairs and to attenuate the first six modes of vibration [6].

The special configuration for placement of finite numbers of sensors and actuators on a continuous flexible structure is investigated by Chemishkian and Arabyan to reduce structural deformation to a minimum in H_∞ sense. The computational overhead is reduced by evaluating the minimum H_∞ norm by each genetic algorithm step and controller H_∞ is designed only when the lower limit is below a predefined value [7, 8]. Optimal placement of two sensor/actuator pairs on a beam is investigated by Hiramoto et al using modified H_∞ based on simple addition and multiplication matrices to reduce the computational time compared with standard norm [9]. Optimal placement of sensors and actuators is investigated by Gawronski for a fixed ended beam based on standard H_2 norm to suppress the first mode individually and collectively. The placement indices for sensors and actuators are explained thoroughly with modelling based on the H_∞, H_2 and H_s norm. The proposed approach consists of determination of the norm of each sensor or actuator for selected modes and then grading them according to their participation in the system norm to reduce the computation time [10]. The spatial H_2 norm is developed by Liu et al to optimise sensor and actuator locations in controlled flexible structures based on dominant modes which have most effect on the structure. They showed that the spatial H_2 norm gives better performance index for structural vibration control [11].

In this paper, the placement of ten sensor/actuator pairs, used to attenuate the first six modes of vibration, is optimised for a cantilever plate using a genetic algorithm. A new objective function, based on modified H_∞, is proposed to minimize an average summation of absolute dB gain for the open loop transfer function for all sensors and modes as a result of applying a unity amplitude disturbance voltage at a predefined single actuator location through all genetic algorithm generations. An isotropic plate with the piezoelectric sensor/actuator pair bonded to its surfaces is modelled using finite element and Hamilton's principle based on first order shear deformation theory. Piezoelectric mass, stiffness and electromechanical coupling effects are taken into account.

2. Modelling

The plate with piezoelectric patches bonded to its surface to form sensor and actuator in pairs is modelled according to first order shear deformation theory to include bending, membrane and shear deformation effects. An isoparametric two dimensional element is chosen with four nodes and five degrees of freedom per node. The full derivation and parameters are explained by Daraji and Hale, the modal dynamic equation and state space matrices are[12];

$$\{\ddot{\eta}\} + 2\xi\omega\{\dot{\eta}\} + \Omega\{\eta\} = [\varphi]^T\{F_u\} - [\varphi]^T[K_{u\phi}]^a\{\phi_a\} \tag{1}$$

$$\{\phi_s\} = -[\varphi]^T[K_{\phi\phi}]^{s^{-1}}[K_{\phi u}]^s\{\eta\} \tag{2}$$

$$\{\dot{X}\} = \begin{bmatrix} 0 & \omega \\ -\omega & -2\xi\omega \end{bmatrix}\{X\} + \begin{bmatrix} 0 \\ -[\varphi]^T[K_{u\phi}] \end{bmatrix}\{\phi_a\} \tag{3}$$

$$\{\dot{X}\} = [A]\{X\} + [B]\{u\}, \quad \{\phi_s\} = [C]\{X\} \tag{4}$$

$$[A_i] = \begin{bmatrix} 0 & \omega_i \\ -\omega_i & -2\xi_i\omega_i \end{bmatrix}, \ [B_i] = \begin{bmatrix} 0 \\ -[\varphi_i]^T[K_{u\phi}]^a \end{bmatrix}, \ [C_i] = \begin{bmatrix} -[\varphi_i]^T[K_{\phi\phi}]^{s^{-1}}[K_{u\phi}]^s\omega_i^{-1} & 0 \end{bmatrix} \tag{5}$$

Where$[A_i]$, $[C_i]$ and $[B_i]$ are the individual modal state, sensor and actuator matrices in which subscript (i) refers to the mode number. The state and actuator matrices for the number of modes n_m and the number of actuators r_a are:

$$\{X\}_{(2n_m\times 1)} = \{\omega_1\eta_1 \quad \dot{\eta}_1 \quad \cdots \quad \omega_{n_m}\eta_{n_m} \quad \dot{\eta}_{n_m}\}^T \tag{6}$$

$$[A]_{(2n_m\times 2n_m)} = \begin{bmatrix} A_1 & \cdots & 0 \\ \vdots & \ddots & \vdots \\ 0 & \cdots & A_{n_m} \end{bmatrix}, \quad [B]_{(2n_m\times r_a)} = \begin{bmatrix} (B_1)_1 & \cdots & (B_1)_{r_a} \\ \vdots & \cdots & \vdots \\ \left(B_{n_m}\right)_1 & \cdots & \left(B_{n_m}\right)_{r_a} \end{bmatrix} \tag{7}$$

Linear quadratic optimal controller design is based on minimization of a performance index J. Values of positive-definite weighted matrices $[Q]$, of dimension$(2n_m \times 2n_m)$, and $[R]$ of dimension $(r_a \times r_a)$ are controlled the value of the performance index, where n_m, r_a represent the number of modes and actuators, Ogata has shown it is possible to follow this derivation to design a linear quadratic controller [13], which leads to the Riccati equation:

$$[A]^T[P] + [P][A] - [P][B][R]^{-1}[B]^T[P] + [Q] = 0 \tag{8}$$

$$[K] = [R]^{-1}[B]^T[P] \quad , \quad \{u\} = -[\mathrm{K}]\,\{X\} \tag{9}$$

Solution of the Reduced Riccati equation (8) gives the value of matrix $[P]$; if matrix $[P]$ is positive definite then the system is stable and the closed loop matrix $[A] - [B][K]$ is stable. The feedback control gain matrix can be obtained after substitution of $[P]$ in Eq. (9).

3. Objective function

A new objective function has been developed to find the optimal sensor/actuator configuration based on the H_∞ norm by measuring the open loop average dB gain for all sensors at all the required modes to be attenuated as a results of unity amplitude disturbance input voltage at a single predefined actuator selected at the position of maximum strain energy for all chromosomes and generations in the genetic algorithm. The open loop transfer function of a system in the frequency domain is

$$G(\omega) = C(j\omega I - A)^{-1}B \tag{10}$$

The peak of the transfer function magnitude is the H_∞ norm of a single mode ω_i .

$$H_\infty = max\big(G(\omega)\big) = G(\omega_i) \tag{11}$$

$$J(x,y) = \frac{1}{n_m n_{sa}} \left| \sum_{i=1}^{n_m} \sum_{j=1}^{n_{sa}} 20Log_{10}(G_j(\omega_i)) \right|_{predefined\ actuator} \tag{12}$$

$$J_{opt}(x,y) = \min\big(J(x,y)\big) \tag{13}$$

Where $x, y \in$ plate dimension 50×50 cm, n_m and n_{sa} refer to number of modes and sensor/actuator pairs respectively.

4. Genetic algorithm

In 1975, Holland invented the genetic algorithm, a heuristic method based on "survival of the fittest" or the principle of natural evolution. It has been continuously improved and is now a powerful method for searching optimal solutions. The fundamental unit in the genetic algorithm is a population of individuals, each defined by a chromosome containing a number of genes. The effectiveness or "fitness" of each individual is calculated according to some rule using the values of the genes. The members of the population with the highest fitness values are allowed to "breed" to form the next generation and the process continues until convergence is achieved. In this case, the ten "genes" are the locations of the ten sensor/actuator pairs, defined by an integer number (integer-coded chromosomes) (1-100), and the fitness function is the linear quadratic index. This process is directly analogous to the survival of the fittest concept in Darwinian natural selection, in which the more successful individuals in a population are inclined to breed and so form the next generation. By this means the genes that code for desirable characteristics, and so give the individuals possessing them a high degree of fitness, are transmitted down the generations at the expense of less useful genes, which die out. A placement strategy using the genetic algorithm including a conditional filter has been investigated by Daraji and Hale and is used in this work [6].

5. Results and discussion

5.1. Research problem

A cantilever flat plate dimensioned $500 \times 500 \times 1.9$mm was mounted rigidly from the left hand edge. The plate is descritised to one hundred elements 10×10 sequentially from left to right and down to up as shown in figure 2. Global optimal placement of ten piezoelectric sensor/actuator pairs is investigated to suppress the first six modes of vibration. This case is chosen to give direct comparison with the published works [4, 6] to test the reliability and robustness of this work. The plate and piezoelectric properties are listed in table1.

Table 1. Plate and piezoelectric properties

Properties	Plate	Piezoelectric PIC225
Modulus, GPa	210	--------
Density, kg/m^3	7810	7180
Poisson's ratio	0.3	--------
Thickness, mm	1.9	0.5
Length, width, mm	500	50
e_{31}, e_{32}, e_{33} , C/m^2	-------	-7.15, -7.15,13.7
$C_{11}^E, C_{12}^E, C_{13}^E, C_{55}^E$, GPa	-------	123,76.7,70.25, 22.226
μ_{33}^{σ} F/m	-------	1.5×10^{-8}

C^E, *e and* μ^{σ} *refer to elasticity, piezoelectric and permittivity constants*

5.2. Natural frequencies

The first six natural frequencies and mode shapes for the cantilever plate were determined using a Matlab m-code program based on the present model and validated by the ANSYS finite element package using two dimensional SHELL63 elements, three dimensional SOLID45 and also experimentally. The results are shown in table 2. They converged with the mesh refining to constant values, showing that the mesh of 10×10 shell63 elements gave good result accuracy for the first six natural frequencies compared with finer meshes and three dimensional solid45 elements, as well as with experimental results as shown in table2.

Table2. ANSYS natural frequencies compared with present model and experimental results

Element type	Mode (Hz)					
	1st	2nd	3rd	4th	5th	6th
ANSYS Shell63 (10×10)	6.59	16.17	41.32	52.37	59.79	104.70
ANSYS Solid45(50×50)	6.59	16.15	40.44	51.68	58.86	103.18
Present model	6.59	16.14	40.62	51.78	58.99	103.29
Experimental Frequency	5.90	16.90	37.30	51.60	58.20	101.00

5.3. Actuator location optimization

The genetic algorithm described in section 4 based on reference [6] was used to find optimal locations for ten piezoelectric actuators on 0.5m square cantilever plate mounted rigidly from the left edge. The progressive convergence of the population onto an optimal solution is shown in figure 1, in which the population is distributed around the circle with radius (**r**) representing the fitness value to be minimised. A unit amplitude voltage disturbance is applied at actuator location (01), forced for all chromosomes in all generations.

At the first generation figure (1a) the population is very diverse with representatives of high and low fitness and the range in between. After thirty generations figure (1b) the population is much less diverse, made up of individuals of high, though not yet optimal, fitness. After 500 generations figure (1c) the population has completely converged to a level of fitness higher than any individual in the first or thirtieth generations.

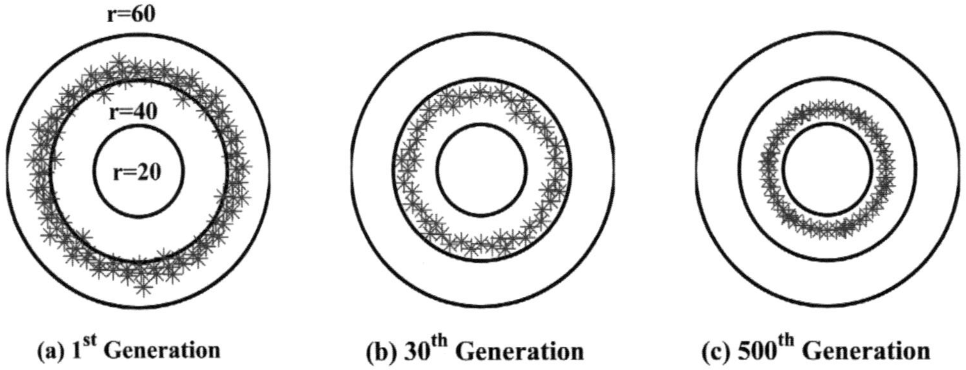

Figure 1. Population fitness progression over 500 generations. Each individual is represented as one of the points distributed around the circle with radii **r** which represent its fitness value to be minimized

5.5. Vibration reduction

The present optimal placement configuration is tested to measure the average closed loop dB gain reduction over all sensor/actuator pairs over the first six modes using an optimal linear quadratic control scheme and the results compared with two references as shown in figure 2. A unit amplitude voltage disturbance is applied at actuator number 11 for all three cases and the present configuration given an improvement of 28.5% reduction at $Q = 10^8$ and 24.35% at $Q = 10^9$ as shown in table3.

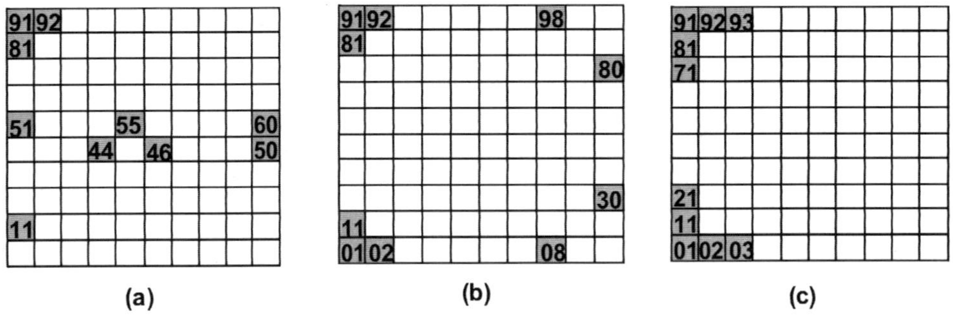

Figure 2. Optimal sensor/actuator pairs configurations for a cantilever plate mounted rigidly from the left edge, **(a)** Ref [4], **(b)** Ref [6], **(c)** present work

Table3 Closed loop average dB gain reduction and number of fitness computational cost for three cases

Case	$Q = 10^8$	$Q = 10^9$	Number of fitness calculation
Ref [4]	6.21dB	12.4dB	$5x10^4$
Ref [6]	7.61dB	14.42dB	$3.19x10^4$
Present work	7.98dB	15.06dB	$0.925x10^4$
Percentage reduction improvement	28.5%	24.35%	81.5%

6. Conclusion

A new objective function is developed to optimize ten sensor/actuator pairs for a cantilever plate using a genetic algorithm and to reduce the computational cost. The objective function is taken as a minimization of an absolute average open loop dB gain for all sensor/actuator pairs and modes as a result of applying disturbance unit volt amplitude at a predefined actuator.

An isotropic plate bonded with a number of piezoelectric patches is modeled using isoparametric four nodes element and Hamilton's principle based on first order shear deformation theory taking the effects of mass, stiffness and electromechanical coupling of piezoelectric pairs. The first six natural frequencies are determined based on the present model and validated with ansys package and experimentally.

The optimal configuration of ten sensor/actuator pairs is found symmetrically distributed about the plate axes of symmetry. The vibration reduction is tested for the present work compared with two references using optimal linear quadratic control. It has been found that the present optimal configuration gives a higher average dB gain reduction of 28.5% and 24.35% than published at $R = 1$ and $Q = 10^8$ and 10^9 respectively. The new objective function achieves a higher vibration reduction and reduced computation cost.

References

[1] Sadri A M, Wright J R and Wynne R J 1999 *Modelling and optimal placement of piezoelectric actuators in isotropic plates using genetic algorithms.* Smart Materials and Structures. **8**(4): p. 490-498.

[2] Han J H and Lee I 1999 *Optimal placement of piezoelectric sensors and actuators for vibration control of a composite plate using genetic algorithms.* Smart Materials and Structures . **8**(2): p. 257-267.

[3] Quek S T , Wang S Y and Ang K K 2003 *Vibration control of composite plates via optimal placement of piezoelectric patches.* Journal of Intelligent Material Systems and Structures. **14**(4-5): p. 229-245.

[4] Kumar K R and Narayanan S 2007 *The optimal location of piezoelectric actuators and sensors for vibration control of plates.* Smart Materials and Structures. **16**(6): p. 2680-2691.

[5] Daraji A H and Hale J M 2012 *Active vibration reduction of a flexible structure bonded with optimised piezoelectric pairs using half and quarter chromosomes in genetic algorithms* in *Modern Practice in Stress and Vibration (MPSVA)*: Glasgow University , UK

[6] Daraji A H and Hale J M 2012 *A conditional filter for reduction of the search space in genetic algorithm optimisation of sensor and actuator location in active vibration control* in *international congress of sound and vibration ICSV19*: Vilnius , Lithuania

[7] Arabyan A and Chemishkian S 1998 *H∞-optimal mapping of actuators and sensors in flexible structures.* Proceedings of the IEEE Conference on Decision and Control. **1**: p. 821-826.

[8] Chemishkian S and Arabyan A 1999 *Intelligent algorithms for H∞-optimal placement of actuators and sensors in structural control.*

[9] Hiramoto K, Doki H and Obinata G 2000 *Optimal sensor/actuator placement for active vibration control using explicit solution of algebraic Riccati equation.* Journal of Sound and Vibration. **229**(5): p. 1057-1075.

[10] Gawronski W K 1998 *Advanced Structural Dynamics and Active Control of Structures.* Springer: New York.

[11] Liu W, Hou Z and Demetriou M A 2006 *A computational scheme for the optimal sensor/actuator placement of flexible structures using spatial H_2 measures.* Mechanical Systems and Signal Processing. **20**(4): p. 881-895.

[12] Daraji A H and Hale J M 2012 *Optimisation of number and placement of piezoelectric sensor/actuator pairs for active vibration reduction of a felxible plate using the genetic algorithm* in *international congress of sound and vibration ICSV19*: Vilnius, Lithuania

[13] Ogata K 1997 *Moderen Control Engineering*. third ed. USA: prentice-Hall. Inc.

The nonstationary, nonlinear dynamic interactions in slender continua deployed in high-rise vertical transportation systems in the modern built environment

S. Kaczmarczyk

School of Science and Technology, The University of Northampton, St. George's Avenue, Northampton NN2 6JD, UK

E-mail: stefan.kaczmarczyk@northampton.ac.uk

Abstract. This paper presents a nonlinear mathematical model and numerical results concerning the nonstationary lateral dynamic behaviour of long low tension slender continua deployed and moving at speed in high-rise vertical transportation systems installed in tall structures. The analysis presented in this study involves the identification of conditions for internal lateral resonances that can readily arise in the system when the slowly varying frequencies approach the fundamental or higher frequencies of the structure. The passage through the fundamental resonance leads to dangerously large displacements in the plane of the excitation. Due to the nonlinear (cubic) coupling, interactions between the in-plane modes and the out-of-plane modes occur. These interactions are studied numerically in order to predict and to examine the non-planar motions that may arise due to the autoparametric resonances. In order to suppress the internal resonance interactions higher speed levels and /or cable tension levels should be applied. Alternatively, an active tension control algorithm can be considered.

1. Introduction

One-dimensional slender structures such as strings, belts, ropes and cables are deployed as stationary and/or moving structural members in many engineering systems. Their dynamic behaviour is inherently non-linear demonstrating complex dynamic responses [1-3]. In the context of the modern built environment nonlinear resonance phenomena occur in suspension ropes, compensating and overspeed governor cables employed in high-speed elevator installations in high-rise buildings [4-5]. Large whirling motions of ropes and cables in elevator installations may result in damage caused by the impact against the elevator equipment located in the shaft and/or against the shaft walls. The dynamic responses of such systems can be predicted and analyzed by modeling their behaviour and solving the resulting equations of motions. Due to the nonstationary and nonlinear nature of the problems involved solving these equations by applying analytical methods is difficult and the most convenient approach is using numerical simulation [6]. In this paper, the nonstationary linear planar model presented in [7] is developed further to accommodate the nonlinear modal interactions in a system comprising a heavy, low tension vertical cable of varying length moving at speed within a tall host structure subjected to a low frequency sway. The nonstationary nonlinear responses of the cable are studied numerically in order to predict and to examine its non-planar motions that arise due to autoparametric resonances. The results presented in the paper can be applied in the design of compensation systems in high-rise vertical transportation systems.

2. Dynamic model

The model of a heavy cable of time-varying length *L(t)* is depicted in Figure 1. The cable has mass per unit length *m*, elastic modulus *E* and an effective cross-sectional area *A*. It passes over a sheave at the bottom of the structure with the upper end fixed to a termination at C moving at speed *V(t)* while the host structure is subjected to bending in-plane and out-of-plane deformations described by the shape functions $\Psi_v(z)$ and $\Psi_w(z)$, respectively, with z denoting a coordinate measured from ground level. The bending deformations result in a sway of the structure producing the in-plane and out-of-plane harmonic motions $v_0(t)$ and $w_0(t)$ at frequencies Ω_v, Ω_w and amplitudes $v_{0\max}$, $w_{0\max}$, respectively, measured at the level defined by the coordinate z_0. The system is equipped with the actuator F acting at the sheave end of the system to control the tension of the cable.

The sway of the structure leads to an inertial dynamic load acting upon the cable. The dynamic response to this load is represented by the lateral in-plane and out-of-plane displacements denoted as $v(x,t)$ and $w(x,t)$, respectively, where x is a spatial coordinate measured from the centre of the sheave as shown in Figure 1. The lateral motions are coupled with axial (longitudinal) motions that are denoted as $u(x,t)$.

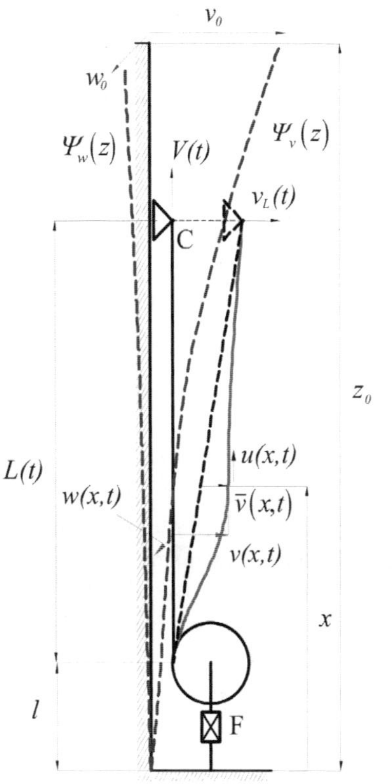

Figure 1. Cable model

The mean tension of the element is expressed as $T^i(x,t) = T_0 + m\left[g + a(t)\right]x$, where T_0 represents a constant tension term, $a(t) = \dot{V}(t)$ is the acceleration of the upper support (an overdot denotes the time derivative) and g is the acceleration of gravity. By using the large displacement approximation Green's axial strain due to stretching of the element given as $\varepsilon(x,t) = u_x + \frac{1}{2}\left(v_x^2 + w_x^2\right)$, where $(\ )_x \equiv \partial(\ )/\partial x$, the equations governing the undamped dynamic displacements $u(x,t)$, $v(x,t)$ and $w(x,t)$ can be developed by applying Hamilton's principle. It is assumed in the derivation that for tensioned members such as steel wire ropes and metallic cables the lateral frequencies are much lower than the longitudinal frequencies. Furthermore, for structures such as high-rise towers and buildings the excitation frequencies Ω_v, Ω_w are much lower than the fundamental longitudinal frequencies of the cable and no interaction will take place between the lateral modes and the longitudinal modes. Thus, the longitudinal inertia of the cable can be neglected and by taking into account the boundary condition $u(0,t) = u\left[\left(L(t),t\right)\right] = 0$ the lateral in-plane and out-of-plane responses are described by the following two equations

$$mv_{tt} - \left\{T_0 + m\left[gx - V(t)^2\right] + EAe(t)\right\}v_{xx} + m\left[a(t) - g\right]v_x + 2mV(t)v_{xt} = 0 \tag{1}$$

$$mw_{tt} - \left\{T_0 + m\left[gx - V(t)^2\right] + EAe(t)\right\}w_{xx} + m\left[a(t) - g\right]w_x + 2mV(t)w_{xt} = 0 \tag{2}$$

where $e(t) = \dfrac{ma(t)L(t)}{2EA} + \dfrac{1}{2L(t)}\displaystyle\int_0^{L(t)}\left(v_x^2 + w_x^2\right)dx$ is the quasi-static axial strain in the cable. The displacements and the spatial coordinate x are defined in a time-variant domain defined as $x \in D(t) = \{0 < x < L(t)\}$. Noting that the lateral displacements at the boundaries are given as

$v(0,t)=0;\ v\left[\left(L(t),t\right)\right]=v_L(t);\ w(0,t)=0;\ w\left[\left(L(t),t\right)\right]=w_L(t)$, where $v_L(t)$ and $w_L(t)$ represent the in-plane and out-of-plane lateral displacements of the structure corresponding to the upper end of the cable, equations (1)–(2) are discretized by expressing the lateral displacements as $v(x,t)=\bar{v}(x,t)+\dfrac{x}{L(t)}v_L(t)$ and $w(x,t)=\bar{w}(x,t)+\dfrac{x}{L(t)}w_L(t)$, respectively. In these expressions the displacements $\bar{v}(x,t)$ and $\bar{w}(x,t)$ are defined as

$$\bar{v}(x,t)=\sum_{n=1}^{N}\Phi_n\left[x;L(t)\right]p_n(t);\ \bar{w}(x,t)=\sum_{n=1}^{N}\Phi_n\left[x;L(t)\right]q_n(t),\ \ 0\le x\le L(t) \tag{3}$$

These are approximate solutions for the in-plane and out-of-plane modes, respectively, that satisfy homogenous boundary conditions. Functions $\Phi_n\left[x;L(t)\right]$, $n=1,2,\ldots,N$ represent the trial functions defined as $\Phi_n\left[x;L(t)\right]=\sin\dfrac{n\pi}{L(t)}x$, and $p_n(t)$, $q_n(t)$, $n=1,2,\ldots,N$ represent the modal coordinates / displacements, with N denoting the number of modes assumed in the expansions (3).

It can be assumed that the length L of the cable is a slowly varying parameter, i.e. that its variation is observed on a slow time scale defined as $\tau=\epsilon t$, where where $\epsilon\ll 1$ is a small quantity [6]. Thus, the derivatives of L with respect to time t is proportional to ϵ

$$\frac{\mathrm{d}L}{\mathrm{d}t}\equiv\dot{L}=\frac{\mathrm{d}L(\tau)}{\mathrm{d}\tau}\frac{d\tau}{dt}=\epsilon\frac{\mathrm{d}L(\tau)}{\mathrm{d}\tau};\ \ddot{L}=\epsilon^2\frac{\mathrm{d}^2L(\tau)}{\mathrm{d}\tau^2} \tag{4}$$

Consequently, the velocity V and acceleration a are also assumed to be slowly varying. By using (3) in the equations of motion (1-2), taking into account (4), orthogonalising with respect to the natural modes, when terms $O(\epsilon)$ and $O(\epsilon^2)$ are neglected, the following set of $2N$ ordinary nonlinear differential equations with slowly varying coefficients results

$$\begin{aligned}&\ddot{p}_r(t)+2\zeta_r\omega_r(\tau)\dot{p}_r(t)+\sum_{n=1}^{N}C_{rn}(\tau)\dot{p}_n(t)+\\&\lambda_r^2(\tau)\left\{\bar{c}^2-V^2(\tau)+\frac{1}{2}a(\tau)L(\tau)+\frac{1}{2}c^2\left[\left(\frac{v_L(t)}{L(\tau)}\right)^2+\left(\frac{w_L(t)}{L(\tau)}\right)^2\right]\right\}p_r(t)+\\&\sum_{n=1}^{N}K_{rn}(\tau)p_n(t)+\left(\frac{\lambda_r(\tau)}{2}c\right)^2p_r(t)\sum_{n=1}^{N}\lambda_n^2(\tau)\left[p_n^2(t)+q_n^2(t)\right]=P_r(t,\tau),\quad r=1,2,\ldots,N\end{aligned} \tag{5}$$

$$\begin{aligned}&\ddot{q}_r(t)+2\zeta_r\omega_r(\tau)\dot{q}_r(t)+\sum_{n=1}^{N}C_{rn}(\tau)\dot{q}_n(t)+\\&\lambda_r^2(\tau)\left\{\bar{c}^2-V^2(\tau)+\frac{1}{2}a(\tau)L(\tau)+\frac{1}{2}c^2\left[\left(\frac{v_L(t)}{L(\tau)}\right)^2+\left(\frac{w_L(t)}{L(\tau)}\right)^2\right]\right\}q_r(t)+\\&\sum_{n=1}^{N}K_{rn}(\tau)q_n(t)+\left(\frac{\lambda_r(\tau)}{2}c\right)^2q_r(t)\sum_{n=1}^{N}\lambda_n^2(\tau)\left[p_n^2(t)+q_n^2(t)\right]=Q_r(t,\tau),\quad r=1,2,\ldots,N\end{aligned} \tag{6}$$

where the modal damping represented by the ratios ζ_r has been added, ω_r, $r=1,2,\ldots,N$, are slow-varying undamped natural frequencies of the system, $\lambda_n(\tau)={n\pi}/{L(\tau)}$, $c=\sqrt{EA/m}$ and $\bar{c}=\sqrt{T_0/m}$

represent the longitudinal wave speed and the lateral wave speed, respectively. $K_{rn}(\tau)$, $C_{rn}(\tau)$ are linear slow time-variant coefficients and $P_r(t;\tau), Q_r(t;\tau)$ are the modal excitation functions. Equations (5-6) accommodate the influence of weight of the cable as well as the Coriolis and centrifugal acceleration effects. It is evident that the host structure motion results in modal external excitation terms and also appears in the stiffness coefficients as parametric excitation. The presence of cubic nonlinear coupling terms arising due to the effect of cable stretching promotes conditions for autoparametric interactions which is investigated in the case study in what follows.

3. Autoparametric interactions: a case study

Numerical simulation tests are applied to study the lateral dynamics of the system in a scenario where the attachment point C of the cable is moving upwards starting from the lowest position and coming to a stop after the cable is fully extended. The cable is a steel wire rope of mass per unit length m = 1.54 kg/m, having the modulus of elasticity $E = 0.7\times10^5$ N/mm^2 and an effective cross-sectional area A = 178 mm^2. The modal damping ratios ζ_r of the rope are assumed to be *3%* across all modes. The rope is being accelerated from rest, when its initial length is $L(0) = l$ = 4.9 m, at the rate of 1.1 m/s^2. It is moving upwards at the maximum speed and is later being decelerated at the rate of 1.1 m/s^2, coming to rest reaching the maximum length of $L_{\max} = 416.2$ m. Two maximum speeds are considered in the simulation tests: $V_{1max} = 2$ m/s and $V_{2max} = 8$ m/s, respectively. The host structure is subjected to a strong in-plane harmonic excitation and sways in its fundamental mode with large bending motions in the in-plane direction. The deformation shapes of the structure $\Psi_v(z)$ and $\Psi_w(z)$ are approximated by cubic polynomial functions [7]. These motions are of frequency 0.0675 Hz ($\Omega_v = 0.4241$ rad/s) with large displacements reaching the amplitude of $v_{0\max} = 0.8$ m at the level corresponding to $z_0 = 422.1$ m. On the other hand, in the out-of-plane direction the fundamental frequency of the structure is assumed to be much higher, it is 0.2 Hz ($\Omega_w = 1.2566$ rad/s), and only very small bending fundamental mode motions of the structure occur, with the maximum amplitude assumed as $w_{0\max} = 0.007$ m recorded at the same level (422.1 m). Thus, in this case scenario the excitation can be considered as nearly planar. The numerical simulations are carried out when the cable is subjected to a constant low mean tension of $T_0 = 61.3$ N. The explicit Runge-Kutta (4,5) formula is used to integrate the equations of motion (5-6) on the fast time scale with the slowly varying coefficients C_{rn} and K_{rn} pre-calculated and evaluated at each integration step. The dynamic displacements relative to the stretched configuration of the cable are then determined from the solution using the expansions (3). The tests conducted show that a good convergence in (3) can be achieved by using relatively small number of modes N. In the case study under consideration N = 5 has been applied.

The natural frequencies $\omega_r(\tau)$, $r = 1,2,\ldots,N$, are determined by freezing the slowly varying parameters in the matrix of linear coefficients in the model (5-6) at each time step and calculating the eigenvalues taking into account the influence of weight, speed and acceleration of the cable. The first three natural frequencies of the cable for the two speeds are plotted vs. time in Figure 2(a) and 2(b), respectively, together with the frequencies of the in-plane the out-of plane motions of the host structure. It is evident that a passage through the fundamental resonance in the in-plane direction takes place at the time instant of approximately 90.4 s and 22.1 s for $V_{1max} = 2$ m/s and $V_{2max} = 8$ m/s, respectively. In the out-of-plane direction the fundamental resonance is reached at the time instant of

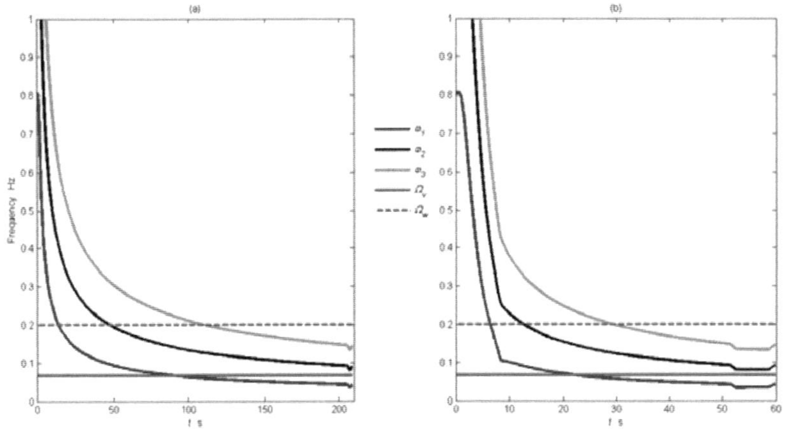

Figure 2. The natural frequencies of the cable shown together with the frequencies of the structure: (a) for $V_{1max} = 2$ m/s , (b) for $V_{2max} = 8$ m/s

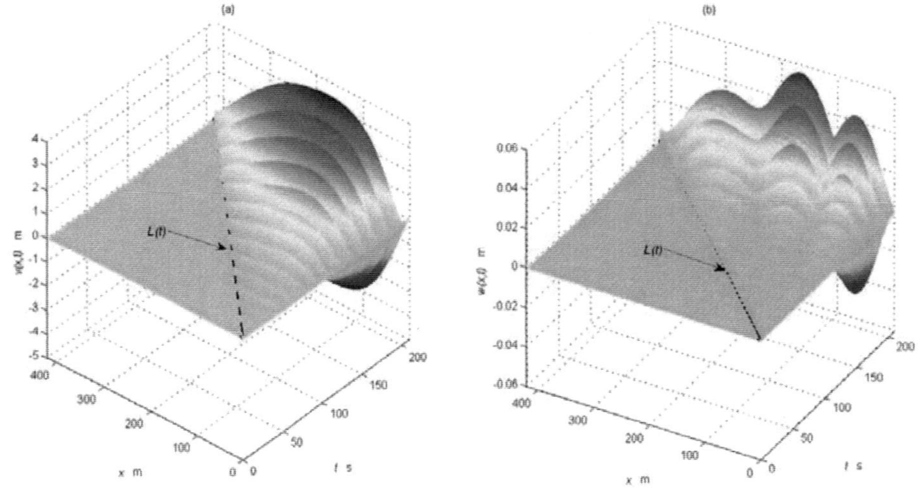

Figure 3. Displacements of the cable corresponding to $V_{1max} = 2$ m/s : (a) *v(x,t)* and (b) *w(x,t)*

approximately 14 s and 6.4 s, respectively. Then, at the time instants of approximately 47.5 s and 12.6 s the frequency of the excitation becomes equal to the second natural frequency of the cable. More importantly, at the time instants of approximately 110.8 s and 29.3 s the frequency of the excitation becomes equal to the third natural frequency of the rope which results in passage through the third out-of-plane resonance. The analysis of responses shown in Figure 3 and Figure 4 and the examination of the corresponding modal coordinates can be summarized as follows. Both the in-plane and out-of-plane responses grow with time. The in-plane dynamics exhibit responses with large amplitudes and the influence of the fundamental resonance is evident. The out-of-plane displacements are smaller with strong 3rd mode contribution being evident. The cubic nonlinearities promote energy exchanges between the modes. The energy from the fundamental in-plane mode is being transferred to the corresponding out-of-plane mode due to autoparametric (internal) 1:1 resonance. This phenomenon is prominent during the final stages of motion when the cable approaches its maximum length. Furthermore, noting that the excitation frequencies of the structure are in the ratio of $\Omega_w / \Omega_v \simeq 3$, in the region corresponding to the time instant of 110.8 s for the lower speed, and 29.3 s for the higher speed, the frequency tuning $\Omega_w = \omega_3 \simeq 3\omega_1$ takes place. Thus, due to a 3:1 internal resonance the energy is being exchanged between the 3rd and the fundamental out-of-plane modes. The responses corresponding to the higher speed exhibit smaller amplitudes due to the fact that the passages

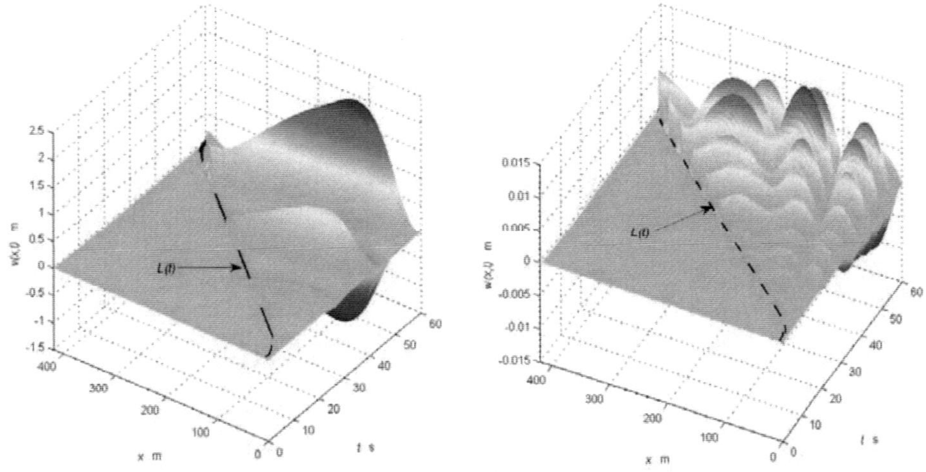

Figure 4. Displacements of the cable corresponding to $V_{2max} = 8$ m/s : (a) *v(x,t)* and (b) *w(x,t)*

through the resonance regions are faster. However, further simulation tests carried out to investigate the behaviour of the system after the cable has stopped reaching its maximum length reveal that the in-plane displacements continue to grow resulting in full non-planar tubular motions and high tensile stresses in the cable. A possible resonance control strategy would involve using the actuator *F* to increase the mean tension in order to shift the natural frequencies above the frequency of the sway excitation. Active methods such as active stiffness control [8] can also be considered.

4. Conclusions

The dynamic behaviour of long, low tension heavy cables and ropes moving at speed in tall civil structures is inherently non-linear. The model presented in this paper accommodates the cubic geometric nonlinearities arising due to the effect of cable stretching and takes into account the influence of slow variation of length, the weight of the cable as well as the Coriolis and centrifugal acceleration effects. The analysis of results presented in the paper shows that due to the nonlinear coupling, important interactions between the in-plane mode and the out-of-plane mode might occur. Ultimately, the planar motions may become unstable resulting in large non-planar tubular motions with the cable being subjected to high tensile stresses. The scenarios discussed in this paper reveal that in order to avoid adverse internal resonance interactions the amplitudes of the excited modes that are involved in 1:1 and 3:1 resonances should not reach their critical threshold levels. This can be achieved if higher travel speeds and/ or cable tensions are applied. Alternatively, a suitable actuator control algorithm can be used to actively suppress the resonance responses.

References

[1] Wickert J A Mote Jr. C D 1990 *ASME J. Appl. Mech.* **57** 738-744
[2] Leamy M J Gottlieb O 2000 *JSV* **236** 683-703
[3] Riedel C H Tan C A 2002 *Int. J. Non-Lin. Mech.* **37** 101-116
[4] Terumichi Y Kaczmarczyk S Turner S Yoshizawa M Ostachowicz M 2003 *Mater. Sci. Forum* **440-441** 497-504
[5] Salamaliki-Simpson R Kaczmarczyk S Picton P Turner S 2006 *Appl. Mech. Mat.* **5-6** 217-224.
[6] Kaczmarczyk S Ostachowicz W 2003 *JSV* **262** 245 - 289
[7] Kaczmarczyk S Iwankiewicz R Terumichi Y 2009 *J. Phys.: Conf. Series* **181** doi:10.1088/1742-6596/181/1/012047
[8] Kaczmarczyk S 2011 *Springer Proc. Phys.* **139** 183 - 188

Modern Practice in Stress and Vibration Analysis 2012 (MPSVA 2012) IOP Publishing
Journal of Physics: Conference Series **382** (2012) 012038 doi:10.1088/1742-6596/382/1/012038

On-line identification of time-varying systems equipped with adaptive control

R. Ceravolo[1,2], E. Matta[1], A. Quattrone[1], C. Surace[1], L. Zanotti Fragonara[1]

[1] Department of Structural, Geotechnical and Building Engineering, Politecnico di Torino, corso Duca degli Abruzzi 24 – 10129 Turin, Italy

[2]E-mail: rosario.ceravolo@polito.it

Abstract. The present study aims at numerically investigating the feasibility of an adaptive TMD control system applied on lightweight, flexible structures characterized by time-varying inertial properties. The case study will consist of a photovoltaic support structure subject to snow drifting and slippage in windy conditions.

1. Introduction

In its basic arrangement, a passive tuned mass damper (TMD) is a single-degree-of-freedom mechanical appendage of the main structure, schematically consisting in a linear mass/spring/damper system. When optimally tuned to the resonant target mode of vibration of the primary structure, a TMD adds significant damping in the dominant vibration modes, thereby reducing the overall vibration response of the structure. Optimal tuning generally requires that, once a proper mass is chosen for the absorber (usually small if compared to the effective mass in the structural mode), its frequency and damping ratios are set to the respective optimal values such as to minimise a desired norm (typically either the H_2 norm or the H_∞ norm) of the input-output transfer functions for the overall dynamical system. De-tuning, including any possible deviation of the TMD parameters from their optimal values, and resulting for example from variations in the structural properties of the primary structure, deterioration of TMD's mechanical parameters, incorrect design forecasts, etc., may lead to a significant loss of control performance. One possible remedy to de-tuning is represented by the "multiple TMD" [1], constituted by lightly damped oscillators having frequencies scattered over a small range around the target natural frequency. An alternative, more efficient solution will follow in this paper, represented by the so-called "adaptive TMD", i.e. a TMD where the stiffness and damping parameters can be adjusted, or re-tuned, in real time to their optimal values at a minimum of energy expense [2], depending on some measures of the structural dynamic response.

Although different re-tuning criteria are available in the literature, including data driven, non-model-based algorithms, a model-based method will be presented in the sequel, in which the optimal control parameters are continuously updated based on an on-line tracking of the actual dynamical properties of the overall system, herein performed through the dynamic identification technique known as the Unscented Kalman Filter (UKF) [3], which has been chosen among the others because of its limited computational cost. The UKF has been firstly proposed in 1995 by Julier *et al* [4] as an improvement of the Extended Kalman Filter (EKF).

2. Model Description

On-line identification for the re-tuning of the TMD, through the recursive algorithm UKF, has been applied to a structure modelled as two degrees of freedom system. This structure is constituted by a steel pillar, clamped at the base, which at the top supports a photovoltaic panel assumed to be infinitely rigid (Figure 1), representing an application of practical relevance that may be subjected to loading by snow in adverse weather conditions.

In practice, the reduced slope of the panel due to efficiency requirements would let snow accumulate on the panel. As the weather conditions change, snow could then slip and detach resulting in a highly time-varying nature of mass loading and hence the inertial properties of the dynamic system.

Figure 1. Solar panels: example of a typical linear arrangement and model scheme.

Generally, snow tends to accumulate on surfaces and on the ground unevenly depending on the action of the wind τ and the threshold level τ^*: for $\tau > \tau^*$ there is erosion; instead for $\tau < \tau^*$ there is deposit or transport. In the latter case, in order to establish which of the two phenomena occur, i.e. either transport or deposit, energetic considerations must be taken into account: if the energy of the particle in motion, which corresponds to the sum of the kinetic and wind energies and depends on τ, is greater than the energy corresponding to τ^*, the snow particle will bounce off the surface and continue to move. Conversely, if the overall energy of the particle is insufficient to break the ties among the snow particles at the moment of impact, deposit will occur [5].

The difficulty of determining precisely the areas and the entity of erosion and deposit is correlated with the multiplicity of physical and geometrical parameters influencing the value of τ * [6,7,8] which in practice depends on the cohesion, dimension and shape of the snow particles, on the density and evolution of the snowdrift. Initial research on the wind transport of snow dates back to the studies of Owen [9]. Referring to the work conducted by Bagnol (1941) on the wind transport of sand, Owen immediately realised that the analysis of snow was more complex [10] since the particles do not have a nominally spherical shape like sand; the dendritic shape of snowflakes affects the elastic properties and the rebound capabilities of the particles on the surface. Furthermore, due to the subsequent rebounds and variations of temperature, these characteristics tend to modify over time and can do so rapidly. Although several studies have been conducted since 1941 concerning snow drifting, analytical relations available in published literature address the determination of the snowdrift evolution in terms of settling and/or exposure to the wind. On the basis of these considerations, a parabolic distribution of the snow on the panel has been assumed according to Yan et al [11].

Concerning the model of snow fall and detachment, reference has been made to the work of Taylor *et al* [12]. According to the model proposed by Taylor, the detachment of a block of snow happens when the component tangent to the surface of the force of gravity equals the snow tensile strength and of the attractive and adhesive forces generated at the snow-surface interface. The tensile strength, which depends on the density and on the degree of humidity, has been conformed to values suggested by Zempachi [13].

Considering a temporal interval of approximately 40s, two subsequent detachments are assumed to occur that alter the condition of the panel from the initial condition of complete covering to the condition of unloaded panel. Figure 2 illustrates the time-variant characteristics of the system, the mass *m(t)* and the polar inertia *I(t)* calculated with respect to the joint between the panel and the pillar.

The mathematical model, describing the temporal evolution of the displacements of the system under wind action, can be expressed in a matrix form as:

$$\begin{bmatrix} m(t) & 0 \\ 0 & I(t) \end{bmatrix} \begin{Bmatrix} \ddot{u} \\ \ddot{\theta} \end{Bmatrix} + \begin{bmatrix} c_{11} & c_{12} \\ c_{21} & c_{22} \end{bmatrix} \begin{Bmatrix} \dot{u} \\ \dot{\theta} \end{Bmatrix} + \begin{bmatrix} k_{11} & k_{12} \\ k_{21} & k_{22} \end{bmatrix} \begin{Bmatrix} u \\ \theta \end{Bmatrix} = \begin{Bmatrix} F \\ C \end{Bmatrix} \tag{1}$$

where u and θ are the translation in the panel plane and its rotation (Figure 1).

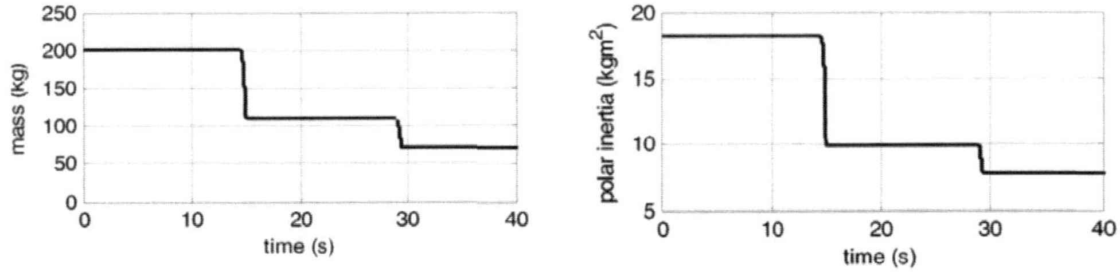

Figure 2. Time-evolution of mass parameters due to progressive snow drifting and slippage.

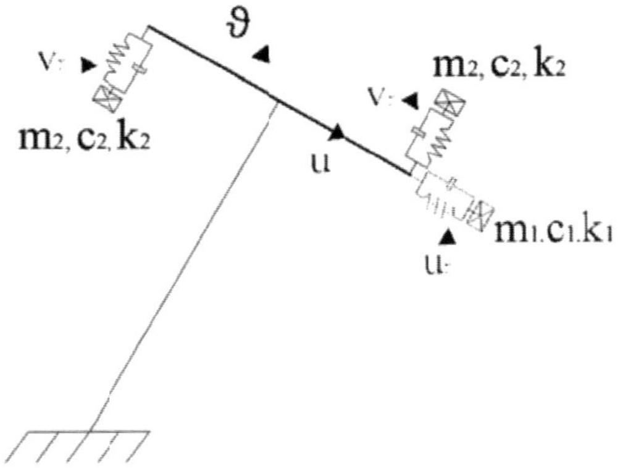

Figure 3. TMDs configuration, the brown TMD controls the translation, the blue TMDs the rotation.

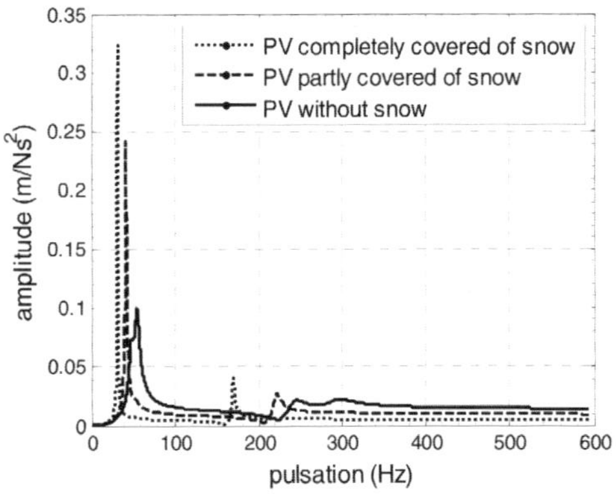

Figure 4. FRFs between force and tangential acceleration.

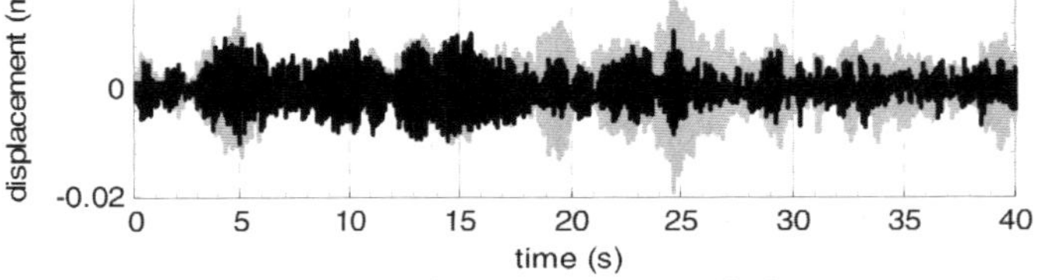

Figure 5. Displacements for the not-controlled structure (grey line) and controlled one (black line).

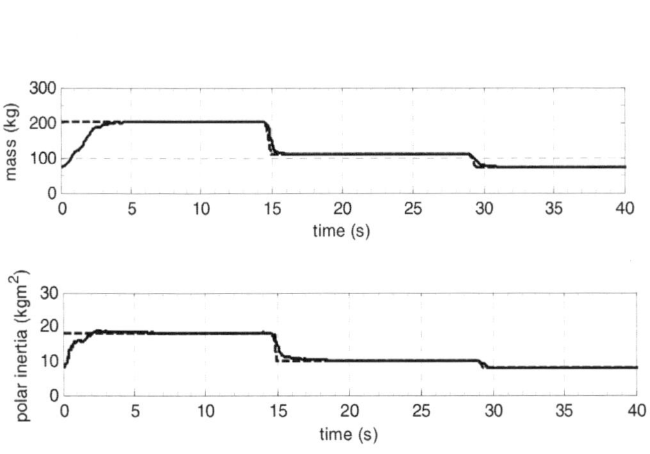

Figure 6. On-line identification of the mass and the polar inertia of the system (real values in dashed line, identified values in continuous line).

3. Passive control with TMD and time-variant identification via UKF

Since the system has two degrees of freedom, three TMDs have been used, placed as illustrated in Figure 3 and tuned to the dynamic characteristics of the panel without snow, which is the reference serviceability condition. In order to ensure an appropriate performance of the passive control system even in presence of snow, it was decided to adopt a relatively high mass ratio equal to 1% of the panel mass completely covered in snow (snow level from the ground assumed equal to 0.5 m). In this case the equations of the controlled dynamic system are:

$$\begin{bmatrix} m(t) & 0 & 0 & 0 \\ 0 & I(t) & 0 & 0 \\ m_1 & 0 & m_1 & 0 \\ 0 & m_2 L & 0 & m_2 \end{bmatrix} \begin{Bmatrix} \ddot{u} \\ \ddot{\theta} \\ \ddot{u}_1 \\ \ddot{v}_2 \end{Bmatrix} + \begin{bmatrix} c_{11} & c_{12} & -c_1(t) & 0 \\ c_{21} & c_{22} & 0 & -c_2(t) \\ 0 & 0 & c_1(t) & 0 \\ 0 & 0 & 0 & 2c_2(t) \end{bmatrix} \begin{Bmatrix} \dot{u} \\ \dot{\theta} \\ \dot{u}_1 \\ \dot{v}_2 \end{Bmatrix} + \begin{bmatrix} k_{11} & k_{12} & -k_1(t) & 0 \\ k_{21} & k_{22} & 0 & -k_2(t) L \\ 0 & 0 & k_1(t) & 0 \\ 0 & 0 & 0 & 2k_2(t) \end{bmatrix} \begin{Bmatrix} u \\ \theta \\ u_1 \\ v_2 \end{Bmatrix} = \begin{Bmatrix} F \\ C \\ 0 \\ 0 \end{Bmatrix} \quad (2)$$

where u_1 and v_2 are the relative displacements of the TMD with respect to the panel in the tangential and normal directions respectively.

The accumulation of snow on the panel, and the subsequent detachment of blocks, causes sudden changes of the natural frequencies of the structure and the consequent de-tuning and efficiency loss of the passive control system. This is clearly visible from the comparison between the Frequency Response Functions obtained for different temporal intervals (Figure 4). Figure 5 shows that when the panel mass diverges from the value used for the TMDs tuning (without snow), the control efficiency reduces significantly.

Having calculated the output data in terms of accelerations at two opposed corners of the panel, the TMD controlled system has been identified using the UKF, selecting appropriate values of the parameters α, β and κ [14]. The algorithm identifies correctly the time-variant inertial characteristics of the principal structure as highlighted in figure 6.

4. UKF identification and control through the adaptive TMD

In contrast to other identification techniques, the UKF enables the determination of the dynamic characteristics of the principal structure by simplifying significantly the device re-tuning. In fact this can be performed by using the same formulas used for the passive TMDs. After measuring the input and output of the system at regular time intervals and applying the unscented filter, it is possible to determine the time-varying characteristics of the system and consequently the natural frequencies and the effective masses of the principal system. Simulations have shown that whilst the UKF identifies in an accurate way the characteristics of the principal structure (see figure 7), the quality of the results is directly related to the appropriate selection of the UKF algorithm parameters. The parameters values have been chosen accordingly to previous studies available in the literature [4,15]. Anyway, the topic may represent an interesting future area of research.

Once identification has been performed, the adaptive TMDs with constant mass should be re-tuned, varying the stiffness and damping. The dynamic behaviour of the panel and of the auxiliary masses continues to be described by a system [2] with time-varying terms c_1, c_2, k_1, k_2.

In figure 7, the optimal characteristics of TMDs, calculated using Warbuton's formulas [16] as a function of the inertial properties instantaneously identified, are shown together with the respective values identified via the UKF.

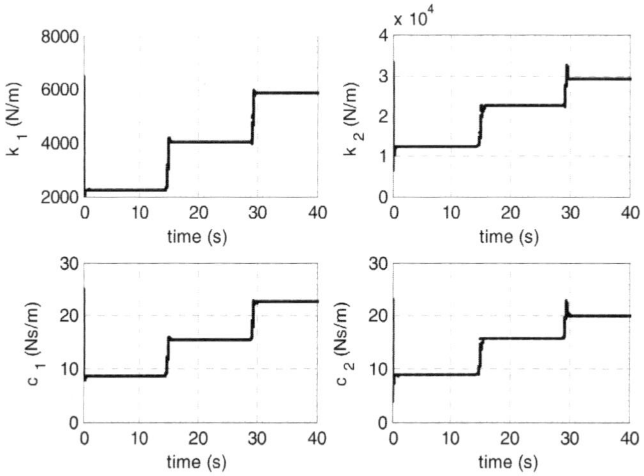

Figure 7. Comparison between the optimal characteristics of translational and rotational TMDs and those obtained with UKF.

Table 1 displays the percentage variation of the inertial characteristics of the system referred to the condition of unloaded panel and the relative variations (ΔM mass, ΔI_p polar inertia), in terms of variation of TMD stiffness (ΔK_{TMD}) and damping (ΔC_{TMD}) in the degrees-of-freedom u and θ.

Table 1. Percentage variation of the inertial characteristics referred to the condition of unloaded panel.

Variation	**ΔM**	ΔK_{TMDx}	ΔC_{TMDx}	ΔI_p	$\Delta K_{TMD\theta}$	$\Delta C_{TMD\theta}$
step 1-3	173.52%	-62.00%	-60.87%	128.57%	-57.31%	-55.00%
step 1-2	48.68%	-31.49%	-30.43%	24.84%	-23.02%	-20.00%

The highest effective level of passive damping can be evaluated through Table 2 which reports the Root Mean Square (RMS) values of the displacements u and θ and their relative reductions due to the introduction of the passive (TMD) or adaptive (STMD) devices.

Table 2. Comparison of results in the cases of structure which is not-controlled, controlled with passive devices, and controlled with adaptive device.

	RMS u					**RMS θ**				
t_i - t_f [s]	without TMDs [m]	with TMDs [m]		with STMDs [rad]		without TMDs [rad]	withTMDs [m]		with STMDs [rad]	
0-15	0.0042	0.0041	98%	0.0025	60%	0.0053	0.0051	96%	0.0032	60%
15-30	0.0058	0.0035	60%	0.0024	41%	0.0072	0.0043	60%	0.003	42%
30-40	0.0048	0.0023	48%	0.0029	48%	0.0061	0.0023	38%	0.0029	48%

5. Conclusions

This paper proposed an on-line procedure for the identification of a time-variant dynamic system by resorting to the UKF algorithm. The aim of the identification was to determine the time-variant characteristic of the system with accuracy adequate to allow for the re-tuning of adaptive TMDs. Through simulations on a system characterised by sharp variations of dynamic properties, it has been shown that STMDs supported by a UKF identification algorithm can recover their optimal performance with delays fully compatible with practical applications. Future developments may involve the numerical investigation of the performances of the UKF with different excitation sources and the application of the methodology to real-scale structure.

Acknowledgments

The authors wish to thank Marica Pecorelli for the precious help and assistance provided.

6. References

[1] Xu K and Igusa T 1992 Dynamic characteristics of multiple substructures with closely spaced frequencies *Earthquake Engineering and Structural Dynamics* **21** 1059-1070

[2] Nagarajaiah S and Varadarajan N 2005 Semi-active control of wind excited building with variable stiffness TMD using short-time Fourier transform *Journal of Engineering Structures* **27** 431-441

[3] Wu M and Smyth AW 2008 Real-time parameter estimation for degrading and pinching hysteretic models *Int. J. of Non-Linear Mech.* **43** 822-833

[4] Julier SJ, Uhlmann JK and Durrant-Whyte HF 1995 A new approach for filtering nonlinear systems *Proc. Am. Control Conference* p. 1628-1632

[5] Kashef A and Baskaran BA 1995 Numerical simulation of air flow with different roof configuration for the prediction of snow accumulation *91CWE*

[6] Zhang J and Huang N 2008 Simulation of snow drift and the effects of snow particles on wind *Modelling and simulation in engineering* 2008

[7] Li L and Pomeroy JW 1996 Estimates of threshold wind speeds for snow transport using meteorological data *J. Appl. Met.* **36** 205-213

[8] Schimidt RA 1986 Transport rate of drifting snow and the mean wind speed profile *Bound. Layer Meteor.* **34** 213-234

[9] Owen PR 1964 Saltation of uniform grains in air *Journal of Fluid Mechanics* **20** 225-242

[10] Doorshot JJ and Lehning M 2002 Equilibrium saltation: mass fluxes, aerodynamic entrainment and dependence on grain properties *Bound. Layer Meteorology* **104** 111-130

[11] Yan K, Zhou X, Gu M and Kasperski M 2010 Wind-drifted snow load simulation on gable roof *The Fifth International Symposium on Computational Wind Engineering (Chapel Hill)*

[12] Taylor DA 1987 Snow on sloping roofs *Proceedings of CSCE Centennial Conference (Montreal)* p. 145-158

[13] Zempachi W Breaking strength of deposited snow. Scientific Report. Fukushima University, Department of Physics. Report No.: n. 25.

[14] Xiong K, Zhang HY and Chan CW 2006 Performance evaluation of UKF-based nonlinear filtering *Automatica* **42** 261-270

[15] Sarkka S 2004 Recursive Bayesian Inference on Stochastic Differential Equations. PhD Thesis. Helsinki University of Technology.

[16] Warburton GB 1982 Optimum absorber parameters for various combination of response and excitation parameters *Earthquake Engineering & Structural Dynamics* **10** 381-401

Journal of Physics: Conference Series **382** (2012) 012039 doi:10.1088/1742-6596/382/1/012039

Active vibration reduction of a flexible structure bonded with optimised piezoelectric pairs using half and quarter chromosomes in genetic algorithms

A H Daraji and J M Hale
School of Mechanical Systems and Engineering, Newcastle University, NE1 7RU, UK

Email: a.h.a.daraji@ncl.ac.uk jack.hale@ncl.ac.uk

Abstract. The optimal placement of sensors and actuators in active vibration control is limited by the number of candidates in the search space. The search space of a small structure discretized to one hundred elements for optimising the location of ten actuators gives 1.73×10^{13} possible solutions, one of which is the global optimum. In this work, a new quarter and half chromosome technique based on symmetry is developed, by which the search space for optimisation of sensor/actuator locations in active vibration control of flexible structures may be greatly reduced. The technique is applied to the optimisation for eight and ten actuators located on a 500×500mm square plate, in which the search space is reduced by up to 99.99%. This technique helps for updating genetic algorithm program by updating natural frequencies and mode shapes in each generation to find the global optimal solution in a greatly reduced number of generations. An isotropic plate with piezoelectric sensor/actuator pairs bonded to its surface was investigated using the finite element method and Hamilton's principle based on first order shear deformation theory. The placement and feedback gain of ten and eight sensor/actuator pairs was optimised for a cantilever and clamped-clamped plate to attenuate the first six modes of vibration, using minimization of linear quadratic index as an objective function.

1. Introduction

Genetic algorithm is a powerful random search method guided by a fitness function, which may be used to detect efficient locations for discrete sensors and actuators on smart structure for active vibration control. The complexity of the genetic algorithm for the placement problem is defined by the number of candidate solutions in the search space, which is the statistical combination of the number of sensors and actuators and the total number of possible locations on the structure.

Flexible structure bounded with distributed sensor and actuator pair investigated thoroughly for vibration attenuation [1-3]. The dynamic equation of piezoelectric electroelasticity is formulated by Allik and Hughes and found to be reducible in form to the well-known equations of structural dynamics [1]. Optimal placement of discrete piezoelectric sensors and actuators was investigated for flexible structures by many researchers using the genetic algorithm based on different types of objective functions. Optimal placement of two actuators was obtained by Sadri et al for a thin plate based on modal [4] and grammian controllability [5] as an objective function using a binary number coded chromosome. Placement of two actuators and piezofilm sensors were optimised used by Han and Lee using controllability, observability and spillover as an objective function to suppress the first three modes of vibration using a binary number coded chromosome [6]. Peng et al studied optimal placement of four sensor/actuator pairs to control the first five modes of vibration based on grammian controllability index maximization as an objective function [7]. Kumar and Narayanan investigated optimal placement of ten sensor/actuator pairs using minimization of linear quadratic index as an objective function to suppress the first six modes of vibration [8]. Roy and Chakraborty studied a composite shell to place six piezoelectric sensor/actuator pairs using a modified genetic algorithm based on maximization of closed loop damping with limited feedback voltage value as an objective function to suppress the first eight modes of vibration [9]. Various published works have investigated symmetrical dynamic structures and variously found asymmetrical [4-6, 8, 9] and symmetrical [7] optimal configurations.

Daraji and Hale have investigated global optimal locations for two, four, six, eight and ten piezoelectric actuators for flat plates and plates stiffened by beams with both symmetrical and asymmetrical geometries and boundary conditions. They have found symmetrical configurations of actuators for symmetrical structures and asymmetrical actuators configurations for asymmetrical structures using minimization of linear quadratic index as an objective function and the symmetrical piezoelectric configuration gave higher vibration attenuation than published asymmetrical configuration [10-13]. Hale and Daraji [14] have researched global optimal piezoelectric sensor/actuator configurations for a cantilever plate based using new fitness function based on H_infinity and found a symmetrical piezoelectric actuator distribution which gives higher vibration reduction than published.

This paper complements the published works [10-15] by reducing the genetic algorithm search space and obtaining the global optimal sensor and actuator locations with few generations using a half and quarter chromosome technique. In this paper, an isotropic plate with a number of discrete piezoelectric sensor/actuator pairs bonded to its surface is modelled using finite element and Hamilton's principle based on first order shear deformation theory taking account of piezoelectric mass, stiffness and electromechanical coupling effects. A matlab m-code computer program has been built for this new technique to locate ten and eight piezoelectric actuators, respectively, for a cantilever and clamped-clamped plate, based on minimization of optimal linear quadratic control as an objective function.

2. Modelling

The plate with number of piezoelectric patches bonded to its surfaces is modelled based on finite element, Hamilton's principle and first shear deformation theory using isoparametric four nodes element, the modal dynamic equation for the plate with a number of discrete piezoelectric pairs bonded to its surfaces was written in state space form as follows[11];

$$\{\dot{X}\} = \begin{bmatrix} 0 & \omega \\ -\omega & -2\xi\omega \end{bmatrix}\{X\} + \begin{bmatrix} 0 \\ -[\varphi]^T[K_{u\phi}]^a \end{bmatrix}\{\phi_a\} \tag{1}$$

$$\{\dot{X}\} = [A]\{X\} + [B]\{u\} \quad , \quad \{\phi_s\} = [C]\{X\} \ , \ \{X\} = \{\omega\eta \quad \dot{\eta}\}^T \tag{2}$$

$$[A_i] = \begin{bmatrix} 0 & \omega_i \\ -\omega_i & -2\xi_i\omega_i \end{bmatrix} , \ [B_i] = \begin{bmatrix} 0 \\ -[\varphi_i]^T[K_{u\phi}]^a \end{bmatrix} , \ [C_i] = \begin{bmatrix} -[\varphi_i]^T[K_{\phi\phi}]^{s^{-1}}[K_{u\phi}]^s\omega_i^{-1} & 0 \end{bmatrix} \tag{3}$$

Where $[A_i]$, $[B_i]$ and $[C_i]$ are state, actuator and sensor matrices in which subscript i refers to mode number. $[K_{u\phi}]$ and $[K_{\phi\phi}]$ refer to piezoelectric electromechanical coupling and capacitance matrices.

3. Control law and objective function

Linear quadratic optimal controller design is based on minimization of a performance index J. Values of positive-definite weighted matrices $[Q]$, of dimension $(2n_m \times 2n_m)$, and $[R]$ of dimension $(r_a \times r_a)$ are controlled the value of the performance index, where n_m, r_a represent the number of modes and actuators, respectively. Ogata has shown it is possible to follow this derivation to design a linear quadratic controller [16], which leads to the following Riccati equation:

$$[A]^T[P] + [P][A] - [P][B][R]^{-1}[B]^T[P] + [Q] = 0 \tag{4}$$
$$[K] = [R]^{-1}[B]^T[P] \quad , \quad \{u\} = -[K]\{X\} \tag{5}$$

Solution of the Reduced Riccati equation (4) gives the value of matrix $[P]$; if matrix $[P]$ is positive definite then the system is stable or the closed loop matrix $[A] - [B][K]$ is stable. Feedback control gain can be obtained after substitution of $[P]$ in equation (5). Minimization of the linear quadratic cost function J is proposed by Kondoh et al as an objective function to optimise gain and piezoelectric actuator location [17].

$$J(x,y) = Trace([P(x,y)]) \tag{6}$$
$$J_{opt}(x,y) = \min(Trace([P(x,y)]), [K]) \tag{7}$$

Where $x, y \in$ plate dimension 50×50 cm

4. Genetic algorithm

In 1975, Holland invented the genetic algorithm, a heuristic method based on "survival of the fittest" or the principle of natural evolution. It has been continuously improved and is now a powerful method for searching optimal solutions. The published works [10-15], for flat and stiffened plate by beams with different boundary conditions based on minimization of linear quadratic index and H_infinity and using the full length chromosome, has shown that the global optimal distribution of piezoelectric sensors and actuators has the same axes of symmetry as the structure. In this work, the chromosome length is reduced to quarter and half based on symmetries of the structure, using quarter chromosome length for a structure having two axes of symmetry and half length for a structure having one axis of symmetry. The placement strategy using genetic algorithm comprising conditional filter proposed by Daraji and Hale is used in this work [10].

5. Results and discussion

5.1. Research problem

A flat plate dimensioned $500 \times 500 \times 1.9$mm was mounted rigidly from the left hand edge to form a cantilever and from left and right hand edges to form a clamped-clamped plate, respectively. The plate is discretised to one hundred elements 10×10 sequentially from left to right and down to up as shown in Figure 2. Table 1 shows plate and piezoelectric material properties.

Table 1. Plate and piezoelectric properties

Properties	Plate	Piezoelectric PIC225
Modulus, GPa	210	--------
Density, kg/m^3	7810	7180
Poisson's ratio	0.3	--------
Thickness, mm	1.9	0.5
Length, width, mm	500	50
e_{31}, e_{32}, e_{33}, C/m^2	-------	-7.15, -7.15,13.7
$C_{11}^E, C_{12}^E, C_{13}^E, C_{55}^E$, GPa	-------	123,76.7,70.25, 22.226
μ_{33}^{σ} F/m	-------	1.5×10^{-8}

C^E, e and μ^{σ} refer to elasticity, piezoelectric and permittivity constants

5.2. Natural frequencies

The first six natural frequencies and mode shapes for the cantilever plate were determined using a Matlab m-code program based on the present model and validated by the ANSYS finite element package using two dimensional SHELL63 and three dimensional SOLID45 elements, and also experimentally. The results converged with mesh refining to constant values, showing that the mesh of 10×10 shell63 elements gives good accuracy for the first six natural frequencies compared with finer meshes and with three dimensional solid45 elements, as well as with experimental results as shown in table2.

Table2. ANSYS natural frequencies compared with present model and experimental results

Element type	Mode (Hz)					
	1st	2nd	3^{td}	4th	5th	6th
ANSYS Shell63 (10×10)	6.59	16.17	41.32	52.37	59.79	104.70
ANSYS Solid45(50×50)	6.59	16.15	40.44	51.68	58.86	103.18
Present model	6.59	16.14	40.62	51.78	58.99	103.29
Experimental Frequency	5.90	16.90	37.30	51.60	58.20	101.00

5.3. Actuators optimization using half chromosome length

The genetic algorithm described in section 4 based on reference [10] was used to find optimal locations for ten piezoelectric actuators on a 0.5m square cantilever plate mounted rigidly from the left hand edge using half chromosome length. The progressive convergence of the population onto an optimal solution is shown in figure 1, in which the population is distributed around the circle with radius **(r)** representing fitness value to be minimised.

At the first generation figure 1(a) the population is very diverse with representatives of high and low fitness and the range in between. After fifteen generations figure 1(b) the population is much less diverse, made up of individuals of high, though not yet optimal, fitness. After 100 generations figure 1(c) the population has completely converged to a level of fitness higher than any individual in the first or fifteenth generations.

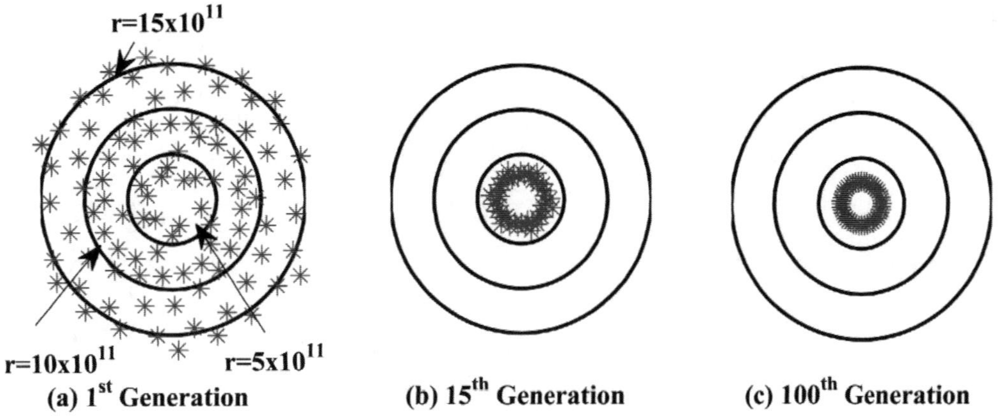

Figure 1. Population fitness progression over 100 generations. Each individual is represented as one of the points distributed around the circle with radii **r** which represent its fitness value to be minimised, using half chromosome

Figure 2 shows optimal piezoelectric configuration for cantilever and clamped-clamped plate using half and quarter chromosome (figures 2(b) and 2(d)) compared with published results using the full chromosome (figures 2(a) and 2(c)). It can observed from figure 2 that same optimal configuration of sensors and actuators was obtained with a smaller number of generations.

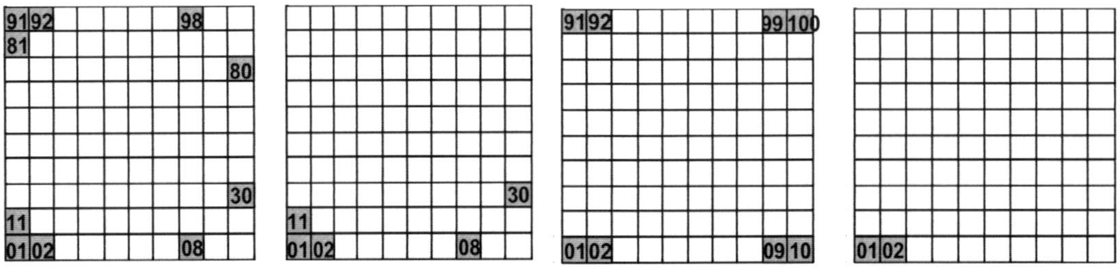

Figure 2. Optimal piezoelectric sensor/actuator configuration for a cantilever plate mounted rigidly at the left hand edge using **(a)** full length chromosome [10], **(b)** present work using half length chromosome. Clamped-clamped plate from left and right hand edge **(c)** full chromosome [13], **(d)** present quarter chromosome

5.4. Genetic algorithm problem reduction

Percentage genetic algorithm problem reduction is calculated using the half and quarter chromosomes compared with the full chromosome. The results are represented in table 3. Two measures of reduction are calculated: size of search space and average number of fitness calculations required to get an optimal solution. The total search space (the number of candidate chromosomes) is the statistical combination of the number of actuators to be optimised and the number of possible locations. This gives a reduction in the number of candidates equal to 99.9999878% for half and 99.99999983% for quarter chromosome length compared with the full chromosome.

The second reduction is calculated on the basis of the actual computational effort required to obtain the optimal solutions. table 3 shows that the half chromosome length technique requires 700 fitness solutions compared with 31900 for the full chromosome, giving a 97.80564263% reduction in computer calculation. Similarly, table 3 shows that the quarter chromosome requires only 25 calculations, compared with 33450 for the full chromosome, which gives 99.2537313% reduction in computer calculation.

Table 3. Percentage genetic algorithm problem reduction obtained using half and quarter chromosomes

	Case	Number of candidates	Number of fitness calculations
cantilever plate (one axis of symmetry)	Full Chromosome [10]	1.73×10^{13}	31900
	Present half chrom.	2.1×10^{6}	700
	Reduction obtained	**>99.99%**	**97.80%**
clamped-clamped plate (two axes of symmetry)	Full chromosome [13]	1.86×10^{11}	3350
	Present quarter chrom.	300	25
	Reduction obtained	**>99.99%**	**99.25%**

6. Conclusion

Half and quarter chromosomes are developed to optimise discrete piezoelectric locations for a flat plate fixed as a cantilever and clamped-clamped, respectively, giving one and two axes of symmetry. Minimization of linear quadric index is used as an objective function to locate ten and eight discrete piezoelectric actuators and attenuate the first six modes of vibration.

An isotropic plate with these discrete actuators bonded to its surface is modelled using finite element and Hamilton's principle based on first order shear deformation theory taking account of piezoelectric mass, stiffness and electromechanical coupling effects. The first six natural frequencies are determined by this model and validated using the Ansys finite element package and experimentally. It has been observed in this work that the configuration of optimal sensor/actuator pairs is in agreement with the modal strain and electric charge distribution over the plate and piezoelectric surface respectively.

A Matlab m-code genetic algorithm program has been built incorporating the new half and quarter chromosome technique and the results for optimal placement of ten and eight actuators are given, validated with published work. The new technique gives >99.99% reduction in genetic algorithm search space and >97.8% reduction in computer calculations compared with the conventional full length chromosome.

References

[1] Allik H and Hughes T J R 1970 *Finite element method for piezoelectric vibration.* International Journal for Numerical Methods in Engineering. **2**(2): p. 151-157.

[2] Tzou H S and Tseng C I 1990 *Distributed piezoelectric sensor/actuator design for dynamic measurement/control of distributed parameter systems: A piezoelectric finite element approach.* Journal of Sound and Vibration. **138**(1): p. 17-34.

[3] Reddy J N 1999 *On laminated composite plates with integrated sensors and actuators.* Engineering Structures. **21**(7): p. 568-593.

[4] Sadri A M, Wright J R , and Wynne R J 2002 *LQG control design for panel flutter suppression using piezoelectric actuators.* Smart Materials and Structures. **11**(6): p. 834-839.

[5] Sadri A M, Wright J R, and Wynne R J 1999 *Modelling and optimal placement of piezoelectric actuators in isotropic plates using genetic algorithms.* Smart Materials and Structures. **8**(4): p. 490-498.

[6] Han J H and Lee I 1999 *Optimal placement of piezoelectric sensors and actuators for vibration control of a composite plate using genetic algorithms.* Smart Materials and Structures. **8**(2): p. 257-267.

[7] Peng F, Ng A and Hu Y R 2005 *Actuator placement optimization and adaptive vibration control of plate smart structures.* Journal of Intelligent Material Systems and Structures. **16**(3): p. 263-271.

[8] Kumar K R and Narayanan S 2007 *The optimal location of piezoelectric actuators and sensors for vibration control of plates.* Smart Materials and Structures. **16**(6): p. 2680-2691.

[9] Roy T and Chakraborty D 2009 *Genetic algorithm based optimal control of smart composite shell structures under mechanical loading and thermal gradient.* Smart Materials and Structures. **18**(11).

[10] Daraji A H and Hale J M 2012 *A conditional filter for reduction of the search space in genetic algorithm optimisation of sensor and actuator location in active vibration control* in *international congress of sound and vibration ICSV19*: Vilnius , Lithuania

[11] Daraji A H and Hale J M 2012 *Optimisation of number and placement of piezoelectric sensor/actuator pairs for active vibration reduction of a felxible plate using the genetic algorithm* in *international congress of sound and vibration ICSV19*: Vilnius, Lithuania

[12] Daraji A H and Hale J M 2012 *Determination of state space matrices for active vibration control using ansys finite element package* in *Proceedings of the ASME international 24th conference (VIB)*: Chicago,IL, USA.

[13] Daraji A H and Hale J M 2012 *The effects of symmetry on optimal transducer location for active vibration control* in *Proceedings of the ASME international 24th conference (VIB)*: Chicago, IL, USA.

[14] Hale J M and Daraji A H 2012 *Optimal placement of sensors and actuators for active vibration reduction of a flexible structure using a genetic algorithm based on modified H_infinity*, in *Modern practice in stress and vibration (MPSVA)*: University of Glasgow, United Kingdom.

[15] Hale J M and Daraji A H 2012 *Active vibration reduction of stiffened plates with optimally placed sesnors and actuators* in *ISMA*: Leuven Belgium.

[16] Ogata K 1997 *Moderen Control Engineering*. third ed. USA: prentice-Hall. Inc.

[17] Kondoh S, Yatomi C and Inoue K 1990 *Positioning of sensors and actuators in the vibration control of flexible systems.* JSME international journal. **33**(2): p. 145-152.

TMD-Based Structural Control of High Performance Steel Bridges

Tae Min Kim[1], Gun Kim[2] and Moon Kyum Kim[3]
[1,2,3] Department of Civil & Environment Engieering, Yonsei University, 50 Yonsei-ro, Seodaemun-gu, Seoul 120-749, Korea

E-mail: taemin@yonsei.ac.kr

Abstract. The purpose of this study is to investigate the effectiveness of structural control using tuned mass damper (TMD) for suppressing excessive traffic induced vibration of high performance steel bridge. The study considered 1-span steel plate girder bridge and bridge-vehicle interaction using HS-24 truck model. A numerical model of steel plate girder, traffic load, and TMD is constructed and time history analysis is performed using commercial structural analysis program ABAQUS 6.10. Results from analyses show that high performance steel bridge has dynamic serviceability problem, compared to relatively low performance steel bridge. Therefore, the structural control using TMD is implemented in order to alleviate dynamic serviceability problems. TMD is applied to the bridge with high performance steel and then vertical vibration due to dynamic behavior is assessed again. In consequent, by using TMD, it is confirmed that the residual amplitude is appreciably reduced by 85% in steady-state vibration. Moreover, vibration serviceability assessment using 'Reiher-Meister Curve' is also remarkably improved. As a result, this paper provides the guideline for economical design of I-girder using high performance steel and evaluates the effectiveness of structural control using TMD, simultaneously.

1. Introduction

High performance steel for bridges (HSB) has emerged a promising material that offers innovative opportunities to design cost-effective steel bridges with respect to tensile strength, yield strength, and so on. HSB800, the most developed grade in Korea, is recently manufactured and has entered the bridge industry. Due to these outstanding factors, the total amount of steel used in bridge construction will be fairly reduced and weight lightening has become a possibility in the construction industry. Moreover, if fabrication costs associated with HSB800 steel are similar to those associated with conventional grade such as HSB600, these decreases in weight will translate into cost reduction. However, in the case of reduced girder by using HSB800, its structure can be exposed to vibrational problems such as deterioration of vibrational serviceability, dynamic displacement, and so on. To resolve these weaknesses of HSB, structural stability should be guaranteed in advance.

In this research, the adaptability of HSB800 is analyzed in comparison with HSB600 in terms of vibration serviceability. And also, TMD-based structural control is conducted in order to improve the serviceability problem induced by adopting HSB with the road-surface roughness. By using Meister

[1] Ph.D. student, E-mail address : taemin@yonsei.ac.kr
[2] Researcher, E-mail address : gunkim@yonsei.ac.kr
[3] Professor, E-mail address : applymkk@yonsei.ac.kr

Curve associated with human response to steady-state vibrations, the damping effects of TMD are finally examined.

2. Three-dimensional dynamic analysis of the steel plate girder with HSB

2.1. Analysis model for vehicle loading

Based on diverse vehicle specification, vehicle models influencing to dynamic behavior of bridges are also diverse. In Bridge Design Specification Interim (Korea), the standard truck load is expressed in DB-Load, while triaxial truck is denoted by HS-loading in ASSHTO LRFD Bridge Construction Specification, as shown in the figure 1. However, because the condition of DB-Load is similar to HS-loading, this study just focuses on HS-loading as a truck model in order to facilitate comparison of DB and HS-loading.

When it comes to modeling for vehicle loading, Li Hong (2005) developed techniques for modeling of triaxial truck as depicted in figure 2. This model is consist of two rigid bodies such as car body (m_1), and trailer (m_2) and also the degree of freedom (DOF) of each wheel is also composed of vertical displacement (Z'_1, Z'_2), pitching rotation (θ_1, θ_2), and rolling rotation (α_1, α_2). In addition, as shown in figure 2, spring-dashpot systems are also attached to each mass in order to simulate a real truck. Based on the principle of virtual work, equilibrium equation of motion for vehicle having 11degrees of freedom is constituted by the relationship from the gravity (M_{1g}, M_{2g}, m_{ig}, i=1…6), inertia force and moment ($M_j\ddot{Z}_j$, $I_{\theta j}\ddot{\theta}$, $I_{\alpha j}$, $\ddot{\alpha}_j$, j=1,2; $m_i\ddot{z}_i$, i=1…6), suspension force (F'_i, i=1…6), to tire force (F_i, i=1…6).

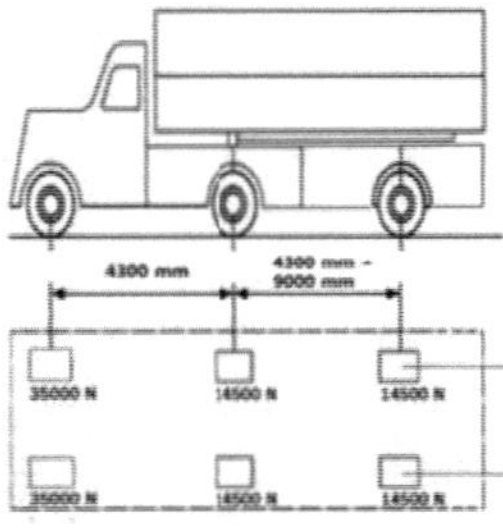

Figure 1. Truck Load (AASHTO, 2002)

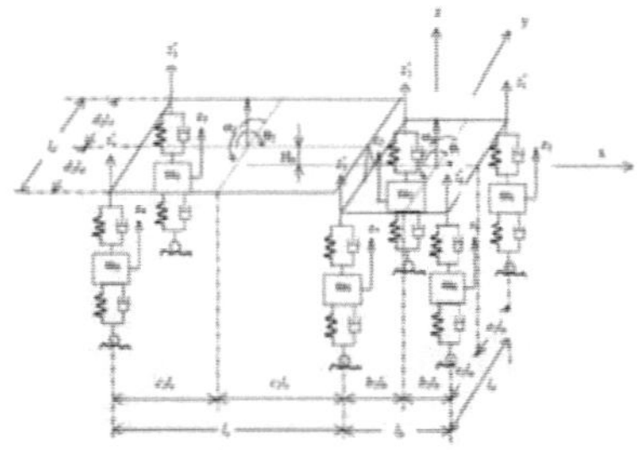

Figure 2. Model for triaxial truck (Li, 2005)

2.2. Dynamic analysis for the vibrational behavior of a steel plate girder

To investigate the application of high performance steel for bridges, a modeling for 1-span steel plate girder bridge arrangement with 54m span length is constructed using ABAQUS/Explicit 6.10, as shown in the figure 3. Based on allowable bending compressive stresses introduced by Bridge Design Specification Interim (2010) regarding HSB800 and 600, the modeling of two girders is performed and analysed, respectively.

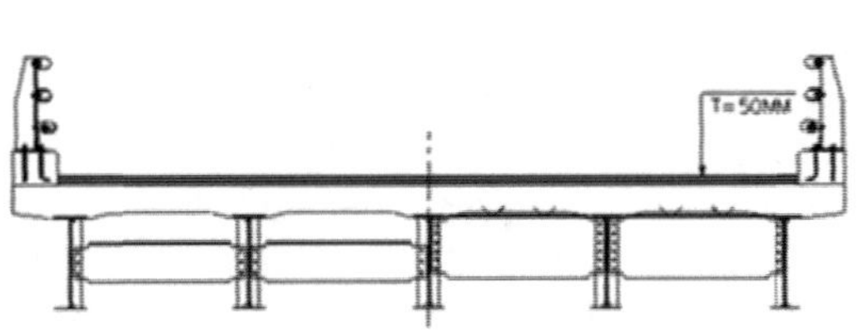

Class: 1st class bridge (DB-24/ DL-24)
Width: 15 m
Span: 54 m
Line : 4 lines, Type : Steel Plate Girder

Figure 3. Specification for design of girder

The design of bridge is based on the first class bridge design load. And also, the fixed and live loads are considered in bridge. In the design of girder, the size of top flange (width is 508 mm and thickness is 29 mm), bottom flange (width is 610 mm and thickness is 29 mm), and web (thickness is 29 mm) of HSB800 is same as the size of HSB600, excepting the height of girder: HSB800 is 1700 mm and HSB600 is 2100 mm, respectively. Throughout adapting the HSB800, the effectiveness in reducing height of girder is about 400mm and the total weight of the structure is reduced about 10%, simultaneously.

2.3. Power Spectral Density theory for the road surface roughness

When it comes to considering the roughness, commonly used method is 'Power Spectral Density (PSD) theory. Dodds and Roboson(1973) have proposed the form of power law, S(r) as shown is equation (1) and (2).

$$\mathrm{S(r)} = \mathrm{L}(r_0)(r/r_0)^{-W_1}, \quad \mathrm{r} \le r_0 \tag{1}$$

$$\mathrm{S(r)} = \mathrm{L}(r_0)(r/r_0)^{-W_2}, \quad \mathrm{r} \ge r_0 \tag{2}$$

Where, $\mathrm{L}(r_0)$ is called the roughness coefficient (m); r represents wave number and r_0 is $1/2\,\pi$ (cycle/m) — both are determined from field experiments. In this research, road surface roughness is generated by using 'Power Spectral Density (PSD)' theory. By using the probability distribution, equation (1) and (2) are changed into combination of amplitude, circular frequency, and phase angel as shown in equation (3).

$$\mathrm{Y(t)} = \sum_{k=1}^{n} \sqrt{4S(r_0)\cdot(\frac{\pi k}{Lw_0})^{-2}\cdot\frac{\pi}{L}}\ \cos(\frac{\pi kx}{L} + \theta_k) \tag{3}$$

Here, Y(t) is the changed surface height using power spectral density: θ_k is independent phase angle from 0 to 2π, L is the length of span, w_0 is $1/2\pi$ 1, and $S(r_0)$ is the roughness coefficient proposed by International Organization for Standardization (ISO). ISO has assorted the road surface roughness into four levels: Very Good, Good, Average, and Poor. In this research, using the principle of PSD theory and the changed surface coordinate, dynamic analysis for the behavior of bridge is conducted by considering Poor level of the roughness.

3. Comparison for dynamic responses of HSB800 and 600

In this project, three-dimensional numerical analysis is performed by using structural analysis program ABAQUS 6.10. The numerical analysis of HSB-girder generally involves three major steps:

- Natural frequency
- Time history analysis for acceleration and vertical displacement
- The evaluation for serviceability by using 'Reiher-Meister curve'

3.1. Time history analysis

Firstly, the modal analysis is conducted in order to analyse the dynamic behavior of the girder. By applying the high performance steel, the height and the total weight of the girder is fairly reduced. Throughout the modal analysis, it is concluded that the natural frequency of HSB 600 indicates 1.635 Hz. On the other hand, the natural frequency is 1.334Hz in case of HSB800. In order to scrutinize the effect of vertical vibration at the center of span, a HS-24 model (80km/h), as mentioned in chapter 2.1, is applied on the middle of the lane in the numerical model and dynamic analysis is performed. During this analysis, the surface of the girder is changed according to the poor level of roughness. As shown in Figure 4, the result of numerical analysis shows that the maximum displacement induced by vehicle model is 14 mm in HSB600. However, the maximum displacement is increased by 25 mm in HSB800 model. Moreover, the residual amplitude in a steady-state is fairly increased in the model of HSB800, compared to HSB600. However, it is hard to distinguish dynamic behavior between HSB600 and HSB800 on the acceleration history as shown in Figure 5. In addition, the maximum acceleration is abnormally taken place in HSB600. According to these results, it is finally concluded that the displacement history is influenced by the height of girder.

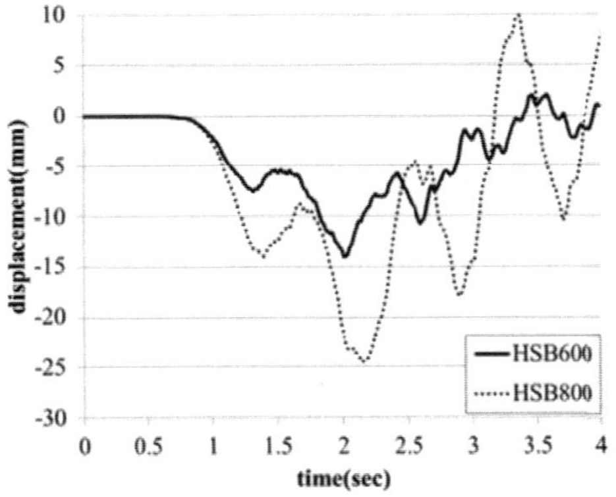

Figure 4. Vertical displacement at the center of span

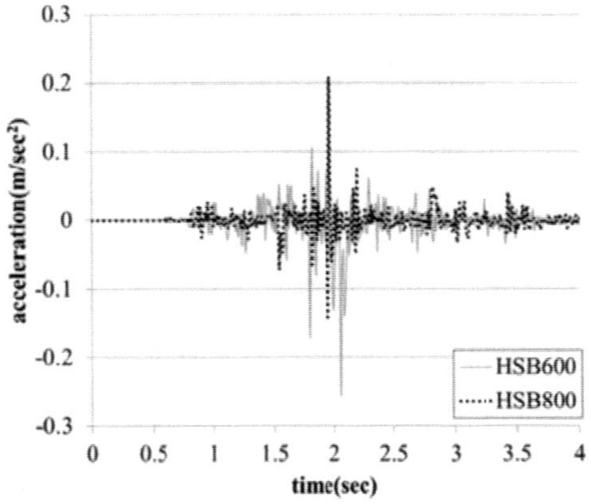

Figure 5. Acceleration history at the center of span

3.2. The evaluation for serviceability by using 'Reiher-Meister curve'

Based on the vertical displacement in subchapter 3.2, the evaluation for dynamic serviceability is performed by using Reiher-Meister curve, as depicted in Figure 6. As a result, the B level (Disturbing) is appeared in case of HSB800 and the C level (Distinctly Perceptible) is occurred in case of HSB 600.

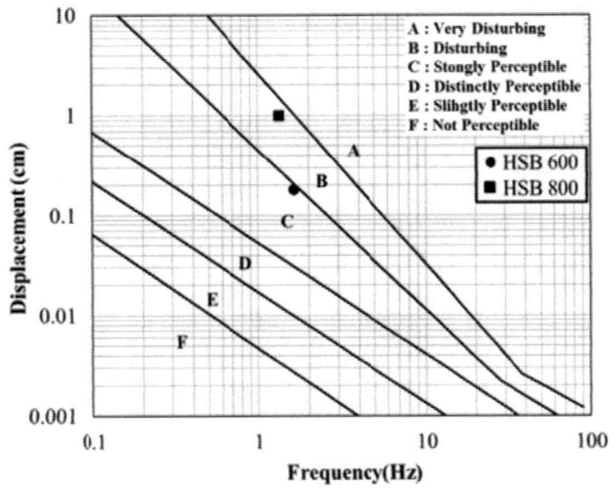

Figure 6. Dynamic Serviceability Evaluation using Reiher-Meister curve

4. The application of TMD for Structural Control of HSB800-Bridges

In order to control the vertical vibration of bridges, control performance, and constructability should be certainly guaranteed in advance. In the perspective of construction cost and constructability, it cannot be said that stiffening girder is also suitable for suppressing traffic induced vibrations. Consequently, when it comes to construction availability, serviceability, and efficiency, it can be generally said that Tuned Mass Damper (TMD) is the most adaptive equipment.

4.1. The principle and design of TMD

TMD is generally consisted of a secondary mass with properly tuned spring and damping elements, providing a frequency-dependent hysteresis that increases damping in the primary structure. To design of TMD, this project introduces the concept of effective mass and calculates optimal damping ratio using Den Hartog theory. Effective mass is the mass which must be added to the suspended mass to correctly predict the behavior of the entire system. It is also expressed in equivalent lumped mass at a particular point. In case that stiffness at a certain point is calculated, then the equation of natural frequency is simplified as expressed in equation (4). The figure 7 describes a second degree of freedom of TMD-adapted bridge deck.

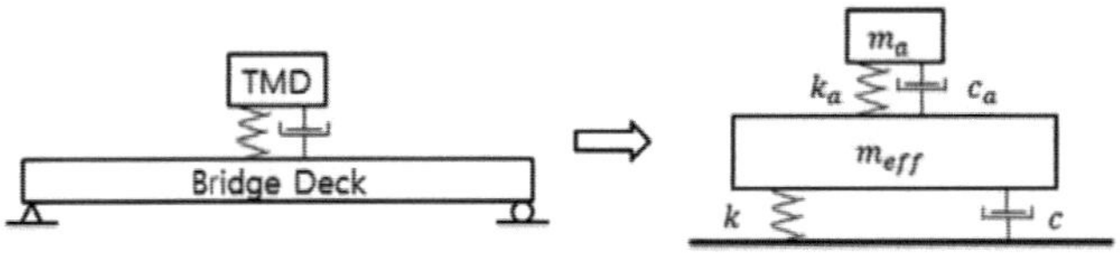

Figure 7. The concept of TMDs for structural applications

$$f_n = \frac{1}{2\pi}\sqrt{\frac{k}{m_{eff}}} \tag{4}$$

Here, f_n means natural frequency, k means stiffness at the center of bridges, and m_{eff} means effective mass. Natural frequency f_n is considered as first natural frequency throughout modal analysis as mentioned in chapter 3. And also, the stiffness is calculated by considering vertical unit load at the center of bridge. Therefore, the determined values of natural frequency and stiffness are 1.334 Hz and 18,191N/mm, respectively. By using these values, the effective mass m_{eff} is finally calculated and the value indicates 259,112kg. Dan Hartog has proposed 'optimum TMD' theory in 1956. Using this theory, optimal mass ratio and frequency of TMD can be calculated as depicted in equation (5) and (6).

$$\mu = \frac{m_a}{m_{eff}} = \frac{2600}{259{,}112} \cong 0.01 \tag{5}$$

$$f_a = \frac{f}{1+\mu} = \frac{1.334}{1+0.01} = 1.321Hz \tag{6}$$

Here, μ indicates the mass ratio in the relationship between the calculated mass m_a and the effective mass m_{eff}, and f_a means optimized frequency of TMD. Finally, optimized damping ratio ζ_{opt} of the bridge can be determined by equation (7) as follow:

$$\zeta_{opt} = \sqrt{\frac{3\times 0.01}{8(1+0.01)}} = 0.0609 \tag{7}$$

Throughout this sequence, variables for mass and damping of TMD are obtained.

4.2. Dynamic behavior of TMD-adopted steel plate girder

In this subchapter, TMD-based structural control of bridges applied HSB800 is evaluated through the three-dimensional dynamic analysis, in order to redeem the vibrational displacement to its original state prior to adopting high performance steel. As mentioned in subchapter 2.3, the poor level of roughness on the surface of bridges is also described in order to inspect the effect of roughness. Consequently, the result of dynamic analysis considering TMD is depicted in figure 8. As shown in figure 8, the dynamic behavior of HSB800 is similar to HSB600 in the field of transient vibration because it involves to the static displacements induced by the vehicle model. However, after the vehicle load is disappeared, it is confirmed that the residual amplitude is appreciably reduced by 85%. Figure 9 shows that vibration serviceability assessment using 'Reiher-Meister Curve' is also remarkably improved. Namely, it is concluded that the control effectiveness of TMD is obviously proved.

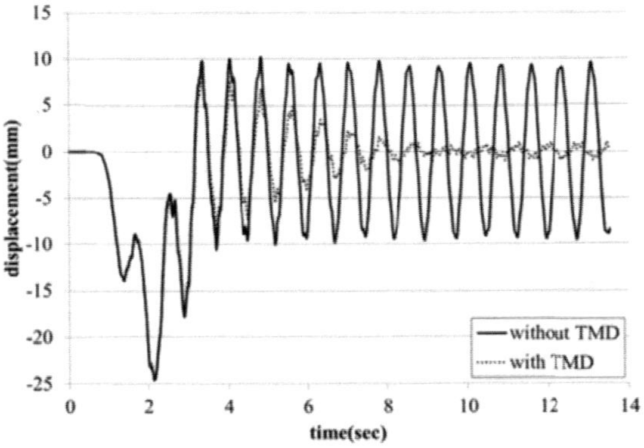

Figure 8. The control effectiveness of TMD

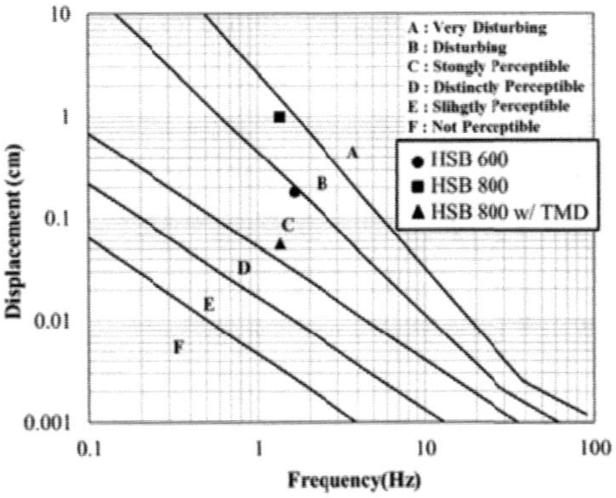

Figure 9. Dynamic Serviceability Evaluation of TMD-adopted bridge

5. Concluding Remarks

This study covers the TMD-based structural control of bridges being built using High performance Steel for Bridges (HSB). The applicability of HSB800, the most recently developed HSB grade is verified by three-dimensional dynamic analysis, serviceability evaluation, and structural control. Throughout this study, several developments and key results were determined as follows:

- This study proposed the method for applying HSB800 and 600 in the design of steel plate girder. When it comes to HSB800, it is verified that the effectiveness in reducing height of girder is about 400mm in comparison with HSB600 and the total weight of the structure is reduced about 10%, simultaneously.
- The vibration serviceability assessment using Reiher-Meister curve is conducted to evaluate the stability of HSB800 and HSB600. As a result, the application of HSB800 and HSB600 induces dynamic serviceability problem.
- The structural control using tuned mass damper (TMD) is implemented in order to alleviate dynamic serviceability problems. In consequent, by using TMD, it is confirmed that the residual amplitude is appreciably reduced by 85% in steady-state vibration..

Acknowledgements

This research was supported by a grant from POSCO

References

[1] AASHTO(American Association of Standard Highway and Transportation Officials) 2002 AASHTO LRFD Bridge Design Specification
[2] Chun P 2010 Skewed bridge behaviors: Experimental, analytical, and numerical analysis (Ph.D. diss.) Wayne state University
[3] Ellis B R and Ji T 1997 Human-structure interaction in vertical vibration, *Proc. Jnstn Civ. Engrs Structs & Bldgs* **122** Feb 1-9
[4] Li H 2005 Dynamic response of highway bridges subjected to heavy vehicles (Ph.D. diss) The Florida State University
[5] Ministry of Land, 2010 Transport and Maritime Affairs, Bridge Design Specification Interim
[6] Shi X and Cai C S 2008 Suppression of vehicle-induced bridge vibration using tuned mass damper *Journal of vibration control* **14** **7** 1036-1054

Journal of Physics: Conference Series **382** (2012) 012041 doi:10.1088/1742-6596/382/1/012041

Optimal control methods for vertical and horizontal beam dynamics

A Mitura[1], J Warminski[1] and M Bochenski[1]

[1]Department of Applied Mechanics, Lublin University of Technology, Nadbystrzycka 36, 20-618 Lublin, Poland

a.mitura@pollub.pl, j.warminski@pollub.pl, m.bochenski@pollub.pl

Abstract. An application of the Macro Fiber Composite (MFC) actuators for damping of a composite beam is presented in this paper. The effectiveness of vibration reduction by a selected control method is tested for vertical and horizontal position of the beam. The original model has been studied numerically by using Galerkin's discretisation method. The numerical results for the vertical and horizontal beams are compared.

1. Introduction

Structures composed of beams are important in civil and mechanical engineering systems. Examples of such structures are L-shaped and T-shaped coupled beams. In these coupled beams the resonance behaviour depends on internal autoparametric coupling and nonlinearities. Theoretical and experimental investigations of vibrations of an autoparametric L-shaped system are presented in paper [4]. The authors proposed an analytical model and verified it by experiments and the finite element method. They show that the natural frequencies and vibration modes obtained by both methods are consistent. The present work is a continuation of those previous investigations and it is the first step to designing the optimal control of an L-shaped beam structure.

The L-shaped beam structure consists of two lightweight, inextensible beams (horizontal and vertical) connected in a right angled shape. To reduce vibration two active elements, fixed to each beam are used. Macro Fiber Composite (MFC) has been selected as the actuator technology. This kind of piezoactuator allows a proper interaction with flexible structures [3]. Different beam positions (vertical or horizontal) determine different dynamic properties. The vertical and horizontal beams are controlled independently but take into account the interactions between them. The complete L-shaped structure is not tested as a whole in this work but its subsystems. Separate research into the horizontal and vertical beam dynamics allows the determination of the initial control parameters. In the next step these parameters will be used as the initial point in the analysis of the optimal control of the full structure.

In the literature several strategies to control beam dynamics are proposed [1,2,5]. They are for example: proportional and cubic displacement feedback, cubic velocity feedback, positive position feedback (PPF), nonlinear saturation control (NSC) and others. In this paper we have selected the PPF algorithm for vibration suppression of the (a) horizontal beam under vertical excitation near the first natural frequency and (b) vertical beam near the principal parametric resonance.

2. Model of the structure

Mathematical models of (a) horizontal and (b) vertical beams are presented in this section. The subsystems are shown in figure 1. Both structures consist of a composite flexible beam with a rectangular cross-section which can perform large flexural oscillations. The tested composite beams are made of glass-epoxy material with fibre orientation 0/90/45/-45/45/90/0.

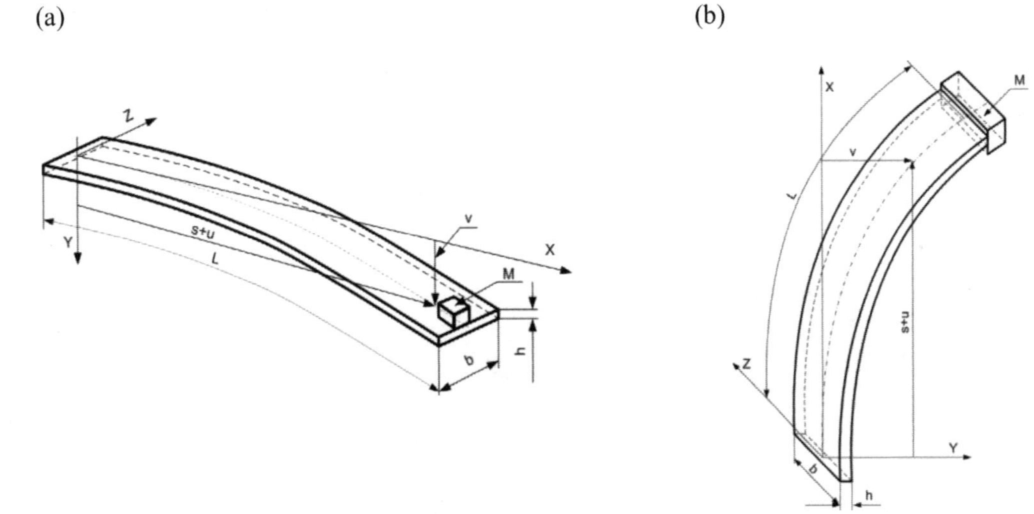

Figure 1. A model of the horizontal (a) and vertical (b) beams.

The horizontal and vertical beams are mounted on an electromagnetic shaker which is a source of excitation along the X and Y axes, respectively. The kinematic excitation can be written as for the horizontal beam

$$y = y_0 \sin(\Omega t) \tag{1}$$

and for the vertical beam

$$x = x_0 \sin(\Omega t) \tag{2}$$

The beam is considered as an Euler-Bernoulli model with an added nonlinear curvature component

$$\rho_\zeta = v'' + \frac{1}{2} v'^2 v'' \tag{3}$$

Kinetic and potential energies of the beam and the added mass M take form:
for the horizontal beam

$$T = \frac{1}{2}\int_0^L \rho A\left(\dot{u}^2 + (\dot{v} + \dot{y})^2\right) ds \qquad V = \frac{1}{2}\int_0^L D_\zeta \rho_\zeta^2 ds \tag{4}$$

$$T_M = \frac{1}{2} M\left(\dot{u}_M^2 + (\dot{v}_M + \dot{y})^2\right) \qquad V_M = -M g\,(v_M + y) \tag{5}$$

for the vertical beam

$$T = \frac{1}{2}\int_0^L \rho A\left((\dot{u} + \dot{x})^2 + \dot{v}^2\right) ds \qquad V = \frac{1}{2}\int_0^L D_\zeta \rho_\zeta^2 ds \tag{6}$$

$$T_M = \frac{1}{2} M\left((\dot{u}_M + \dot{x})^2 + \dot{v}_M^2\right) \qquad V_M = M g\,(u_M + x) \tag{7}$$

where $u=u(s,t)$ and $v=v(s,t)$ describes the position of the arbitrary beam's cross section.

Following the procedure presented in paper [4] the dimensionless equations of motion take the form:
for the horizontal beam

$$\ddot{v}-\left(\lambda v'\right)'+v^{IV}\left(1+v'^2\right)+4v'''v''v'+v''^3=y_0\Omega^2\sin\left(\Omega t\right) \tag{8}$$

for the vertical beam

$$\ddot{v}-\left(\lambda v'\right)'+v^{IV}\left(1+v'^2\right)+4v'''v''v'+v''^3=0 \tag{9}$$

Considering the lumped mass, we get dynamical boundary conditions in the form:
for the horizontal beam

$$v(0,t)=0\,,v'(0,t)=0\,,v''(L,t)=0\,,\ v'''(L,t)=\lambda v'_M+M\left(\ddot{v}_M+g^*\right) \tag{10}$$

for the vertical beam

$$v(0,t)=0\,,v'(0,t)=0\,,v''(L,t)=0\,,\ v'''(L,t)=\lambda v'_M+M\ddot{v}_M \tag{11}$$

where $g^*=g\rho AL^3/D_\varsigma$ is the dimensionless gravitational acceleration and λ is the Lagrange multiplier defined
for the horizontal beam

$$\lambda=\int_L^s \rho A\ddot{u}ds-M\ddot{u}_M \tag{12}$$

for the vertical beam

$$\lambda=\int_L^s \rho A(\ddot{u}+\ddot{x})ds-M\left(\ddot{u}_M+\ddot{x}-g^*\right) \tag{13}$$

In order to transfer the partial differential equations (8), (9) into ordinary differential equations the classical Bubnov–Galerkin procedure is applied. Only the first bending mode is taken into account. Then the differential equations can be written:
for the horizontal beam

$$\ddot{v}+2\mu\omega_s\dot{v}+\omega_s^2v+\beta v^3+\delta\left(v\dot{v}^2+v^2\ddot{v}\right)=\xi y_0\,\Omega^2\sin\left(\Omega t\right) \tag{14}$$

for the vertical beam

$$\ddot{v}+2\mu\omega_s\dot{v}+\omega_s^2v+\beta v^3+\delta\left(v\dot{v}^2+v^2\ddot{v}\right)=\xi x_0 v\,\Omega^2\sin\left(\Omega t\right) \tag{15}$$

The dimensionless parameters of the composite flexible beam are taken from paper [5] and they have the following values: for the vertical beam: $\xi=1.49929$, $\omega_s^2=9.3728$, $\mu=0.05$, $\beta=14.4108$, $\delta=3.27463$ and for the horizontal beam $\xi=0.89663$, $\omega_s^2=9.3728$, $\mu=0.05$, $\beta=14.4108$, $\delta=3.27463$.

The MFC actuator is bounded on one side of the composite beam as shown in figure 2 while on the opposite side a sensor - strain gauge is glued. The PPF algorithm has been programmed in a DSP system [5]. Differential equations of motion of the considered composite beam with PPF controller take the form:
for the horizontal beam

$$\begin{cases}\ddot{v}+2\mu\omega_s\dot{v}+\omega_s^2v+\beta v^3+\delta\left(v\dot{v}^2+v^2\ddot{v}\right)=\xi y_0\Omega^2\sin\left(\Omega t\right)+\gamma u_c\\ \ddot{u}_c+2\zeta\omega_c\dot{u}_c+\omega_c^2u_c=\alpha v\end{cases} \tag{16}$$

for the vertical beam

$$\begin{cases}\ddot{v}+2\mu\omega_s\dot{v}+\omega_s^2v+\beta v^3+\delta\left(v\dot{v}^2+v^2\ddot{v}\right)=\xi x_0 v\,\Omega^2\sin\left(\Omega t\right)+\gamma u_c\\ \ddot{u}_c+2\zeta\omega_c\dot{u}_c+\omega_c^2u_c=\alpha v\end{cases} \tag{17}$$

The difference between the horizontal and vertical beam dynamics consists of a different excitation mechanism, in equation (16) the system is forced by an external harmonic force while in equation (17) it is by a parametric excitation.

a) b)

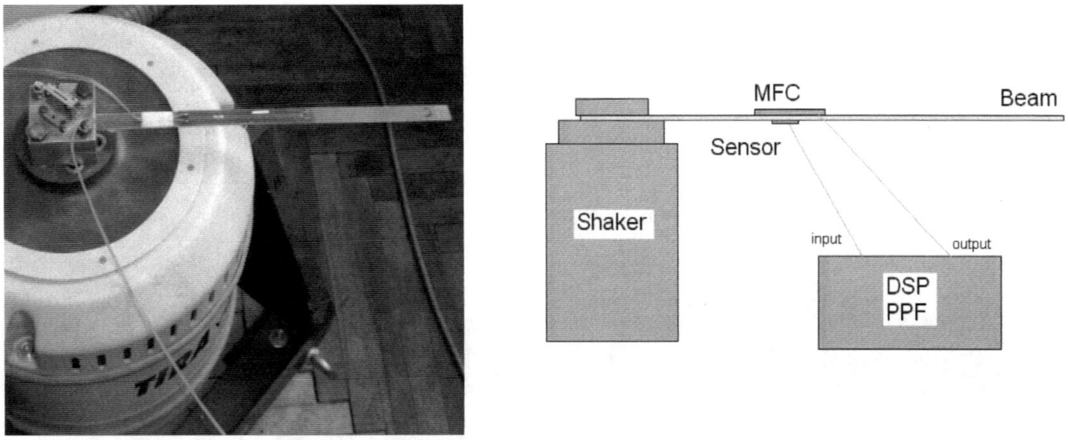

Figure 2. A view (a) and a sketch (b) of an experimental setup with a horizontal beam.

3. Numerical results

Numerical simulations of the horizontal and vertical beam without (14), (15) and with control (16), (17) are performed in the MATLAB package. Selected results are presented in figure 3 and figure 4. All simulations have been started from the same initial conditions $v(t=0)=0.1$ and $\dot{v}(t=0)=0$.

a) b)

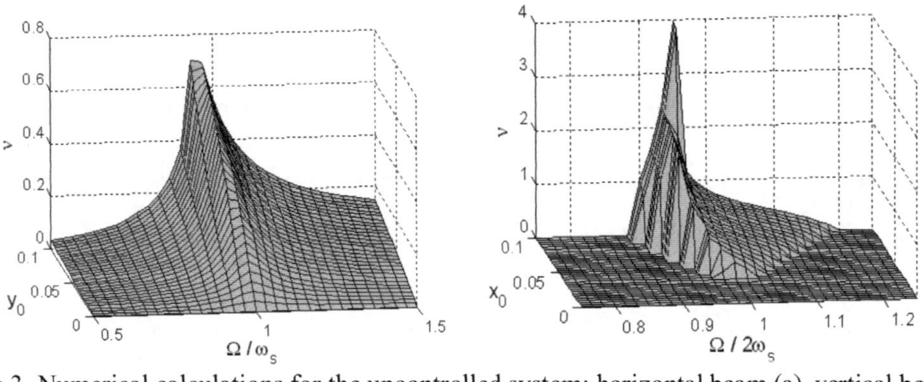

Figure 3. Numerical calculations for the uncontrolled system: horizontal beam (a), vertical beam (b).

3D characteristics, beam deflection versus amplitude and frequency of harmonic excitation for the uncontrolled system are shown figure 3. We can see a fundamental difference in the position of the resonance region for both beams. The horizontal beam has a resonance region around the natural frequency of the linear system ω_s. In the second case, for the vertical beam, the region is about the principal parametric resonance frequency of the linear system $2\omega_s$.

The PPF method is employed to damp the high amplitude vibration of the systems. The other parameters of the controller are $\omega_c=\omega_s$, $\gamma=6$, $\alpha=6$, $\varsigma=0.05$. These parameters give a smaller amplitude in the resonance region than for the uncontrolled system. The disadvantage of the PPF controller is that a single degree of freedom beam model is extended to a two degree of freedom system. Thus, the system with PPF has two resonance regions (figure 4). In the case of the vertical beam, two resonance regions are observed for large excitation amplitudes $x_0>0.05$ (5% of the beam length). The areas below this threshold are characterized by the absence of vibration $v=0$ (figure 4b), therefore the PPF controller works very effectively. For the horizontal beam the PPF controller can only be used to minimize the vibration in the resonance region. Outside of this area, the dynamics of the beam are worse than for the uncontrolled system.

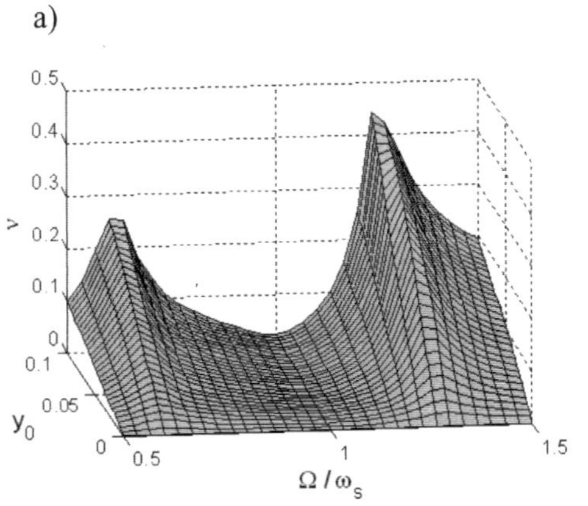

Figure 4. Numerical calculations for the controlled system: horizontal beam (a), vertical beam (b).

4. Conclusions

Numerical results for the positive position control (PPF) method applied to strongly nonlinear horizontal and vertical beam models are presented in this paper. The results show a possibly very good vibration reduction of the vertical beam for low level of excitation amplitudes. In the case of the horizontal beam this control method has to be modified. In future research we will propose more effective model based algorithms for control with a study of the parameters' influence on the system's response.

Acknowledgments

The research leading to these results has received funding from the European Union Seventh Framework Programme (FP7/2007-2013), FP7-REGPOT-2009-1, under grant agreement No: 245479. The support by Polish Ministry of Science and Higher Education – Grant No 1471-1/7 PR UE/2010/7 – is also acknowledged.

References

[1] Ashour O N and Nayfeh A H 2002 Adaptive control of flexible structures using a nonlinear vibration absorber *Nonlinear Dynamics* **28** 309-322

[2] Oueini S S and Nayfeh A H 1999 Single-mode control of a cantilever beam under principal parametric excitation *Journal of Sound and Vibration* **224** 33-47

[3] Sodano H A, Park G and Inman D J 2004 An investigation into the performance of macro-fiber composites for sensing and structural vibration applications *Mechanical Systems and Signal Processing* **18** 683-697

[4] Warminski J, Cartmell M P, Bochenski M and Ivanov I 2008 Analytical and experimental investigations of an autoparametric beam structure *Journal of Sound and Vibration* **315** 486-508

[5] Warminski J, Bochenski M, Jarzyna W, Filipek P and Augustyniak M 2011 Active suppression of nonlinear composite beam vibrations by selected control algorithms *Communications in Nonlinear Science and Numerical Simulations* **16** 2237-2248

The design of impact absorbing structures for additive manufacture

J Brennan-Craddock, D Brackett, R Wildman and R Hague

Additive Manufacturing Research Group, Loughborough University, Loughborough, Leicestershire, LE11 3TU, UK

E-mail: D.J.Brackett@lboro.ac.uk

Abstract. Additive manufacturing (AM) is increasingly becoming a viable manufacturing process due to dramatic advantages that it facilitates in the area of design complexity. This paper investigates the potential of additively manufactured lattice structures for the application of tailored impact absorption specifically for conformal body protection. It explores lattice cell types based on foam microstructures and assesses their suitability for impact absorption. The effect of varying the cell strut edge design is also investigated. The implications of scaling these cells up for AM are discussed as well as the design issues regarding the handling of geometric complexity and the requirement for body conformity. The suitability of AM materials for this application is also discussed.

1. Introduction

This paper details work on the design of lattice structures for energy absorbing applications as well as methods to conform them, with the ultimate aim of manufacturing body protection tailored to an individual's body. The complexity of such parts means that Additive Manufacturing (AM) is likely to be one of the few processes that can realise these designs.

1.1. The potential of AM lattice structures in conformal body protection

The geometric freedom afforded by additive manufacturing processes allows the design of products of a complexity not achievable through any other means [1]. Without the need for physical tooling (such as the moulds used in injection moulding or the machine tool used in CNC machining) a product does not need to be designed with compromises to facilitate removal from the tooling. Parts of previously unattainable complexity can be fabricated through AM, utilising the complexity to consolidate assemblies into the minimum number of necessary components, or building assemblies of parts when necessary. Weight and size can be minimised, and products can be designed purely with function in mind [1].

Although covering a wide range of technologies, AM processes are united through the method in which parts are built: material is deposited, fused or cured in a succession of two dimensional slices, stacked and combined with the layer below to form a three dimensional shape. As AM parts are fabricated layer-by-layer, the material deposition mechanism (such as a print head, nozzle or directed laser) never encroaches on the part as it is being built. One particularly novel avenue that is opened by this is the fabrication of lattice structures, with physical properties that can be tuned to a particular application. Lattice structures as defined by this work are a tessellation of a unit cell, itself composed of struts, as shown in figure 1. When a lattice structure is deformed in some way to fit a shape, it is then termed a conformal lattice structure.

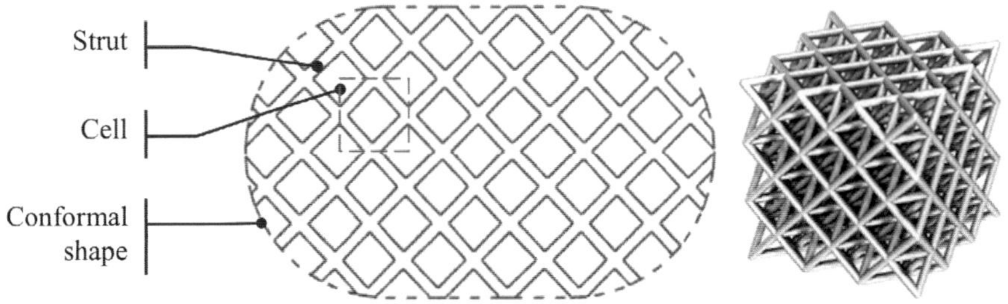

Figure 1. Naming conventions for a conformal lattice structure.

By utilizing body scan data of an individual's body, personalized body-fitting apparel can be designed. Personal protective equipment (PPE) used in sport in particular is an area that tailored lattice structures could be of use, and were the driving force behind the work detailed in this paper.

1.2. Suitability of AM processes for conformal body protection

The materials available for processing in AM are currently fairly limited and are specific to each AM process. It was determined that due to the relative strength and resilience of the range of polymers available, laser sintering would be the process used to build conformal lattice structures for this application. Laser sintering uses a reflected laser to selectively fuse layers of powdered material together. The part is supported by the surrounding un-sintered powder as it is being built, which is removed once the build is finished with a compressed air source, shot blast or water bath. This support mechanism is also advantageous to lattice structures; most other AM processes require an actual support structure to be built alongside the part, which would be difficult and time-consuming to remove from a lattice structure.

2. The design of conformal energy absorbing AM lattice structures

Conventional materials were first investigated as a basis for the design of AM lattice structures. As the most commonly used material in energy absorbing applications, polymeric foams in particular were the focus of study.

2.1. The structure of foam

Foams are suited to absorbing energy due to a low density, cellular structure that deforms readily under load [2]. There are two broad categories of foam: open cell and closed cell. A closed cell structure consists of walls and struts, forming isolated cells of gas, while an open-cell structure consists solely of struts. Open cell foam absorbs energy through strut buckling and the work done in pushing air out of the weaving structure. Closed cell foams absorb energy through strut buckling, cell wall bending and the compression of trapped gas. Foams generally undergo compressive behavior of three regimes. Initially, the foam exhibits a linear elastic region where very little energy is absorbed. A wide plateau region follows, with a densification phase where stress rises steeply [2].

During the manufacture of foams, the foaming process generates a three-dimensional cellular structure that is controlled by the principle of minimal surface energy [2,3]. Thomson (later Lord Kelvin), investigating the minimum surface qualities of aqueous foams, developed a repeatable unit cell for a perfectly ordered, regular foam. This later became known as the 'Kelvin cell'. The Kelvin cell is a modified truncated octahedron (a polyhedron composed of six square and eight hexagonal faces) and has been used extensively more recently in simulations of foam behavior [4-7].

2.2. Implications of scale

While foams can consist of individual cells less than a millimetre in diameter, a key constraint of current laser sintering technology is resolution. Laser sintering is limited to a minimum feature size of 0.4mm [8]. In terms of lattice structure design, this means that practically the smallest feasible strut diameter imposed is 0.4mm. With this as a consideration, cell diameter must be roughly an order a magnitude higher than strut diameter, making minimum cell diameters in the order of 5 to 10mm. Another constraint of laser sintering is the removal of un-sintered powder after fabrication. The lattice structure design must facilitate this. For a structure inspired by foams, only an open cell foam could be replicated by AM.

As Closed-cell structures cannot be manufactured, the bending of cell walls and compression of trapped air that contribute to energy absorption in closed-cell foams cannot be replicated. Also, the scale possible in AM is an order of magnitude higher than the micro-scale of conventional foam cellular structure. Air flow will not be significantly impeded by an AM structure at the macro scale like a micro-scale foam, so will not contribute to energy absorption to any useful level. This leaves cell strut bending as the only remaining mechanism for energy absorption in an AM lattice structure.

2.3. AM lattice structure designs

As a regular representation of the cellular structure of foam, the Kelvin cell makes a promising candidate for the basis of energy absorbent AM lattice structures. Two such designs are shown in figures 2 and 3. The first is a straight strut, triangular cross-section design, with fillets between the nodes to imitate the natural structure of foam. Both have a cell diameter of 15mm, and at two cells thick are approximately 30mm in height (geometric features of the struts slightly increase overall height). Five samples of the first design were manufactured in a blend of 50% virgin / 50% recycled PA 2200 material (nylon-12) from EOS on a Formiga P100 laser sintering machine.

Figure 2. Straight strut structure (inset: SEM image of open cell polymeric foam [6])

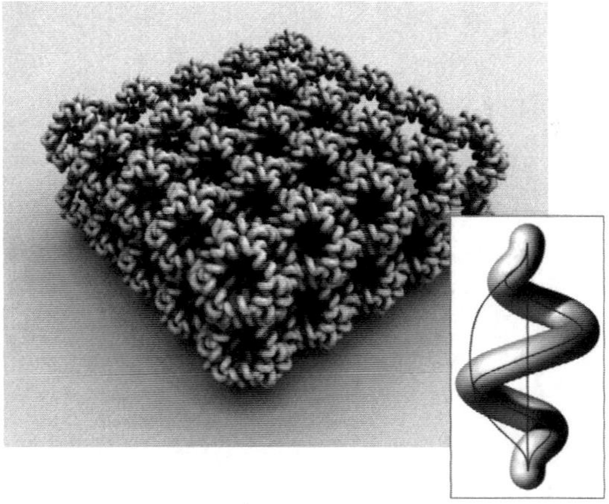

Figure 3. Helical strut structure (inset: helical strut)

When compressed, the samples exhibited the characteristic three stage compression seen in foams. Samples were compressed on an Instron 3366 Dual Column Testing System at a strain rate of 100mm/min. Figure 4 shows an initial linear increase in stress as the samples were compressed (A), followed by a wide plateau observed (B) before another sharp increase as the structure compresses to the point that struts start to make contact with each other (C).

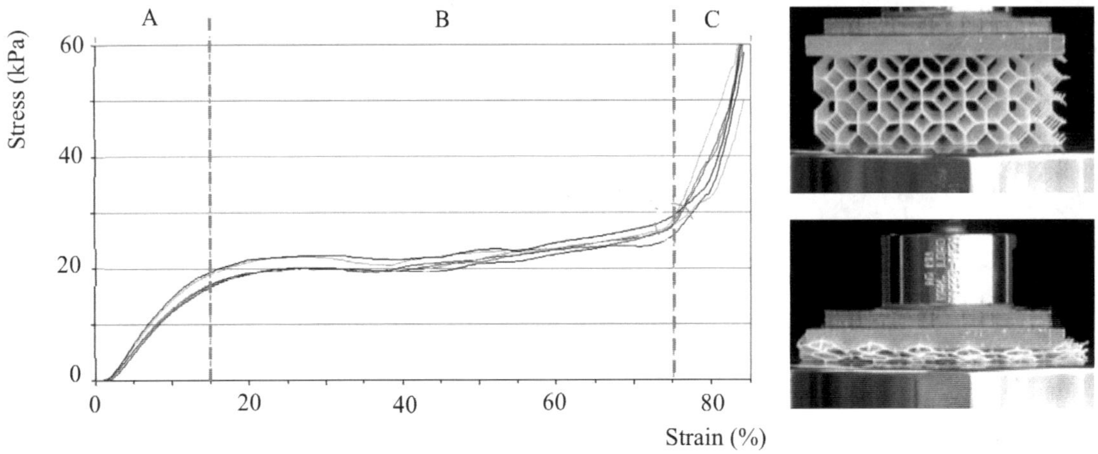

Figure 4. Compressive behavior of the straight strut samples

Taking advantage of the geometric freedom that AM provides, the second design was augmented with helical struts. The helix is constrained to a 'law curve' that reduces the radius of the helix to zero at each end. This allows helical struts to connect without intersecting each other. The reasoning behind this helical design was that increasing the overall length of each strut would allow more deformation within the structure (both for struts in tension and compression) and thus increase energy absorption.

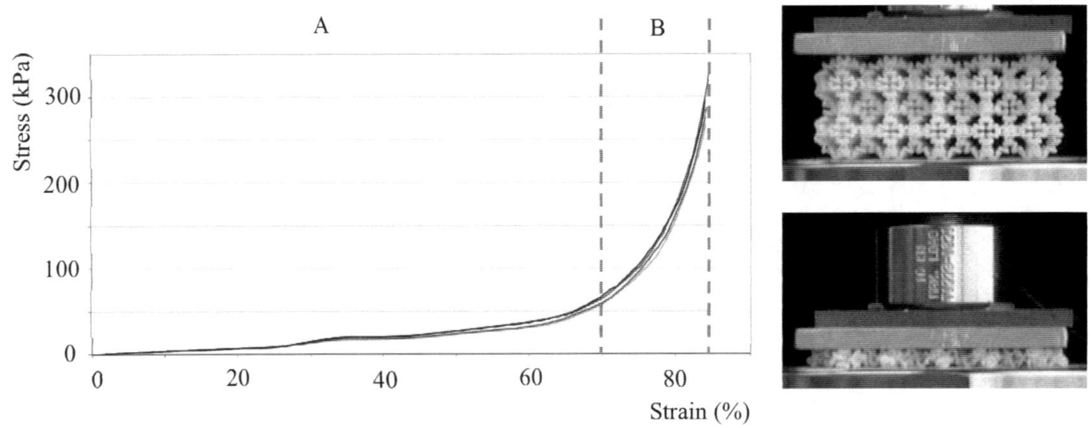

Figure 5. Compressive behavior of the helical strut samples

As shown in figure 5, the helical strut samples exhibited a different compressive response to the straight strut samples. A two stage compression is apparent, an approximately linear region (A) followed by a steep densification region (B) where stress quickly rises. Again, this is due to struts starting to make contact. The helical struts store or absorb more energy than the straight strut alternative, but due to the more bulky design, the densification stage starts slightly earlier (at around 70% strain rather than 75%).

3. Conformal methods for fitting a lattice structure to the human body

There are several methods that have been identified of fitting a structure to shape, i.e. generating a conformal structure. Two dimensional examples of these methods are shown in figure 6 for comparison.

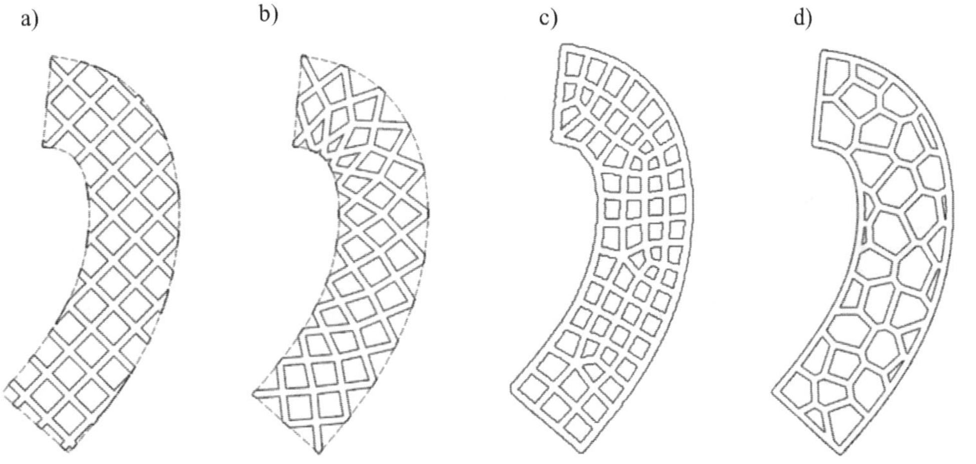

Figure 6. Conformal methods: a) trimmed b) swept c) meshed d) random

The 'trimming method' trims or cuts a regular tessellation to fit a particular conformal shape. The structure retains its orientation, and only the cells at the boundary of the structure are affected by the conformal method. Mullen *et al* use trimming to generate lattice structures for medical applications [9]. A structure can be trimmed to any shape, although there is no guarantee that the result will be useable. The 'sweeping method' deforms a structure to follow a surface of the conformal shape and as such is only suitable for use on sheet-like shapes, as used by Chu *et al* [10]. Mesh-based methods are a two-step process where a finite element meshing algorithm is used to generate a volumetric mesh of, for example, tetrahedral elements. This mesh is used as a skeleton to map lattice structure struts to (for example, see Gervasi *et al* [11]). Although a robust method, the structure will usually exhibit variation in cell size and shape. Finally, a randomised structure can be applied to a conformal shape, such as a 3D Voronoi tessellation. However at the macroscopic scale that cellular structures can be replicated at by current laser sintering technology, the ratio between cell size and part size is such that a randomised structure only serves to randomise its properties.

Each of these methods deform a structure in some way to fit a conformal shape. The sweeping and mesh-based methods consider the surface of the conformal shape as the basis of the structure generation, and 'fill in' the volume with a warped structure. As a result, these methods form 'closed' structures of intact (but varied) cells. In contrast, the trimming method populates a volume with structure and cuts it to fit the surface of the conformal shape. This results in a perfectly regular structure with 'open' and incomplete cells at the boundary.

The trimming method is compatible with any three dimensional tessellation. With this greater variety of tessellations brings a wider range of structure types, each with potentially different physical properties when manufactured. The trimming method is the most viable method of conforming a structure to a shape as it is robust and flexible.

3.1. Skinning a conformal structure

The main disadvantage of the trimming method is that the boundary regions of the structure are weakened where cells have been cut. A common method of strengthening the boundary of a trimmed structure is to wrap the structure (or a portion of the structure) with a sheet of solid material, commonly termed a 'skin'. This is shown in figure 7. A solid skin such as this can also provide some added rigidity for a lattice structure in weight-saving applications, and may provide some useful spreading of an impact in an energy absorbing role, but a different type of skin was developed by the authors to retain the flexibility of an energy absorbing structure. This skin type was termed a 'net skin' and is also shown in figure 6. Unlike a solid skin, the topology of a net skin is defined by the underlying lattice structure; cut cells are connected with skin struts to reconnect trimmed struts.

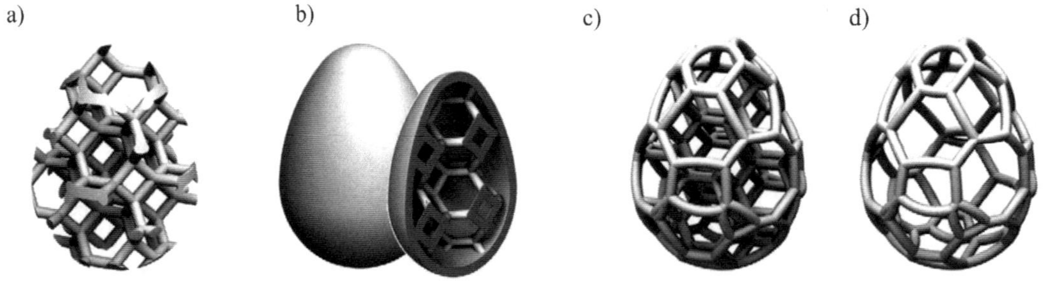

Figure 7. Skin types: a) no skin b) solid skin c) net skin d) net skin only

4. Handling geometric complexity

Additive manufacturing allows the fabrication of complex lattice structures, however conventional computer aided design (CAD) software becomes a limiting factor to size and complexity. Conventional CAD utilises a method of boundary-representation (b-rep) to represent three-dimensional geometry [12]. 'Solid' models are actually represented as a 'watertight' shell of zero-thickness surfaces. If the model is not watertight, it is usually not possible to tell what is 'inside' and 'outside', making the part a challenging proposition to manufacture.

While this is an elegant way to represent conventional products or components, this model structure is not suited to representing lattice structures. A single strut may only be comprised of a few surfaces, but even a moderately sized lattice structure of a few thousand struts becomes so computationally demanding that it makes the use of b-rep unfeasible. Additionally, the sheer number of struts increases the risk of non-manifold errors occurring; gaps between polygons or intersecting or overlapping polygons can compromise the watertight boundary [12].

Other geometry representation methods have been investigated for the fabrication of lattice structures. Voxels – a 3D analogy to pixels – are a volume representation that describes a model as a series of discrete cubes [13]. Function representation, developed by Pasko *et al* uses functions to implicitly describe geometry and has been used to great effect to represent a range of lattice structures [14].

Slice files represent a model as a sequence of 2D cross-sections and are used by AM processes to build parts. A CAD model must be converted to a slice file as this is the format that an AM machine reads data. Each slice of the model corresponds to a built layer. A novel method developed by the authors processes slice files to generate trimmed structures on a layer by layer basis. This is both a less computationally expensive method of generating trimmed geometry and also forgoes any need for further format conversions required by other geometry representations. A concept conformal chest protector designed by the author and built through the process developed is shown in figure 8. As well as a net skin, it contains in excess of 50,000 helical struts.

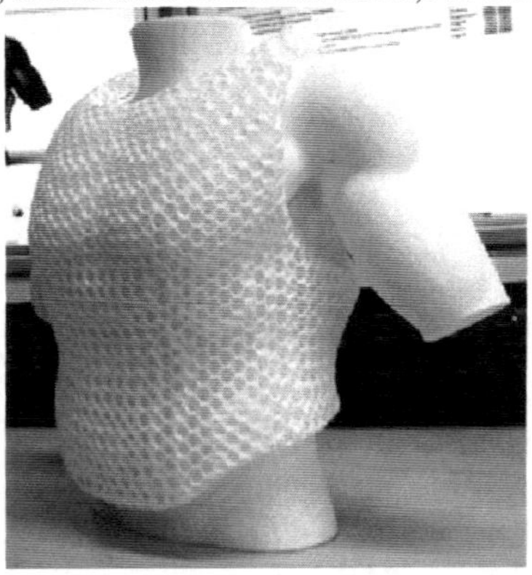

Figure 8. Chest protector concept

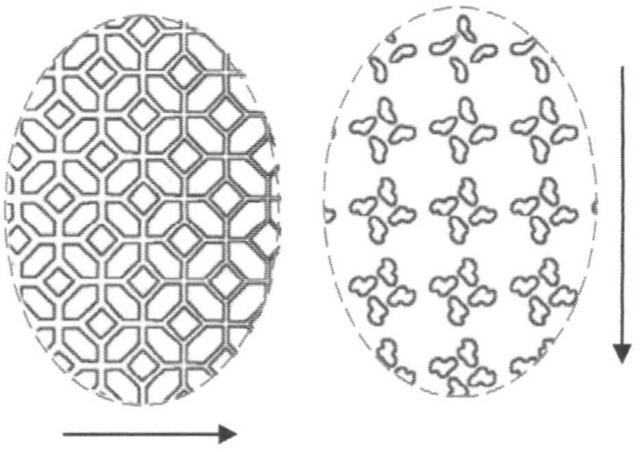

Figure 9. Functionally graded structures

The method developed also allows for functionally graded lattice structures, where structure geometry varies across the part. The base tessellation must remain the same, as one tessellation cannot morph into another without severe warping, but strut geometry can be varied as required. Example layers of functionally graded structures are shown in figure 9.

5. Conclusions

A methodology for generating conformal lattice structures based on the structure of polymeric foams has been discussed. The preliminary tests discussed in this paper have shown that AM lattice structures can be designed to exhibit similar compressive properties as conventional foams, as well as the ability to dramatically transform the compressive response by taking advantage of the geometric complexity AM affords. The tests conducted do not replicate an environment for sports PPE. A combination of both limited material options for laser sintering and the absence of many of the mechanisms of energy absorption available in foams mean that the energy absorption capabilities of the structures are limited. However, the tests do show the potential for tailored lattice structures, and as materials and processes improve so too will the potential for energy absorbent AM lattice structures.

References

[1] N. Hopkinson, R. J. M. Hague, and P. M. Dickens, Eds., Rapid Manufacturing: An Industrial Revolution for the Digital Age. Wiley, 2006.

[2] L. J. Gibson and M. F. Ashby, *Cellular Solids: Structure and Properties*, 2nd ed. Cambridge University Press, 1997.

[3] W. Thompson, "On the division of space with minimum partitional area," *Philosophical Magazine*, vol. 24, no. 151, p. 503, 1887.

[4] K. Li, X. L. Gao, and G. Subhash, "Effects of cell shape and strut cross-sectional area variations on the elastic properties of three-dimensional open-cell foams," *Journal of the Mechanics and Physics of Solids*, vol. 54, pp. 783-806, 2006.

[5] L. Gong, S. Kyriadekes, and N. Triantafydllidis, "On the stability of Kelvin cell foams under compressive loads," *Journal of the Mechanics and Physics of Solids*, vol. 53, no. 771-94, 2005.

[6] N. J. Mills, "The high strain mechanical response of the wet Kelvin model for open-cell foams," *International Journal of Solids and Structures*, vol. 44, pp. 51-65, 2007.

[7] W. E. Warren and A. M. Kraynik, "Linear elastic behavior of a low-density Kelvin foam with open cells," *Journal of Applied Mechanics*, vol. 64, pp. 787-94, 1997.

[8] "EOS Literature." [Online]. Available: http://www.eos.info/en/news-events/press-material/literature.html. [Accessed: 24-May-2012].

[9] L. Mullen, R. C. Stamp, W. K. Brooks, E. Jones, and C. J. Sutcliffe, "Selective Laser Melting: a regular unit cell approach for the manufacture of porous, titanium, bone in-growth constructs, suitable for orthopedic applications.," *Journal of biomedical materials research. Part B, Applied biomaterials*, vol. 89, no. 2, pp. 325-34, May. 2009.

[10] C. Chu, G. Graf, and D. W. Rosen, "Design for Additive Manufacturing of Cellular Structures," *Computer-Aided Design and Applications*, vol. 5, no. 5, pp. 686-696, 2008.

[11] V. R. Gervasi and D. C. Stahl, "Design and fabrication of components with optimized lattice microstructures," in *Proceedings of the Solid Freeform Fabrication Symposium*, 2004,

[12] C. R. Alavala, *Computer Graphics*. PHI Learning Private Ltd., 2009.

[13] S. Gibson, "Beyond Volume Rendering: Visualization, Haptic Exploration, and Physical Modeling of Voxel-based Objects," *Proceedings of Eurographics Workshop on Visualization in Scientific Computing*. pp. 10–24, 1995.

[14] A. Pasko, V. Adzhiev, A. Sourin, and V. Savchenko, "Function representation in geometric modeling: concepts, implementation and applications," *The Visual Computer*, vol. 11, no. 8, pp. 429-446, Aug. 1995.

Journal of Physics: Conference Series **382** (2012) 012043 doi:10.1088/1742-6596/382/1/012043

Finite element analysis of the high strain rate testing of polymeric materials

C. V. Gorwade[1], A. S. Alghamdi[2], I. A. Ashcroft[1], V. V. Silberschmidt[1], M. Song[2]
[1] Wolfson School of Mechanical and Manufacturing Engineering, Loughborough University, UK, [2] Departments of Materials, Loughborough University, UK

E-mail: C.Gorwade@lboro.ac.uk

Abstract. Advanced polymer materials are finding an increasing range of industrial and defence applications. Ultra-high molecular weight polymers (UHMWPE) are already used in lightweight body armour because of their good impact resistance with light weight. However, a broader use of such materials is limited by the complexity of the manufacturing processes and the lack of experimental data on their behaviour and failure evolution under high-strain rate loading conditions. The current study deals with an investigation of the internal heat generation during tensile of UHMWPE. A 3D finite element (FE) model of the tensile test is developed and validated the with experimental work. An elastic-plastic material model is used with adiabatic heat generation. The temperature and stresses obtained with FE analysis are found to be in a good agreement with the experimental results. The model can be used as a simple and cost effective tool to predict the thermo-mechanical behaviour of UHMWPE part under various loading conditions.

1. Introduction

Advanced polymeric materials are of interest to researchers, owing to their versatile applications in aerospace, defence, medical sectors as well as the limited current understanding of these materials in terms of manufacturing processes and the complex behaviour of these materials at different loading and temperature conditions [1]. One of the important applications of polymeric materials in the defence sector is as a 'liner' for armour material. The need for efficient, light-weight armour is necessary because of the increase in military personnel and civilians being placed in life threatening situations. Typical armour consists of a bullet proof metal or ceramic front plate, backed by a liner material, such as polymer [2]. A liner is necessary to reduce the risk of shock impact behind the bullet proof layer.

Impact response is strongly dependent on the strain hardening, strain rate sensitivity and temperature sensitivity of a material. In high speed impact, high temperature levels can be induced in polymeric materials, which lead to thermal softening and degradation of their mechanical properties [3]. This is a result of irreversible thermodynamic processes which convert the plastic energy into heat. Adiabatic temperature increase is a precursor of plastic instabilities and consequent failure of a material. In this work, UHWMPE samples are tensile tested, to investigate the temperature effects at high strain rate and obtain experimental results to be used as a reference for the development of a suitable numerical modeling method. A high resolution thermal camera was used to record the temperature increase in sample during the test and an elastic-plastic material model with adiabatic conditions was used to develop a finite element model of the experimental test.

2. Materials and preparation

Nascent ultra-high molecular weight polyethylene (UHMWPE) powders (Sarbic®UHMWPE3548) were purchased from SABIC with an average molecular weight of 3×10^6 mol/g. Compression moulding was used to mould the material using a square mould (10 X 10 X 1.65 mm) heated to 190 °C, which is higher than the melting point of the UHMWPE (approximately 135 °C). Various pressures (154, 232, 309, and 386 MPa) were investigated to optimise the properties of the material, such as hardness and crystallinity [4]. Various holding times at maximum pressure (10, 15 and 30 minutes) were also used to identify the most appropriate moulding parameters. The mould was cooled to room temperature using water. All specimens and testing methods were prepared according to ASTM D638 (2010).

3. Mechanical testing and characterisation

Tensile tests were carried out using an Instron 3366 tensile testing machine at room temperature (~22 °C). Three crosshead speeds were applied, 0.2, 1 and 5 mm/s and at least 5 samples were tested at each condition. The specimen dimensions are shown in figure 1. During tensile testing, a thermal camera (FLIR SC3000) was used to measure the internal heat changing in the sample, with a temperature range from -20 to +150 °C and accuracy of ± 1 °C. 'FLIR ResearchIR' software was used for temperature analysis. This was used to create point, line and area profiles combined with histogram charts for the temperature – time changes. Figure 2 shows example of thermal contour plot at different times in the testing of a UHWMPE sample.

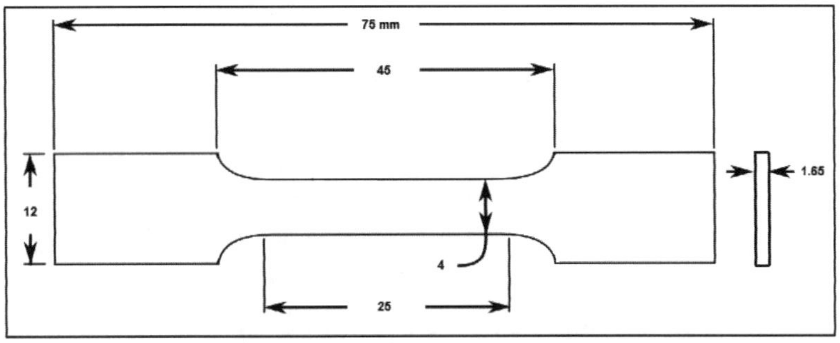

Figure 1. Specimen dimensions (in mm)

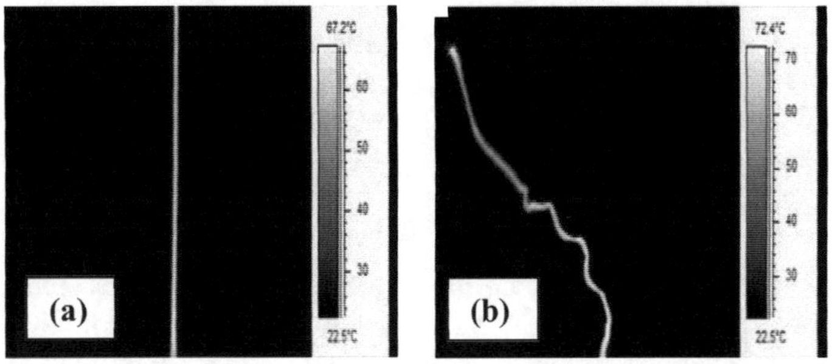

Figure 2. Thermal images show temperature contours (a side view of the test section) during the tensile testing of UHMWPE at 5 mm/s crosshead speed a) 620 % strain and b) failure

4. Experimental results and discussion

In order to demonstrate the influence of internal heat generation on the stress-strain behaviour of UHMWPE, crosshead speeds of 1 mm/s and 5 mm/s were applied using uniaxial tensile testing machine and a thermal camera was used to measure the temperature changes. Figure 3 shows the plot of engineering strain vs. engineering stress and temperature change. A significant temperature increase in the sample can be seen with the 5 mm/s crosshead speed, with the temperature increasing after the yield point. The results indicate that the heat generated during the plastic deformation is strongly dependent on the crosshead speed. This heat can be generated due to the high friction during the rearrangement and mobility of UHMWPE chains. High heat generation cause material softening as appears in the strain hardening region for the specimen tested at 5 mm/s crosshead speed, and this reduces the tensile strength. It appears from the figure 3 that approximately 50℃ can be considered as a critical temperature to significantly affect UHMWPE mechanical behaviour. At 1 mm/s crosshead speed, the temperature is 40 % less than the critical temperature value, and hence there is little thermo-mechanical effect at this crosshead speed.

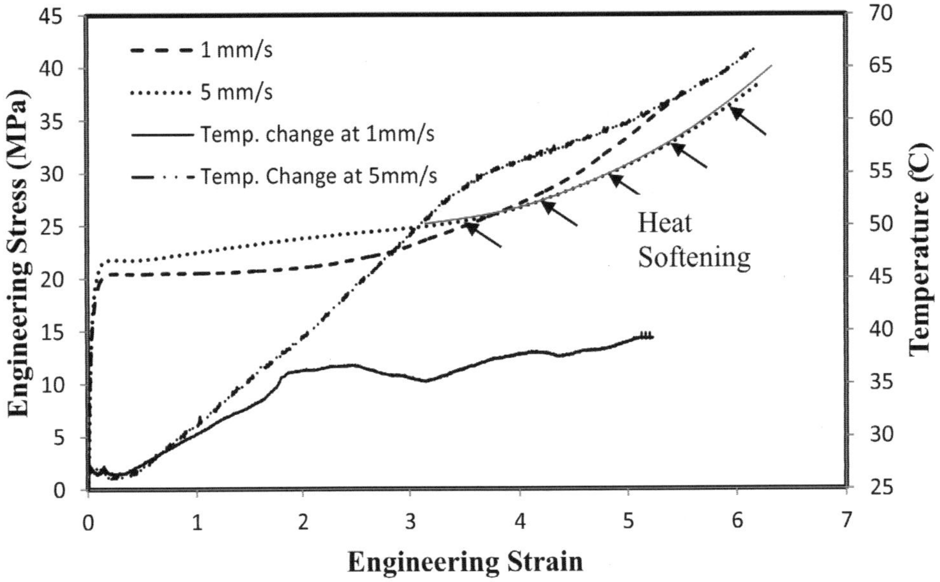

Figure 3. Effect of temperature on the tensile behaviour of UHMWPE at high strain rate.

5. Finite element modeling

A numerical analysis was carried out to study the details of the underlying mechanics during the deformation process in the tensile test experiment. The finite element model was used to simulate the plastic deformation and heat generation in the UHMWPE and its effect on test sample.

5.1 Methodology and Numerical simulation

The simulation was performed using the ABAQUS explicit finite element code. The first approach was to assume an elastic-plastic material model with adiabatic conditions, i.e. no heat lost to the environment and all energy released in plastic deformation converted into internal heat generation. The elastic-plastic material model is a simple and effective way to investigate the effect of internal heat generation during plastic deformation [6]. The numerical model uses an inelastic heat fraction parameter, η, to define heat generation caused by mechanical dissipation associated with plastic straining. The model assumes that plastic straining gives rise to a heat flux. The heat flux r^{pl} is written as,

$$r^{pl} = \eta\sigma : \dot{\varepsilon}^{pl} \quad (1)$$

where σ is the stress and $\dot{\varepsilon}^{pl}$ is the rate of plastic straining. The plastic strain increment is written from the flow potential as $\dot{\varepsilon}^{pl} = \dot{\varepsilon}^{pl} n$, where n is the flow direction. We assume $n = n(\sigma, \dot{\varepsilon}^{pl}, \theta)$ where θ is the temperature.

5.2 Parameters of model

The finite element model is modelled with the same dimensions as shown in figure 1. Table 1 shows that mechanical and thermal properties of the material model used in the finite element model [7]. The model is meshed with element C3D8RT, an 8 node thermo-mechanical brick, element with reduced integration. A displacement rate of 5 mm/s was applied to one end of the test sample and the other end was fixed in all directions. An initial temperature of 26℃ was applied.

Table 1. Mechanical and thermal properties of UHWMPE

Material	Elastic Modulus (MPa)	Poisson's Ratio	Coefficient of expansion (/°C)	Thermal conductivity (W/m-c)	Specific heat (J/Kg-c)	Inelastic fraction	Density Kg/ m^3
UHMWPE	450	0.45	0.0002	0.51	1900	0.9	934

5.3 Simulation results

The figure 4 shows a comparison of experimental and simulation results for the tensile test. The value of inelastic heat fraction used was 1, which represents as 100% of the plastic strain energy used to increase the internal temperature of test specimen. In this case an elastic-plastic material model overestimates the thermo-mechanical response of the UHMWPE.

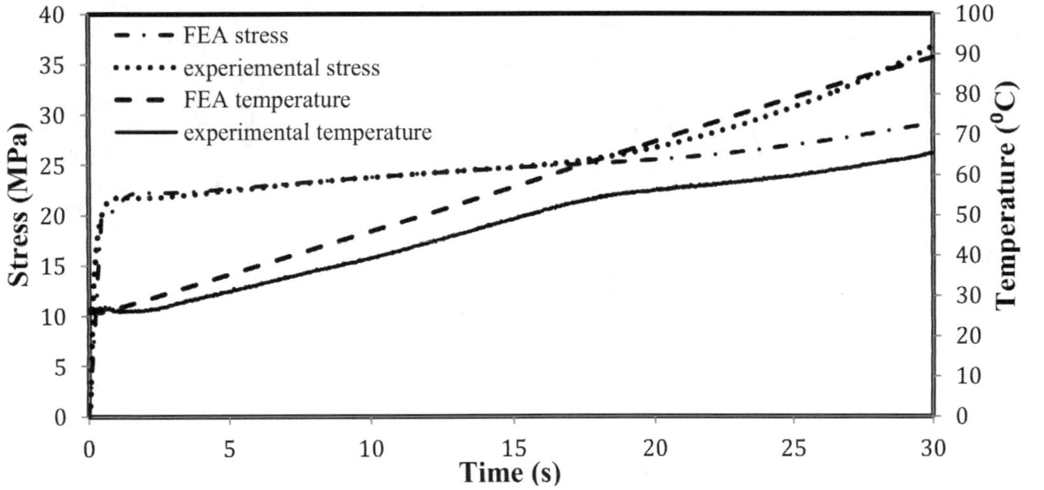

Figure 4. Experimental and simulation stress and temperature plot (inelastic heat fraction = 1)

The value of inelastic heat fraction was reduced to 0.7 in finite element model to optimise the overestimated temperature. The figure 5 shows a comparison of experimental and simulation results for the tensile test. The value of inelastic heat fraction used was 0.7, which represents 70% of the plastic strain energy used to increase the internal temperature of test specimen. Figure 5 shows that an elastic-plastic material model can predict the thermo-mechanical response of the UHMWPE in a good agreement with experimental results. The material model used has limitation to predict the accurate stress values after the curve reaches the stress value of 25 MPa as shown in figure 5.

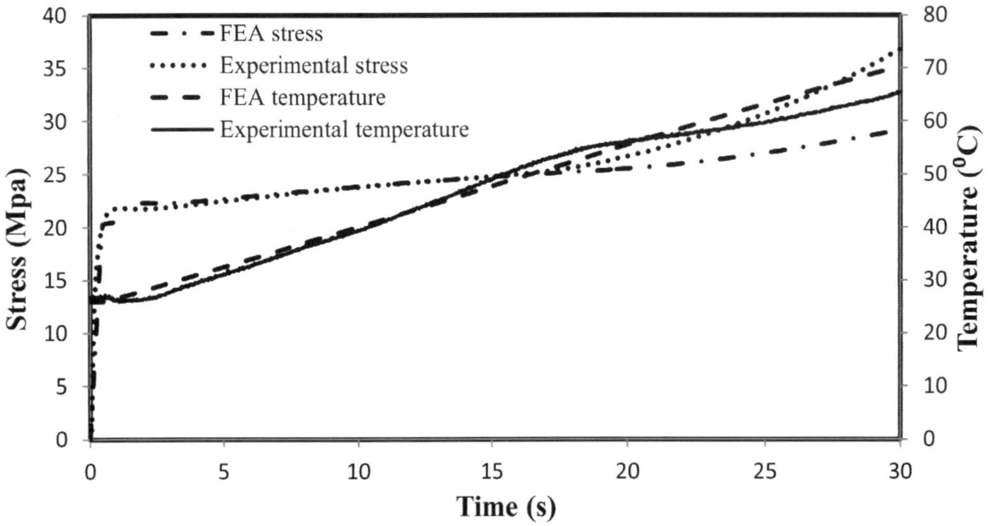

Figure 5. Experimental and simulation stress and temperature plot (inelastic heat fraction = 0.7)

The figure 6 shows a comparison of temperature contour for experimental and simulation at approximate 84% and 309% strain. The predicted FEA temperature contour plot shows approximately same temperature profile in necking area of the sample as observed in the experimental test by thermal camera.

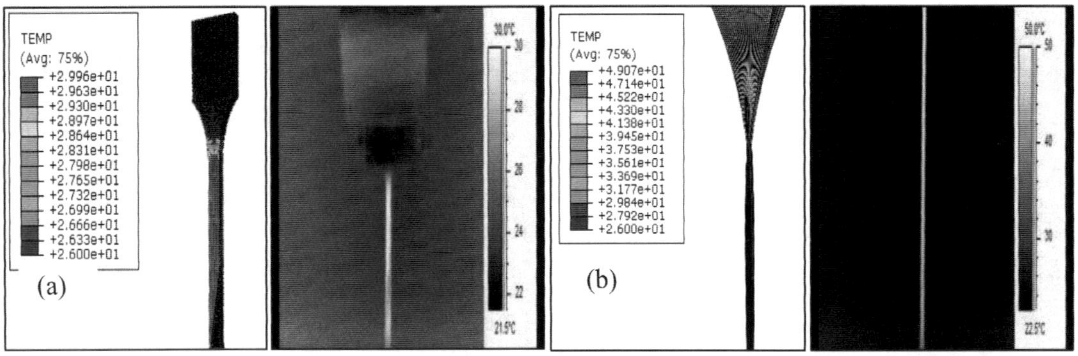

Figure 6 Comparison of (a) simulation and (b) experimental temperature contour at approximate 85% and 309 % strain

6. Conclusion

The finite element model of the tensile test for UHMWPE has been developed and validated with experimental work. A good agreement was observed between experimental results and optimised FEA results. The tensile experimental results show that cross head speed parameter has influence on the rate of temperature increase and strength of UHMWPE. The proposed material model can be used as a simple cost and time effective tool to predict the thermo-mechanical effects in UHMWPE at high strain rates. Further experimental and simulation work is in process for other polymer blends and polymer nanocomposites material to understand their mechanical and thermal behaviour under high strain rate.

Acknowledgement

Authors would like to acknowledge EPSRC and DSTL for funding support for the project and EPSRC instrument loan pool for the loan of thermal camera instrument.

Reference

[1] T. Erik et. al., Review nanocomposites in context, Composites Science and Technology 65 (2005) 491–516, doi:10.1016/j.compscitech.2004.11.003
[2] Ong. C. W. et. al., Advanced layered personal armour, International Journal of Impact Engineering, doi:10.1016/j.ijimpeng.2010.12.003
[3] Rodrı´guez-Martı´nez J. A. et. al., Temperature measurements on ES steel sheets subjected to perforation by hemispherical projectiles, International Journal of Impact Engineering, doi:10.1016/j.ijimpeng.2009.12.005
[4] Parasnis N.C., Ramani K., J. of Materials Sci.: Materials in Medicine 1998; 9: 165-172.
[5] Park J., Eom K., Kwon O., Woo S. Microsc Microanal 2001; 7: 276-286.
[6] Abaqus 6.11, Abaqus Theory manual, Dassault Systèmes Simulia Corp., Providence, RI, USA, 2011.
[7] Spiegelberg S., The UHMWPE handbook, Chapter 12- Characterisation of physical, chemical and mechanical properties of UHMWPE, doi.org/10.1016/B978-012429851-4/50013-9

Modern Practice in Stress and Vibration Analysis 2012 (MPSVA 2012) IOP Publishing
Journal of Physics: Conference Series **382** (2012) 012044 doi:10.1088/1742-6596/382/1/012044

Inspiration from Victorian times in Ultrasonic Surgical Tool Design

O. A. Ganilova[1], M. Lucas[1], Z. Pan[1], A. Y. Muir[2], H. Simpson[2]
[1] School of Engineering, University of Glasgow, Glasgow, G12 8QQ, UK
[2] Royal Infirmary of Edinburgh, Department of Orthopaedics, University of Edinburgh, Edinburgh, EH16 4SA, UK

E-mail: olga.ganilova@glasgow.ac.uk

Abstract. This work is devoted to the investigation of performance of surgical tools used in orthopaedics in terms of the occurrence of signs of necrosis, the accuracy of the cut and cutting tool design. For the comparison of the surgical tool performance different types of cutting devices were studied in a series of experiments. A Victorian surgical saw, its copy, a contemporary surgical saw, a surgical scalpel and an ultrasonic blade designed for a surgical application were chosen for the performance assessment. Such geometrical parameters as cutting edge shape, angle of teeth inclination, and sharpness of the cutting tools were analysed in terms of the quality of the cut and signs of necrosis. As a result of the analysis of experimental data obtained and theoretical insight the authors have come up with a creative solution for a novel design for a surgical ultrasonic blade which benefits from the design advantages of each of the analysed surgical tools and eliminates their drawbacks.

1. Introduction

The first report of the use of ultrasound in dentistry was published in 1952 when an industrial ultrasonic grinder was used to prepare cavities in extracted human teeth [1], and further applications in medical practice started to emerge rapidly. In 1970, Balamuth patented use of ultrasonic frequency vibratory forces for removal of layers of highly compliant biological tissue [2] and in 1967, Kelman reported the use of a combination of low frequency ultrasound and aspiration for removal of human tissue [3]. In 1955, another high-frequency longitudinal vibrating surgical cutting instrument for bone surgery was introduced and in 1974, Volkov reported on the simplification of orthopaedic procedures using an ultrasonic cutting device [4]. His device used a 25–30 kHz mechanical longitudinal vibration to drive an end-effector (scalpel or saw), which was modified depending on the task performed. It was emphasized that simple orthopaedic operations could be performed through a smaller incision than before. Further success and development took place in the area of ultrasonic bone scalpels for precise bone osteotomy [5] and for sawing and uniting bony fractures. Advantages of ultrasonic devices applications were also claimed in the removal of callus without damage from titanium osteosynthetic material, in particular callus from the slots of screws allowing fixation plate removal, [6]. The success of an ultrasonic device application can be traced to the ability of ultrasound to cut bone under finer control and with less pressure than for rotary or hand chisels. Nevertheless any development of an ultrasonic cutting instrument must account for associated vibrational energy, and the use of such a device may be very different from the cutting process associated with rotary burrs. An ultrasonic surgical hand piece usually consists of four basic elements: a generator or power supply, an ultrasonic motor (transducer), a mechanical wave amplifier (referred to as an acoustic horn) and a sonotrode (or probe) [1]. The generator converts electrical signals at low-frequency into high-frequency current to drive the transducer for conversion into longitudinal or transverse low amplitude mechanical vibrations, at 20–60 kHz for the applications reviewed [1]. The vibration amplitude is usually of the order of a few microns and so this is transmitted to a horn which is used to increase or decrease the amplitude. These vibrations are finally transmitted longitudinally along the length of the probe/tip. The horn and tip are normally profiled to achieve a specified mechanical gain [1]. An ultrasonic chisel, as developed for surgically cutting bone in contrast to a curved insert, has a straight ultrasonic tip and is used and shaped as a chisel that oscillates in longitudinally. The ultrasonic chisel differs in its ability to remove bone compared to a conventional rotary drill. Further development and miniaturisation of the ultrasonic hand piece has evolved novel tip or blade designs for the facilitation of certain medical and surgical procedures. Although the success of ultrasound in orthopaedics was first noted in 1974, surgical devices still have not been routinely replaced by ultrasonic hand pieces. Therefore this research represents the continuation of years of research in the area of ultrasonic blades at the University of Glasgow and is concentrating on the investigation and experimental testing of an Victorian surgical saw, a copy, a contemporary surgical saw, a surgical scalpel, and an ultrasonic blade designed for a surgical application (Fig. 1) in terms of performance assessment, i.e. cutting edge shape, angle of tooth inclination, sharpness of the cutting, the quality of the cut, and signs of necrosis.

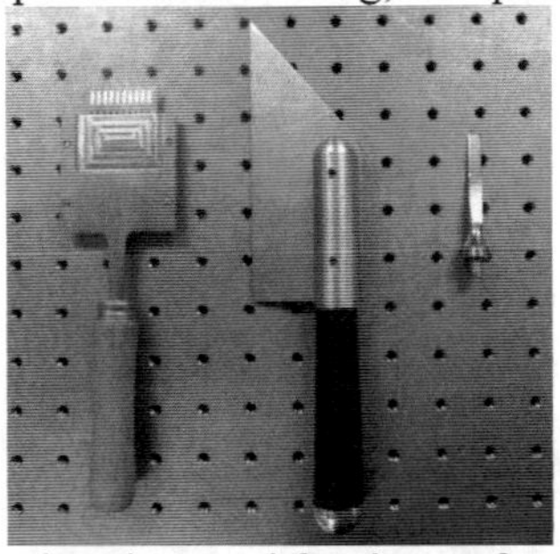

Figure1. Surgical tools tested for the performance analyses

2. Analysis of the Victorian saw shape

In the search for some inspiration for a novel and more efficient ultrasonic blade design for orthopaedic surgery, an antique Victorian saw used for an orthopaedics in the XIX century (Figure 2) was borrowed from the Museum of the Royal College of Surgeons of Edinburgh for a thorough scientific investigation and performance assessment.

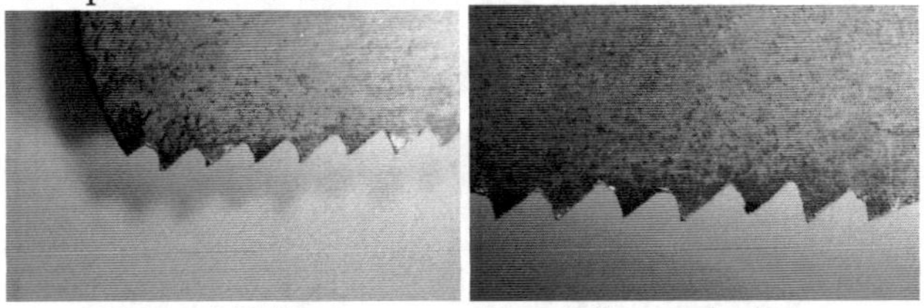

Figure 2 XIX century orthopaedic surgical saw

As a result of a series of tests conducted at the Royal Infirmary of Edinburgh to compare the performance of the orthopaedic Victorian and contemporary surgical saws, in terms of signs of necrosis, and by using a thermo-imaging camera, it was observed that the Victorian saw, in contrast with the contemporary surgical saw which reached the level of 90°C and higher, did not cause any dangerous temperature accumulation due to friction the level of temperature at the cutting site was well below 36°C. Therefore the shape of the antique saw rake and teeth was analysed in order to project the shape into an ultrasonic blade design. Photographs of the saw were taken with a macro-camera at a scale of 4.5 microns per pixel and processed using GIMP software (Figure 3).

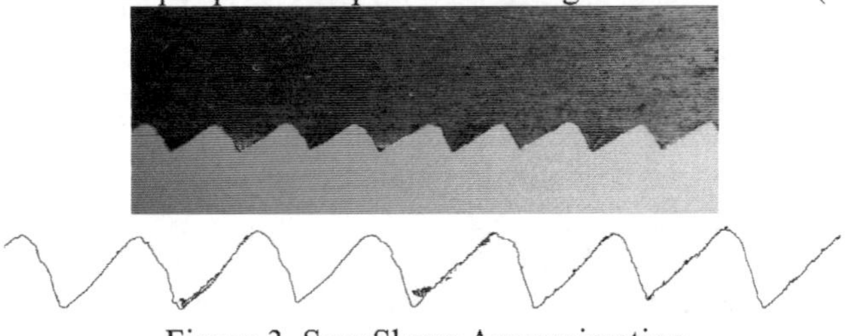

Figure 3. Saw Shape Approximation

The coordinates of the saw tooth were found (Figure 4) and the average angle of the tooth inclination was obtained (Figure 5).

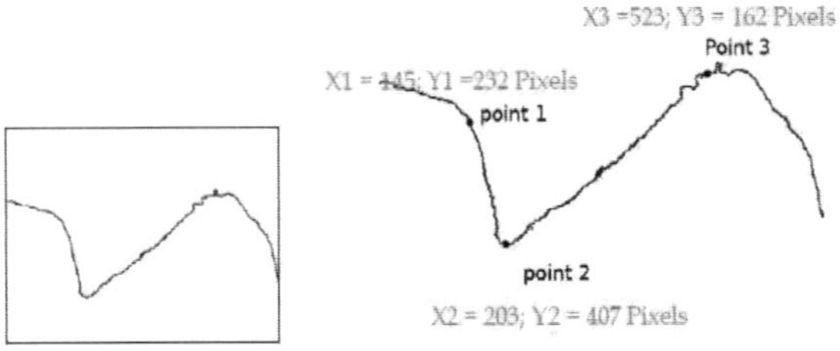

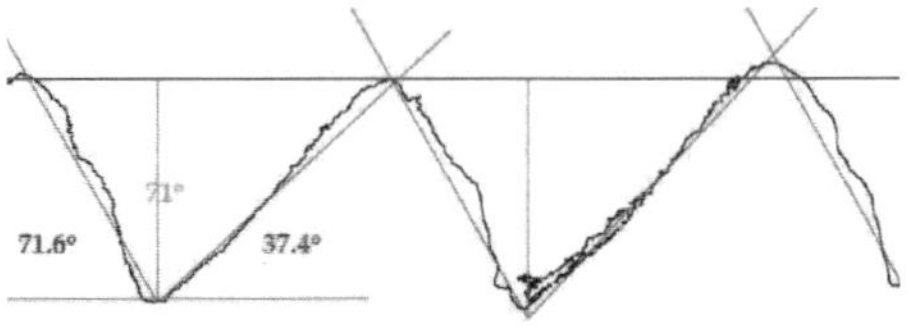

Figure 5. Tooth coordinates Figure 6. Average saw tooth angle of inclination

It should be pointed put that the cutting edge is normally considered to have a finite sharpness and to be constructed of two intersecting flat planes formed by the grinding process, with typical wedge angles in the region of 20° [7]. Herbert's experiments [8] showed that the main factor influencing durability and work output is the clearance angle or slope of the back of the tooth. This controls the space for debris, but too great a clearance results in the tip of the cutting edge breaking off. A clearance of about 25° on a file with zero rake (a vertical front face) gave best results. Figure 7 shows that the saw teeth are bent sideways alternately, and therefore the saw is expected to cut a path slightly wider than the thickness of the blade. This reduces the contact between the blade and work piece, and as consequence reduces friction, heat generation, possible overheating and damage of the blade. It prevents the saw being pinched and possibly jamming in the slot; it also allows for debris clearance. However, the sideways splay of the teeth (the kerf) also determines the width of the slot and volume of material lost. Therefore it is desirable to have the smallest kerf possible in order to increase the volume recovery of the bone.

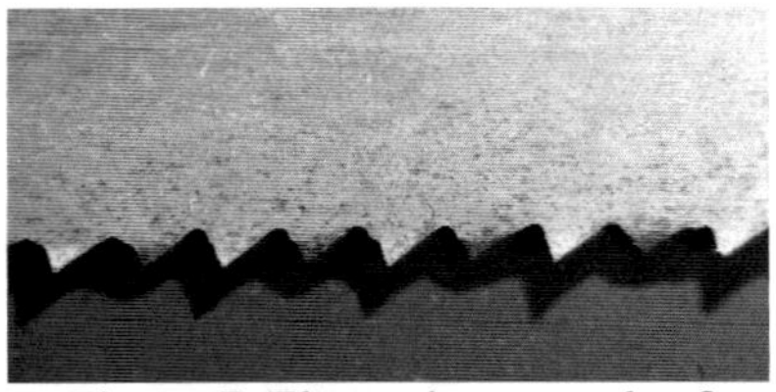

Figure 7. The antique saw kerf

Some research is discussed next on saw design in order to highlight the advantages of the Victorian design.

3. An insight into the performance of saws

Generally in ultrasonic procedures an indentation cut with a single pointed cutting instrument takes place. The cutting instrument cuts by perpendicular penetration into the work piece, so that the wedge angle of the cutting instrument is positioned symmetrically about the line of cutting action. Based on the research presented in [7] it was stated that the factors affecting the cutting process are: tool geometry, tool material, work material and cutting conditions and that the magnitude of the interaction is greatest between the first three factors while cutting conditions only marginally affect the instrument geometry [9]. According to [8] the factors affecting the cutting efficiency (i.e. the material removal per stroke) of files are: sharpness of tooth slope of the front face of the teeth (the rake); slope of the back face of the teeth (the clearance); the angles at which the two cuts lie relative to the axis of the file; the different pitches of the two sets of teeth; and the ratio between the two pitches. The cutting edge can be defined as a Fine edge; Burr edge and Blunt edge. The saw under consideration (Fig. 4) was characterised as a Fine edge, a cutting edge optimised for cutting as a result of the sharpening process and exhibiting no characteristics associated with bluntness such as cutting edge deformation, cutting edge dulling, and cutting edge cratering [10].

Some studies developed so far may be considered as macro level work concerned principally with the effect of the wedge-cutting angle and other cutting approach angles. However they do not account for the effect of the cutting edge radius on the cutting forces encountered in orthogonal cutting, and so neglect the effect of edge sharpness on cutting performance. In a study on ultra precision machining [11], the effect of the tool edge on the cutting process is considered and it was confirmed that there is a significant size effect in the measurement of cutting forces, which can be directly related to the variation in size of the cutting edge radius. Furthermore, the research indicated that this size effect is also linked to the localised stresses at the tool tip and the temperature gradient along the tool. Molecular dynamics studies on the effect of the tool radius on the nanometric cutting process carried out in [12] confirms that the size ratio of the depth of cut to the radius of the cutting edge is an important factor in the cutting process. They established that this relationship is linear. These studies confirm the importance of the cutting angles and the cutting edge radius on the performance of the cutting instrument, and emphasise the factors affected by these parameters as part of the cutting process. It has been shown that the shape of the tip profile is important in determining the puncture or penetration force and the load/displacement profile of the penetration force [7]. Although quite a few models do not account for variations in the cutting edge radius on the tool, and its effect on the cutting process, there have been attempts to refine the model for cutting with respect to the limitations due to the chip produced during the cutting process. The model described in [13] allows the advance prediction of chip breakage. It was also demonstrated that the tool coating is an important factor in the chip breaking problem. In the study [14], devoted to experimental work on cutting edges of different radii, 10, 50 and 60 μm, an advantage of the honing process in the form of having homogeneous and reliable cutting edges was demonstrated in contrast to ground inserts, the cutting edges of which failed rapidly. It was also emphasised that the lack of coating in the tool-chip contact area leads to a high wear rate.

The role of tool material properties on chip control has neither been fully understood nor considered as significant in the past. Nevertheless, the study presented in [13] presents a formula which describes the energy transformed into heat within the temperature field in the contact area, i.e. between the cutting tool chip and the work piece. This amount of thermal energy generated in the machining process that is transferred into the chip depends on the thermal conductivity of the cutting tool. When the thermal conductivity of the cutting tool is very low, the temperature in the chip area, and especially in the lower side of the chip, will be very high. However, cutting tool materials with low thermal conductivity show short contact lengths and produce chips with small radii. For this reason, the chips immediately leave the cutting tool and the crater wear on the tools is very close to the cutting edge.

It should be pointed out that thermal conductivity of most cutting tool materials, coated as well as uncoated, increases with increasing temperature. Only a few materials, such as aluminium oxide or silicon nitride, behave in the opposite way, i.e., as temperatures increase, their thermal conductivity decreases.

Therefore it can not be expected that a purely mechanical model would be valid since we know the thermal properties have a significant influence on the chip formation process. Several attempts have been made in the past for making analytical predictions of chip breakability and chip forms/shapes. It was not possible to do so owing to the complexity of the interactions among the work material, cutting tool material and geometry, and the cutting conditions. In view of the difficulty involved in developing purely analytical predictive models for chip breaking, as a practical approach, the feasibility of predicting the approximate levels of chip breakability through relevant experimental databases and knowledge-based systems has been demonstrated [13]. This method is an effective compromise between the time-consuming extensive experimental work and the totally non-scientific operator-based or handbook-dependent methodologies for selecting conditions that would warrant an effective chip breaking. Generally there are many different varieties of saw tooth profile for cutting different materials but all aim to give a blade along which the teeth do not break while cutting, that does not wear too quickly and remains sharp, that permits dust to escape, which does not jam by friction during cutting the slot, and where cutting can be performed with least effort [8]. Clearance or relief angles on saw teeth are different for different materials: very dense material require bigger reliefs than dry soft material. Serrated cutting edges changed force direction along the cutting edge, improving stiffness and resistance to vibration, and increasing tool-life [13]. It the teeth are spaced closely the rate of material removal is reduced. The quality of the sawn surface (roughness, etc.) is, however, improved with finer teeth [8]. With regard to the conditions of cutting it should be emphasised that dry machining leads to higher mechanical and thermal loading on cutting edges.

4. Experimental set up and results discussion

In [15] a study of in vitro the direction and force of applied loads applied by clinicians when using both a conventional slow surgical hand piece (CH) and an ultrasonic chisel (USC) for cutting bone was presented. Five clinicians took part in the cutting of a bovine bone using either an USC or a CH and the rate of cut was calculated over a fixed time-period and the depth of cut measured using a penetratometer. The magnitude of the longitudinal forces generated was demonstrated as a variable between 1.48 and 3.22 N (USC) and 0.04 and 4.56 N (CH). The CH had a pulling force directed towards the operator ranges from 20.11 to 20.72 N. Both instruments produced a similar range of downward forces, although there was intra- and inter-operator variability. The rate of cut varied in a similar manner, however, the CH produced a significantly greater depth of cut. The forces measured for the ultrasonic chisel varied between each cut made by the clinician and between each individual. The depth of cut ranges from 0.12 to 0.36 mm and 0.21 to 0.71 mm for the ultrasonic chisel and the conventional burr, respectively.

Following this study an experimental test of a copy of the Victorian surgical saw, a contemporary surgical saw, and an ultrasonic blade designed for a surgical application was performed with a clinician handling the blade. For this a copy of the Victorian saw was manufactured based on the average geometrical propertied identified in the Section 2. To test all the blades under the same conditions the blade of the contemporary surgical saw (Stryker) was equipped with a handle for testing under 'hand sawing' conditions. The ultrasonic blade, operating in the longitudinal mode of vibration at 35 kHz, was designed and tested using ABAQUS software. The view of the cut site and the temperature effect of the ultrasonic tool and the copy of Victorian saw application can be observed in Figure 8.

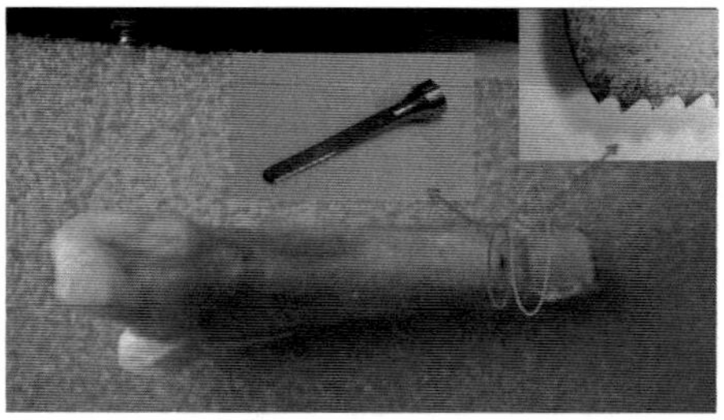

Figure 8. Cutting area after the test with an ultrasonic blade and a copy of the Victorian saw

The temperature accumulated during the test with an ultrasonic blade was measured using a thermal imaging camera FLIR and reached 115°C for the blade 1(a) (Figure 9). Investigation of the blade condition revealed the existence of a manufacturing defect which contributed substantially into the excessive temperature accumulation by increasing the amount of friction between the blade and the bone (Blade 1 in Figure 9). Therefore, one of the reasons for such a rapid temperature accumulation during the test with an ultrasonic blade could be the excess of material at the tip of the blade (manufacturing defect) or simply the large contact area of the blade with the bone at the cut sight. The manufacturing defect was eliminated by modifying the technique of blade fabrication (Figure 9, Blade 2) which allowed a reduction in temperature at the cut site down to 55 °C (Blade 2 in Figure 9). Nevertheless, the contact area reduction requires significant design or cutting technique modification. The technique which introduces short delays and applies only the tip of the blade to bone cutting, reduces in this way the contact surface and a reduces the temperature accumulated down to 54 °C (Figure 9, Blade 1(b)). Nevertheless to avoid any sign of necrosis some further minimisation of the contact area is required. This effect was also confirmed in [16] for the saw tooth radius, the maximum temperature increases with increase in the tool edge radius, which occurs as a heat transfer along the tool edge due to the increased size of the contact region.

Results of the experiments with a copy of Victorian saw and Stryker blade saw confirmation that the temperature at the cut site never exceeded 36°C. However it should be pointed out that normally Stryker blade is used as a contemporary orthopaedic surgical tool operating at very high speed causing very high rate of temperature accumulation (well above 90°C). For the saw based on the Stryker blade, the clinician confirmed that operating it by hand, at low speed similar to that for an ultrasonic blade and the copy of the Victorian saw, the clinician confirmed that a much higher load should be applied as well as a higher pulling force.

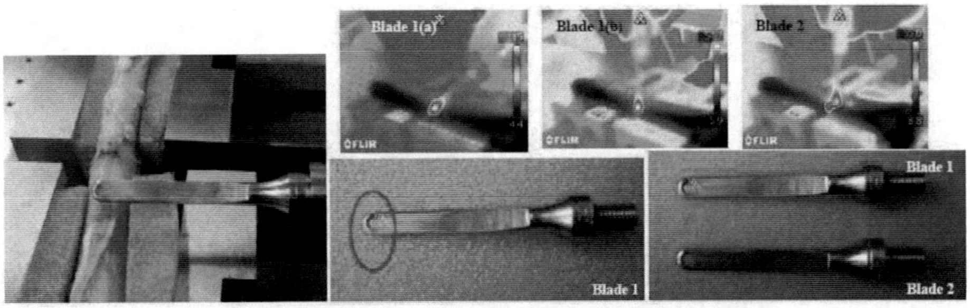

Figure 9. Thermal effect and manufacturing defect

Nevertheless, if now we investigate the shape of a surgical scalpel, as studied at the Royal Infirmary of Edinburgh in terms of performance in bone cutting during orthopaedic surgery (Figure 10), it is possible to notice that the cross-section of the scalpel reduces towards the top of the blade.

Figure 10. Surgical scalpel and its cross-section

Therefore inspired by the Victorian saw performance in terms of saw tooth shape and the possibility for reduction of the contact area through the reduction of the thickness of the blade alone the cross-section, adopting the scalpel design, it has been possible to come up with novel blade design presented in Figure 11.

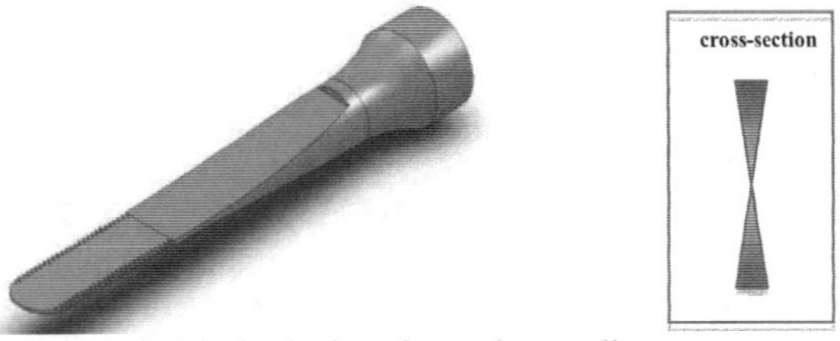

Figure 11. Novel ultrasonic blade design for orthopaedic surgery

In contrast to the scalpel shape the blade was reinforced at the top to reduce any significant influence of the coupled modes of vibration. This was achieved by adopting a 'bowl' type of cross-section (or combination of two scalpels) with the minimal thickness of the blade along the middle longitudinal line.

5. Conclusions

The work presented is concentrating on the investigation of the performance of surgical tools used, or can be used potentially in orthopaedics in terms of the occurrence of signs of necrosis, the accuracy of the cut and cutting tool design. For the comparison of the surgical tool performance different types of cutting devices were studied in a series of experiments. A copy of the Victorian surgical saw, a contemporary surgical saw blade (Stryker), a surgical scalpel, and a 35 kHz ultrasonic blade designed for a surgical application were chosen for the performance assessment. Such geometrical parameters as cutting edge shape, angle of tooth inclination, and sharpness of the cutting tools were discussed in terms of the quality of the cut and signs of necrosis. As a result of the analysis of the experimental data obtained it was concluded that minimisation of the contact area due to the optimal design and surgical tool handling technique affects the friction at the cut site in the most pronounced way, reducing the temperature accumulated from 115 °C down to 54 °C. This is an improvement comparing to the results obtained at the Royal Infirmary of Edinburgh (90 °C) for a different design of ultrasonic blade tested at 20 and 35 kHz. Manufacturing problems resulting in the excess of material or roughness of the blade surface contribute to the parasitic temperature increase. The maximum temperature also increases with increase in the tool tooth radius due to the increased size of the contact region [16].

Therefore to avoid any sign of necrosis some further minimisation of the contact is required. That is why the authors have suggested a creative solution for a novel design for a surgical ultrasonic blade which benefits from the design advantages of each of the analysed surgical tools, and eliminates their drawbacks (Fig.11). According to the finite element analyses the performance of the blade presented should reduce the contact area at the cut site and therefore the temperature accumulated. Additionally a reinforced symmetrical design will help to reduce undesirable effect of the coupled modes of vibration. Results of the experiments with a copy of the Victorian saw and the Stryker blade saw confirmed the temperature at the cut site never exceeded 36°C. Future work will be devoted to the test of a novel ultrasonic blade in comparison with the performance of the surgical scalpel and Victorian saw in terms of the cut accuracy, and signs of necrosis. The test will be performed under the same conditions, blade sharpness, under the same unified perpendicular loading, and on the same bone sample. The saws will be tested following [14] with a cutting speed of 50 m/min and a feed rate $f = 0.1$ mm/rev. Further work will also involve the consideration of the result obtained in [17] stating that a spherical pit proves to be superior to a groove in meeting the various directions of chip flow and can provide a positive rake angle to the main part of the cutting edge, regardless of the chip flow direction.

Acknowledgment
The research reported here has been supported by EPSRC Grant EP/G046948. The authors gratefully would like to acknowledge this support.

References

[1] O'Daly B J, Morris E, Graham P G, O'Byrne J M, McGuinness G B 2008 *J. of materials processing technology* **200** 38–58.
[2] Balamuth L 1970 Removing Tissue from a Biological Organism. *US Patent* 3 526 219.
[3] Kelman C D 1967 *Am. J. Ophthalmol.* **64** 23–35.
[4] Volkov MV, Shepeleva I S 1974 *Reconstr. Surg. Traumatol.* **14** 147–152.
[5] Hoigne D J, Stubinger S, Von Kaenel O, Shamdasani S, Hasenboehler P 2006 *Musculoskelet. Disord.* **7** 36.
[6] Eggers G, Klein J, Blank J, Hassfeld S, 2004 *J. Oral Maxillofac. Surg.* **42** 451–453.
[7] Reilly G A , McCormack B A O, Taylor D 2004 *J. of Materials Processing Technology* **153–154** 261–267.
[8] Atkins T 2009 *The Science and Engineering of Cutting — The Mechanics and Processes of Separating, Scratching and Puncturing Biomaterials, Metals and Non-Metals.* (electronic material) Chapter 7 pp 171-188.
[9] Kalder S, Venuvinod P K 1997 *J. Manuf. Sci. E-T ASME* **119** 1–9.
[10] Balevi B, *J. Peridont.* **67** 374–378.
[11] Kim K W, Lee W Y, Sin H C *J. Mater. Process. Technol.* **86** 45–55.
[12] Komanduri R, Chandrasekaran N, Raff L M 1998 *Wear* **219** 84–97.
[13] Jawahir I S and van Luttervelt C A 1993 *Annals-Manufacturing Technology* **42(2)** 659-693.
[14] Rech J, Yen Y-C, Schaff M J , Hamdi H, Altan T, Bouzakis K D 2005 *Wear* **259** 1168–1176.
[15] Khambay B S, Walmsley A D 2000 *J. of Dentistry* **28** 31–37.
[16] Kim K W, Lee W Y, Sin H C 1999 *J. of Materials Proc. Tech.* **86** 45–55.
[17] Nakayama K, Arai M, Kondo T 1981 *Annals-Manufacturing Technology* **30(1)** 5-8.

Variability and Anisotropy of Fracture Toughness of Cortical Bone Tissue

Adel Abdel-Wahab[1], Norhaziqah Nordin[2] and Vadim Silberschmidt[3]

Wolfson School of Mechanical and Manufacturing Engineering, Loughborough University, Loughborough, Leicestershire, LE11 3TU, UK

[1] a.a.abdel-wahab@lboro.ac.uk, [2] N.Nordin-10@student.lboro.ac.uk, [3]V.silberschmidt@lboro.ac.uk

Abstract. Bones form protective and load-bearing framework of the body. Therefore, their structural integrity is vital for the quality of life. Unfortunately, bones can only sustain a load until a certain limit, beyond which they fail. Therefore, it is essential to study their mechanical and fracture behaviours in order to get an in-depth understanding of the origins of its fracture resistance that, in turn, can assist diagnosis and prevention of bone's trauma. This can be achieved by studying mechanical properties of bone, such as its fracture toughness. Generally, most of bone fractures occur for long bones that consist mostly of cortical bone. Therefore, in this study, only a cortical bone tissue was studied. Since this tissue has an anisotropic behaviour and possesses hierarchical and complex structure, in this paper, an experimental analysis for the fracture toughness of cortical bone tissue is presented in terms of J-integral. The data was obtained using single-edge-notch bending (SENB) cortical specimens of bone tested in a three-point bending setup. Variability of values of fracture toughness was investigated by testing specimens cut from different cortex positions of bovine femur called *anterior*, *posterior*, *medial*, and *lateral*. In addition, anisotropy ratios of fracture toughness were considered by examining specimens cut from three different orientations: longitudinal, transverse and radial. Moreover, in order to link cortical bone fracture mechanisms with its underlying microstructure, fracture surfaces of specimens from different cortices and along different orientations were studied. Experimental results of this study provide a clear understanding of both variability and anisotropy of cortical bone tissue with regard to its fracture toughness.

1. Introduction

Bone is a natural composite material with hierarchical organization at different length scales. At the nano-scale, it consists of a collagen matrix impregnated with ceramic nano-particles known as carbonated hydroxyapatite [1, 2]. At the micro-scale, cortical bone is laid down in lamellar layers of 5 μm thickness. Similar to plywood composite structure, inside a layer, collagen fibers are parallel; however, their orientations are different for different layers. Across a bone section, not all lamellae are arranged in the same way, for instance, near the outer and inner surfaces, circumferential lamellae are parallel and arranged along the cortical bone's circumference. On the other hand, the outside and inside circumferential lamellae pack a region made of circular structures called osteons, formed from concentric lamellae impregnated in old remnants of a bone's remodeling process called interstitial matrix. The interface between osteons and interstitial matrix is called cement line; it is a collagen-free and highly mineralized layer. Cement lines have a paramount effect on bone's behavior, especially its fracture. Osteons are, on average, 200 μm in diameter and 1 cm long and parallel to the bone's longitudinal axis [3]. In addition, a network of canals and channels is formed across the bone's section and along its axis; these canals accommodate blood vessels and called Haversian canals. Moreover, bone has living cells called osteocytes that live within an interconnected network of microscopic channels called canaliculi. The latter are responsible for exchange of nutrients and waste between osteocytes [3]. At the millimeter length scale, bone consists of a dense and thick outer layer called cortical bone and a sponge-like structure called trabecular bone [4]. All these hierarchical levels work together in symphony to enhance macroscopic mechanical properties of bone tissue in the meter range [4]. Microarchitecture of the cortical bone tissue is quite complex and has a significant effect on its mechanical and fracture properties. Moreover, the preferential alignment of both collagen fibrils and nano-scaled mineral crystals causes anisotropy in both mechanical and fracture properties of the tissue [4]. From a fracture toughness perspective, the cortical bone tissue has different fracture resistance for various crack-propagation directions relative to the long bone axis that is called fracture toughness anisotropy. Various toughening mechanisms were reported for the cortical bone tissue including microcracks in the vicinity of the main crack due to stress concentrations ahead of the crack tip [5-7], and crack deflection and blunting at cement lines that are weak interfaces at the boundaries of secondary osteons [8]. Recently, it was reported that ligament bridging of crack in the wake zone is a dominant toughening mechanism in cortical bone as it reduces a driving force at the crack tip [9-11]. Several authors reported that toughening mechanisms are highly dependent on the crack propagation direction; therefore, fracture toughness of long bones is significantly higher in transverse and radial directions compared to the longitudinal one [11-13]. Despite interest for many researchers to fracture toughness of the cortical bone tissue, understanding of the causes of bone fracture is still not fully developed. Therefore, in this paper, fracture toughness of cortical bone tissue was studied as a function of both crack propagation direction and cortex position to promote our understanding for the origins of its fracture resistance a step further.

2. Materials and Method

2.1. Specimen preparation

The specimens in this study were cut from three fresh bovine femora (aged 1.5-2 years). The mid-part of three femurs (diaphysis) was extracted using a fine teeth band-saw. Then, the diaphysis part of each femur was sliced into four cortices – anterior, posterior, medial and lateral. Twenty-one specimens were cut from each cortex to allow crack growth along three different orientations relative to bone axis – longitudinal, transverse and radial as shown in Fig. 1. After cutting, specimens were ground under tap water using a series of grinding papers Standard ANSI grit: 240, 600, and 1200 to make sure that the surface is clean, without any scratches or irregularities. After preparation, the test specimens were put in a 0.9% physiological saline solution until tested. All specimens were prepared with the same dimensions for comparison according to the British Standard [14]: 25 mm x 5.43 mm x 2.72 mm (length × width × thickness). Also, a very fine slit of 2.7 mm x 5.43 mm was produced using a low-speed diamond saw for all specimens according to British Standard [14]. In this paper, specimens are labeled based on the crack propagation direction: longitudinal, transverse, or radial. Hence, specimens with crack propagating parallel to the bone axis is called longitudinal, perpendicular to it is called transverse and in the radial direction is called radial, see Fig. 1.

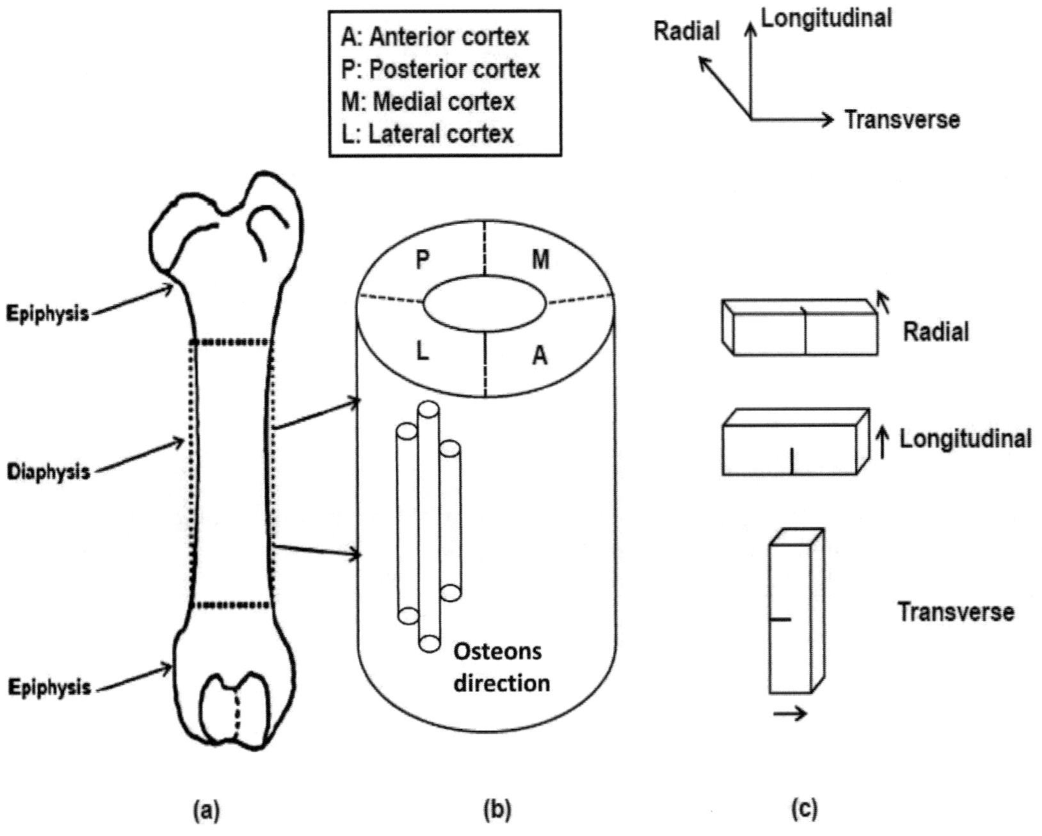

Figure 1. (a) Schematic illustration of bovine femur; (b) cortex positions in cortical bone; (c) specimens with different crack propagation directions. Specimens are labeled based on crack propagation directions depicted by arrows.

2.2. Fracture toughness measurements

The fracture toughness testing was performed using single-edge-notch bending on an Instron 3345 machine with a 5 kN load cell. All specimens were loaded to failure with a displacement rate of 1 mm/min. Specimens were loaded in three-point bending with load measured and recorded using the machine's load cell and the corresponding load-line displacement was simultaneously measured using a linear variable differential transducer (LVDT). The obtained load-displacement curves were analyzed according to the British Standard [14]. Specimens cut from the diaphysis part of bovine femur were tested with pre-notch in transverse, longitudinal and radial orientation for different cortex positions called *anterior*, *lateral*, *posterior* and *medial*. After fracture tests, fracture surfaces of all the specimens were investigated using scanning electron microscopy (SEM). Since cortical bone is not a conductive material, before investigation specimens were air dried and gold coated.

Plane strain fracture toughness, K_{Ic}, crack opening tip displacement (CTOD), or J-integral values can be determined using the specimen dimensions, depth of notch, 0.2% proof strength (σ_{YS}) and specific data from the force-displacement record of the fracture test. When the fracture follows elastic-plastic conditions, it is not possible to determine a valid K_{Ic} value to represent fracture toughness of a material. However, either critical CTOD or critical J-integral values can be calculated. Obtaining a valid K_{Ic} value depends on the shape of the force versus displacement record, the specimen size and form, and the 0.2% proof strength and toughness of the material at the temperature of interest. For a valid K_{Ic} measurement, the specimen dimensions (a nominal crack length (a), thickness (B) and the uncracked ligament (W – a)) each has not to be less than: $2.5\left(\frac{K_{Ic}}{\sigma_{YS}}\right)^2$ [14]. In this study, the behavior of all specimens was predominantly non-elastic and all specimens failed to satisfy the validation criterion of K_{Ic}. Therefore, elastic plastic fracture mechanics (EPFM) parameter, J-integral, was calculated based on British Standard [14].

3. Results and Discussion

This study focuses on evaluating fracture-toughness values for specimens with cracks growing parallel to bone axis, perpendicular to it and in the radial orientation. In addition, the anisotropy ratios of the fracture toughness values were calculated. Results of this study show that all specimens exhibited a non-linear elastic-plastic behavior. Therefore, based on the British Standard [14], the critical stress intensity factor K_{Ic} was not valid for all specimens, and the J-integral was used to quantify its fracture toughness. Table 1 lists the average levels and standard deviation of critical values of J-integral and their cortex dependency for all crack growth directions.

Table 1. Critical J-integral values (N/m) for specimens with longitudinal, radial and transverse crack growth in bovine femoral cortical bone tissue at different cortex positions.

	Anterior		Medial		Posterior		Lateral	
	mean	SD	mean	SD	mean	SD	mean	SD
Long.	1033.90	±254.5	1768.47	±98.8	1165.65	±340.1	2034.27	±509.9
Radial	1199.09	±153.1	1418.22	±97.2	982.97	±369.5	2664.16	±554.4
Trans.	4509.08	±422.1	5925.47	±802.9	3876.66	±847.3	5661.61	±452.7

It can be noticed from Table 1 that significant differences were found among the resistance to fracture of specimens cut from different cortices of bovine femur cortical bone. In general, bovine femoral cortical bone shows higher resistance to fracture when a crack grows perpendicular to osteons direction, see Fig. 1. Resistance to fracture is lower for cracks grow in both radial and longitudinal directions for all specimens. For cracks grow in the transverse direction, specimens cut from medial cortices showed the highest fracture toughness when those cut from posterior cortex exhibited the lowest resistance to fracture. The mean fracture toughness value for specimens cut from medial cortices is higher by some 34.58%, 23.9% and 4.45% relative to those cut from posterior, anterior and lateral, respectively. On the other hand, among specimens with radially extending cracks, those cut from lateral cortices demonstrated the highest resistance to fracture while those cut from posterior exhibited the lowest resistance. The former's mean fracture toughness value is some 54.99%, 46.77% and 63.1% higher compared to those of anterior, medial and posterior, respectively. Finally, for specimens with cracks extending parallel to osteons, similar to those with radial cracks, the mean fracture toughness value was the highest for specimens cut from lateral cortices. Dislike specimens with radially extending cracks, the mean fracture resistance value was the lowest for specimens cut from anterior cortices. The ratios of average fracture toughness values of specimens cut from anterior, medial and posterior cortices relative to that had the highest value were 49.18%, 13.1% and 42.6%, respectively. This implies that bovine femur has a non- uniform fracture resistance for cracks extend parallel to osteons, perpendicular to it and in the radial direction. Obviously, microstructure of cortical bone has a predominant effect on its resistance to fracture, which is the case for other mechanical properties, such as elastic modulus, yield stress, and ultimate strength [16]. Due to a natural loading regime exerted by animal's weight and muscle forces, long bones are exposed to combined loading conditions that are non-uniform. As it is well known from literature, bone is a dynamic tissue that reacts to mechanical loading by alternating its shape, internal microstructure and properties to meet external loading environment [17]. Therefore, the findings of this study demonstrated different crack resistance for specimens cut from different cortices with various underlying microstructures. Toughening mechanisms depend on crack growth orientation [17] for osteonal cortical bone. For cracks emerging in the transverse direction, toughening mechanisms such as crack deflection and twist are dominant, whereas for cracks extending in both longitudinal and radial directions, the dominating toughening mechanism is uncracked-ligament bridging [17]. In this study, it was found that the underlying microstructure of bovine cortical bone changed from one cortex to another and included primary, secondary and plexiform bones (see Fig. 2). Thus the interaction between the crack and different microstructures triggers even more toughening mechanisms that, in turn, were reflected in the form of different values of fracture toughness for specimens from different cortices. Also, this study demonstrated fracture toughness anisotropy ratios of specimens with cracks growing longitudinally, transversely and radially, see Table 2.

Table 2 introduces anisotropy ratios of fracture toughness values for specimens with cracks growing in longitudinal, transverse and radial orientations and cut from different cortex positions.

Table 2. Values of anisotropy ratios of fracture toughness values of longitudinal, transverse and radial crack growth directions of bovine femoral cortical bone tissue at different cortex positions

	Anterior	Medial	Posterior	Lateral
Trans/Long	4.36	3.35	3.33	2.78
Trans/ Radial	3.76	4.18	3.94	2.13

It was noticed from table 2 that bovine cortical bone demonstrated high anisotropy ratios of J-integral values. Anisotropy ratios were calculated by relating fracture toughness values of specimens with transverse cracks to those with cracks extending longitudinally and radially. It was also found that anisotropy ratios were different for different cortices. Specimens cut from lateral cortices with cracks growing radially showed the lowest anisotropy ratio, while those cut from anterior cortices exhibited the highest ratio.

At the microstructure level, specimens cut from lateral and medial cortices revealed secondary osteonal bone with cement-line layers that trigger the deflection-of-crack toughening mechanism. This is a possible interpretation of the higher J-integral values of lateral and medial cortices specimens. Liang et al. [15] suggested that osteons, which were newly formed and had low stiffness, strengthened the cortical bone tissue via promoting crack propagation toward osteons, thus causing the crack arrest in the cement lines.

Fracture surfaces were studied using scanning electron microscopy (SEM) for the crack growth regions for all the specimens. Generally, higher energy consumption during crack growth was linked with rough and uneven surfaces while flat surfaces were indication of low levels of energy (see Fig. 2). The most common mode of bone fracture is the transverse fracture where a crack cuts across osteon direction [18]. Figure 2 shows the fracture surfaces of specimens from four different cortex positions with cracks growing perpendicular to dominant orientation of osteons (bone axis). This study demonstrates that different fracture resistances are underpinned by microstructure adaptations of cortical bone to applied mechanical loadings.

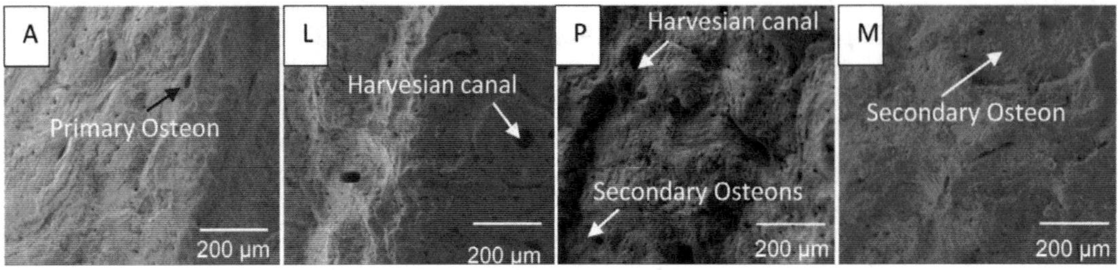

Figure 2. Fracture surfaces of specimens cut from different cortex positions with cracks growing in transverse direction (⊥ bone axis). A: anterior, L: lateral, M: medial and P: posterior.

In summary, bone is designed to resist cracks to different extent and by employing diverse mechanisms for various crack-growth orientations. In addition, bone's underlying microstructure plays an important role in the fracture process. In the future research, we will try to reveal the fracture mechanisms caused by the interaction of cracks with different microstructures of the cortical bone tissue.

4. Conclusions

In the present study, the fracture toughness of bovine femoral cortical bone was evaluated, and the effect of its microstructure on fracture toughness values was examined. Based on the current study, the following conclusions were made:

- Bovine femoral cortical bone demonstrated non-uniform fracture toughness for specimens cut from different cortices and for cracks extending along bone axis, perpendicular to it and in the radial direction. Different fracture resistances values are linked to cortical bone's underlying microstructure.
- Bovine cortical bone demonstrated high anisotropy ratios of J-integral values. It was also found that anisotropy ratios were different for different cortices.

Acknowledgment

The authors acknowledge support from EPSRC (grant no. EP/G048886/1).

References

[1] Currey JD. J Exp Biol. 1999;202:3285-94.
[2] Fratzl PG, H. S.; Paschalis, E. P.; Roschger, P. Structure and mechanical quality of the collagen–mineral nano-composite in bone. J Mater Chem 2004;14:2115-23.
[3] Ethier CR, Simmons, C. A. . Introductory biomechanics: from cells to organisms Cambridge Unversity Press; 2007.
[4] Peterlik H, Roschger, P. , Klaushofer, K., Fratzl, P. . Orientation dependent fracture toughness of lamellar bone. Int J Fract. 2006;139:395-405.
[5] Zioupos P aCJ. J Mater Sci. 1994;29:978-86.
[6] Zioupos P WXT, Currey JD. Clin Biomech. 1996;11:365-75.
[7] Vashishth D TKE, Bonfield W J Biomech. 2003;36:121-4.
[8] Liu D WS, Wanger HD. J Biomech. 1999;32:647-54.
[9] Nalla R K KJH, Ritchie R O. Nat Mater. 2003;2:164-8.
[10] Nalla R K KJJ, Ritchie R O. Bone. 2004;34:790-8.
[11] Nalla R K KJJ, Kinney J H, Ritchie R O. Biomater. 2005;26:217-31.
[12] Martin R B B, D L. J Biomech. 1993;26:1047-54.
[13] Behiri JC B, W. J Biomech. 1989;22:863-72.
[14] BRITISH STANDARDS INSTITUTE B-FMTTMfDoK, Critical CTOD and Critical J Values of Metallic Materials. London: British Standards Institute.
[15] Liang L, Guo, XE, Goldstein, SA. J Biomech Eng. 1998;1:112-7.

Cointegration and why it works for SHM

Elizabeth J. Cross and Keith Worden

Department of Mechanical Engineering, University of Sheffield, Mappin Street, Sheffield, U.K.

E-mail: e.j.cross@sheffield.ac.uk

Abstract. One of the most fundamental problems in Structural Health Monitoring (SHM) is that of projecting out operational and environmental variations from measured feature data. The reason for this is that algorithms used for SHM to detect changes in structural condition should not raise alarms if the structure of interest changes because of benign operational or environmental variations. This is sometimes called the data normalisation problem. Many solutions to this problem have been proposed over the years, but a new approach that uses cointegration, a concept from the field of econometrics, appears to provide a very promising solution. The theory of cointegration is mathematically complex and its use is based on the holding of a number of assumptions on the time series to which it is applied. An interesting observation that has emerged from its applications to SHM data is that the approach works very well even though the aforementioned assumptions do not hold in general. The objective of the current paper is to discuss how the cointegration assumptions break down individually in the context of SHM and to explain why this does not invalidate the application of the algorithm.

1. Introduction

One of the largest stumbling blocks to the implementation of structural health monitoring (SHM) in practice outside of laboratory conditions is the confounding influence of changing environmental and operational conditions on damage sensitive features. Recently the idea of cointegration, from the field of econometrics, has been introduced by the authors as a way to remove trends induced by changing environmental and operational conditions [1]. Cointegration is a property of nonstationary time series which admit to a stationary linear combination. The main ideas behind the use of cointegration for SHM will be discussed briefly here, but see [1] for more details.

To use cointegration in the way suggested by the authors, a nonstationary multivariate training set of feature data from a structure operating in its normal condition is employed. If the variables in this training set are cointegrated, then a linear combination of them may be found that is stationary. Following common practice in econometrics, the Johansen Procedure [2] is implemented to find the most stationary linear combination of the variables possible. The Johansen procedure is a maximum likelihood approach for finding stationary combinations (culminating in a cointegrating *vector*) of nonstationary variables whose first difference is stationary. Briefly, the variables of interest $\{y_i\}$ are arranged in a vector error correction model form (VECM), that may be seen as a rearrangement of the more familiar vector autoregressive model (VAR). This VECM takes the form:

$$\{\Delta y_i\} = [\Pi]\{y_{i-1}\} + \sum_{j=1}^{p-1} [B_j]\{\Delta y_{i-j}\} + \{\epsilon_i\} \tag{1}$$

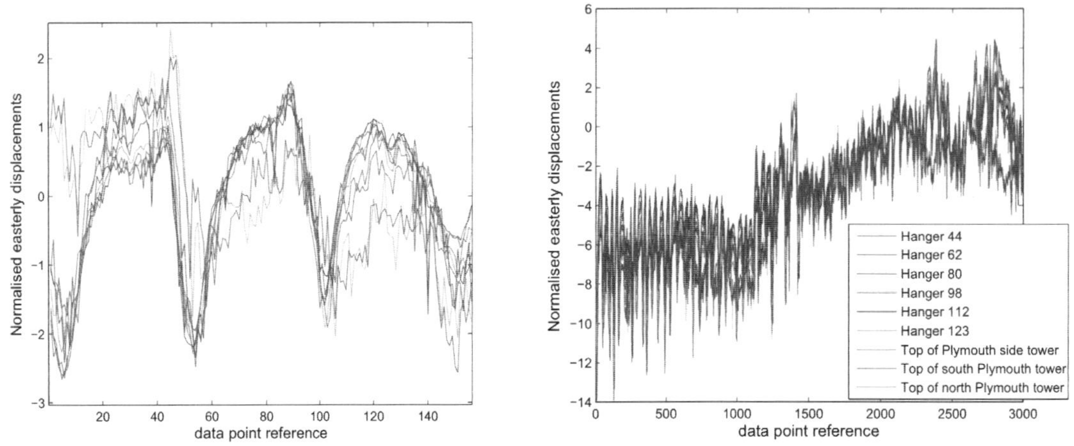

(a) (b)

Figure 1: TPS measurements of bridge displacements (a) over a three day period, (b) over a two month period.

where y_i denotes an n-vector including all n variables to be analysed, with the subscript i relating to time, $i = 1, \dots N$, p represents the model order, or the number of lags to be included in the model, and ϵ_i is a noise process. In this form, the matrix $[\Pi]$ contains the coefficients that will create the most stationary linear combination of the original variables, in other words, $[\Pi]$ contains the cointegrating vectors. Unfortunately, if the variables are cointegrated, $[\Pi]$ must be of reduced rank, which means that parameter estimates for (1) must be obtained using a reduced rank regression. The reduced rank regression involves the decomposition of the matrix $[\Pi]$, and parameter estimation through maximising the likelihood of observing the correct noise sequence $\{\epsilon_i\}$, however, as the process is rather in depth no further details are given here, instead readers are again referred to [1] or [2].

If a stationary linear combination can be found, the continuing stationarity of the combination (commonly referred to as a residual) when new data is projected onto it can be used as an indicator that the structure continues to operate in its normal condition. An example of data from the Tamar Bridge, monitored by the Vibration Engineering Section at the University of Sheffield, is included here for illustrative purposes.

Figure 1 shows the displacement measurements of nine points on the deck and towers of the Tamar Bridge over a three day period and over a two month period respectively, measured by a total positioning system (TPS). The daily and seasonal trends visible are induced mainly by the changing temperature. As the variables share common trends it is likely that they will submit to a stationary linear combination (i.e. they are cointegrated). Figure 2(a) shows the linear combination suggested by the Johansen Procedure when the data shown in Figure 1(a) is used as a training set. From this figure one can see that the temperature induced trend has successfully been removed. If the data over the longer time period is then projected onto this combination the residual remains stationary, as shown in Figure 2(b) indicating without influence from the temperature that the structure continues to respond in its normal condition, as defined by the training set.

The remainder of the paper aims to give a general discussion about why cointegration is applicable to engineering data. In the following discussion the authors will argue that measured responses from healthy structures exhibit nonstationary behaviour over relatively short time

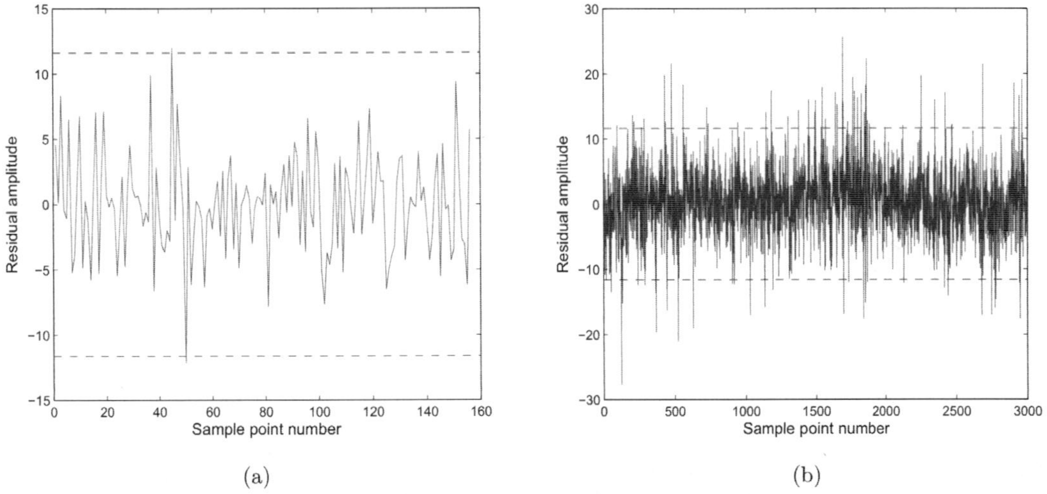

Figure 2: Cointegrated residual of variables (a) for three day training period, (b) over two months.

periods, but should generally be stationary in the long term. As one would wish to detect any occurrence of structural degradation swiftly, these shorter time periods are of great interest to SHM. The discussion below will argue that such nonstationarity exhibited by structural response variables may be well represented by econometric theory, which models nonstationarity with unit root processes or with a deterministic time trend (or both). While under the time periods of interest such models of nonstationarity are valid, and therefore, too, the applied theory of cointegration, the acceptance of a unit root generating process for a structural response is not a comfortable one. The authors will argue that so long as the econometric models fit the data well, the philosophical question of whether a unit root assumption is valid is not really of interest. Indeed, to the engineer, so long as the combination of variables found via the Johansen procedure, or any other cointegrating method, does the job of removing confounding influences, the underlying assumptions of the process used to arrive at this combination may not be of any interest. None the less, a good understanding of such assumptions may help to avoid any suboptimal use of the sophisticated theory imported from econometrics.

2. The assumptions behind cointegration applied to engineering data

In previous works by the authors [1, 3], cointegration has been adopted from econometrics for the purposes of removing environmental and operational trends in damage sensitive features. The theory applied in these works, and in the example above, assumes that the variables of interest are nonstationary and, for the Johansen procedure, integrated of order one (nonstationary with first difference stationary, i.e. a unit root process). A unit root process is one whose characteristic equation has at least one root equal to one, it is these roots that will determine the stability/stationarity of a time series. For an auto-regressive model (AR) of order p:

$$y_i = a_1 y_{i-1} + a_2 y_{i-2} + \cdots + a_p y_{i-p} + \varepsilon_i \tag{2}$$

where ε_i can be considered to be a Gaussian white noise process driving the model, its characteristic equation is as follows;

$$1 - a_1 z - a_2 z^2 - \cdots - a_p z^p = 0 \tag{3}$$

where z are the roots of the characteristic equation. This process will remain stable as long as $|z_i| > 1$. If any $|z_i| < 1$ the process will behave explosively. Now if any $z_i = 1$, (unit root), then the process has marginal stability. This section aims to discuss whether the unit root assumption is valid in the context of an engineering application and whether structural response can be nonstationary in the long term.

A nonstationary time series is defined as one whose statistics change with time. A weakly stationary process has constant mean and variance, and an autocovariance that depends only on the lag length considered. In the presence of nonstationarity standard regression techniques and inferences made on such regressions have been found to be unreliable, and since Yule's seminal paper [4], much research has been carried out on nonstationary processes.

Of the variables of interest to SHM, many of them appear to exhibit nonstationarity. The most relevant example to this paper is the easterly displacements of the deck of the Tamar bridge, as plotted in Figure 1. Studying Figure 1, the time series certainly appear nonstationary over the time of available measurements.

To an engineer, when considering a dynamic response the usual modelling approach would be to employ the classic second order differential equation of motion. The nonstationarity of a response is then explained by the changing physical parameters such as mass and stiffness, or a change in the nature of the excitation. In the context of applying cointegration theory, the interest is in modelling the entire process (response) with fixed parameters that cannot change. For this task, the most comfortable way of thinking about a structural response is probably as a function of a number of external conditions which are themselves fluctuating. One might expect that if all external conditions could be accounted for, a structural response would be stationary around the trends introduced by these external conditions. Unfortunately, it may be very infeasible to account for all external conditions driving the response of a structure, as it certainly is in the field of econometrics where relations between variables of interest are uncertain. This is where an AR type model can be useful and why they are regularly applied to econometric time series, as one no longer needs information about each driving factor to be able to describe a dynamic process. The econometric cointegration theory commonly used was developed in the context of such models, and so to utilise this theory one must adopt them too.

In econometrics and for the cointegration procedure, an error correction model (ECM) is used to model a process variable, y_i of interest:

$$\triangle y_i = \rho y_{i-1} + \sum_{j=1}^{p-1} b_j \triangle y_{i-j} + \mu + \nu t + \varepsilon_i, \tag{4}$$

When considering such models, nonstationarity of a time series can be prescribed to two different mechanisms; either a deterministic trend νt or a unit root. Where a deterministic trend is the cause of nonstationarity, econometricians refer to the time series as a trend stationary process; fluctuations around this deterministic trend are stationary in nature. Where a unit root is the cause of nonstationarity, the time series is referred to as difference stationary processes, due to the fact that difference operations will render the series stationary.

Now considering structural response variables that exhibit nonstationarity, any term in a descriptive model that continually increases or decreases in time is unlikely to mimic the behaviour of a stable structural response. This implies that to continue using an autoregressive type model, the nonstationarity of the variable must be described by a unit root process.

Many debates surround the idea of unit root modelling and its suitability to real life applications, within and outside economics (for example, there is currently much debate around using unit root processes to model the Earth's temperature change (see for example [5, 6, 7])).

The idea of a variance that increases with time, which is inherent to a unit root process, certainly sits uncomfortably when considering a structural response over time. However, while it may be that unit root processes do not ideally suit the dynamics of a structural response, the framework in which the cointegration theory has been borrowed from has been established on these types of models, and for the present it seems sensible to utilise what developed theory one can. If the ECM model structure can describe the variables of interest in SHM well enough, then the philosophical question on unit roots becomes unimportant, and the theory in the previous section can be applied without further ado. As, for the present, the interest is not in accurate forecasting of a structural response variable a potential model misspecification is less important.

For the purposes of utilising the cointegration theory, the authors suggest that so long as the variables of interest to an engineer follow a similar behaviour to a unit root process over a given time interval, then cointegration theory is applicable. In this case, the theory can happily be applied to process variables which are nonstationary with first difference stationary, which at the very least may be roughly checked visually with ease.

Returning to the previous example, the first difference of the easterly deck displacements at one point on the deck of the Tamar Bridge are plotted in Figure 3. Studying this figure the first difference appears to be a stationary process. As the variable is nonstationary with first difference stationary, the Johansen procedure is applicable.

In fact, it is only necessary that the first difference be *effectively* stationary i.e. that the longer time scale structure in the difference be highly suppressed relative to the shorter time scale behaviour characteristic of the structural dynamics. A relatively simple example can be used to argue the case, certainly for civil engineering structures. Suppose that the signal of interest has the form,

$$y(t) = A_s \sin(\omega_s t) + A_d \sin(\omega_d t) + A_y \sin(\omega_y t) \tag{5}$$

where ω_s is a characteristic frequency of the structural dynamics, ω_d is a daily frequency perhaps associated with daily temperature variations and ω_y is a frequency of one year and is associated with seasonal variations. Differentiating (5) trivially yields,

$$\dot{y}(t) = A_s\omega_s \cos(\omega_s t) + \omega_d A_d \cos(\omega_d t) + \omega_y A_y \cos(\omega_y t) \tag{6}$$

Now supposing ω_s to be of the order of 6.82 rad/s (1 Hz) for a typical civil structure, one has $\omega_d = 7.27 \times 10^{-5}$ rad/s and $\omega_y = 2 \times 10^{-7}$ rad/s. One obtains from (6),

$$\dot{y}(t) = 6.82 A_s \cos(\omega_s t) + 7.27 \times 10^{-5} A_d \cos(\omega_d t) + 2 \times 10^{-7} A_y \cos(\omega_y t) \tag{7}$$

and one sees that, even if the daily (trend) component of the original signal is 1000 times bigger than the structural component, in the differenced signal the trend component represents only about 1% of the total. This analysis seems to beg the question as to why one wishes to use cointegration at all rather than simply work with differenced signals. There are two good reasons why one should not differentiate in general:

(i) It is usually a bad idea to numerically differentiate experimental data as this will greatly amplify any high frequency noise components. In contrast, cointegration will at worst generate a weighted average of the noises.

(ii) Damage itself will often manifest as a trend in the data and differentiation would remove this also. In contrast, the appearance of a new trend will still (usually) destroy the balance imposed by the cointegrating vector and leave the damage exposed.

The authors believe that, although SHM variables may exhibit nonstationary behaviour over observable windows, in the nature of engineering, these response variables should be stationary

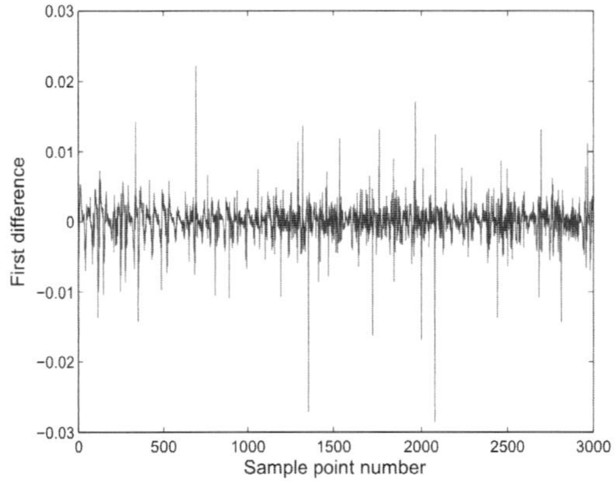

Figure 3: First difference of easterly deck displacements at one point as plotted in Figure 1.

when considering long periods of time. By design, engineering variables from stable healthy structures should be describable by a long run mean and a variance. In the long term one can therefore describe them as stationary, which enables one to use the regression techniques commonly applied in attempt to model a features dependence on an external condition, for example [8, 9]. The interests of SHM, however, are focused on response variables over shorter time periods, as any useful detection of structural degradation should be swift. One must therefore rely upon nonstationary theory and, in this work, cointegration. In the circumstance where a set of variables of interest include stationary and nonstationary measurements, the inclusion of the stationary variables doesn't invalidate the Johansen Procedure.

3. Conclusions

Econometricians understand nonstationarity in terms of trend stationary and difference stationary processes. This paper has argued that although neither approach may be philosophically suited for SHM feature variables, these feature variables do mimic the behaviour of difference stationary processes, for which the Johansen procedure was developed. With this in mind, no problems should arise with applying the theory presented in the previous chapter to SHM data.

References

[1] Cross E J, Worden K and Chen Q 2011 *Proceedings of the Royal Society: Part A* **467** 2712–2732
[2] Johansen S 1995 *Likelihood-based Inference in Cointegrated Vector Autoregressive Models* (Oxford: Oxford University Press)
[3] Cross E J, Worden K, Sturgeon R and Chen Q 2010 A tutorial on cointegration for engineers - a tool for non-stationary time series analysis *Proceedings of the 10th International Conference on Recent Advances in Structural Dynamics*
[4] Yule G 1926 *Journal of the royal statistical society* **89** 1–63
[5] Kaufmann R and Stern D 2002 *Journal of Geophysical Research* **107** 4012
[6] Verbeke T and De Clercq M 2006 *Environmental Modelling & Software* **21** 1447–1454
[7] Stern D and Kaufmann R 2000 *Climatic Change* **47** 411–438
[8] Peeters B and De Roeck G 2001 *Earthquake engineering & structural dynamics* **30** 149–171
[9] Sohn H, Dzwonczyk M, Straser E, Law K, Meng T and Kiremidjian A 1998 *Smart Systems for Bridges, Structures, and Highways* 127–138

Modern Practice in Stress and Vibration Analysis 2012 (MPSVA 2012) IOP Publishing
Journal of Physics: Conference Series **382** (2012) 012047 doi:10.1088/1742-6596/382/1/012047

Novelty detection applied to vibration data from a CX-100 wind turbine blade under fatigue loading.

[1]N. Dervilis, [3]M. Choi, [1]I. Antoniadou, [2]K.M. Farinholt, [2]S.G. Taylor, [1]R. J.Barthorpe, [2]G. Park, [1]K. Worden and [2]C.R. Farrar

[1]Dynamics Research Group, Department of Mechanical Engineering, University of Sheffield, Mappin Street, Sheffield S1 3JD, England.
[2]The Engineering Institute, Los Alamos National Loboratory (LANL), Los Alamos, NM 87545, USA
[3]Department of Aerospace Engineering, Chonbuk National University

E-mail: N.Dervilis@sheffield.ac.uk

Abstract. The remarkable evolution of new generation wind turbines has led to a dramatic increase of wind turbine blade size. In turn, a reliable structural health monitoring (SHM) system will be a key factor for the successful implementation of such systems. Detection of damage at an early stage is a crucial issue as blade failure would be a catastrophic result for the entire wind turbine. In this study the SHM analysis will be based on experimental measurements of Frequency Response Functions (FRFs) extracted by using an input/output acquisition technique under a fatigue loading of a 9m CX-100 blade at the National Renewable Energy Laboratory (NREL) and National Wind Technology Center (NWTC) performed in the Los Alamos National Laboratory. The blade was harmonically excited at its first natural frequency using a Universal Resonant Excitation (UREX) system. For analysis, the Auto-Associative Neural Network (AANN) is a non-parametric method where a set of damage sensitive features gathered from the measured structure are used to train a network that acts as a novelty detector. This traditionally has a highly complex "bottleneck" structure with five layers in the AANN. In the current paper, a new attempt is also exploited based on an AANN with one hidden layer in order to reduce the theoretical and computational difficulties. Damage detection of composite bodies of blades is a "grand challenge" due to varying aerodynamic and gravitational loads and environmental conditions. A study of the noise tolerant capability of the AANN which is associated to its generalisation capacity is addressed. It will be shown that vibration response data combined with AANNs is a robust and powerful tool, offering novelty detection even when operational and environmental variations are present. The AANN is a method which has not yet been widely used in the structural health monitoring of composite blades.

1. Intoduction

New generation offshore wind turbines are introducing a demanding increase in the blade size which has brought together a chain of structural challenges. Reliability remains on issue for such structures and it is apparent that the making of a robust and online system of Structural Health Monitoring (SHM) is a dominant concern in the successful implementation of such alternative systems in the energy arena. Reinforced composite materials such as carbon fibre reinforced polymer (CFRP) or glass fibre reinforced polymer (GFRP), are leading the industry in the manufacture of blades as they are characterised by high strength–to-weight and stiffness-to-weight ratio [1].

The purpose of this study is to examine a damage detection approach for a 9m CX-100 blade under fatigue loading such as may be used for new generation wind turbine blades. The overall approach taken is to apply auto-associative neural networks with different architectures to FRF data.

2. Novelty detection analysis methods

The premise of novelty detection techniques is to seek the answer to a simple question; given a newly presented measurement from the structure, does one believe it to have come from the structure in its undamaged state? The objective of this study is to demonstrate the technique of novelty detection in the context of auto-associative neural networks (AANNs). Artificial Neural Networks (ANNs) offer a holistic, nonlinear parameterised mapping between a set of inputs and a set of outputs [2, 3].

The AANN is a type of multilayer perceptron whose target outputs are the same as the inputs. Generally, the auto-associative neural network consists of five layers including the input, mapping, "bottleneck", de-mapping and output layers. Also, here for comparison reasons outlier analysis based on the Mahalanobis-Squared Distance (MSD) and monte-carlo simulation, is used as a robust and simple statistical method. The theoretical analysis behind these methods is not described in this study. For this, as well as the critical issue of calculating the thresholds, the reader is referred to previous works [4, 5, 6, 7, 8, 9, 10] in turn, the definition of MSD is briefly given by the following equation:

$$D_i^2 = (x_i - \mu_x)^T \Sigma^{-1} (x_i - \mu_x). \tag{1}$$

where x_i is the potential outlier, μ_x is the mean of the samples observations and Σ is the sample covariance matrix. The mean and covariance matrix could be inclusive or exclusive measures.

In order to find the best network architecture the training data was tested for several different architectures, by applying at the same time an early-stopping criterion in order to avoid over-fitting problems and achieve a better generalisation. For this purpose a percentage of the input data was used for validation purposes and testing purposes in order to implement the early-stopping criterion.

3. Experimental overview of the testing blade

The full-scale fatigue test was performed between 11/08/2011 and 9/11/2011. The test was running continuously for approximately 8.5 million cycles until a visible crack was observed on 9/11/2011. The crack nature was a through-thickness crack appeared on the surface of the blade in the root area near to the leading edge. The blade is made of a fibre glass body (shell) and a carbon fibre spar cap with balsa wood core (small percentage of spar cap of root section is made of glass fibre with some carbon fibre layers in the thick skin). The specimen was excited at its first natural frequency at 1.8 Hz. Two saddle positions were implemented at 1.6 m and 6.75 m from the root and on 13/09/2011, 10/10/2011 and 18/10/2011 extra mass was added on the first saddle at 1.6 m, leading to increase of mass from 582.4 Kg initially to 642.3, 702.16, 762.04 Kg respectively. The structural health monitoring sensor system consisted of several different systems implementing active sensing and passive sensing [21], [22].

In this study active sensing measurements are used for the novelty detection methods. Using this active sensing system (LASER sensing system) two different sensor arrays were implemented (figure 2) called the INNER and OUTER sensor arrays; they consisted of 6 and 7 sensors respectively and an actuator was used in each of them (thick blue dot). The excitation frequency bandwidth was between 5 KHz and 40 KHz with a sampling rate of 96 KHz giving a resolution of 7200 spectral points for the FRFs that were measured and used for this study. The observations are between 11/08/2011 and 9/11/2011 corresponding to 565 for the INNER sensors and 534 for the OUTER sensors (crack observed at observation 560 for INNER and 529 for OUTER array).

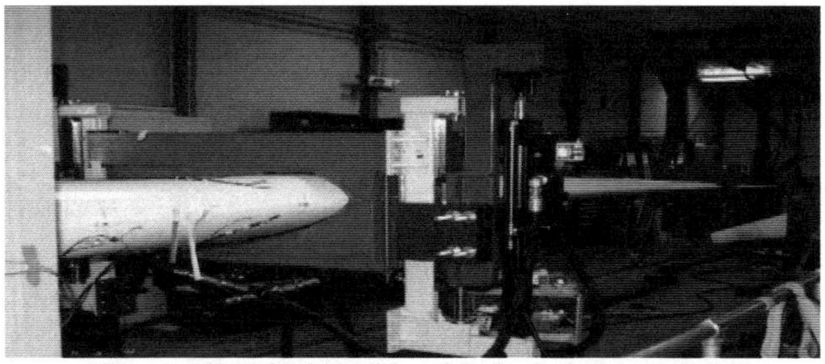

Figure 1. Wind Turbine blade fatigue testing.

Figure 2. INNER (on the left) and OUTER (on the right) sensor arrays.

4. Feature selection

In order to implement the novelty detection methods and overcome the "curse of dimensionality" a classification of features according to their ability to separate the normal from any other condition was assumed. For the purpose of this work only "strong" 50 dimensional features were selected. For a detailed analysis the reader is referred to the previous work [10, 15, 18, 19]. As was described in Section 3, the boundary conditions regarding the added mass were changed. During the analysis it was found that this change had influenced a limited number of the features (peaks) in the FRF spectrum. For this reason, a method to overcome the influence of external factors on normal conditions is introduced in Section 6.

5. Novelty detection results

After selecting the training, validation and testing data for each of the 13 sensors labelled INNER and OUTER, a well-trained AANN was used via unsupervised learning to generalise the normal condition of 50 dimensional features. Then a novelty distance (Euclidean distance [7,10]) was calculated by feeding the network with testing data [10]. In Figure 3, there is representative image of two sensors in the INNER and OUTER families showing the novelty results and in Figure 5, there is a summary graph of all sensors results. The blue bar shows when initially novelty was observed like Figure 4 and in some sensors there is a red bar that represents a monotonic different pattern as the days pass indicating a noticeable novelty change about 10 days before a visible crack was observed as in Figure3.

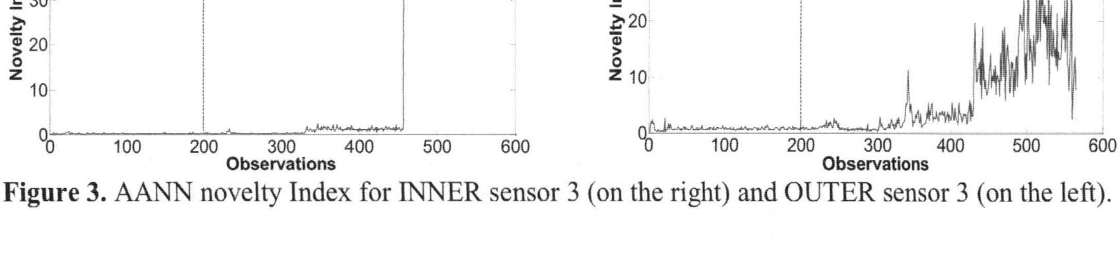

Figure 3. AANN novelty Index for INNER sensor 3 (on the right) and OUTER sensor 3 (on the left).

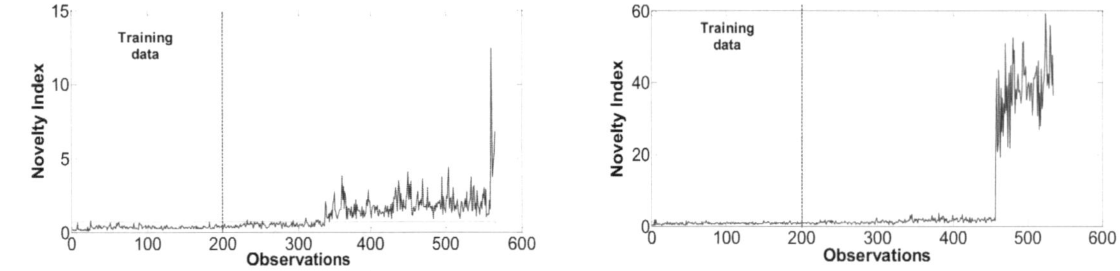

Figure 4. AANN novelty Index for INNER sensor 6 (on the left) and OUTER sensor 5 (on the right).

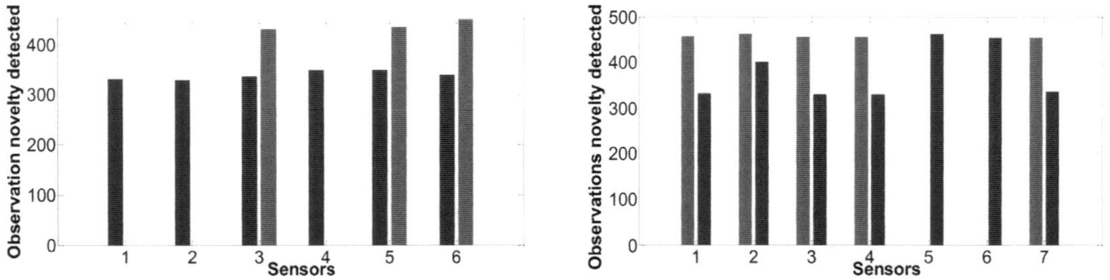

Figure 5. Summary of all sensor novelty results in INNER array (on left) and OUTER array (on right).

6. Features affected by change in boundary conditions

Through an extensive analysis of the FRF data it was found that a limited number of specific features (peaks on FRF) were affected by the mass change on the saddle during the fatigue test. Figure 6, represents an outlier analysis where the normal/trained data was selected directly after the 13/09/2011 which is the first added mass condition up to 05/10/2011 before the change of the second added mass, in order to visualise how the boundary conditions (black arrows) influenced this specific feature, see Figure 6 (left). In Figure 6 (right), are the results of a well-trained AANN and this shows the ability in overcoming the difficulties of the boundary changes by "learning" how to adapt on these external factors using only one node in the hidden bottleneck layer [9]. For this reason the trained data was selected so as to include some data after the second added mass on 10/10/2011. For comparison reasons with the same trained data as the AANN, the results from auto-association with only three layers with linear and non-linear transfer functions are presented in Figure 7.

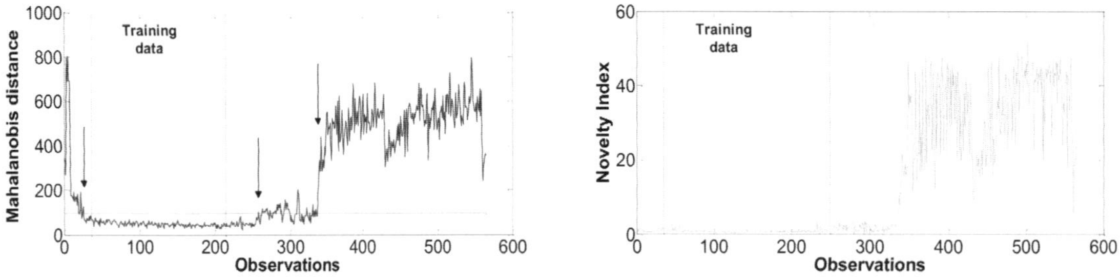

Figure 6. Outlier novelty index (on the left) and AANN novelty improved index (on the right).

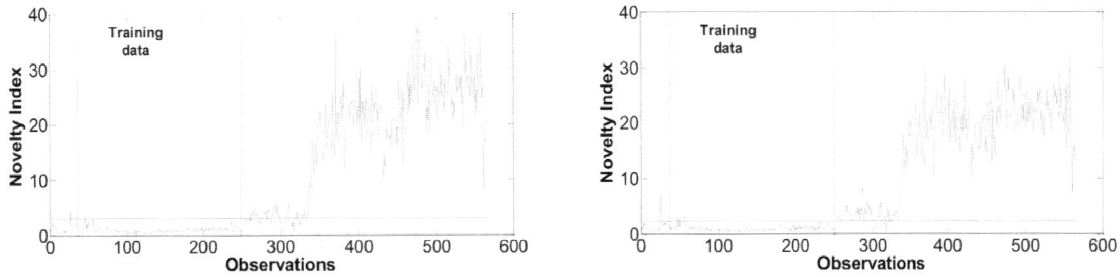

Figure 7. AANN with only three layers with linear (on the right) and non-linear (on the left) transfer functions.

7. Conclusion

In this paper a novel approach applied to wind energy systems was addressed. The ability of autoassociation of a three layer network with nonlinear and linear transfer function and their difference to generalise on simple changes in boundary conditions was investigated. The use of the AANN as a novelty detection algorithm has been shown to be effective in detecting alternate mechanisms during the continuous fatigue test. Differences between the sensors were noticeable in detecting the fault mechanisms. Active sensing measurements combined with novelty detection methods point out that the integration of AANN enables a quantitative and qualitative damage detection even when the system exhibits a range of normal conditions. The paper demonstrated the differences between linear and nonlinear activation function. It is noticeable that the one layer auto-association is doing a better classification than the linear auto-association, as it is able to generalise better. Feature selection was found not to be a trivial process and the number of dimensions of each feature warrants further discussion.

The authors believe that the initial damage was introduced internally and started from the main carbon spar. In the experiment, the excitation of the fatigue test was the first bending moment with a natural frequency at 1.8 Hz. This specific load is carried from the main spar and in high proportion at the blade root. If a failure occurs in the spar components then the blade's shell is not able to carry the high loading and usually damage will appear inevitably to the shell's body (as it happened). The shell mainly is for aerodynamic reasons but also its design is playing a small structural role by "helping" in stiffening and strengthening the spar.

The results are very encouraging for classification and pattern recognition purposes. The study raised many issues that warrant further attention. As discussed above, further considerations include factors such as variability, structure of the real blade, loading and environmental conditions, boundary conditions, feature selection and AANN architecture that will all affect the performance of the classifier.

Acknowledgments

The research was funded by the Department of Energy through the Laboratory Directed Research and Development program at Los Alamos National Laboratory. The authors would also like to acknowledge Scott Hughes and Mike Desmond from the National Renewable Energy Laboratory, and Mark Rumsey and Jon White from the Sandia National Laboratory for their support and guidance on this study. The authors gratefully acknowledge the support of the EU Marie Curie scheme through the Initial Training Network SYSWIND.

References

[1] B. D. Agarwal, L. J. Broutman and K. Chandrashekhara, Analysis and performance of fibre composites, Wiley, 2006.

[2] Deraemaeker, K. Worden, New trends in vibration based structural health monitoring, CISM Courses and Lectures, Vol.520, Springer-Verlag, 2010.

[3] K. Worden, G. Manson, N. R. Fieller, Damage detection using outlier analysis, Journal of Sound and Vibration, 229, 647–667, 1999.

[4] C. M. Bishop, Neural networks for pattern recognition, Oxford University Press, 1995.

[5] C. M. Bishop, Pattern recognition and machine learning, Springer Press, 2006.

[6] M. A. Kramer, Nonlinear principal component analysis using auto-associative neural networks. AIChE Journal, 37(2):233–243, 1991.

[7] K. Worden, Structural fault detection using a novelty measure, Journal of Sound and Vibration, 201(1), pp. 85-101, 1997.

[8] L. Tarassenko, A. Nairac, N. Townsend, I. Buxton, Z. Cowley, Novelty detection for the identification of abnormalities, International Journal of Systems Science, 31(11), pp.1427-1439, 2000.

[9] H. Sohn, K. Worden, C.F. Farrar, Novelty detection under changing environmental conditions, SPIE's Eighth Annual International Symposium on Smart Structures and Materials, Newport Beach, CA. (LA-UR-01-1894), 2001.

[10] N. Dervilis, R. Barthorpe, I. Antoniadou, W. J. Staszewski, K. Worden, Damage detection in carbon composite material typical of wind turbine blades using Auto-Associative Neural Networks, SPIE's Annual International Symposium on Smart Structures and Materials, 8348, 2012.

[11] M. Scholz and R. Vig´ario. Nonlinear PCA: a new hierarchical approach. In M. Verleysen, editor, Proceedings ESANN, pages 439–444, 2002.

[12] Bourlard, H., & Kamp, Y., Auto-association by multilayer perceptrons and singular value decomposition. Biological Cybernetics, 59, 291–294, 1988.

[13] Cottrell, G. W., & Munro, P., Principal component analysis of images via back propagation. In Proceedings of the Society of Photo-Optical Instrumentation Engineers. Cambridge, MA, 1988.

[14] N. Japkowicz, S.J. Hanson, M.A. Gluck, Nonlinear autoassociation is not equivalent to PCA, Neural Computation 12, 531-545, Massachusetts Institute of Technology, 2000.

[15] G. Manson, K. Worden, D. J. Allman. Experimental validation of a structural health monitoring methodology: Part III. Novelty detection on an aircraft wing, Journal of Sound and Vibration, 259 , 345–363, 2003.

[16] P. Filzmoser, A multivariate outlier detection method, Proceedings of the 7th International Conference on Computer Data Analysis and Modeling, Minsk, Belarus,18-22, 2004.

[17] N. M. M. Maia, J. M. M Silva, Theoretical and experimental modal analysis, Research Studies Press, 1998.

[18] G. Manson, K.Worden, D. J. Allman, Experimental validation of a structural health monitoring methodology: Part II. Novelty detection on an aircraft wing, Journal of Sound and Vibration. 259, 345–363, 2003.

[19] K. Worden, G. Manson, D. J. Allman, Experimental validation of a structural health monitoring methodology: Part I. Novelty detection on a laboratory structure, Journal of Sound and Vibration, 259, 323–343, 2003.

[20] Stuart G. Taylor et al., Full-scale fatigue tests of CX-100 wind turbine blades. Part II: analysis, SPIE's Annual International Symposium on Smart Structures and Materials, 8348, 2012.

[21] Kevin M. Farinholt et al., Full-scale fatigue tests of CX-100 wind turbine blades. Part I: testing, SPIE's Annual International Symposium on Smart Structures and Materials, 8343, 2012.

Modern Practice in Stress and Vibration Analysis 2012 (MPSVA 2012) IOP Publishing
Journal of Physics: Conference Series **382** (2012) 012048 doi:10.1088/1742-6596/382/1/012048

Damping estimation of a loosely supported single U-bend tube used in a steam generator

S Kaneko[1], R Kobayashi[1], T Watanabe[1] and T Nakamura[2]
[1] The University Tokyo, Hongo7-3-1, Bunkyo-ku, Tokyo
[2] Osaka Sangyo University, 3-1-1, Nakagaito, Daitou-shi, Osaka

E-mail: kaneko@mech.t.u-tokyo.ac.jp

Abstract. An evaluation method for estimating the damping of loosely supported single U-bend tube used in a steam generator colliding with a support plate is proposed. First, we performed experimental modal analysis and obtained natural frequencies, modes and damping without collision by Impulse Modal Test and then analysed natural frequencies, modes and damping with collision employing FEM analysis taking account of the collision force. In modelling the characteristics of the collision force, we applied Bijlaard's model for the spring constant and assumed hysteresis. After that, we performed experiments for measuring the damping ratio by changing the gap size and the support plate position. Comparison between calculated and experimental results is made which shows good agreement. Experimentally observed fact showing damping coefficient increases with the initial amplitude is well explained by theoretical model.

1. Introduction

It is well known that heat exchanger tubes used in a steam generator start to vibrate beyond critical speed. This phenomenon is called fluid-elastic vibration and the critical speed is usually estimated by Connors' equation(1) ,

$$\frac{U_c}{fD} = K\left[\frac{m\delta}{\rho D^2}\right]^{1/2} \tag{1}$$

where U_c, f, D, m, δ, ρ denote critical velocity, natural frequency, outer diameter, mass per unit length, damping ratio of single tube, density of the fluid and K denotes coefficient called critical factor. In order to estimate critical velocity accurately, evaluation of damping ratio under collision is a key issue(2),(3).

To evaluate damping ratio under collision, restitution factor is often introduced and model target selected is a cantilever pipe(4). The idea of restitution coefficient is convenient however it lacks energy dissipation process during collision. In this study, to propose more accurate method, we propose a method using energy loss coefficient proposed by Bijlaard(5),(6) and single U bend tube as a target.

First, we performed experimental modal analysis and obtained natural frequencies, modes and damping without collision by Impulse Modal Test and then analysed natural frequencies, modes and damping with collision employing FEM analysis taking account of the collision force. Finally, analytical results are compared with experiments.

---

[1] Department of Mechanical Engineering, The University Tokyo, Hongo7-3-1, Bunkyo-ku, Tokyo, Japan

2. Time history analysis of single U bend tube

As shown in Fig.1, we first performed experimental modal analysis and obtained natural frequencies, modes and damping without collision by Impulse Modal Test. Information on target object is tabulated in Table 1.

Fig. 1 Single U bend tube and Impulse Modal Test

Table 1 Major dimensions

Material	Stainless Steel
Height	1510mm
Outside Diameter of the tube (D_T)	22.23mm
Inside Diameter of the plate (D_B)	24.23mm
Thickness of the tube	1.27mm
Radius of curvature of bending beam	381.1mm
Clearance between tube and hole	1.0mm

Then, time history under collision is analysed by Finite Element Method taking collision effect into consideration. In our model, collision effect is represented by a linear spring shown in Fig.2.

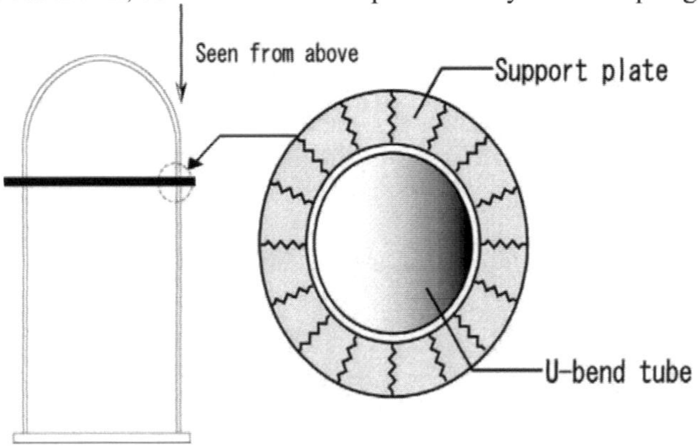

Fig.2 Collision model

Introducing virtual displacement as shown in Fig.3 and assuming hysteresis characteristics of collision force as shown in Fig.4, then radial force acting on the surface can be described by Eq.(2).

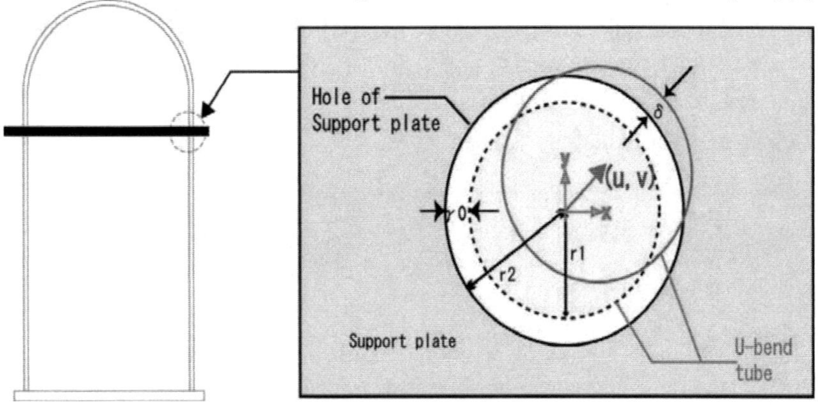

Fig.3 Virtual displacement

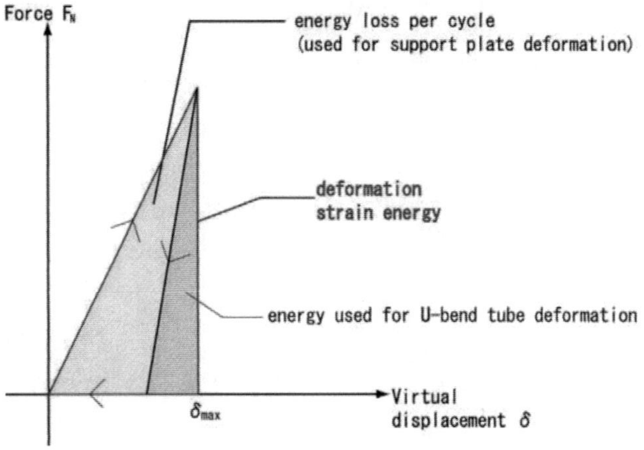

Fig.4 Hysteresis characteristics

$$F_N = \begin{cases} k \times \delta \quad (\dot{\delta} \geq 0) \\ \max[k\delta_{\max} - k(\delta_{\max} - \delta)/(1-\alpha)\ ,\ 0] \quad (\dot{\delta} < 0) \end{cases} \tag{2}$$

In the above equation, k denotes spring constant determined based on Bijlaard[(5),(6)], δ denote virtual displacement defined by

$$\delta = \gamma - \left(D_B - D_T\right)/2 \tag{3}$$

where γ, D_B and D_T denote radial displacement of a tube, inside diameter of the plate and outside diameter of the tube, respectively.

In addition, $\delta_{\max}$ in Fig.4 is defined as corresponding virtual displacement when time derivative of δ is equal to zero and α in Eq.(2) is defined as energy loss factor representing the ratio between strain energy and hysteresis damping energy.

Then, equation of motion for a single U bend tube taking collision effect into account can be written as

$$[\mathbf{M}]\{\ddot{\mathbf{U}}\}+[\mathbf{C}]\{\dot{\mathbf{U}}\}+[\mathbf{K}]\{\mathbf{U}\}=\{\mathbf{F}\} \quad (4)$$

where $[\mathbf{M}],[\mathbf{K}]$ denote mass and stiffness matrices and $\{\mathbf{U}\},\{\mathbf{F}\}$ denote nodal displacement and nodal force obtained by FEM analysis. When we calculate time history, we inserted the data obtained by experimental modal analysis without collision to damping matrix $[\mathbf{C}]$.

In Fig.5, one example of calculated transient virtual displacement is shown.

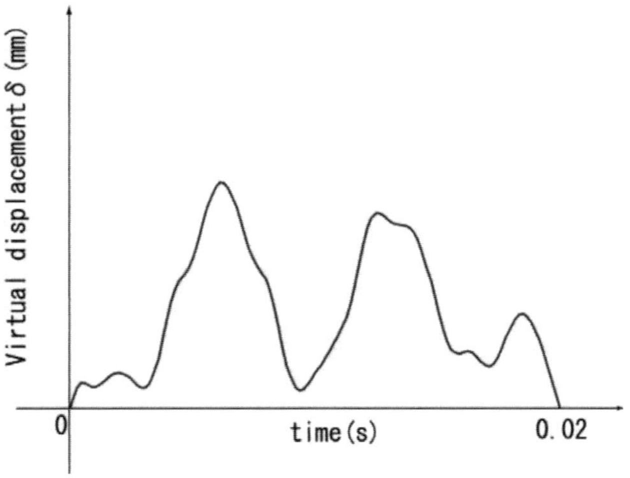

Fig.5 Transient virtual displacement

3. Experiment with single U bend tube under collision

In Fig.6, experimental setup and coordinate system is shown. Single U bend tube is fixed by two blocks and support plate is equipped in between the top of U bend tube and blocks. Gap between support plate and U bend tube is set 1mm.

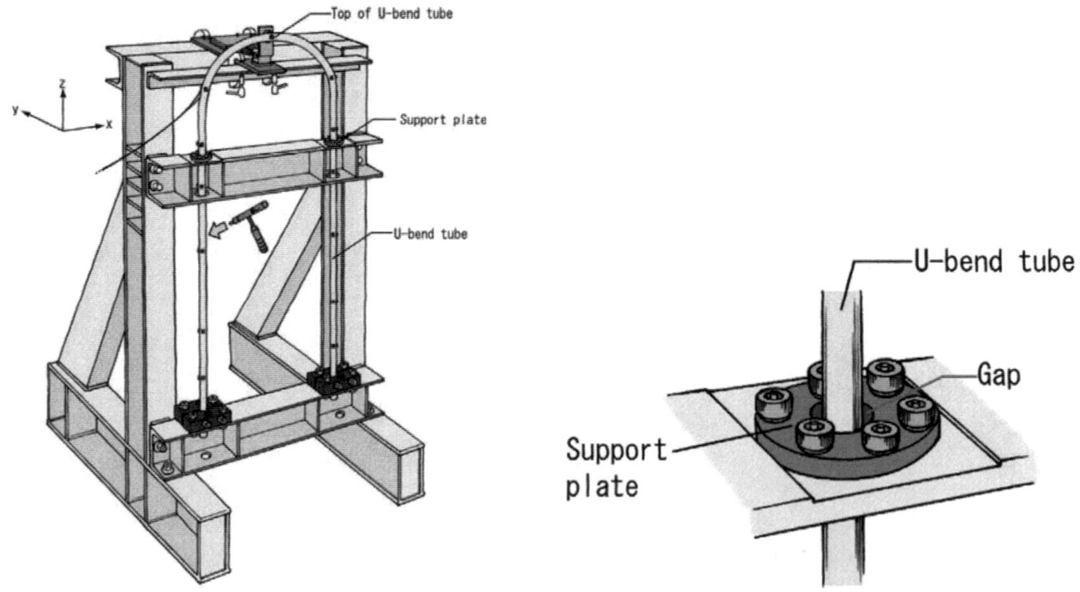

Fig.6 Experimental setup

Fig.7 shows transient wave form in the direction of y measured on the top of U bend tube when applied initial amplitude is 2mm. One can find that the envelope of decay during collision corresponding to the first three wave forms in Fig.7 is almost linear but not exponential. In addition, higher mode is included as shown in enlarged wave form of Fig.7.

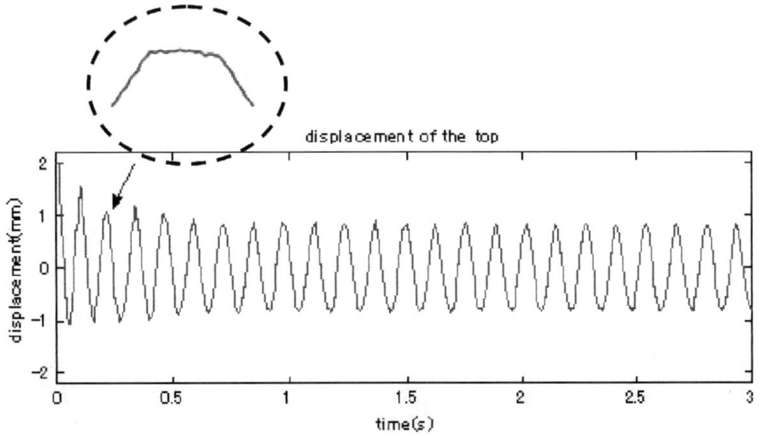

Fig.7 Experimentally obtained time history of U bend tube (Initial Disp.=2.0mm)

4. Analytical results

In Fig.8, calculated results are shown in two cases with energy loss factor α is 0.25 and 0.5. In the calculation, the value of α was determined by logarithmic decay of time history. Comparing Fig.7 and Fig.8, similar trend on the envelope can be found. In case of low energy loss factor which means duration time of collision is longer than the case of high energy loss factor, fundamental period of vibration becomes shorter than the case of high energy loss factor. Therefore, we can conclude the trend that the fundamental natural frequency becomes higher is exerted by the increase of strain energy due to the local deformation when a tube collides with a support plate.

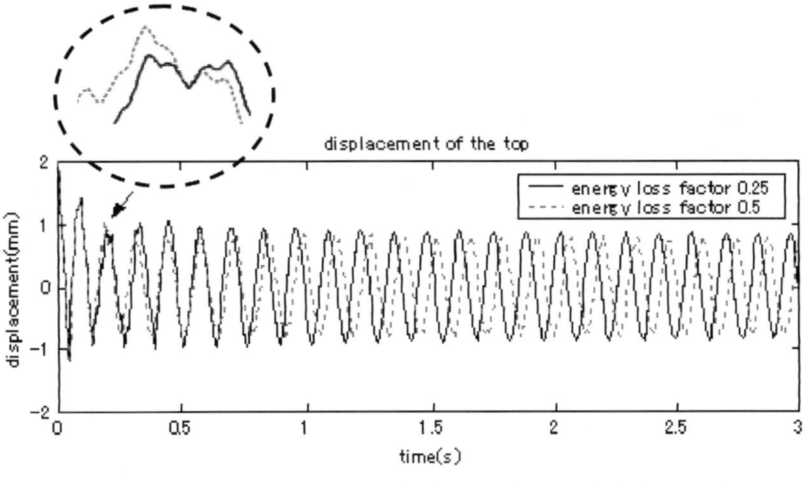

Fig.8 Calculated time history of U bend tube (Initial Disp.=2.0mm)

Finally, in Fig.9, calculated and experimentally obtained damping ratio are shown against initial displacement. Both results show good agreement.

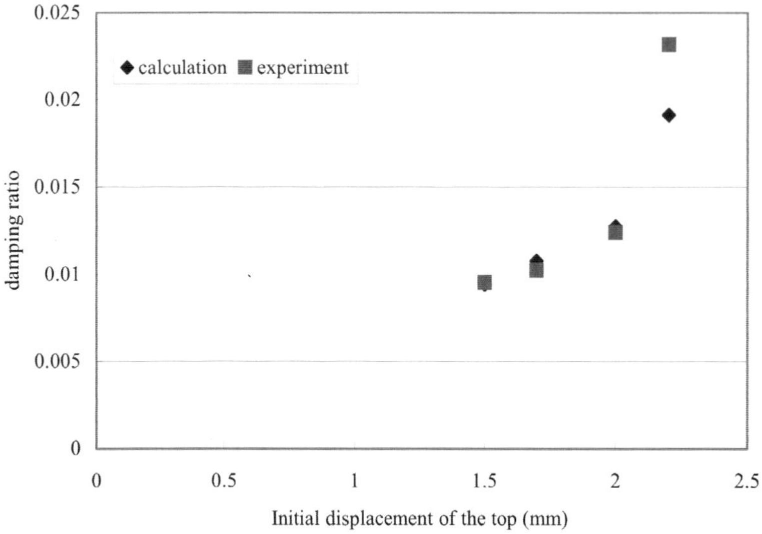

Fig.9 Average damping ratio - initial displacement dependency

Let us explain more on the detail of vibration. In the case that initial displacement is 1.5mm and 1.7mm, U bend tube collides with a wall with one end and in the case of 2.0mm and 2.2mm, U bend tube collides with a wall with both ends. Therefore, higher damping ratio is realised in the case of larger initial amplitude.

5. Conclusion

In this paper, the method for evaluating damping ratio of single U bend tube vibration taking account of collision against support plate is proposed. Comparing calculated results with experimental results, this method is found to be effective for the purpose of accurate damping evaluation.

Reference

[1] Connors H J 1980 ASME PVP **41** 93
[2] Rogers R J and Pick R J 1975 Nuclear Eng. and Design **36** 81
[3] Nakamura T and Fujita K 1989 Japan Trans JSME Ser.C **55-516** 1878
[4] Endo M and Oka T 1986 Japan Trans JSME Ser.C **52-477** 1520
[5] Bijlaard P P 1955 Trans ASME 805
[6] Bijlaard P P 1954 Welding Research Supplement 615

On the Development of a Damage Detection System using Macro-fibre Composite Sensors

R Pullin[1], M J Eaton, M R Pearson, C Featherston, J Lees, J Naylon, A Kural, D J Simpson and K Holford

Cardiff School of Engineering, Cardiff University, The Parade, Newport Road, Cardiff, Wales, UK, CF24 3AA

Abstract. Macro-fibre composite (MFC) sensors, originally developed as actuators by NASA, have been investigated for three components of a damage detection system for composite structures; actuation, sensing and energy harvesting. MFC sensors are constructed from piezoelectric fibres embedded in an epoxy matrix and offer greater flexibility than traditional sensors for embedding due to their low profile and low weight. It is proposed that embedded MFCs could be used to act as damage detectors, whilst energy either transmitted ultrasonically or harvested ambiently could be used to power the system. To assess the applicability of the MFCs a scale A320 composite wing was manufactured. Ten MFC sensors were embedded within the wing structure. Through a series of investigations on the wing the use of MFCs as part of an acousto-ultrasonic (AU) and Acoustic Emission (AE) damage detection system were investigated. Utilising AE source location and an AU cross-correlation techniques damage induced by impact was identified. In a further experiment the capability of transmitting and harvesting energy with the same embedded MFC actuators was completed. By impedance matching it was possible to improve the transmitted power. Furthermore an analysis of the MFCs ability to capture ambient vibrations, associated with aircraft structures, was completed. The completed experimental work demonstrated that it would be possible to embed sensors, energise them through active or passive vibration, and detect damage.

1. Introduction

Structural Health Monitoring (SHM) systems offer asset managers a distinct advantage over routine maintenance programmes. The ability to complete prognosis and diagnosis during service allows increased asset use, reducing maintenance costs and furthermore structures can be designed closer to the critical limit which in the example of aircraft can reduce weight and hence emissions.

The application of SHM is critical to composite structures where Barely Visible Impact Damage (BVID) can lead to catastrophic failures. The use of composites is increasing in aerospace applications and in the manufacture of wind turbine structures where the ability to remotely monitor damage is a necessity.

Acoustic Emission (AE) and Acousto-Ultrasonics (AU) offer a real possibility for an SHM system. AE is the most sensitive form of Non-Destructive Testing (NDT) and utilises an array of sensors to detect and locate the stress wave emitted due to damage propagation. AE can be regarded as a passive NDT technique. The AE technique has been applied previously to identify impact damage [1], matrix cracking [2] and delamination and damage growth in composites. AU utilises a pulse and receive

[1] pullinr@cardiff.ac.uk

Published under licence by IOP Publishing Ltd

approach for damage detection. The approach uses the same AE sensor array and is used to validate any detected and located AE sources. AU is often referred to as an active technique.

Any SHM system should ideally be embedded. The structure itself then acts as protection to environmental damage and limits the possibility of being damaged during routine service. Traditional piezoelectric sensors are not suitable for this application due to their geometry. In addition to prevent extra weight being added to a structure through cabling energy harvesting (EH) can be incorporated. EH can be active or passive depending if ambient vibrations are captured and convereted to energy or whether energy is transferred through the structure using ultrasonic vibrations.

Macro-fibre composite (MFC) transducers are a piezoelectric transducer manufactured by Smart Materials Corp [3]. Unlike traditional piezoceramic sensors they are constructed from piezoceramic fibres with a square cross-section. The fibres are aligned and held in a structural epoxy that is sandwiched between a polyimide film with printed interdigitated electrodes (Figure 1). To reduce electrical noise in sensing applications the MFCs are supplied with a very thin film of a copper/tin alloy to provide shielding.

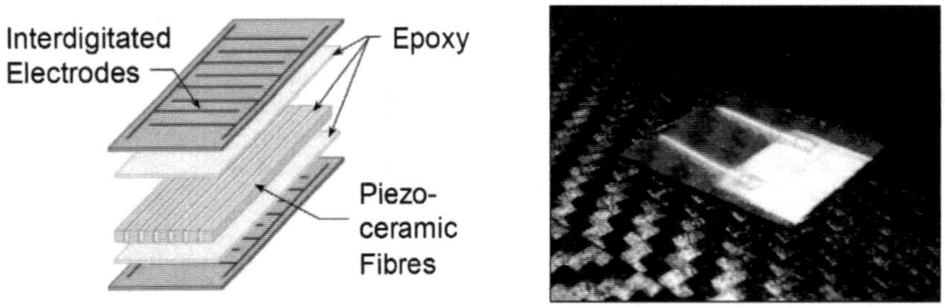

Figure 1: Schematic of MFC sensor and MFC embedded in glass fibre and carbon composite laminate.

2. Embedded MFC Carbon Wing Manufacture

A two part wing mould was designed and manufactured at Cardiff School of Engineering. The two parts of the wing were fabricated using four layers of woven carbon fibre (CFRP) and glass fibre (GFRP) composite with a 0-90° layup. Ten shielded MFC sensors (28x14 mm) were embedded between two GFRP plys to give a stacking sequence of CFRP, GFRP, sensors, GFRP and CFRP. Embedding the MFC between the GFRP avoids grounding loop issues. The wing parts were then cured in the University autoclave. Post manufacture, the embedded wing was c-scanned to ensure no damage was present. Finally stiffeners made from CFRP/Nomex honeycomb were added.

Figure 2: Schematic of wing showing embedded sensors (rectangles) and impact location (circle)

3. Experimental Procedure

Prior to impact testing background AU baseline data was captured. A series of signals were pulsed from each MFC and captured at the remaining sensors [4]. In addition a novel AE source location technique, AIC-delta-T [5,6], was used to identify damage. The technique utilises a 'touch and learn' mapping approach to identify the damage position. To assess the performance of the novel approach against the traditional approach artificial sources [7] were created at various points. The wing was then subjected to increasing impact events (2.3-6J) using an instrumented impact test machine. Impact testing was stopped once damage was observed on the inner face of the wing. AE was recorded throughout all impacts and post damage a further set of AU results were captured and the wing C-scanned again. A disbond of $100m^2$ was identified. To assess the ability of the embedded MFCs to harvest both active and passive energy two further experiments post impact were completed. Initially a 20 V peak to peak signal over a frequency range of 10-200 kHz was used to drive one transducer whilst energy was captured at the reaming sensors. The use of inductors to impedance match the devices at 65kHz was also used. Finally, based on real flight data, the wing was subjected to vibrations of 10-400 Hz and the energy harvested.

4. Results and Discussion

The results of the AE source location investigation using the embedded sensors is shown in Figures 3 and 4. It is evident from the results that in the majority of cases the novel approach outperforms the traditional method. The traditional approach assumes a single propagating velocity which is not the case in composite materials.

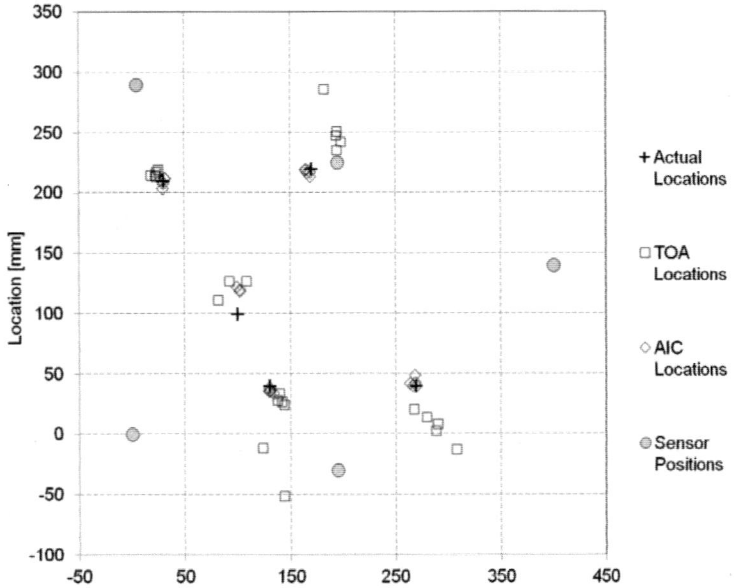

Figure 3: Comparison of AIC delta-T and time of arrival AE source location of artificial sources

Figure 5 presents waveform traces captured by the closest MFC from non-damage and damage inducing impact events. Superimposed over the waveforms is the load trace from the impact event. The load trace on the right hand image shows distinct drops, beyond the noise, indicating damage which coincides with an increase in the waveform amplitude. It is therefore clear from the plots that it is possible to identify when damage has occurred.

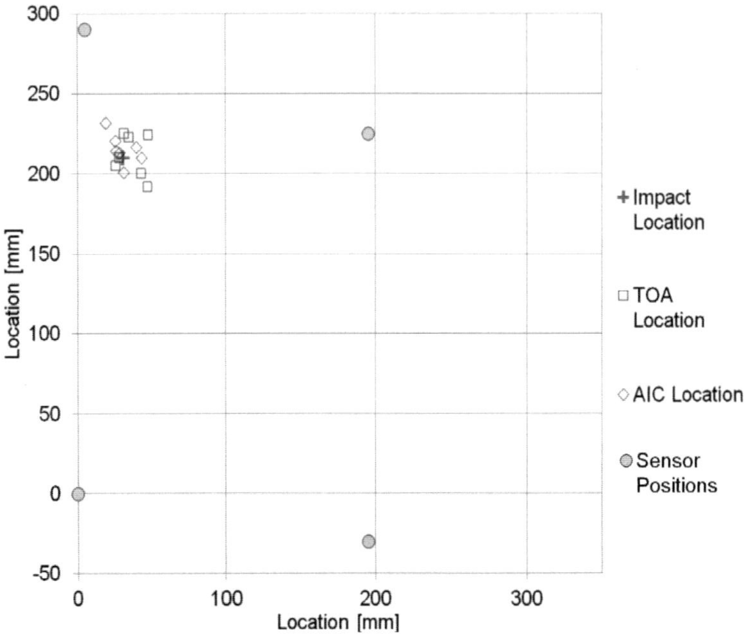

Figure 4: Comparison of AIC delta-T and time of arrival AE source location of impact

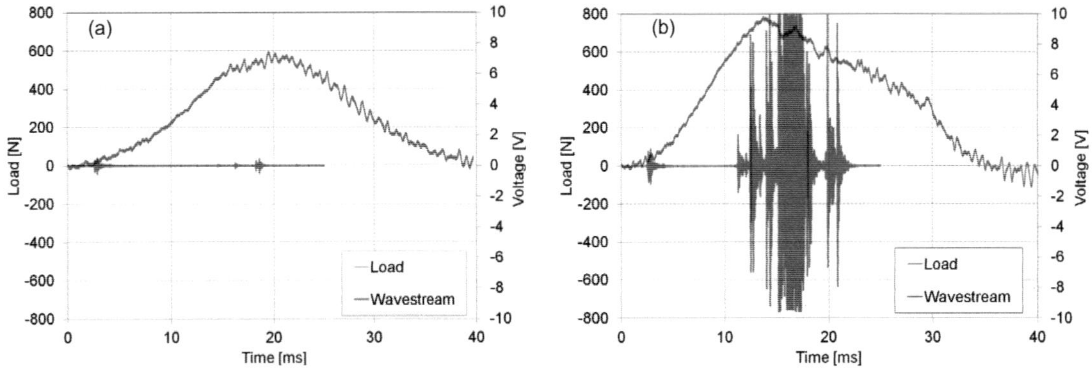

Figure 5: Comparison of waveforms recorded during non-damage (a) and damage impact events (b)

Figure 6 presents the AU results using the sensor closest to the impact position pulsing and the nearest four embedded sensors receiving. Channel 6 is a commercial sensor surface mounted for comparison. Three sets of data are presented, the cross correlation coefficient of the pre damage data set (black), the post damage data set (grey) and the post damage compared with the pre damage data set (white). A value of unity for the cross correlation coefficient indicates that two data sets are identical, whilst a value less than unity indicates a difference in the two sets [4]. By comparing the initial waveform recorded before impact with the waveform post impact an indication of damage can be found. It is evident from Figure 5 that damage has occurred in the structure.

Figure 7 shows the results of the active energy harvesting experiment. The power transmitted through the MFC is presented alongside the received power. It can be seen that the experimental results closely matches a simulation model developed at Cardiff, that forms part of a larger research programme [8].

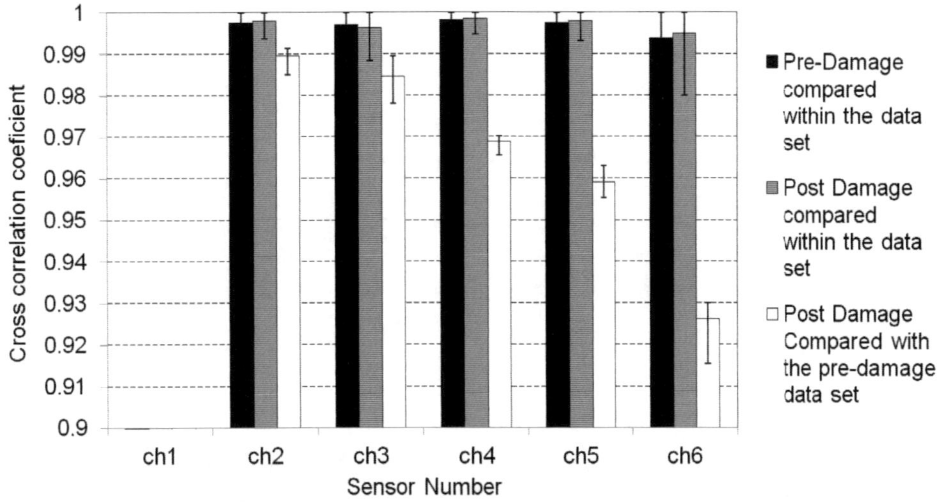

Figure 6: AU results after final impact

The results demonstrate that 118 μW can be transmitted for a drive of 20V peak to peak. Although this initially appears low it should be remembered that each MFC is receiving power of this magnitude and that the embedded devices have not yet been optimised. Furthermore the drive voltage could be significantly increased.

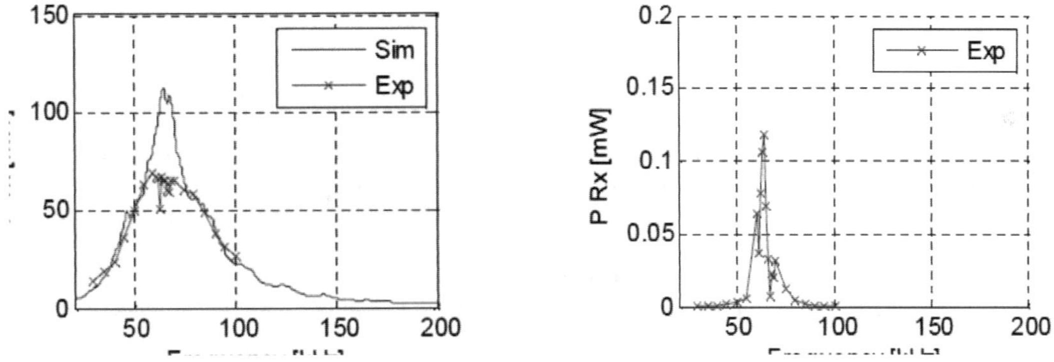

Figure 7: Active energy harvesting

Figure 7 presents the results of the passive harvesting experiment. As stated the wing was excited at 10-400Hz (aircraft data). It can be seen from the figures that 12μW peak and 5μW rms power was harvested from the MFC closest to the impact due to the vibration source. Again the MFC position has not been optimised for this geometry; however a numerical method to achieve this is currently being investigated. Currently 170 mW of power is required to use the MISTRAS Group wireless AE system. There is clearly a power gap between what has been investigated within this experimental program and what is currently required. However the MISTRAS system has been designed for bridge structures and could be further optimised electronically to increase efficiency. Furthermore the powers found here are worst cases. The position and size of the harvesters have not been optimised and the drive voltage could be increased. To ensure that a system could be installed within a wing structure a number of other aspects need investigating. Wireless technologies will be integrated to ensure data can be transmitted. In addition to vibrational sources, thermal gradients exist during flight cycles and these could be used to generate further power to reduce any gap. However to complete this an effective power management systems would be required. This is currently being investigated at Cardiff.

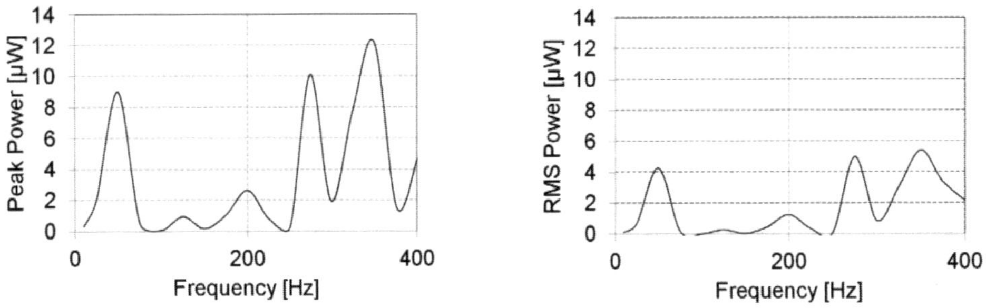

Figure 7: Passive energy harvesting

5. Conclusions

The completed research program on a composite wing with embedded MFCs has demonstrated true potential. Impact locations can be accurately located with a novel AE method whilst damage can be identified using both AE and AU. Furthermore the powering of an SHM system has been explored through active and passive harvesting. Further work needs to completed on power management and wireless technologies to ensure a full embedded system can be created.

6. Acknowledgements

The authors would like to acknowledge the support of the Cardiff School of Engineering Technicians.

7. References

[1] Wang, CS and F-K Chang 2000 Diagnosis of Impact Damage in Composite Structures with Built-in Piezoelectroc Network Proceedings of SPIE - *The Society Photo-Optical Instrum. Engineers* **3990** 13–19

[2] Michael May, Rhys Pullin, Mark Eaton, Carol Featherston, Stephen R. Hallett. An advanced model for initiation and propagation of damage under fatigue loading – part II: Matrix cracking validation cases, **93**, Issue 9, August 2011, Pages 2350–2357

[3] Williams, R.B. and D.J. Inman. An Overview of Composite Actuators with Piezoceramic Fibers. Smart Materials Corp Publication 2003 30\11\2009]; Available from: http://www.smart-material.com/Smart-choice.php?from=Whatis.

[4] M R Pearson, M J Eaton, C A Featherston, K M Holford and R Pullin. Impact Damage Detection and Assessment in Composite Panels using Macro Fibre Composites Transducers. 2011 *J. Phys.: Conf. Ser.* **305** 012049

[5] Baxter MG, Pullin, R., Holford, K.M., and Evans, S.L. Delta T Source Location for Acoustic Emission. *Mechanical Systems and Signal Processing* 2007;**21**:1512-20.

[6] Kurz JH, Grosse, C.U., and Reinhardt, H-W. Strategies for Reliable Automatic Onset Time Picking of Acoustic Emission and of Ultrasound Signals in Concrete. *Ultrasonics* 2005;**43**:538-46.

[7] Hsu, N.N. and F.R. Breckenbridge, Characterisation and Calibration of Acoustic Emission Sensors. *Materials Evaluation*, 1981. **39(1)**: p. 60 - 68.

[8] A Kural, R Pullin, C Featherston, C Paget and K Holford. Wireless power transmission using ultrasonic guided waves, *2011 J. Phys.: Conf. Ser.* **305** 012088

Identification of the Onset of Cracking in Gear Teeth Using Acoustic Emission

R Pullin[1], **A Clarke, M J Eaton, M R Pearson and K M Holford**
Cardiff School of Engineering, Cardiff University, The Parade, Newport Road, Cardiff, Wales, UK, CF24 3AA

Abstract. The development of diagnostic methods for gear tooth faults in aerospace power transmission systems is an active research area being driven largely by the interests of military organisations or large aerospace organisations. In aerospace applications, the potential results of gear failure are serious, ranging from increased asset downtime to, at worst, catastrophic failure with life-threatening consequences. New monitoring techniques which can identify the onset of failure at earlier stages are in demand. Acoustic Emission (AE) is the most sensitive condition monitoring tool and is a passive technique that detects the stress wave emitted by a structure as cracks propagate. In this study a gear test rig that allows the fatigue loading of an individual gear tooth was utilised. The rig allows a full AE analysis of damage signatures in gear teeth without the presence of constant background noise due to rotational and frictional sources. Furthermore this approach allows validation of AE results using crack gauges or strain gauges. Utilising a new approach to AE monitoring a sensor was mounted on the gear and used to continuously capture AE data for a complete fatigue load cycle of data, rather than the traditional approach where discrete signals are captured on a threshold basis. Data was captured every 10th load cycle for the duration of the test. A developed fast fourier transform analysis technique was compared with traditional analytical methods. In this investigation the developed techniques were validated against visual inspection and were shown to be far superior to the traditional approach.

1. Introduction

Detection and diagnostic methods for gear tooth faults is a fundamental research activity for military and aerospace organisations such as NASA [1-3]. In addition Health and Usage Monitoring Systems (HUMS) are mandatory for helicopter operators servicing, for example, the North Sea oil industry. Currently methods for detecting damage in transmission systems are predominantly based on vibration [4,5].

Any gear failure results in increased asset downtime and maintenance expenditure and at worst can cause catastrophic life threatening failure. It is essential therefore that any new techniques that can identify damage at earlier stages than vibration methods should be investigated and employed as part of a global Structural Health Monitoring (SHM) system.

Acoustic Emission (AE) monitoring is a possible alternate technique to vibration analysis. It is one of the most sensitive damage detection methods and is used widely in other applications such as bridge structures and pressure vessels. Furthermore it offers significant advantages in terms of early fault detection and diagnosis compared to other techniques [6]. AE is the detection of stress waves, released as a result of damage advancement, that propagate through a solid material as it undergoes stress.

[1] pullinr@cardiff.ac.uk

Compared with the reasonably mature applications of AE to structural monitoring, use of AE to monitor rotating machinery in general, and high speed, heavily loaded aerospace transmission gears in particular, is at the developmental stage. Researchers have investigated AE from spur gears [7-10] predominantly using measures of RMS AE levels, with some success in detecting gross changes. However a much clearer understanding of AE sources in gear contacts and radically improved signal analysis and interrogation methods are necessary to detect the onset of damage and to provide viable automated techniques for identifying damage onset in these applications.

The aim of this paper therefore is to build on previous investigations [11] and to further assess novel methods for identification in gear structures using AE.

2. Experimental Procedure

A bespoke test rig as described in [11] was manufactured to allow the static loading of an individual gear tooth pairing (Figure 1). A PANCOM acoustic sensor (50-450 kHz) was attached and coupled to the gear using cyanoacrylate. The response of the system to a Hsu-Nielsen source [12, 13] was used to ensure the correct sensitivity of the installed monitoring equipment. AE data was captured using a MISTRAS group PCI 2 data acquisition system. The system was used to record both AE transients and wave-streams for the life of a tooth subjected to fatigue loading (0.35-3.5 kN at 1Hz).

Figure 1. CAD Model of developed test rig.

Traditional AE data was captured throughout the test whilst a wave-stream was captured every 10th cycle using a micro-switch positioned under the loading arm and a decade counter connected to the external trigger of the acquisition system. Wave-streams, unlike traditional waveforms are independent of threshold and AE data is sampled over a user defined period, in this example one complete cycle. In contrast the traditional approach uses timing strategies to determine when a signal has surpassed and dropped below a threshold and a waveform is then recorded. Figure 2 presents an example wave-stream covering two cycles with the corresponding sinusoidally varying load superimposed. Visual observation of the onset of cracking was used to validate any findings.

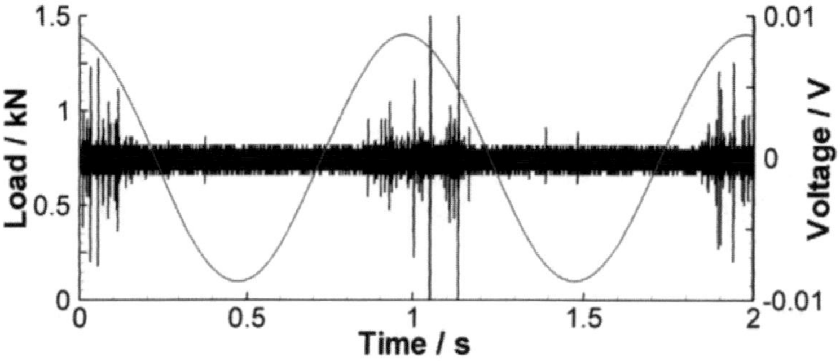

Figure 2. Example AE Wave-stream and discrete signal

3. Results and Discussion

The completed analysis has been separated into two distinct sections – *viz.* a traditional analysis that examines the amount of acoustic energy and the frequency of the detected acoustic signals and a novel approach that utilises wave-streams. Periodic visual observation showed that the crack started shortly after 22’000 cycles.

3.1 Traditional Analysis

Conventional analysis of AE data relies on the assessment of parameters that describe detected acoustic waves that exceed and drop below a user defined threshold as previously described. These parameters include measures of energy, amplitude, timing features such as rise to peak and duration and counts (number of threshold crossings). In theory, during a fatigue investigation, the amount of detected energy should remain linear when there is no crack present. Energy will be detected from load machine and test rig noise but this should remain constant on a cycle per cycle basis. However as a fatigue crack starts to initiate a new energy source will be observed, changing the rate of energy detected and hence an indication of the onset of cracking is noted. A similar pattern should be seen for the number of detected signals that pass the threshold, more commonly known as hits. Figure 3 shows the energy and hits detected above 900 N for the entire test.

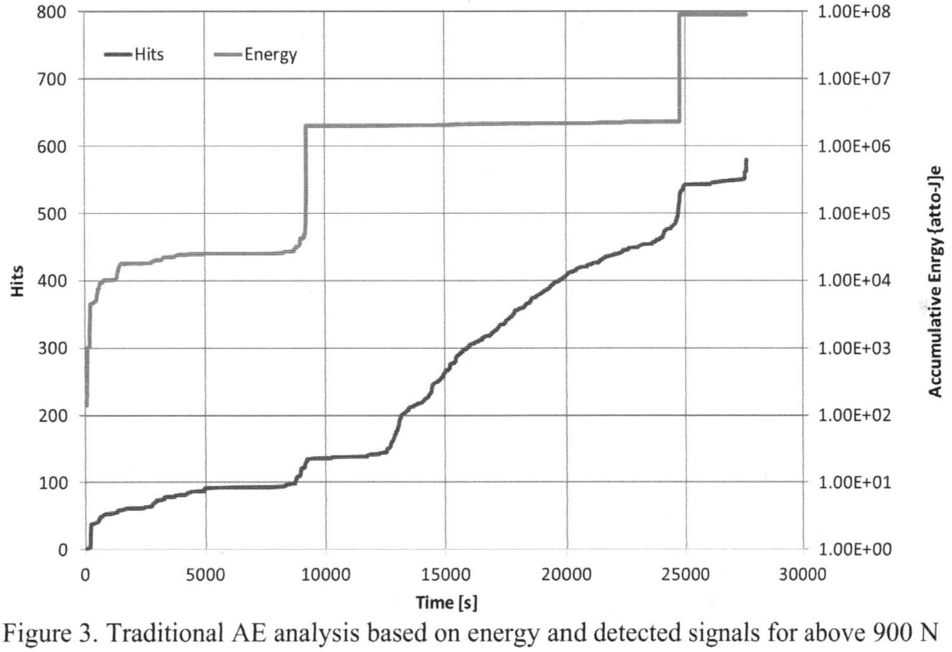

Figure 3. Traditional AE analysis based on energy and detected signals for above 900 N

It can be seen from the traditional approach that there is no clear change in detected signals after visual observation until just prior to 25’000 cycles. An initial assessment of all the detected signals showed no clear identification and therefore the signals detected above 900 N were considered. However there are also rises at approximately 0, and 9’000 and 13,000 cycles making automatic identification difficult.

3.2 Novel Approach

The captured wave-streams were used to perform further analysis. Initially, the RMS approach favoured by other researchers [7-9] was adopted. The RMS level of each wave-stream was calculated and can be seen in Figure 4.

Journal of Physics: Conference Series **382** (2012) 012050 doi:10.1088/1742-6596/382/1/012050

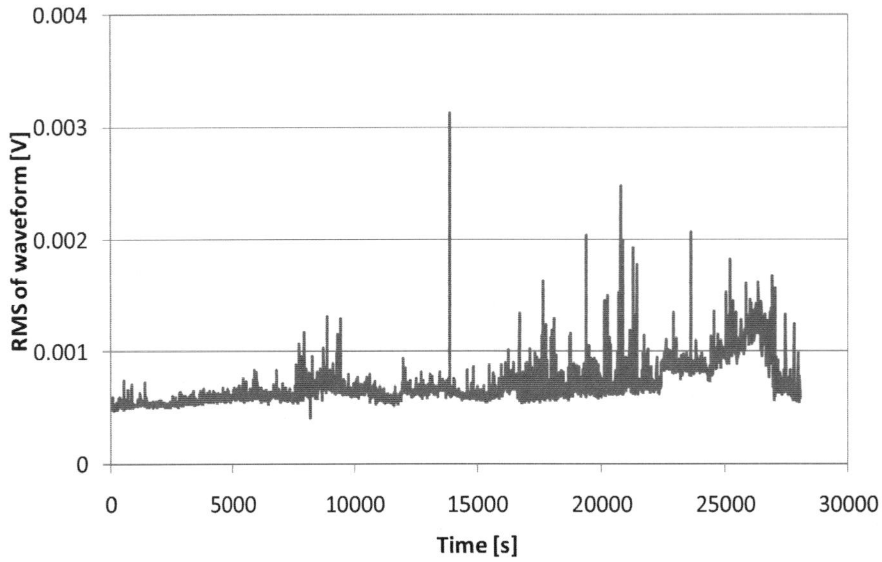

Figure 4: RMS level of wave-stream

It may be seen that whilst the RMS approach does demonstrate an overall increase in Acoustic Emission towards the later stages of the test, it does not clearly discern the onset of cracking in the gear tooth. Subsequently, a fast fourier transform (FFT) of an entire wave-stream was completed for every cycle. This FFT was then sub-banded into distinct frequency ranges/sub-bands. In theory noise due to rotating signals will occur at a distinct frequency band (in this example load machine noise) whilst damage will occur at higher frequencies, typically 100-150 kHz in metallic materials. The sub-banding process was designed to visually show how different noises dominate frequency bands throughout the test. Figure 5 shows the peak amplitude of the FFT of every wave-stream, captured through the duration of the test, separated into eleven sub bands. Each band was 20kHz wide, equally spaced between 80 and 300 kHz.

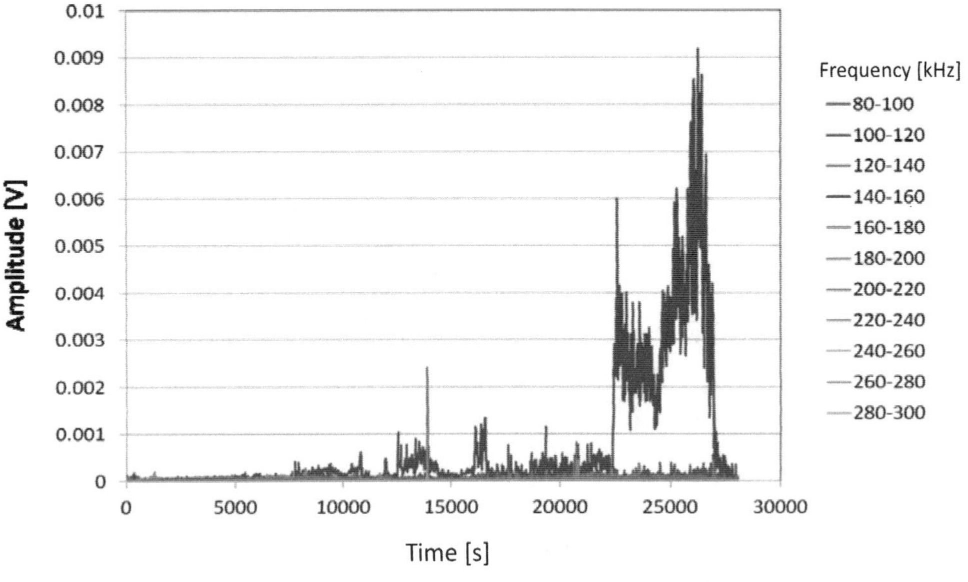

Figure 5. Peak value of FFT in distinct frequency bands

Figure 5 clearly shows a distinct rise in the peak amplitude of the FFT of the wave-stream in the 100-120 kHz sub-band. The plot shows that in this frequency range several peaks occur but particularly after the visual observation at 22,000 cycles there is a clear rise that continues to the end of the test, clearly crack growth. A further demonstration of the technique is presented in Figure 6.

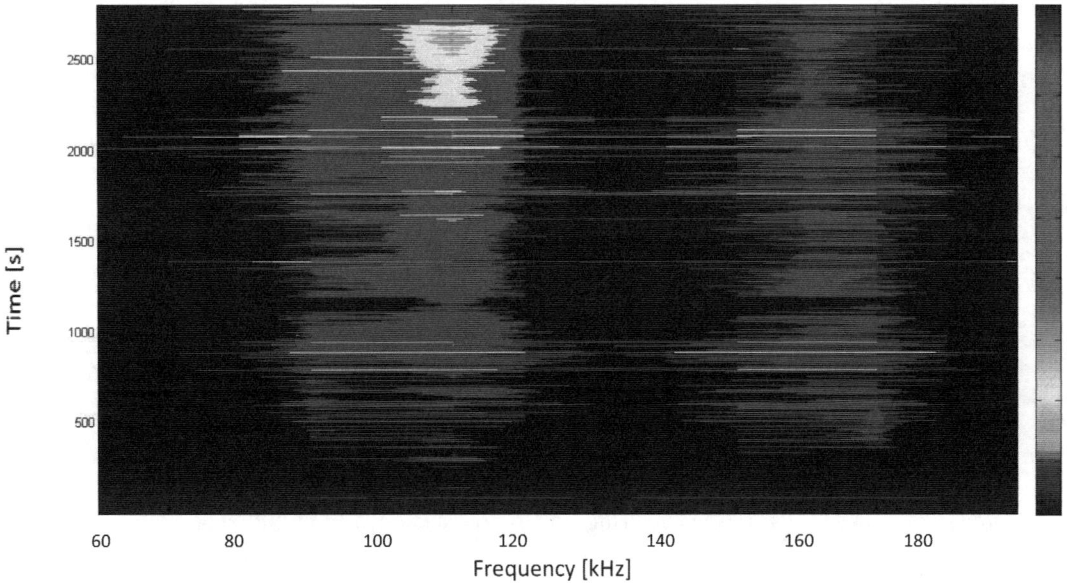

Figure 6. Intensity of FFTs in 100-120 kHz frequency bands

Figure 6 shows the intensity of each individual FFT in the 60-190 kHz frequency band plotted against test cycles. In this analysis, intensity is the sum of the amplitudes of the FFT at each frequency within the band. It is proposed that this banded approach is more suitable for complex shapes such as gear teeth where signal paths to the sensor may be subject to attenuations, that merely tracking a single frequency. An alternate approach using mean levels within each band identified similar trends.

The plot clearly demonstrates that something novel occurs shortly after 22'000 cycles and is therefore identifying the crack in the tooth. This demonstrates the sensitivity of the technique but importantly is visual, therefore a non-AE expert could easily identify a new mechanism happening in the gear system. In addition the sub-banding process allows background noise to be filtered, in this static gear rig, leaving only crack signatures. The same approach should function as effectively in rotating machinery, subject to further development. Figure 6 also demonstrates higher levels of intensity in the 100-120 kHz band from about 200 seconds onwards.

The novel technique presented has clearly demonstrated damage detection where a traditional approach failed. In addition the method outlined could offer further damage assessment opportunities including the ability to assess gear run in and possibly the detection of the breakdown in lubricant in gear boxes which will cause asperity contacts in the gears to develop. The techniques developed need to be assessed in rotating structures to validate these findings.

Although the process has focussed on gear structures it is evidently applicable to damage detection in any fatigue test where the identification of a novel source such as fatigue crack is important. The process will be trialled on other aerospace materials such as composites and will be the focus of future applications.

4. Conclusions

A novel test rig, designed to fatigue load an individual gear tooth, has been utilised to assess the performance of traditional AE analysis compared with a wave-stream approach. A traditional analysis could not easily identify the onset of cracking. Wave-streams captured very 10th cycle were used to identify fatigue cracking of the gear tooth. An assessment of the peak FFT value for distinct sub-bands of the captured data allowed a very visual approach to crack detection to be realised. Furthermore an intensity analysis of the one sub-band further demonstrated the technique developed. All results were validated against visual observation.

5. Acknowledgements

The authors would like to acknowledge the support of MISTRAS Group Ltd and the technical staff of the Cardiff School of Engineering. In particular the support of Paul Farrugia and Denley Slade for the design and manufacture of the decade counter micro switch trigger is acknowledged.

6. References

[1] Wong AK, "Vibration-based helicopter health monitoring – an overview of the research programme in DSTO", DSTO-HUMS2001, 2001.

[2] Decker HJ, "Gear crack detection using tooth analysis", NASA-TM2002-211491, 2002

[3] Decker HJ, "Crack detection of aerospace quality gears", NASA-TM2001-211492, 2002

[4] Larder BD, "Helicopter HUM/FDR: Benefits and Developments", *American Helicopter Society 55th Annual Forum*, Montreal, Canada, 25-27th May 1999.

[5] McColl J, "Overview of transmissions HUM performance in UK, North Sea Helicopter Operations", *in proceedings of the IMechE Seminar on Rotor Transmission System Health Monitoring* – experience update, November, 1997

[6] Loutas TH *et al.*, "Condition monitoring of a single-stage gearbox with artificially induced gear cracks utilizing on-line vibration and acoustic emission measurements", *Applied Acoustics*, 2009, **70**, 1148-1159.

[7] Toutountzakis T, Tan CK, Mba D, "Application of acoustic emission to seeded gear fault detection", *NDT E Int*, 2005, **38(1),** 27-36

[8] Tan CK and Mba D, "Investigation of the acoustic emission source during a comparative study on diagnosis of a spur gearbox", *Tribol. Int.*, 2005, **38(5)**, 469-480

[9] Tan CK and Mba D, "Experimentally established correlation between acoustic emission activity, load, speed, and asperity contact of spur gears under partial elastohydrodynamic lubrication", *Proc Inst Mech Eng Part J, J.* Eng Tribol, 2005, **219**, 401-409

[10] Loutas TH *et al.*, "A novel approach for continuous acoustic emission monitoring on rotating machinery without the use of slip ring", *Trans ASME J Vibration and Acoustics*, 2008, **130**, 064502-2

[11] Pullin R, Clarke A, Eaton M, Holford K M, Evans S L, and McCory J P," Detection of Cracking in Gear Teeth Using Acoustic Emission", *Applied Mechanics and Materials*, 2010, **24-25**, 45-50

[12] Hsu NN, Breckenridge FR. Characterization and Calibration of Acoustic Emission Sensors. *Mater Eval.* 1981, **39(1)**:60-8.

[13] ASTM. A standard guide for determining the reproducibility of acoustic emission sensor response. *American Society for Testing and Materials*, 2010;E976.

Non-linear characterisation of the physical model of an ancient masonry bridge

L Zanotti Fragonara[1], **R Ceravolo**[1,2], **E Matta**[1], **A Quattrone**[1], **A De Stefano**[1], **M Pecorelli**[1]

[1] DISEG, Politecnico di Torino, C.so Duca degli Abruzzi 24, 10129 Torino, Italy

Abstract. This paper presents the non-linear investigations carried out on a scaled model of a two-span masonry arch bridge. The model has been built in order to study the effect of the central pile settlement due to riverbank erosion. Progressive damage was induced in several steps by applying increasing settlements at the central pier. For each settlement step, harmonic shaker tests were conducted under different excitation levels, this allowing for the non-linear identification of the progressively damaged system. The shaker tests have been performed at resonance with the modal frequency of the structure, which were determined from a previous linear identification. Estimated non-linearity parameters, which result from the systematic application of restoring force based identification algorithms, can corroborate models to be used in the reassessment of existing structures. The method used for non-linear identification allows monitoring the evolution of non-linear parameters or indicators which can be used in damage and safety assessment.

1. Introduction

Non-linear identification is still scarcely applied to the identification of real structures, due to the large quantity of data required and to the difficulty of the problem. However, information about the nonlinear behaviour of a structure can be essential in the assessment of the seismic behaviour of a real structure. Moreover, changes in the structure's dynamic properties, including deviations from linearity, may provide quantitative evidence that damaging phenomena are underway.

The Restoring Force Surface (RFS) method was first developed by Masri & Caughey (1979) in order to achieve a non-parametric identification of non-linearity in structural systems. The basic idea behind the method is that the restoring force of a non-linear system can be expressed in terms of the state variables of the system itself. Once the restoring force surface is defined, it is possible to set the parameters of an analytical model in terms of displacements and velocities using a least square approximation. Masri & Caughey [1] at first used an orthogonal Chebyshev polynomial representation to construct a model of the restoring force.

This paper deals with the non-linear characterisation of an experimental masonry bridge. The structure underwent various damage steps, allowing for the collection of dynamic response data at each step. In particular, an electro-mechanical shaker has been used to apply different excitation levels in order to investigate the behaviour of the bridge at resonant frequencies, also in order to perform a non-linear identification of the structure based on a simple Single Degree of Freedom (SDoF) equation form. Thus, the shaker was placed so as to apply harmonic excitations at sensible points of the structure.

[2] Corresponding author. E-mail: rosario.ceravolo@polito.it

2. The experimental masonry bridge model

The 1:2 scaled model of the masonry arch bridge shown in figure 1 was built in the laboratory of the Department of Structural, Building and Geotechnical Engineering at the Politecnico di Torino [2,3] consisting of a twin-arch bridge with a length of 5.90m, a width of 1.60m and it is 1.75m high. The two arches are segmental arches with a radius of 2.00m and an angular opening of 30°. Each span is 2.00m long between the supports and the thickness of the arch is equal to 0.20m. The model was built with handmade clay bricks also scaled to 130x65x30mm to respect the adopted modelling scale law. Low compressive strength elements were chosen and a mortar with poor mechanical properties was used to bound them in order to reproduce the typical materials of historical constructions.

Figure 1. The scaled masonry bridge: notice the settlement application device under the central pier.

The mid-span masonry pier, which was cut at a hypothetical middle-height section to allow the insertion of a settlement application system, is imagined to be placed inside the streambed and subjected to the scour of its foundation.

Foundation settlements and rotations were applied on the bridge model by means of the four independent screws installed at the extremities of the settlement application system. The spherical plain bearings placed at the head of the screws allow the rotations of the plate which supports the central pier about axes parallel to the longitudinal and transversal directions of the bridge.

2.1. Dynamic experimental test

The experimental study involved three different testing campaigns. The first campaign regarded the undamaged structure: an extensive set of dynamic tests was carried out on the bridge model in order to characterise its "healthy" state. The second campaign started after the application of additional masses on the central pier, in order to take in account the weight of the missing part of the pier. In the same campaign the first four settlement steps were applied on the upstream side of the pier. Dynamic tests were conducted in correspondence of each settlement step. During the third campaign five further settlement steps were applied (11 mm total settlement).

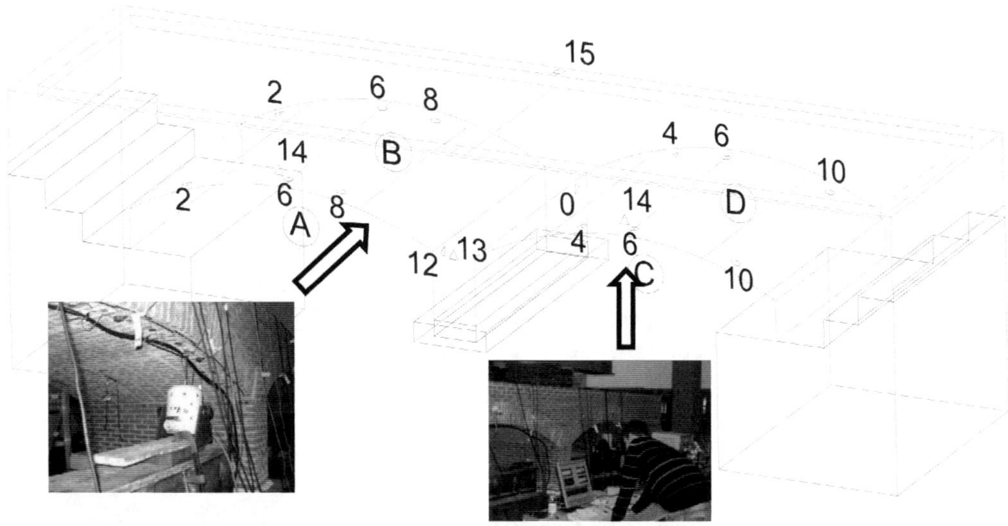

Figure 2. Shaker application in positions 9A and 6C (right), experimental setups for vibration tests (left).

Measuring points were selected in order to achieve a sufficient mode shapes resolution. The arch barrels were subdivided in 11 segments whose ends were assumed as measuring points for both the edge and the middle lines. Other 6 positions at the springing sections of the pier were materialised to capture the longitudinal displacements. The 4 mid-span sections of the arch barrels lateral faces and the 2 pier frontal faces were considered for the lateral and torsional modes. Finally, the 2 positions on the longitudinal spandrel walls at the middle section of the deck were added to identify the vertical modes.

Forced vibration tests were performed by using a shaker TIRA TV 51220, capable of supplying a rated peak force of 200 N. The force applied was acquired by using a mechanical impedance sensor PCB Piezotronics 288D01 (measurement range ±222.4 N pk). Figure 2 shows both the location where the shaker has been applied and the accelerometers' location on the structure.
Through a first linear identification, at each damage step, the modal frequencies were identified, in order to plan the forced vibration tests. In fact, at each modal frequency a harmonic excitation has been applied by using different force levels, usually 33 N, 66 N and 100 N for one minute. Both the force and the acceleration input to the system were acquired. Using the accelerometers, dynamic measurements were collected at 18 different locations, as shown in figure 2.

3. Nonlinear identification

3.1. Non-linear identification in principal coordinates

Classical linear theory allows the decoupling of motion equations in terms of modal or principal coordinates. On the other hand, several authors investigated the non-linear vibration of mechanical systems also in the principal coordinates - for instance see Bellizzi & Bouc [4] or Spottswood & Allemang [5] - though in such a case the equations of motion are coupled, as it is shown by the equation referred to the *k*-th coordinate:

$$\ddot{p}(t)_k + 2\xi_k\omega_k\dot{p}(t)_k + \omega_k^2 p(t)_k + \theta_k\left(p(t)_1, p(t)_2, \dots, p(t)_n\right) = U_{ki}F_i(t) \tag{1}$$

where p_k represents the principal coordinate, ξ_k the modal damping, ω_k the *k*-th natural frequency, $F_i(t)$ the force applied at position *i* in physical coordinates, U_{ki} the modal or basis vector and θ_k the non-linear component that couples the equations of motion.

In light of what has been shown, whenever the excitation is high enough to induce a non-linear response in the system, modal interaction and internal resonances between different modes are expected.

In some circumstances, especially in the case of resonant excitation applied to a weakly non-linear system, most of the energy is carried by a single resonant component, p_k, the other components being negligible. This means that the coupling in Equation (1) vanishes and the system approximately behaves like a SIMO (Single Input Multiple Output) system. In the specific case of the resonance tests conducted on the masonry bridge, the latter circumstance has been confirmed by experimental evidence [6]. In fact, while at higher excitation levels the presence of internal resonances was evident (peculiarly in the 100 N excitation case), all resonance peaks were still related to the super-harmonics of the excited frequency, the coupling components being excluded [6].

For a non-linear characterisation purpose, one can neglect the interactions among the principal responses and non-linearity may be ultimately studied through the following equation:

$$\ddot{p}(t)_k + 2\xi_k\omega_k\dot{p}(t)_k + \omega_k^2 p(t)_k + f_k\left(p(t)_k\right) = U_{ki}F_i(t) \tag{2}$$

where the non-linear modal component f_k is decoupled.

4. Non-linear polynomial identification from shaker tests

As previously stated, the bridge dynamic response has been treated as a series of SISO systems, using resonant excitation. In particular, the model proposed in equation (2) may be expressed also in physical coordinates. For an excitation applied at position i and a response measured at position j one has:

$$\begin{cases} \ddot{x}_{ki} + f_{nl}\left(x_{ki},\dot{x}_{ki}\right) + f_{hyst}\left(\dot{x}_{ki},f\right) = U_{kj}U_{ki}F_i(t) \\ f_{nl}\left(x_{ki},\dot{x}_{ki}\right) = \sum_{m=0}^{3}\sum_{n=0}^{3}\alpha_{mn}x_{ki}^{m}\dot{x}_{ki}^{n} \end{cases} \tag{3}$$

The lower excitation levels are expected to show a predominance of the linear behaviour. In the optimisation process, linear modal parameters, including a viscous equivalent damping, have been initialised to those available from a previous identification from ambient response signals (SSI algorithm) [2]. An initial estimate of the factor $U_{kj}U_{ki}$ has been conveniently obtained from the classical formula of viscous damping at resonant excitation.

4.1. Non-linear identification results

The results of the fittings are reported in figure 3 for one single response point that is representative of the dynamic behaviour of the structure relative to the first mode at 17.4 Hz (point 6C – see figure 2). Several nonlinear terms assume relevant values with the increase of the excitation level. The higher excitation level (100 N) reported in figure 3, shows clearly a nonlinear behaviour. Red lines in figure 3 correspond to the nonlinear fitting carried out by minimising the α coefficients in equation (3).

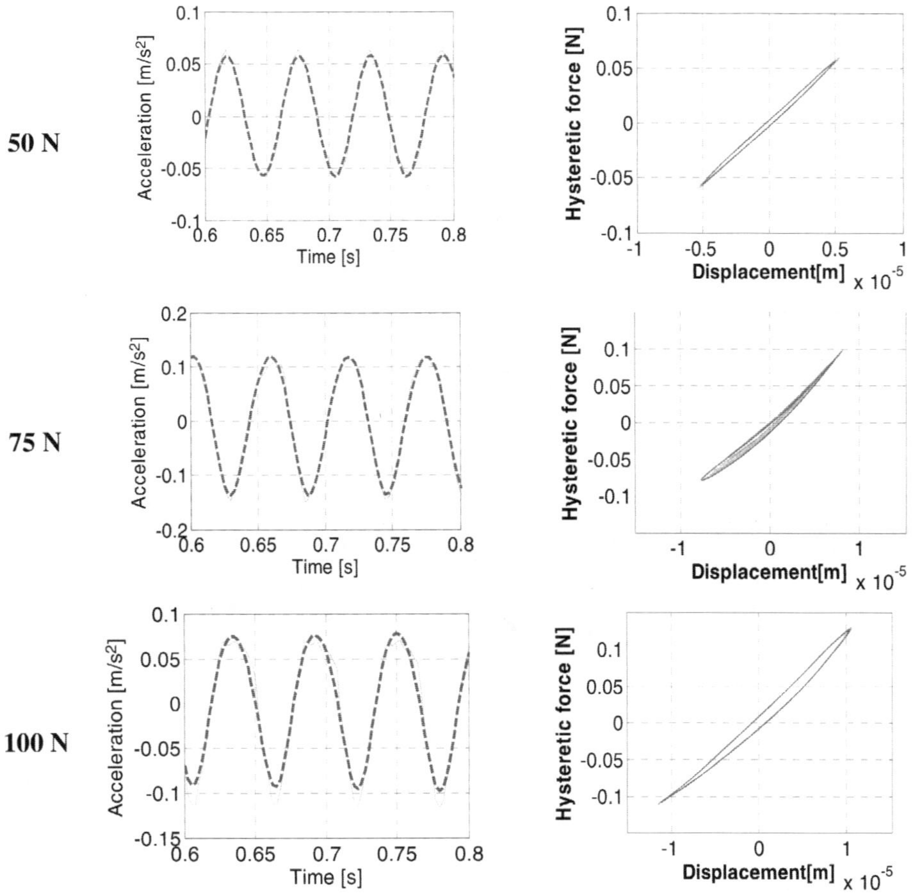

Figure 3. Comparison between experimental data and identified results for three excitation levels in terms of acceleration. The identified restoring force vs displacement plot is reported on the right side.

Figure 4 quantifies the polynomial coefficients as estimated for the various excitation levels. When the system is excited by a 50 N resonant force, only the viscous damping term is activated by the minimisation process, while at higher excitation levels the whole set of nonlinear terms is activated. Estimates for various coefficients are normalised to their maximum value. The value estimated for viscous damping is similar at different excitation levels (in Figure 4 this term seems to vanish, but this is due to normalization with respect to nonlinear coefficients). In fact, at 100 N the most significant nonlinear terms resulted to be the x^2, $x\dot{x}$ and $x^2\dot{x}$.

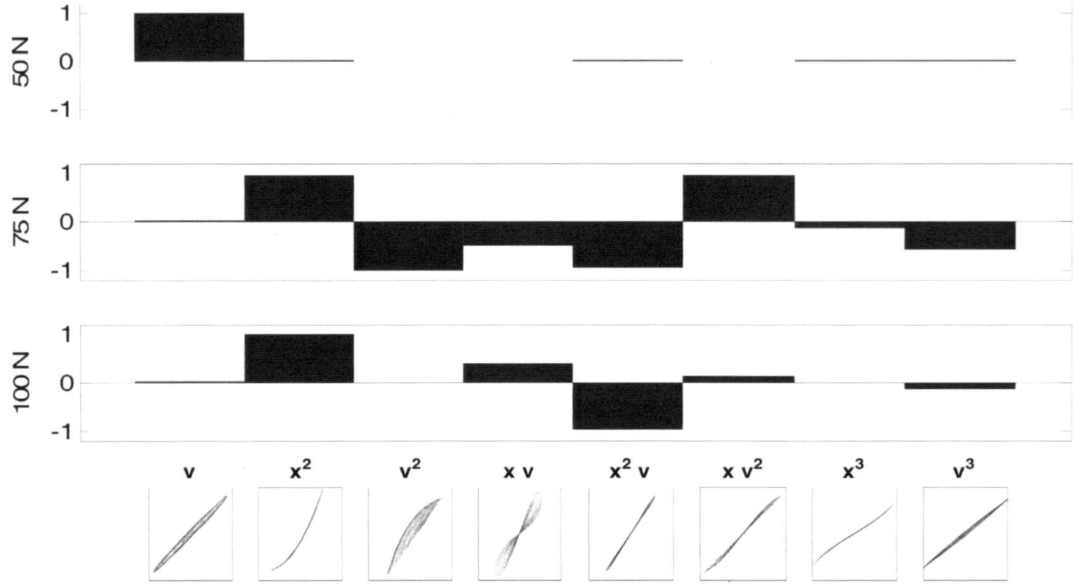

Figure 4. Nonlinear coefficients results for three different excitation levels (sixth damage step applied to the bridge). In the bottom part the corresponding nonlinear terms laws is highlighted.

5. Conclusions

In the last years, non-linear identification has been an area of increasing research interest. In the specific context of cultural heritage structures, this paper has presented a polynomial identification of the non-linear restoring force associated to the resonant response of a scaled model of a masonry arch bridge subjected to shaker dynamic tests at three different excitation levels. Research on the dynamics of masonry offers some perspectives for increasing the knowledge in the fields of structural reassessment of ancient structures, as well as damage detection procedures.

6. References

[1] Masri, S.F., Caughey, T.K., (1979); *A nonparametric identification technique for nonlinear dynamic problems*. J. of Appl. Mech., 46, pp. 433-447.

[2] Ruocci, G., (2010); *Application of the SHM methodologies to the protection of masonry arch bridges from scour*. Politecnico di Torino, PhD Thesis.

[3] Ruocci, G..D.L., Quattrone, A., Zanotti Fragonara, L., Ceravolo, R., De Stefano, A., (2011); *Experimental testing of a masonry arch bridge model subject to increasing level of damage*. Proceedings of 4th International Conference on Advances in Experimental Structural Engineering, Ispra.

[4] Bellizzi, S., Bouc, R., (1999); *Analysis of multi-degree of freedom strongly non-linear mechanical systems with random input Part I: Non-linear modes and stochastic averaging*. Probabilistic Engineering Mechanics, 14, pp. 229-244.

[5] Spottswood, S.M., Allemang, R.J., (2006); *Identification of nonlinear parameters for reduced order models*. Journal of Sound and Vibration, 295, pp. 226-245.

[6] Ceravolo, R., De Stefano, A., Quattrone, A., Zanotti Fragonara, L., Ruocci, G., (2011); *Using non-linear identification from shaker tests for structural reliability assessment*. Proceedings of EVACES 2011, Experimental Vibration Analysis for Civil Engineering Structures, Varenna, pp. 801-808.

Journal of Physics: Conference Series **382** (2012) 012052 doi:10.1088/1742-6596/382/1/012052

Nonlinear normal vibration modes in the dynamics of nonlinear elastic systems

Yu V Mikhlin[1], N V Perepelkin[1], A A Klimenko[1], E Harutyunyan[2]

[1]National Technical University "KhPI", Kharkov, Ukraine
[2]National Academy of Science of Armenia, Yerevan, Armenia

Yuri_Mikhlin@mail.ru

Abstract. Nonlinear normal modes (NNMs) are a generalization of the linear normal vibrations. By the Kauderer-Rosenberg concept in the regime of the NNM all position coordinates are single-values functions of some selected position coordinate. By the Shaw-Pierre concept, the NNM is such a regime when all generalized coordinates and velocities are univalent functions of a couple of dominant (active) phase variables. The NNMs approach is used in some applied problems. In particular, the Kauderer-Rosenberg NNMs are analyzed in the dynamics of some pendulum systems. The NNMs of forced vibrations are investigated in a rotor system with an isotropic-elastic shaft. A combination of the Shaw-Pierre NNMs and the Rauscher method is used to construct the forced NNMs and the frequency responses in the rotor dynamics.

1. Introduction

Kauderer [1] was a forerunner in developing quantitative methods to analyze the NNMs in two-DOF conservative systems. Rosenberg [2] defined NNMs as synchronous periodic motions during which all coordinates of the system vibrate equiperiodically, reaching their maximum and minimum values at the same instant of time. He selected a few classes of essentially nonlinear systems allowing NNMs with rectilinear trajectories (modal lines) in a configuration space. The first formulation of the NNMs can be named **the Kauderer-Rosenberg approach**. In general, the NNM modal lines are curvilinear. The power series method to construct these trajectories is proposed in [3,4]. Shaw and Pierre [5,6] proposed the other formulation of NNMs for a general class of nonlinear discrete conservative or non-conservative systems. This analysis is based on the computation of invariant manifolds on which the NNM oscillations take place. This NNMs formulation can be named **the Shaw-Pierre approach**.

Basic results on NNMs are presented in the book [7-9] where quantitative and qualitative analyses of NNMs in conservative and non-conservative systems are considered.

An efficiency of the NNMs method is showed in some applied problems. In particular, NNMs in vibro-absorption problems are investigated in [10-12]. An essentially nonlinear oscillator, a snap-through truss with three equilibrium positions, and a vibro-impact oscillator are considered as absorbers. Construction and stability analysis of the localized and non-localized nonlinear normal modes are developed. If the localized mode is realized, the system energy is concentrated in the nonlinear absorber. This situation is the most appropriate to absorb vibrations of a linear subsystem. NNMs are used to analyze the cylindrical shell nonlinear dynamics [13]. Initial imperfections are taken into account. The Shaw-Pierre NNMs are used to analyze a 7-dof model for a double tracked road vehicle [14]. Nonlinear response of the suspension is taken into account.

The paper is organized as follow. In Section 2 the Kauderer-Rosenberg and Shaw-Pierre approaches are presented. In Section 3 an investigation of the NNMs in some pendulum systems is described. In Section 4 the Shaw-Pierre NNMs and the modified Rauscher method are used to construct the NNMs of forced vibrations in a one-disk rotor system.

2. Two approaches to the nonlinear normal modes

The Kauderer-Rosenberg approach of the NNMs is based on a construction of trajectories in the system configuration space. If some generalized coordinate is chosen as the independent one, for example, $x_1 \equiv x$, then the following functions defines the NNM:

$$x_i = x_i(x); \quad (i = 2,3,..., n) \tag{1}$$

For a finite-dof conservative system with the potential energy, $\Pi = \Pi(x)$, which is a positive definite analytical function, and the kinetic energy is transformed to the canonical form, the equations to obtain trajectories (modal lines) in the system configuration space can be written as [2, 7-9]

$$2x_i''(h-\Pi)/(1+\sum_{k=2}^{n} x_k'^2)+x_i'(-\Pi_x) = -\Pi_{x_i} \; (i=2,3,...,n) \tag{2}$$

Here the prime means a differentiation by the variable x. These equations have singular points on the maximal equipotential surface $\Pi(x_1,...,x_n)=h$. An analytical continuation of the NNM trajectories to the surface is possible if the next boundary conditions are satisfied [2, 7-9], namely,

$$x_i'(X)[-\Pi_x(X,x_2(X)...,x_n(X))] = -\Pi_{x_i}(X,x_2(X)...,x_n(X)) \; (i = 2,3,..., n) \tag{3}$$

The relations (3) are natural boundary conditions for the variational principle in the Jacobi form. Here $x=X, x_i(X)$ are the trajectory return points lying on the maximal equipotential surface $\Pi(x_1,...,x_n)=h$. If the NNM (1) is obtained from the boundary problem (3)-(4), then the system under consideration is reduced to a single-dof nonlinear oscillator. So, the NNM is a two-parametric (by energy and phase of the motion) family of periodic solutions with smooth modal lines. To construct the NNM, a power series can be used [3,4].

Shaw and Pierre [5,6] reformulated the method of NNMs for a general class of quasilinear dissipative systems. Their analysis is based on the computation of invariant manifolds of motion on which the NNMs take place. One chooses a couple of new independent variables (u, v), where u is some *dominant generalized coordinate,* and *v is the corresponding generalized velocity*. By the Shaw-Pierre concept, the nonlinear normal mode is such a regime when all phase coordinates are univalent functions of the selected couple of variables. It is possible to obtain the solution in the power series by new independent variables u and v using a system of partial differential equations.

Generalization of the NNMs approach to non-autonomous and self-exited systems can be found in Refs. [7-9].

3. NNMs by Kauderer-Rosenberg in the pendulum dynamics

One considers free vibrations of the spring pendulum (Figure 1). Here m is a mass of the pendulum, l is length of the linear spring in the unstressed state. The dynamics of the system are described by two positional coordinates, ρ and φ.

One introduces a new variable $z = \rho - \rho_0$, where $\rho_0 = l + gm/c$ is a static extension of the spring in the equilibrium position. In the system two vibration modes can be selected: a) longitudinal vibration mode when $\varphi=0$, $z=z(t)$; b) coupled vibration mode when $\varphi=\varphi(t)$, $z=z(t)$.

Journal of Physics: Conference Series **382** (2012) 012052 doi:10.1088/1742-6596/382/1/012052

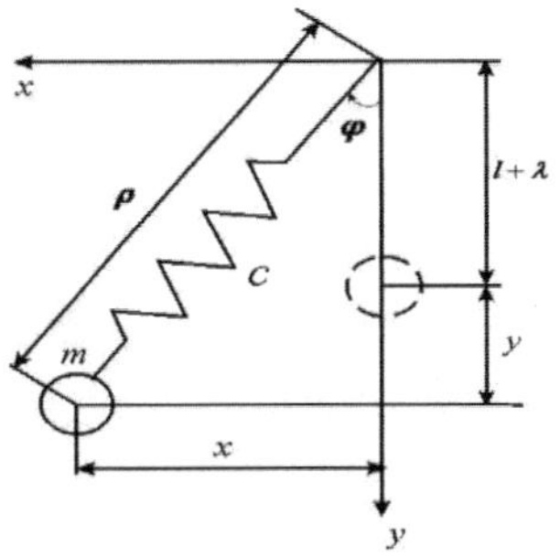

Figure 1. The model of the spring pendulum

The coupled vibration mode can be constructed in the form of the trajectory in the configuration space as $z = z(\varphi)$. Corresponding equation and boundary conditions are similar to equation (2) and (3). The NNM is determined as a power series by the independent variable φ. Coefficients of the series are obtained from algebraic equations. A numerical simulation needed for checking shows a good agreement with the obtained analytical solution.

The longitudinal vibration mode stability was studied previously [15] and results of such investigations are known. The corresponding linearized equation for variations in the horizontal direction can be reduced to the Mathieu equation. A more exact method for the stability analysis is based on reduction to the Hill equation. Analysis of stability of the coupled vibrations mode is made by the analytical-numerical criterion which is proposed in [16]. In this case current values of perturbations are compared with initial ones.

Free vibrations of the two-DOF system, which is shown in Figure 2, are considered too. The anchor spring is linear having a stiffness coefficient k. The linear oscillator of the mass m_1 is connected with the pendulum absorber, having the mass m_2; the length of the pendulum is equal to l. The system motions are determined by two generalized coordinates x and θ.

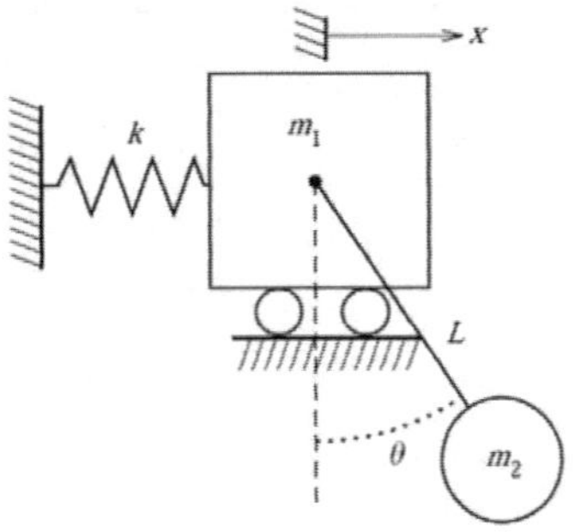

Figure 2. The mechanical system having the pendulum absorber

Two nonlinear vibration modes can be selected in this system: a) the coupled vibrations mode, $x = x(t), \theta = \theta(t)$, when vibration amplitudes of two generalized coordinates have the same order; b) the localized vibration mode, which is the most appropriate for absorption of the linear subsystem vibrations, when amplitudes of $\theta = \theta(t)$ are essentially larger than ones of $x = x(t)$.

The method of the NNMs construction presented in Section 2, is utilized here. The coupled vibration mode, $\theta = \theta(x)$, is found in a power series in x. The trajectory of the localized vibration mode is determined of the form: $x = x(\theta)$. The NNM trajectory is obtained as a power series in the variable θ. The near rectilinear trajectory of the localized mode is presented in Figure 3.

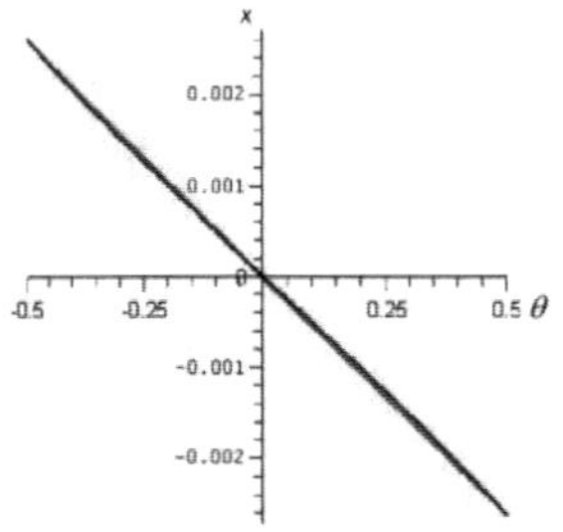

Figure 3. Trajectory of localized vibration mode in configuration space

Again the numerical simulation shows a very good agreement with the analytical solution.

The non-localized and localized NNMs trajectories are near rectilinear. Stability of the mode is determined by variations, which are orthogonal to the trajectories. The Mathieu and Hill equations are used in this investigation. The region of the localized mode instability is very narrow. So, the localized mode is very effective for absorption of elastic vibrations.

The forced NNMs in the system containing the pendulum absorber under the external periodic excitation are obtained too.

4. Nonlinear normal modes of forced vibrations in rotor systems

Different nonlinear effect must be taken into account in the analysis of the dynamical behaviour of rotor systems [17,18]. Note that in many publications mostly the simplest models, for example, the Jeffcott rotor, are considered due to a complexity of the rotor system dynamics. Here nonlinear forced vibrations of the rotor are considered. Gyroscopic effects, nonlinear flexible base, inertial forces in supports, an asymmetrical disposition of the disk in the shaft and internal resonances are all taken into account. Equations of motion of the one-disk unbalanced rotor with the linearly isotropic elastic shaft and nonlinear elastic bearings are the following [9]:

$$
\begin{aligned}
& m\ddot{x}+\rho_1\dot{x}+k_{11}\left(x-h_1x_2-h_2x_1\right)+k_{12}\left(\theta_2-\frac{x_2-x_1}{l}\right)=\varepsilon\Omega^2 m\cos\Omega t \\
& m\ddot{y}+\rho_1\dot{y}+k_{11}\left(y-h_2y_1-h_1y_2\right)-k_{12}\left(\theta_1+\frac{y_2-y_1}{l}\right)=\varepsilon\Omega^2 m\sin\Omega t \\
& I_e\ddot{\theta}_1+\rho_2\dot{\theta}_1+I_p\Omega\dot{\theta}_2+k_{22}\left(\theta_1+\frac{y_2-y_1}{l}\right)-k_{12}\left(y-h_2y_1-h_1y_2\right)=0 \\
& I_e\ddot{\theta}_2+\rho_2\dot{\theta}_2-I_p\Omega\dot{\theta}_1+k_{22}\left(\theta_2-\frac{x_2-x_1}{l}\right)+k_{12}\left(x-h_1x_2-h_2x_1\right)=0 \\
& m_1\ddot{x}_1+\beta\dot{x}_1+\left(\frac{k_{12}}{l}-h_2k_{11}\right)\left(x-h_1x_2-h_2x_1\right)+\left(\frac{k_{22}}{l}-h_2k_{12}\right)\left(\theta_2-\frac{x_2-x_1}{l}\right)+c_x^{(1)}x_1+c_x^{(2)}x_1^3=0 \\
& m_1\ddot{y}_1+\beta\dot{y}_1+\left(\frac{k_{12}}{l}-h_2k_{11}\right)\left(y-h_2y_1-h_1y_2\right)+\left(h_2k_{12}-\frac{k_{22}}{l}\right)\left(\theta_1+\frac{y_2-y_1}{l}\right)+c_y^{(1)}y_1+c_y^{(2)}y_1^3=0 \\
& m_1\ddot{x}_2+\beta\dot{x}_2+\left(-\frac{k_{12}}{l}-k_{11}h_1\right)\left(x-h_1x_2-h_2x_1\right)+\left(-\frac{k_{22}}{l}-h_1k_{12}\right)\left(\theta_2-\frac{x_2-x_1}{l}\right)+k_x^{(1)}x_2+k_x^{(2)}x_2^3=0 \\
& m_1\ddot{y}_2+\beta\dot{y}_2+\left(-\frac{k_{12}}{l}-h_2k_{11}\right)\left(y-h_2y_1-h_1y_2\right)+\left(\frac{k_{22}}{l}+h_1k_{12}\right)\left(\theta_1+\frac{y_2-y_1}{l}\right)+k_y^{(1)}y_2+k_y^{(2)}y_2^3=0
\end{aligned}
\tag{4}
$$

where l is the shaft length; l_1, l_2 are distances of the disk up to left and right supports, correspondently; $h_1=l_1/l; h_2=l/l_2$; $c_x^{(1)}, c_y^{(1)}$ are coefficients which characterize linear terms in the left support restoring force; $k_x^{(1)}, k_y^{(1)}$ are similar coefficients for the right support; $c_x^{(2)}, c_y^{(2)}$ are coefficients which characterize cubic terms in the left support restoring force; $k_x^{(1)}, k_y^{(1)}$ are similar coefficients for the right support; β is a coefficient of damping in the supports; ρ_1, ρ_2 are coefficients of damping during the disk motion; m is the disk mass; ε is an eccentricity of the disk mass center.

One has an 8-dof nonlinear system, describing the displacements and rotations of the disc, and displacements of the nonlinear supports. The NNMs are constructed here in the forced rotor system having an internal resonance. This situation is always realized in the rotor dynamics with the isotropic-elastic shaft and supports. In a case of internal resonance it is possible to obtain a 2-dof nonlinear system for each vibration mode of the forced vibrations.

The Rauscher method was first proposed for the single-DOF system [19]. Generalization of the method to construct NNMs in general non-conservative systems is proposed in [20]. One considers the nonlinear dynamical system under the external harmonic excitation in normal coordinates.

Let $\bar{q} = \{q_1, q_2, ..., q_N\}^T$, $\bar{s} = \{s_1, s_2, .. s_N\}^T$ be the principal coordinates and corresponding velocities. It is assumed that two linearized frequencies ν_1 and ν_2 are close, and they are close to the external frequency, Ω, that is $\Omega \approx \nu_1 \approx \nu_2$. In this case two active coordinates, $q_{1,2}$, and two corresponding velocities, $s_{1,2}$, may be taken as independent master coordinates to construct the forced NNMs. One uses from the zero approximation a representation of the active coordinates as a Fourier series when the other coordinates are essentially smaller than the master ones:

$$\begin{aligned} q_1 &= A_1 cos(\Omega t) + B_1 sin(\Omega t) + A_2 cos(2\Omega t) + B_2 sin(2\Omega t) + A_3 cos(3\Omega t) + B_3 sin(3\Omega t) + ... \\ s_1 &= \Omega(B_1 cos(\Omega t) - A_1 sin(\Omega t) + 2B_2 cos(2\Omega t) - 2A_2 sin(2\Omega t) + 3B_3 cos(3\Omega t) - 3A_3 sin(3\Omega t) + ...) \\ q_2 &= ... \quad s_2 = ... \end{aligned} \tag{5}$$

One has from here, after some trigonometric transformations, that

$$cos(\Omega t) = \alpha_1 q_1 + \alpha_2 s_1 + \alpha_3 q_2 + \alpha_4 s_2 + \alpha_5 q_1^2 + \alpha_6 s_1^2 + ... \tag{6}$$

When the coefficients of the expansions (6) are determined from algebraic equations, then an n-dof "pseudo-autonomous" system is obtained instead of the initial non-autonomous system. It corresponds to the principal idea of the Rausher method. In the obtained autonomous system, having the internal resonance, the NNMs can be constructed as functions of *four* independent coordinates q_1, s_1, q_2, s_2. Corresponding partial differential equations, similar to equations (8), are used. Solutions of these equations are obtained in the form of the power series by *four master coordinates*. It permits the reduction of the n-dof system to the two-dof one. Four active phase coordinates are obtained from this reduced system in a form of the Fourier series. The recurrent process is constructed, and the pointed out series of operations can be repeated some times to reach the necessary condition of exactness. So, the steady-state resonance regimes are constructed in the form of the NNMs.

The procedure of the NNMs construction is used in the rotor dynamics. Frequency responses near the first resonance are presented in Figure 4 for some values of the system parameters. Figures (a) and (b) represent frequency responses for principal coordinate q_1 (first and third harmonic of excitation frequency respectively). Trajectories of the resonance vibrations in the system configuration space are constructed too. Numerical simulation shows a good efficiency of the proposed approach.

Journal of Physics: Conference Series **382** (2012) 012052 doi:10.1088/1742-6596/382/1/012052

a)

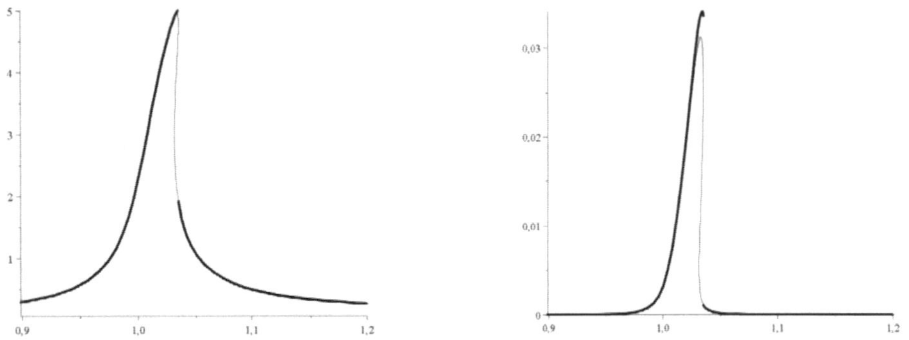

(b)

Figure 4. Frequency responses near the first resonance. Bold curves correspond to the NNM approach, and thin curves correspond to the calculations by the harmonic balance method. Here frequencies on the horizontal axis are dimensionless; all amplitudes are multiplied by a scaling coefficient of 1000.

5. Conclusions

Nonlinear normal modes (NNMs) are typical regimes realized in different conservative or near-conservative finite-dof systems. The method of nonlinear normal modes is one of the approaches for dimension reduction in nonlinear systems. The Kauderer-Rosenberg concept, when all positional coordinates are single-valued functions of some of them, is associated with trajectories in configuration space. The Shaw-Pierre concept is based on the computation of invariant manifolds of motion. In this case the NNMs can be obtained as single-valued functions of two selected phase coordinates. An efficiency of the NNMs theory is shown in some applied problems, in particular, in pendulum and rotor dynamics.

References

[1] Kauderer H 1958 *Nichtlineare Mechanik* (Berlin: Springer-Verlag)
[2] Rosenberg R M 1966 *Adv. of Appl. Mech.* **9** 156
[3] Manevich L I and Mikhlin Yu V 1972 *Prikl. Matem. i Mekh. (PMM USSR)* **36** 1051
[4] Mikhlin Yu 1996 *Nonl. Dyn.* **11** 1
[5] Shaw S and Pierre C 1991 *J. Sound and Vibr.* **150**, 170
[6] Shaw S and Pierre C 1993 *J. of Sound and Vibr.* **164** 85
[7] Vakakis A, Manevitch L, Mikhlin Yu, Pilipchuk V and Zevin A 1996 *Normal Modes and Localization in Nonlinear Systems* (New York: Wiley–Interscience)
[8] Mikhlin Yu and Avramov K 2010 *Appl. Mech. Rev.* **63** 4
[9] Avramov K V and Mikhlin Yu V 2010 *Nonlinear Dynamics of Elastic Systems* (Moscow-Izhevsk: Regular and Chaotic Dynamics)
[10] Mikhlin Yu V and Reshetnikova S N 2005 *J Sound and Vibr.* **283** 91
[11] Avramov K V and Mikhlin Yu V 2004 *J. of Vibr. and Control* **10** 291
[12] Avramov K V and Mikhlin Yu V 2006 *J Sound and Vibr* **29** 705
[13] Avramov K V, Mikhlin Yu V and Kurilov E 2007 *Nonl. Dyn.* **47** 331
[14] Mikhlin Yu and Mitrokhin S 2008 *J. of Theoretical and Applied Mechanics*, **46** 581
[15] Starzhinskii V M 1977 *Applied Methods of Nonlinear Vibrations* (Moscow: Nauka).
[16] Mikhlin Yu V, Shmatko T V and Manucharyan G V 2004 *Comp. and Str.* **82** 2733
[17] Tondl A 1965 *Some Problems of Rotor Dynamics* (London: Chapman and Hall)
[18] Genta G 2005 *Dynamics of Rotating Systems* (Berlin: Springer)
[19] Rauscher M 1938 *J. of Appl. Mech.* **5** A169
[20] Mikhlin Yu 1974 *Prikl. Matem. i Mekh. (PMM USSR)* **38** 425

Journal of Physics: Conference Series **382** (2012) 012053 doi:10.1088/1742-6596/382/1/012053

Green's function and Bloch theory for the analysis of the dynamic response of a periodically supported beam to a moving load

R Lassoued[1], M Lecheheb[1], G Bonnet[2]

[1]Laboratory of Materials Durability of Construction, Department of Civil Engineering Faculty of Engineering sciences, University Mentouri Constantine Algeria
[2] University Paris East Marne la Vallée, Laboratory of modelling and simulation Multi Echelle, CNRS UMR 8208, 5 boulevard Descartes, 77454 Marne la Vallée Cedex, France

E-mail: rachid_lassoued@yahoo.fr

Abstract. This paper describes an analytical method for the wave field induced by a moving load on a periodically supported beam. The Green's function for an Euler beam without support is evaluated by using the direct integration. Afterwards, it introduces the supports into the model established by using the superposition principle which states that the response from all the sleeper points and from the external point force add up linearly to give a total response. The periodicity of the supports is described by Bloch's theorem. The homogeneous system thus obtained represents a linear differential equation which governs rail response. It is initially solved in the homogeneous case, and it admits a no null solution if its determinant is null, this permits the establishment the dispersion equation to Bloch waves and wave bands. The Bloch waves and dispersion curves contain all the physics of the dynamic problem and the wave field induced by a dynamic load applied to the system is finally obtained by decomposition into Bloch waves, similarly to the usual decomposition into dynamic modes on a finite structure. The method is applied to obtain the field induced by a load moving at constant velocity on a thin beam supported by periodic elastic supports.

1. Introduction

The dynamic response analysis of infinite periodic beams under moving loads has been one of the main of the research interests in railway engineering. Genin and Chung have proposed a numerical algorithm for calculation of the response of a moving vehicle on a continuous periodically supported beam [1]. Smith and Wormley [2] use the Floquet principle and the Fourier transform of the beam displacement and then numerically inverting it. The free vibration propagation and forced vibration induced by stationary harmonic loads is analyzed by [3] . A mathematical model is presented by [4] to predict the vibration of a rail excited by a rolling wheel. The track is modelled as an infinitely long beam supported over a finite section by discrete support systems for a non moving load. Vibration of

[1] rachid_lassoued@yahoo.fr.

an infinite periodic beam subject to a moving harmonic load has been investigated by [5]. In this study, the author considered a single segment only by using boundary conditions derived according to the Euler beam theory. The rail is modelled as an Euler beam with flexible supports in the work of [6]. The supports of the beam represent the pad/sleeper/ballast system of the railway track; they are spaced regularly. The exact solution, in the frequency domain, of the linear differential equation is presented. The method used to obtain the response to the dynamic load includes the computation of transfer matrices [7, 8], the use of Bloch waves [9, 6] and finite element solutions [10, 11]. For a Timoshenko beam, [12] produced the dispersion curves and showed that different modes appear, related to the coupling between different degrees of freedom of the beam: flexion, torsion, shear.... The response of a Timoshenko beam to a static (non moving) dynamic load was produced by [13]. A method is proposed to calculate the response of periodic structures to subject to a moving load based on the Floquet decomposition [14].The aim of this paper describes the response of a periodically supported beam using the Green function, the superposition principle and the Bloch transform [15,16]. The model consists of a moving load on a discretely supported rail. The track is reduced to a rail modelled as an infinite Euler beam.

2. Response to an harmonic load on the infinite beam :Green function

The vertical beam displacement is the solution of the dynamic beam equation Euler model is considered in equation (1):

$$EI\frac{\partial^4 U(x,t)}{\partial x^4} + m\frac{\partial^2 U}{\partial t^2} = F(x,t) \tag{1}$$

where m is the mass of the beam by unit of length and E the Young's modulus, I an inertia section moment and $F(x,t)$ a moving load.
If the harmonic concentrated load is applied at the point x_0, in the Fourier space, the movement equation becomes:

$$EI\frac{\partial^4 \tilde{U}(x,\omega)}{\partial x^4} - m\omega^2\tilde{U}(x,\omega) - \delta(x-x_0) = 0 \tag{2}$$

Where $\tilde{U}$ is the Fourier transform of U.
The solution of this equation is given by:

$$G(x,x_0) = \frac{1}{4EIk_h^4}\left[ie^{ik_b|x-x_0|} - e^{-k_b|x-x_0|}\right] \tag{3}$$

Where:

$$k_b = \sqrt[4]{\frac{m\omega^2}{EI}} \tag{4}$$

3. Bloch wave function

We consider a beam periodically supported with N supports placed between $-\infty$ and $+\infty$. The period is given by the regular spacing l. The rigidity of the Nth discrete elastic support, corresponding to the deformation of the system (elastic support, mass support...), is Z_n .It is considered that all the discrete supports are characterized by the same rigidity $Z_n = Z$. The global response of the structure can be described by applying the superposition principle, which consists of the linear summation of all the answers due to the various supports. Then, the displacement with the position x is given by:

$$\eta(x) = \sum_{-}^{+} G(x,x_n)F_n \tag{5}$$

Where: $F_n = -Z_n U(n,l)$ is the force transmitted by the discrete support.
Introducing the value of $G(x,x_n)$ is given by the equation (3) for the Euler beam.
The relation (6) is thus valid for any point x_m.

$$\eta(x_m) + \sum_-^+ Z\eta(x_n)\, G(x_m, x_n) \tag{6}$$

The application of the Floquet theorem gives the relation between two adjacent supports:

$$\eta(x_{m+1}) = \eta(x_n) e^{-\gamma l} \tag{7}$$

With:

$$\gamma = \alpha + ik \tag{8}$$

γ is the Bloch constant propagation. It is generally complex. α is the wave attenuation and k is the Bloch wave number .

By substituting the expression of the Green function given by equation (3) in equation (5) and by multiplying this equation by $e^{\gamma ml}$, changing the index of summation $m-n$ to n gives the following equation which defines the relation between γ and the displacement η_0 .

$$\eta_0 + Z\eta_0 \sum_-^+ \left[\frac{1}{4EIk_b^4}\left(ie^{ik_b|n|l} - e^{k_b|n|l}\right)e^{\gamma nl}\right] = 0 \tag{9}$$

In the following, only the waves not attenuated will be considered, which means that $\alpha = 0$ and $\gamma = -ik$. The sum (9) can be evaluated by the usual formulas for the geometric series.
For the Euler model, we can have:

$$1 + \frac{Z}{4EIk_b^3}\left[\frac{\sin(k_b l)}{\cos(k_b l) - \cos(kl)} - \frac{\sinh(k_b l)}{\cosh(k_b) - \cos(kl)}\right] = 0 \tag{10}$$

This is a second degree polynomial for $\cos(kl)$. The solution produces the values of the wave number k which is between 0 and π. It is a function of k_b which depends on the radial frequency equation (4). It leads therefore to the dispersion equation.

The shape of the Bloch wave can be obtained by using a similar process. Let us consider the free motion related to a couple k , k_b applying to equation (10).
The displacement at position x between two supports is related to the displacements of all supports by using Green's function G , leading to:
Introducing Green's function and the propagating constant γ leads to:

$$\eta(x) = -Z\eta_0 \frac{1}{4k_b^4 EI}\left[\sum_{m=-}^{+}\left(ie^{ik_b|x-x_m|} - e^{-k_b|x-x_m|}\right)e^{-\gamma x_m}\right] \tag{11}$$

Taking into account that the Bloch function must be normalized, it can be defined as:

$$\eta(x) = \left[\frac{e^{ikl}\sin(k_b z) - \sin(k_b(z-l))}{\cos(k_b l) - \cos(kl)} - \frac{e^{ikl}\sinh(k_b z) + \sinh(k_b(l-z))}{\cosh(k_b l) - \cos(kl)}\right] / C \tag{12}$$

The value of η_0 and therefore the value of C must be chosen to build from the Bloch waves an orthonormal basis of $L^2(Y)$ where Y is the periodic cell obtained after transforming the space variable x into the non dimensional variable $X^* = x\frac{2\pi}{l}$. Through this process, the period $[0, l]$ is transformed into $[0, 2\pi]$.

The constant C is chosen to ensure that the L^2 norm of $\eta(X^*)$ $(or\ \eta_1(X^*))$ on $[0,2\pi]$ is equal to 1, leading to:

$$C^{-1} = \left[\frac{2\pi}{l}\int_0^l \left|\frac{N_1}{D_1} + \frac{N_2}{D_2}\right|^2 dx\right]^{1/2} \tag{13}$$

Where:

$$N_1(x,k) = e^{ik}sin(k_b x) - sin\big(k_b(x-l)\big)\ N_2(x,k) = -\left(e^{ikl}sinh(k_b x)\right) - sinh\big(k_b(x-l)\big)$$

$$D_1(k) = cos(k_b l) - cos(kl) \qquad D_2(k) = cosh(k_b l) - cos(kl)$$

4. The response of the railway structure to a moving load

As for the space coordinate, it is convenient, to define the Bloch transform, by using a no-dimensional wave-number K^* defined by: $K^* = \frac{kl}{2\pi}$

The Bloch transform of a function $F(x)$ is then defined by:

$$\widehat{F_n}(K^*) = \int_0^{2\pi} F(X^*)\overline{\eta_n}(X^*,K^*)dX^* \tag{14}$$

Where $\overline{\eta_n}(X^*,K^*)$ is the complex conjugate of the Bloch wave contained in the nth band obtained by equation (5). The Bloch decomposition theorem, whose proof and conditions of validity can be found in [15, 16], states that.

$$F(X^*) = \sum_{n=1}\int_0^1 \widehat{F_n}(K^*)\eta_n(X^*,K^*)dK^* \tag{15}$$

The Bloch transform of equation (1) , considering that the Bloch wave $\eta_n(x,k)$ is the solution of the homogeneous equation for an harmonic equation at radial frequency $\omega_n(k)$, leads to:

$$m\omega_n^2\widehat{U_n}(K^*,t) + m\frac{\partial^2}{\partial t^2}\widehat{U_n}(K^*,t) = \widehat{F_n}(K^*,t) \tag{16}$$

This equation shows that each Bloch component is a solution of the dynamic equation for a '1 DOF' system.

Let us use this result to obtain the response of the beam to a load having a constant intensity F_0 which is moving at the velocity V. It means that F is given by:

$$F(x,t) = \delta(x - Vt)F_0$$

Where δ is the Dirac distribution.

The Bloch transform of $F(t)$ is:

$$F_n(K^*,t) = \overline{\eta_n}\left(X^* = Vt\frac{2\pi}{l}, K^*\right)F_0\frac{2\pi}{l} \tag{17}$$

The Bloch components of the displacement induced by the moving load are therefore solutions of:

$$m\omega_n^2\widehat{U_n}(K^*,t) + m\frac{\partial^2}{\partial t^2}\widehat{U_n}(K^*,t) = \overline{\eta_n}\left(X^* = Vt\frac{2\pi}{l}, K^*\right)F_0\frac{2\pi}{l} \tag{18}$$

Its solution is given by:

$$\widehat{U_n}(K^*,t) = \frac{2F_0C^2}{l}\left[\frac{e^{-2iK^*}\sin(k_bVt) - \sin\big(k_b(Vt-l)\big)}{1[\cos(k_bl) - \cos(2K^*)]} - \frac{e^{-2iK^*}\sinh(k_bVt) - \sinh\big(k_b(Vt-l)\big)}{2[\cosh(k_bl) - \cos(2K^*)]}\right] \tag{19}$$

Where γ_1 and γ_2 are given by:

$$\gamma_1 = m\left(\omega_n^2 - k_b^2V^2\right) \qquad \gamma_2 = m\left(\omega_n^2 + k_b^2V^2\right)$$

The displacement induced by the moving force is finally obtained by a composition of all the Bloch components, leading to:

$$U(X^*,t) = \sum_n \int_{K=0}^{1} \widetilde{U_n}\,(K^*,t)\eta_n(X^*,K^*)dK^* \tag{20}$$

Coming back to x , k and taking into account that k is involved in η only by $cos(kl)$, leads to:

$$U(x,t) = \sum_n \int_{k=0}^{\pi/l} V_n(k,x,t)\,dk \tag{21}$$

The function $V_n(k,x,t) = \frac{1}{\pi}\widetilde{U_n}(k,t)\eta_n(x,k)$ is given by:

$$V_n(k,x,t) = 2F_0C^2\left[\frac{N_1(Vt,k)}{\gamma_1 D_1} + \frac{N_2(Vt,k)}{\gamma_2 D_2}\right]\left[\frac{N_1(x,k)}{D_1} + \frac{N_2(x,k)}{D_2}\right] \tag{22}$$

This solution can be transformed by using new dimensionless variables T, X, B, K defined by $x = Xl$, $t = Tl/V$, $k = K/l$, $k_b = B/l$.
Using these variables U is given by:

$$U(X,T) = \frac{F_0}{\pi\beta l}\sum_n \int_{K=0}^{\pi} \frac{P}{\alpha_1\alpha_2 Q}dK \tag{23}$$

where: $\alpha_1 = B^4 - B^2\phi$; $\alpha_2 = B^4 + B^2\phi$; $\phi = \frac{mV^2l^2}{EI}$; $\beta = \frac{EI}{l^4}$

From the previous results, it can be seen that the displacement depends on the non dimensional position and time and also on the three following parameters:

4.1 Example of application

Let us consider a rail whose properties are given in (table 1).

Table 1. Physical properties of the beam and support

Young's modulus	Section's inertia	Mass per un. length	Support stiffness	Spacing
GPa	m^4	kg/m	GN/m	m
200	8.10-5	158	30 and 0.02	0.6

The dispersion curves are show on figure 1 for the first three passing bands for the stiffer support (Z=30 GN/m). Similar results are obtained for the softer support at lower frequencies.

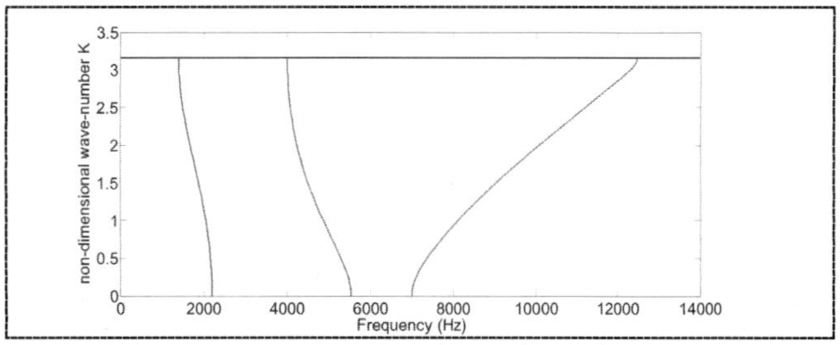

Figure 1. Dispersion curve for the first band tree bands: case for stiffer support

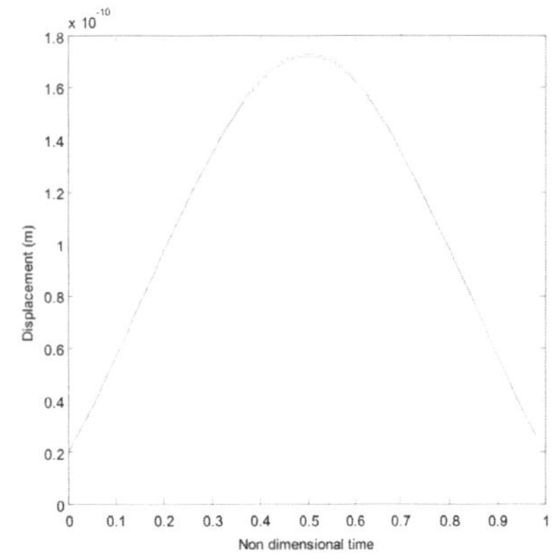

Figure 2. non dimensional time

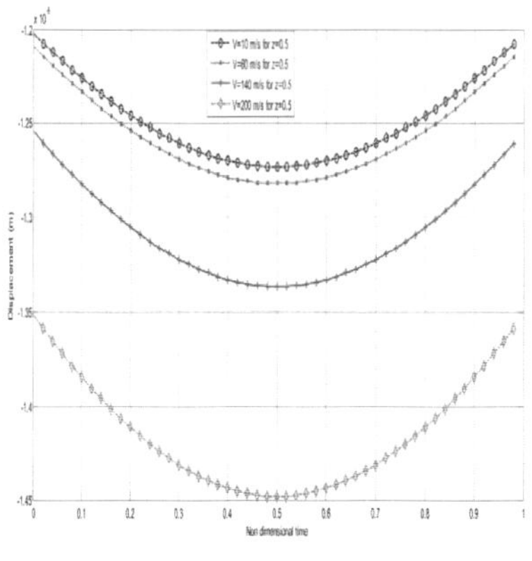

Figure 3. Displacement vs non-dimensional time with various speeds

The displacement induced by a unit moving load at the center of a given period is next computed when the load is moving over that period with two speeds: a low speed 10m/s and a high speed 140m/s. Results are given in the case of a stiff support in figure 2 and in the case of a soft support in figure 3. It appears that the displacement is the same for low speed and high speed, in the case of the stiffer support, while the displacement depends significantly of the speed for the softer support figure 3 where the speeds for the second case are: 10m/s; 60m/s; 140 m/s; 200m/s.
This result is due to the fact that the passing bands are at higher frequencies for the stiffer supports.
The figure 4 show the analysis displacement of the beam with various positions (z=0; z=0.25; z=0.5), for this case the speed is constant v=140 m/s.

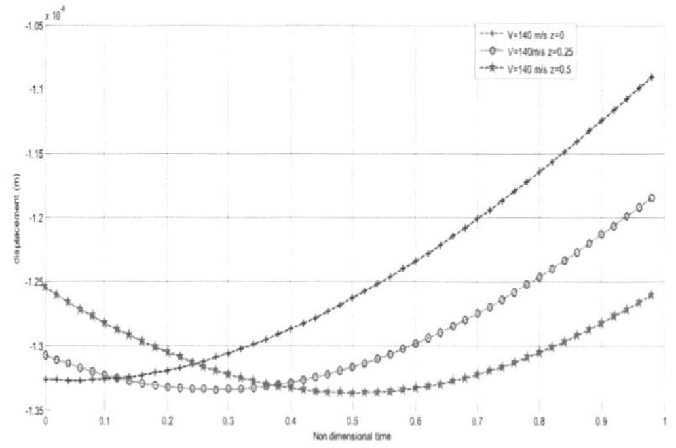

Figure 4. Displacement vs non-dimensional time : Analysis with various positions on the beam

5. Conclusions

The Green function and Bloch theory were used for analysis of dynamic response of periodically supported beam to moving load. The proposed method uses the superposition principle with these two theories. It allows the computation of the response of the beam to any dynamic loading by computing the Bloch transform of the displacement induced by the loading. It only involves the solution of a (continuous) set of decoupled one degree of freedom dynamic equations. The method is applied to the dynamic response of a thin beam.

References

[1] J.Genin, Y.I.Chung 'Response of a continuous guideway on equally spaced supports traversed by moving load' J.S.V (1979) 67,245-251
[2] C.C. Smith, N. Wormley, 'Response of continuous periodically supported guideway beams to travelling vehicle loads' American society of mechanical Engineers Journal of dynamic system.(1975),97,21-29.
[3] Mead, D.J. (1996), Wave propagation in continuous periodic structures: research contributions from Southampton, Journal Sound Vibration. 190 (3) 495-524.
[4] Heckl, M. A. (1995) , rail Noise-can random sleeper spacing's help, Acoustica. 81 559-564.
[5] Belotserkovskiy P.M., (1996), on the oscillations of infinite periodic beams subject to a moving concentrated force, Journal Sound Vibration. 193 (3) 705-712.
[6] Nordborg A, (1998) ,vertical rail vibration: point force excitation, Acoustica, 84 280-288.
[7] Langley, R.S. (1994), on the modal density and energy flow characteristics of a periodic structure, Journal Sound Vibration. 172(4), 491-511.
[8] Munjal, M.L. & Heckl, M. (1982) ,Vibrations of a periodic rail-sleeper system excited by an oscillating stationary transverse force, Journal Sound Vibration. 81(4), 491-500.
[9] Langley, R.S. (1996) ,The response of two-dimensional periodic structures to point harmonic loading, Journal Sound Vibration. 197 (4), 447-469.
[10] Nguyen, V.H. & Duhamel, D. (2006) , Finite element procedures for nonlinear structures in moving coordinates, Part 1: Infinite bar under moving axial loads. Computers and Structures, 84, 1368-1380.
[11] Duhamel, D. B.R. Mace, & M.J. Brennan, (2006), Finite element analysis of the vibration of waveguides and periodic structures, Journal Sound Vibration. 294, 205-220.
[12] Heckl, M.H. (2002) , Coupled waves on a periodically supported Timoshenko beam,. 252(5) Journal Sound Vibration, 849-882.
[13] Hamet, J.F. (1999) , Railway noise: use of the Timoshenko model in rail vibration studies, Acoustica, 85, 1-12.
[14] Chebli. H, Othman. R, Clouteau.D,'response of periodic structures due to moving load' , Comptes rendus mécanique 334 (2006) 347-352.
[15] Allaire, G. (1998), The Bloch transform and applications, ESAIM Proceedings, 3, 65-84.
[16] Sanchez. E, Sanchez Palencia Vibration and coupling of continuous systems, Springer-verglag, Berlin 1989

Static and dynamic analysis of isotropic shell structures by the spectral finite element method

A Żak[1], M Krawczuk[1,2]

[1] Polish Academy of Sciences, Gdansk, Poland
[2] Gdansk University of Technology, Gdansk, Poland

E-mail: `arkadiusz.zak@imp.gda.pl`

Abstract. The paper deals with certain problems related to static and modal analysis of isotropic shell structures by the use of the approach known in the literature as the time-domain spectral finite element method. Although recently this spectral approach has been widely reported as a very powerful numerical tool used to solve various wave propagation problems, its properties make it very well suited to solve static and modal problems. The robustness and effectiveness of the spectral approach has been successfully demonstrated by the authors in the case of a thin-walled spherical shell structure representing a pressure vessel. Static and modal responses of the structure have been investigated by the use of transversally deformable shell-type spectral finite elements and the results of this investigation have been compared to known analytical solutions as well as those obtained by the use of commercially available software for the finite element method.

1. Introduction

The spectral finite element method, also known as the time-domain spectral finite element method, as a numerical technique has been known for a few decades. Its origins come back to the first publications by Patera [1] who by its use investigated a two-dimensional problem of a laminar flow in a channel expansion. Since that time the spectral finite element has been successfully adopted by various physical and applied sciences and nowadays its application fields cover not only fluid dynamics [2] but also heat transfer [3], acoustics [4], seismology [5], and more recently mechanical engineering [6, 7, 8, 9, 10]. The method originates from the application of spectral series for solution of partial differential equations, while at the same time its base ideas are analogous to the classical finite element approach. Its main assumption is the application of orthogonal Lobatto polynomials as approximation functions defined at appropriate Gauss-Lobatto-Legendre integration points. As a consequence of that the inertia matrix obtained in this spectral approach is diagonal making the total cost of numerical calculations much less demanding. Additionally, thanks to the orthogonality of the approximation polynomials the spectral finite element method is characterised by exponential convergence.

In the field of mechanical engineering the time-domain spectral finite element method has established a firm position as a modelling tool used to investigate wave propagation related problems especially for SHM purposes [11, 12]. On the other hand the applicability of this spectral approach to solve static and modal problems has not been widely demonstrated and reported in the literature.

The aim of this paper is to show the robustness and effectiveness of the time-domain spectral finite element method in the case of thin-walled isotropic elastic spherical shell structures representing a pressure vessel and hemispherical dome. Their static and/or modal responses have been investigated by the authors by the use of transversally deformable shell-type spectral finite elements [9] and the results of this investigation have been compared to the known analytical solutions [13, 14, 15] as well as to the solutions obtained by the use of commercially available software for the finite element method.

2. Transversally deformable shell-type spectral finite element

A key factor of the time-based spectral finite element method is an appropriate selection of element nodes. Their coordinates are a key factor that strongly influences the element performance and is directly linked with the type of assumed approximation within the element [10]. The coordinates must be also appropriately selected to avoid Runge's phenomenon associated with the use of uniform grids of element nodes together with higher orders of approximation polynomials.

In the normalised (curvilinear) coordinate system of the element $\xi\eta\zeta$ the coordinates of element nodes can be defined as the roots of the following well-known polynomial expression:

$$(1-t^2)P_n'(t) = 0, \quad t \in [-1,+1] \tag{1}$$

where $P_n'(t)$ is the nth order Legendre polynomial and the symbol $'$ denotes its first derivative. For the 5th order approximation polynomials used by the authors the nodal coordinates of the elements in the normalised (curvilinear) coordinate system $\xi\eta\zeta$ can be found as:

$$\xi_m, \eta_n \in \left\{\pm 1, \pm\sqrt{\tfrac{1}{3} \pm \tfrac{2}{3\sqrt{7}}}\right\}, \quad m,n = 1,\ldots,6 \tag{2}$$

Next a set of elemental shape functions can be built in the local coordinate system spanned on the specified nodes. A Lagrange interpolation function $f(\xi,\eta)$ supported on the nodes can be defined in the following manner:

$$f(\xi,\eta) = \sum_{m=1}^{6}\sum_{n=1}^{6} N_m(\xi)N_n(\eta)f_{mn}, \quad m,n = 1,\ldots,6 \tag{3}$$

where $N_m(\xi)$ and $N_n(\eta)$ are one-dimensional shape functions of the element, while f_{mn} are the nodal values of the function $f(\xi,\eta)$. The approximation shape functions $N_m(\xi)$ and $N_n(\eta)$ are orthogonal in a discrete sense:

$$\int_{-1}^{+1} N_m(\zeta)N_n(\zeta)d\zeta = \sum_{k=1}^{6} w_k N_m(\zeta_k)N_n(\zeta_k) = w_m\delta_{mn}, \quad k,m,n = 1,\ldots,6 \tag{4}$$

where w_m is the Gauss-Lobatto weight and δ_{mn} is the Kronecker delta.

Assuming small strains the displacement field in the mid-plane of the element in the local coordinate system xyz can be expressed in a manner typical for shell-type elements as:

$$\begin{cases} u(x,y,z) = u_0(x,y) + z\cdot\phi(x,y) \\ v(x,y,z) = v_0(x,y) + z\cdot\psi(x,y) \\ w(x,y,z) = w_0(x,y) + z\cdot\theta(x,y) \end{cases} \tag{5}$$

It can be noticed that the element defined in such a manner is transversely deformable ($\epsilon_{zz} \neq 0$), has six degrees of freedom per node including three membrane displacement components $u_0(x,y)$, $v_0(x,y)$ and $\theta(x,y)$ and three flexural displacement components $\phi(x,y)$, $\psi(x,y)$ and $w_0(x,y)$.

Based on the assumed displacement field given by (5) the strains within the element can be easily expressed according to the first order shear deformation theory [16, 17] leading to the well-known definitions of the characteristic inertia (diagonal) and stiffness matrices of the element in the global coordinate system XYZ, as shown in [10].

3. Numerical computations

Numerical simulations were divided into two parts and followed modal analysis and static responses of thin-walled elastic and isotropic spherical shell structure representing a pressure vessel and hemispherical dome. Initially it was assumed that the spherical shell structure is made of aluminium alloy (Young's modulus $E = 72.7$ GPa, Poisson ratio $\nu = 0.33$, density $\rho = 2700$ kg/m^3). The thickness t and the diameter D of the structure were assumed as $t = 5$ mm and $D = 2$ m, respectively.

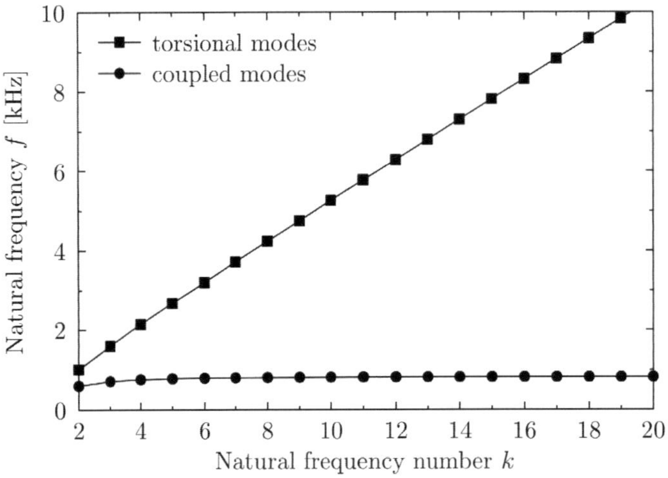

Figure 1. Natural frequencies of a thin-walled elastic, isotropic spherical shell.

It is interesting to note that in the case of a perfectly symmetrical and isotropic elastic sphere there are two main branches of multiple (degenerated) modes of vibrations. The first branch represents purely torsional behaviour ($u_r = 0$), while the second branch corresponds to coupled behaviour due to shearing and stretching [18]. These branches are supplemented by a single purely extensional natural vibration mode ($u_\theta = u_\phi = 0$). The multiplicity (degeneration degree) of particular natural vibration modes (either torsional or coupled) increases linearly with the natural frequency number as $k = 2m + 1$, where $m = 2, 3, \ldots$. For this reason for a given value of the natural frequency (either torsional or coupled) there exists a set of substantially different modes of vibrations (either torsional or coupled), as presented in figure 1.

Table 1. Natural frequency relative error as a function of the mesh density calculated based on the time-domain spectral finite element method (d_1 – 152/6, d_2 – 602/24, d_3 – 1352/54, d_4 – 2402/96, d_5 – 3752/150, d_6 – 60002/2400 nodes/elements).

Mode number	Mode multiplicity	Mesh density					
m	k	d_1	d_2	d_3	d_4	d_5	d_6
1	5	0.00%	0.00%	0.00%	0.00%	0.00%	0.00%
2	7	0.00%	0.01%	0.01%	0.01%	0.01%	0.01%
3	9	-1.41%	0.04%	0.04%	0.04%	0.04%	0.04%
4	11	-1.69%	0.10%	0.10%	0.10%	0.10%	0.10%
5	13	-1.25%	0.21%	0.20%	0.20%	0.20%	0.20%

First a convergence analysis was performed. During this analysis the first 45 natural frequencies of the structure under investigation were calculated assuming free type boundary conditions. The results obtained by the use of the time-domain spectral finite element method and the transversally deformable shell-type spectral finite elements [9] were compared to the known analytical solutions [13, 14, 15] as well as to the solutions obtained by the use of commercially available software (Patran/Nastran) for the finite element method. In that case 6-node triangle CTRIA6 shell finite elements were used.

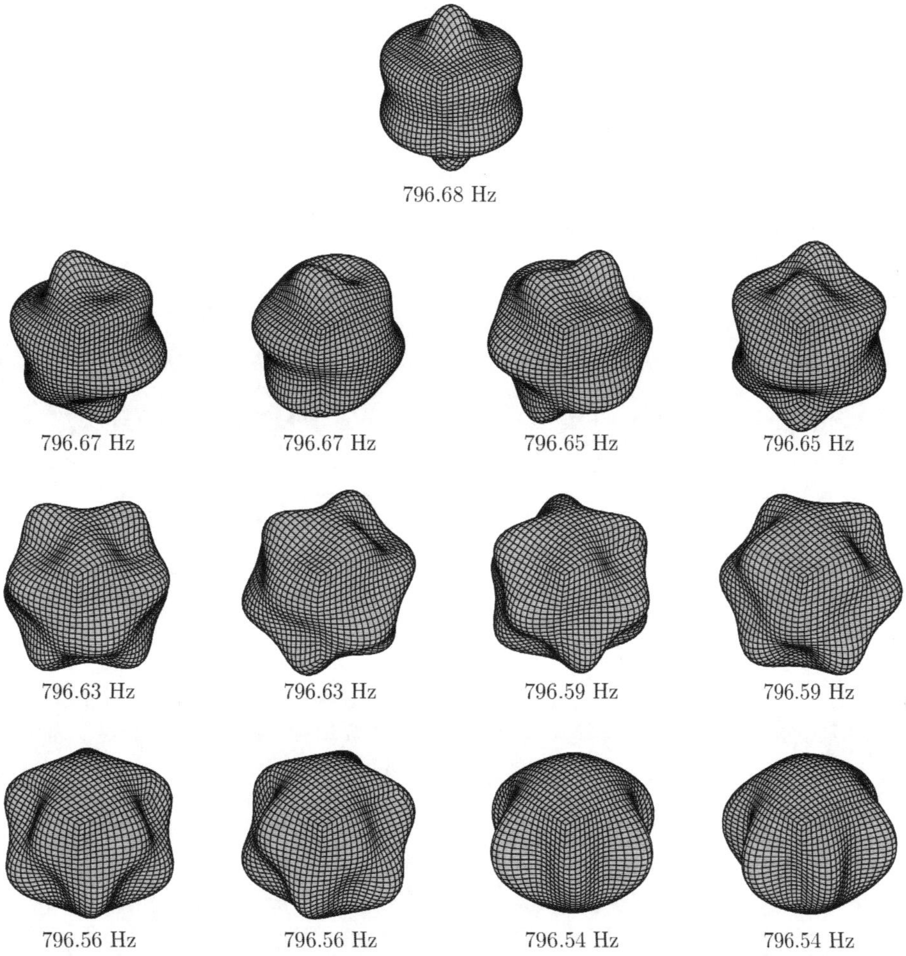

Figure 2. Multiple (degenerated) modes of natural vibrations of a thin-walled elastic, isotropic spherical shell for $k = 6$, calculated based on the time-domain spectral finite element method.

In the case of numerical calculations similar mesh densities were used in order to keep the number of nodes at the same level, but in favour of the finite element method. The obtained results are presented in table 1 and table 2. They express the relative errors between the average values of selected natural frequencies, calculated numerically, and the corresponding values calculated analytically. It should be mentioned that the stabilised and small constant values of the relative errors are a direct consequence of the multiplicity of particular natural vibration modes as well as applied numerical procedures used to calculate them. The multiplicity of natural vibration modes presents a serious numerical challenge to all natural frequency extraction procedures. This results in the fact that the values of the calculated natural frequencies, and corresponding to higher multiplicity numbers, are spread around certain average value, as illustrated in figure 2.

Table 2. Natural frequency relative error as a function of the mesh density calculated based on the finite element method (CTRIA6 triangle shell elements) (d_1 – 258/128, d_2 – 768/392, d_3 – 1602/800, d_4 – 2706/1352, d_5 – 4098/2048, d_6 – 61506/30752 nodes/elements).

Mode number	Mode multiplicity	Mesh density					
m	k	d_1	d_2	d_3	d_4	d_5	d_6
1	5	-10.6%	-5.87%	-2.08%	-0.79%	-0.31%	0.01%
2	7	-15.5%	-7.54%	-2.40%	-0.87%	-0.34%	0.02%
3	9	-18.5%	-8.39%	-2.51%	-0.88%	-0.32%	0.05%
4	11	-20.1%	-8.98%	-2.51%	-0.83%	-0.26%	0.10%
5	13	-20.8%	-9.64%	-2.43%	-0.72%	-0.15%	0.20%

Static analysis concerned a hemispherical thin-walled elastic and isotropic shell structure representing a dome. In this case it was assumed that all material properties as well as the geometry of the structure follow the classical benchmark test data for the shell type of finite elements [20] (Young's modulus $E = 68.25$ MPa, Poisson ratio $\nu = 0.3$). The thickness t and the diameter D of the structure were assumed as $t = 40$ mm and $D = 20$ m, respectively.

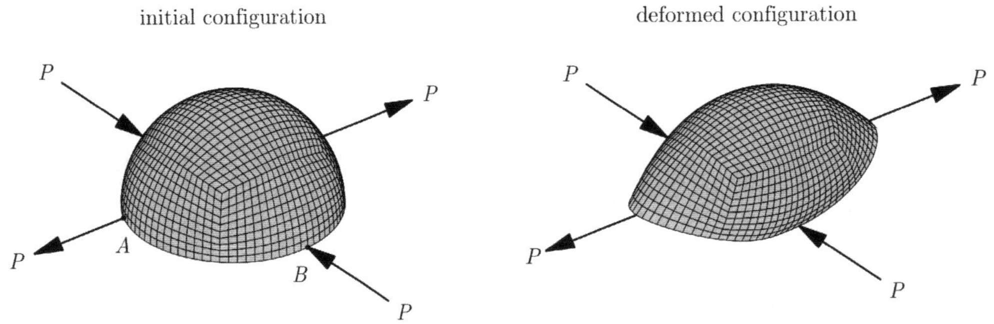

Figure 3. Initial and deformed configurations of a hemispherical thin-walled elastic, isotropic spherical dome calculated based on the time-domain spectral finite element method.

In this static test the structure was loaded by a set of inward and outward symmetrically acting concentrated forces P of 2 N placed on the structure free edge at points A and B, as shown in figure 3. Boundary conditions were provided by fixing the pole point of the structure.

Numerical calculations were carried out for various mesh densities and compared against the known analytical solution, as presented in table 3. It can be clearly seen also in this case that the results obtained thanks to the application of the time-domain spectral finite element method and the transversally deformable shell-type spectral finite elements are characterised by very high accuracy even for relatively low mesh densities.

Table 3. Deflection at point A as a function of the mesh density calculated based on on the time-domain spectral finite element method (d_1 – 321/12, d_2 – 1241/48, d_3 – 2761/108, d_4 – 4881/192, d_5 – 7610/300, d_6 – 30201/1200 nodes/elements).

Mesh density	Deflection	Relative error	Mesh density	Deflection	Relative error
d_1	90.13 mm	-2.46%	d_4	92.02 mm	-0.41%
d_2	91.68 mm	-0.78%	d_5	92.04 mm	-0.39%
d_3	91.98 mm	-0.45%	d_6	92.09 mm	-0.34%

4. Conclusions

Based on the results obtained by the authors and presented in this work it can be stated that the time-domain spectral finite element method is a very effective and robust numerical tool. It can be successfully used not only to solve various wave propagation problems, but also can be applied for static and modal problems. It has been shown that the application of the transversally deformable shell-type spectral finite element, developed by the authors, helps to reduce significantly errors of numerical investigation providing excellent accuracy, especially in comparison to the finite element method and typical finite elements commonly used for that purpose.

Acknowledgements

The authors of this work would like to gratefully acknowledge the support for this research provided by the Polish Ministry of Science and Higher Education through the European Funds System under the Sectoral Operational Programme Improvement of the Competitiveness of Enterprises via MONIT project (Monitoring of Technical State of Construction and Evaluation of Its Lifespan) nr POIG.01.01.02-00-013/08.

References

[1] Patera A T 1984 *Journal of Computational Physics* **54** 468
[2] Canuto C, Hussaini M Y, Quarteroni A and Zang T A 1991 *Spectral Methods in Fluid Dynamics* (Berlin: Springer-Verlag)
[3] Spall R 1995 *International Journal of Heat Mass Transfer* **15** 2743
[4] Dauksher W and Emery A F 1996 *Review of Progress in Quantitative Nondestructive Evaluation* **15** 97
[5] Seriani G 1998 *Computational Methods Applied in Mechanical Engineering* **164** 235
[6] Wilcox P D 2003 *IEEE Transactions on Ultrasonics, Ferroelectrics, and Frequency Control* **50** 699
[7] Giurgiutiu V and Bao J J 2004 *Structural Health Monitoring* **3** 121
[8] Pozdrikidis C 2005 *Introduction to Finite and Spectral Element Methods using MATLAB* (Boca Raton: Chapman & Hall/CRC)
[9] Żak A 2009 *Finite Elements in Analysis and Design* **45** 650
[10] Ostachowicz W, Kudela P, Krawczuk M and Żak A 2012 *Guided Waves in Structures for SHM. The Time-domain Spectral Element Method* (Singapore: John Wiley & Sons Ltd.)
[11] Żak A and Ostachowicz W 2011 *Structural Health Monitoring 2011* **2** 2560
[12] Żak A, Radzieński M, Krawczuk M and Ostachowicz W 2012 *Smart Materials and Structures* **21** 035024
[13] Ulitko A F 1968 *Prikladnaya Mekchanika* **4** 38
[14] Duffey T A, Pepin J E, Robertson A N, Steinzig M L and Coleman K 2007 *Journal of Vibration and Acoustics* **129** 363
[15] Hasheminejad S M and Mirzaei Y 2011 *Journal of Sound and Vibration* **330** 229
[16] Reddy J N 1993 *An Introduction to the Finite Element Method* (Singapore: McGraw-Hill)
[17] ZienkiewiczO C 1989 *The Finite Element Method* (London: McGraw-Hill)
[18] Blevins R D 1979 *Formulas for Natural Frequency and Mode Shape* (Malabar: Krieger Publishing Company)
[19] Barrera R G, Estevez G A and Giraldo J 1985 *European Journal of Physics* **6** 287
[20] Mac Neal R H and Harder R L 1985, *Finite Elements in Analysis and Desing* **1** 3.

On the oscillation death phenomenon in a double pendulum system with autoparametric interaction

I Kovacic[1], M Zukovic[1] and M P Cartmell[2]
[1]Department of Mechanics, Faculty of Technical Sciences, University of Novi Sad, Novi Sad, Serbia
[2] Systems, Power and Energy Division, School of Engineering, University of Glasgow, Glasgow, Scotland, United Kingdom

Abstract. This study is concerned with autoparametric interaction in a four degree of freedom damped mechanical system consisting of two identical pendula fitted onto a horizontal massive rod which can oscillate vertically and rotationally. One pendulum is harmonically excited. The equations of motion indicate that autoparametric interaction is possible by means of typical external and internal resonance conditions involving the system natural frequencies and excitation frequency. An intriguing phenomenon is demonstrated when the unforced pendulum is decoupled and no energy goes into it, as a result of which it stops oscillating. Numerical simulations are carried out to determine when and why this phenomenon occurs for a different excitation magnitude as well as for zero and non-zero initial conditions of the unforced pendulum.

1. Introduction

A pendulum is one of the simplest mechanical systems and one of the most distinctive scientific paradigms. Although its applications date back deeply to the past, a profound change in epistemology was related to the 17th century analyses of pendulum motion and its utilisation in clockwork by Galileo Galilei and Christiaan Huygens [1]. Since then a wealth of phenomena has been associated with its behaviour in different sets of elements, structures or devices. This includes, for example, whirling motion, chaotic behaviour, synchronisation, etc. An intriguing phenomenon is also *oscillation death* (*amplitude death*), which corresponds to the case when oscillations decay to zero in a coupled system, whereas each subsystem can oscillate when isolated. This phenomenon has recently been reported in a reconsideration of Huygens's observations [2]. Therein, an updated version of the two-clock system with pendula attached to a common frame free to move in one dimension is re-examined. In the clock system, the pendulum is attached to an escapement mechanism, which alternately blocks and releases an escape wheel as the pendulum oscillates and is modelled as a simple impulse rule. Pantaleone [3] suggested that the oscillation death phenomenon may also exist in a system of metronomes placed on a common freely moving base, which was demonstrated by the lead author in the fourth demonstration of the video shown in [4]. Metronomes have an escapement mechanism as well, which Pantaleone modelled as a van der Pol term, thus representing metronomes generally as van der Pol oscillators, i.e. self-excited oscillatory systems. Self-excited autonomous oscillators with stronger coupling or time delay in the coupling were thought for a while to be the only systems associated with the oscillation death phenomenon, with no analogy to systems with periodic forcing and direct coupling [5, 6]. However, Pisarchik showed theoretically [6] that oscillation death can also occur in a nonautonomous, parametrically excited system of two coupled pendula each of which is represented as a two-well potential oscillator. In addition, he envisaged further investigations in this area and their experimental realisation as a worthwhile endeavour. The following study is seen as an attempt to contribute to these investigations. A practically realisable mechanism consisting of two pendula attached to an interconnecting rigid rod which can oscillate vertically and rotationally is proposed as a nonautonomous system with autoparametric interaction in which one of the pendula becomes uncoupled and stops oscillating.

2. Mechanical and mathematical model

The mechanical system under consideration is shown in Figure 1. It consists of a massive rod, a slider, two pendula and two springs.

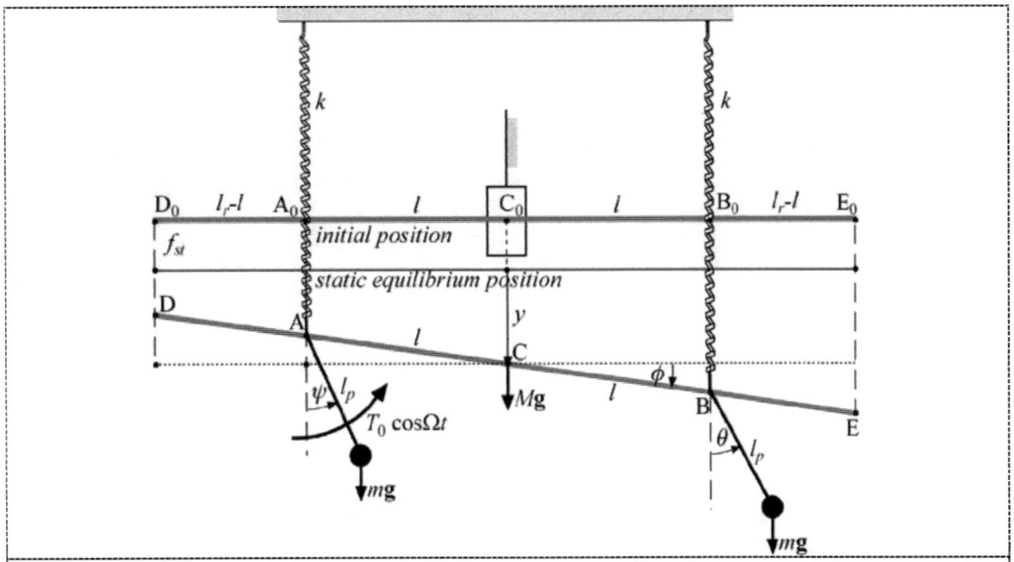

Figure 1. Mechanical system under consideration in motion (after Figure 2 in [7])

The half-length of the rod is l_r and its mass is M. The initially horizontal rod is pivoted on a very low friction bearing at its centre C, while the outer race of this bearing is mounted on a linear bearing allowing rotation of the bar around the axis that is perpendicular to the rod and passes through its centre. This bearing is mounted on the slider whose mass is negligible. The slider enables the rod to perform vertical translation, too. Two springs of stiffness k are equi-positioned at points A and B at distance l from the centre C. Two identical pendula are fitted below points A and B. It is assumed that the mass m within each pendulum is effective as a lump at each end of the rod the length of which is l_p. The system has four degrees of freedom and the generalised coordinates are: y, ϕ, ψ and θ, all labelled in Figure 1. It is assumed that the rotations through ϕ are small so that points A and B just move vertically. A torque is applied to the left pendulum, as defined in Figure 1. In practice, this pendulum would be excited by a brushless dc servomotor directly driving the pendulum at the pivot. The servomotor would be driven by a power amplifier with feedback velocity control, and the input to the amplifier would be provided by a variable frequency function generator operating in its sinusoidal mode.

The total kinetic energy of the system is:

$$T = \frac{1}{2}M\dot{y}^2 + \frac{1}{2}J\dot{\phi}^2 + \frac{1}{2}m\left[\left(l_p\dot{\psi}\cos\psi\right)^2 + \left(l_p\dot{\theta}\cos\theta\right)^2 + \left(\dot{y} - l\dot{\phi}\cos\phi - l_p\dot{\psi}\sin\psi\right)^2 + \left(\dot{y} + l\dot{\phi}\cos\phi - l_p\dot{\theta}\sin\theta\right)^2\right] \tag{2.1}$$

where J is the moment of inertia of the rod with respect to the axis that passes through the centre of mass C, and is perpendicular to the plane in which the system is placed.

The system potential energy is given by:

$$V=\frac{1}{2}k\left(f_{st}+y-l\sin\phi\right)^2+\frac{1}{2}k\left(f_{st}+y+l\sin\phi\right)^2- \\ mg\left(f_{st}+y-l\sin\phi+l_p\cos\psi\right)-Mg\left(f_{st}+y\right)-mg\left(f_{st}+y+l\sin\phi+l_p\cos\theta\right), \tag{2.2}$$

where the static deflection of the springs is $f_{st}=(M+2m)g/(2k)$. Since the torque acts on the left pendulum, as shown in Figure 1, the generalised force corresponding to the coordinate ψ is $Q_\psi=T_0\cos\Omega t$, while the rest of the generalised forces are zero. By using this fact as well as equations (2.1) and (2.2), the equations of motion are derived from Lagrange's equations of the second kind. Then, by approximating the trigonometric function of the angles by the first term of their respective MacLaurin series, the equations of motion are written down in the following non-dimensional form:

$$\tilde{y}''+2\zeta_y\tilde{\omega}_y\tilde{y}'+\tilde{\omega}_y^2\tilde{y}-r\left(\tilde{\theta}''\tilde{\theta}+\tilde{\theta}'^2+\tilde{\psi}''\tilde{\psi}+\tilde{\psi}'^2\right)=0,$$
$$\tilde{\phi}''+2\zeta_\phi\tilde{\omega}_\phi\tilde{\phi}'+\tilde{\omega}_\phi^2\tilde{\phi}-p\left(\tilde{\theta}''\tilde{\theta}+\tilde{\theta}'^2-\tilde{\psi}''\tilde{\psi}-\tilde{\psi}'^2\right)-s\tilde{\phi}'^2\tilde{\phi}=0,$$
$$\tilde{\psi}''+2\zeta_\psi\tilde{\omega}_\psi\tilde{\psi}'+\left(1-\frac{p}{r}\tilde{y}''+\tilde{\phi}''\right)\tilde{\psi}-\frac{r^2}{p^2}\tilde{\phi}'^2\tilde{\phi}\tilde{\psi}=\Gamma\cos\tilde{\Omega}\tau, \tag{2.3a-d}$$
$$\tilde{\theta}''+\zeta_\theta\tilde{\omega}_\theta\tilde{\theta}'+\left(1-\tilde{\phi}''-\frac{p}{r}\tilde{y}''\right)\tilde{\theta}+\frac{r^2}{p^2}\tilde{\phi}'^2\tilde{\phi}\tilde{\theta}=0,$$

where the following notation has been introduced:

$$y=\tilde{y}l,\quad \phi=\tilde{\phi}\frac{l_p}{l},\quad \psi=\tilde{\psi}\frac{l_p}{l},\quad \theta=\tilde{\theta}\frac{l_p}{l},\quad t=\frac{\tau}{\omega},\quad \omega^2=\frac{g}{l},$$
$$\tilde{\omega}_y^2=\frac{2kl_p}{(2m+M)g},\quad \tilde{\omega}_\phi^2=\frac{2kl^2l_p}{\left(2ml^2+J\right)g},\quad \tilde{\omega}_\psi^2=\tilde{\omega}_\theta^2=\frac{\omega_\psi^2}{\omega^2}=\frac{\omega_\theta^2}{\omega^2}=1, \tag{2.4a-n}$$
$$\tilde{\Omega}=\frac{\Omega}{\omega},\quad r=\frac{ml_p^3}{(2m+M)l^3},\quad p=\frac{ml_p^2}{2ml^2+J},\quad s=\frac{2ml_p^2}{2ml^2+J},\quad \Gamma=\frac{T_0l}{ml_p^3\omega^2},$$

with the primes denoting differentiation with respect to τ. Linear viscous damping has also been added to equations (2.3a-d), where the corresponding ratios are defined by ζ_q, $q\in\{y,\phi,\psi,\theta\}$.

It is seen that equations (2.3a-d) represent a set of coupled, non-homogenous ordinary differential equations in which the response of one element parametrically excites the other and where quadratic nonlinearities exist. These features are typical for autoparametric interaction [8, 9]. In general, autoparametric interactions involve subsystems which interact with a 2:1 ratio between principal natural frequencies, so that nonlinear coupling and frequency tuning affects the actual response, causing, for example, large exchanges of energy between the subsystem. Thus, the system under consideration belongs to the class of autoparametric interaction. The natural question arising is how it behaves under typical external and internal resonance conditions involving the system natural frequencies and excitation frequency.

To satisfy the 2:1 frequency ratio condition, the parameters existing in $\tilde{\omega}_y$ are chosen to be as listed in Table 1. The rod is assumed to be a 1'' (25.4 mm) square section stock rod of 6082T6 or 2014T4 aluminium alloy of mass per unit length 1.76 kg/m. Its half-length is 1 m and its mass is M = 3.52 kg. The length of the pendulum rods is l_p = 0.2 m. The pendula are assumed to be made of

$5/16''$(7.94 mm) square section aluminium alloy to the same material specification, but with a mass per unit length of 0.17 kg/m and the bob mass is m = 0.034 kg. The spring stiffness is k = 352 N/m.

Table 1. Physical data

Elements	Rod		Pendula		Spring
Parameters	l_r	M	l_p	m	k
Values	1 m	3.52 kg	0.2 m	0.034 kg	352 N/m

Knowing that $J = M(2l_r)^2/12$, it follows that $\tilde{\omega}_\phi$ equals $\tilde{\omega}_y$ if $l^* = l_r/\sqrt{3}$. This resonance condition defines the position of points A and B (Figure 1), where the springs and the pendula should be attached to the rod. However, imprecise fitting in practice can cause this position to be different than calculated, which would cause the frequency mismatch. The aim of the following section is to explore how the change of the length l with respect to l^* influences the system behaviour.

3. Behaviour due to the frequency match/mismatch

In order to study the behaviour of the pendula for the case when the condition for the internal resonance 2:1 is slightly violated, the ratio between the lengths l and l^* is introduced

$$\Delta_l = l/l^*, \tag{3.1}$$

and substituted into the equations of motion (2.3a-d) as $l = \Delta_l \cdot l^*$. The left pendulum is assumed to be in a perfect external resonance, i.e. the condition $\tilde{\Omega} = 1$ is used and all the damping ratios are assumed to be of the same value $\zeta_y = \zeta_\phi = \zeta_\psi = \zeta_\theta = 0.015$. While changing the magnitude T_0, it has been found that, when in perfect resonance, or around it, the right pendulum can perform qualitatively different motion, which includes the case when the oscillations die out, while the left pendulum stays in motion. Numerical results corresponding to three different values of T_0 are given in Figure 2. They show the amplitude of the left pendulum $A_{\tilde{\psi}}$ as a function of Δ_l obtained numerically by using the following two approaches: i) the system of equations (2.3a-d) is solved with a non-zero initial value of $\tilde{\theta}$ and $\tilde{y}(0) = 0.01, \tilde{\phi}(0) = \tilde{\psi}(0) = \tilde{\theta}(0) = \pi/72$ with initial generalised velocities all being zero; ii) the system (2.3a-c) is solved given the fact that $\tilde{\theta} = 0$ represents a particular integral of (2.3d) for the same initial conditions listed earlier, but excluding the non-zero initial displacement for $\tilde{\theta}$. In both cases the region of Δ_l in which the amplitude of the right pendulum ceases is found to exist. It is shown as a thick solid black line in Figure 2, which indicates that for the parameters used, this region becomes narrower with the increase of T_0. To illustrate typical responses of the right pendulum, three cases C1, C2 and C3 with respect to different values of Δ_l are chosen (Figure 2b) and plotted in Figure 3a - c. In the cases C1 and C3, although having a very small initial amplitude, the amplitude of the right pendulum increases considerably, reaching a large value. In the case of a small region close to the perfect internal resonance condition C2, the oscillations die out. This case underpins the use of autoparametric interaction to achieve oscillation death in this system.

To explain the reason for the occurrence of this phenomenon, the parametric excitations for the right pendulum $\tilde{\phi}''$ and $p\tilde{y}''/r$ existing in equation (2.3d) are plotted separately for the cases C1 (Figure 3d), C2 (Figure 3e) and C3 (Figure 3f) and their sum is shown, too. They show that in the cases C1 and C2 the sum of these parametric excitations is different from zero, while in the case C2,

Journal of Physics: Conference Series **382** (2012) 012055 doi:10.1088/1742-6596/382/1/012055

they have the opposite phase and, thus, a zero sum, as a result of which the right pendulum is not parametrically excited, but performs damped oscillatory motion, with the amplitude tending to zero.

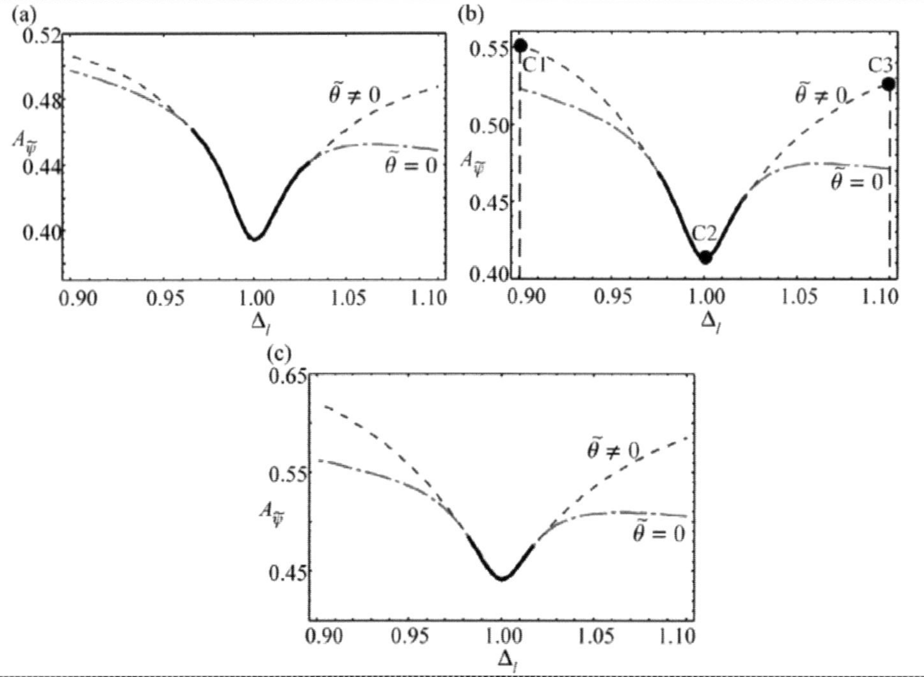

Figure 2. Amplitudes of the left pendulum $A_{\tilde{\psi}}$ for: (a) $T_0 = 0.00208\,\text{Nm}$, (b) $T_0 = 0.00231\,\text{Nm}$, (c) $T_0 = 0.0027\,\text{Nm}$. Dashed lines - numerical solutions of equation (2.3a-d) for $\tilde{\theta}(0) \neq 0$, dashed-dotted lines - numerical solutions of equation (2.3a-c) for $\tilde{\theta} = 0$ and thick solid lines - the case when the oscillations of the right pendulum die out. The response corresponding to C1 – C3 is plotted in Figures 3a - c.

4. Conclusions

In this study, the phenomenon of oscillation death is shown to exist under certain conditions in a four-degree-of-freedom autoparametric mechanical system that consists of two pendula attached to a rod that can oscillate vertically and rotationally. When this phenomenon occurs the externally forced damped pendulum oscillates and the unforced damped pendulum stops oscillating as it becomes uncoupled. The reason for that is that the autoparametric excitations stemming from translatory and rotational motion of the rod are in opposite phases and cancel each other.

Future work will be directed towards answering the question related to the combinations of the magnitude of the external excitation and viscous damping ratios that yield oscillation death and other qualitatively different behaviour as well as to the investigation of this phenomenon in the same mechanism, but without working under the restrictions of small oscillations as in this study.

Acknowledgments

This study has financially been supported by the Provincial Ministry of Science and Technological Development, Autonomous Province of Vojvodina, Republic of Serbia (Project 'Oscillators with parametric excitation: behaviour, phenomena and applications').

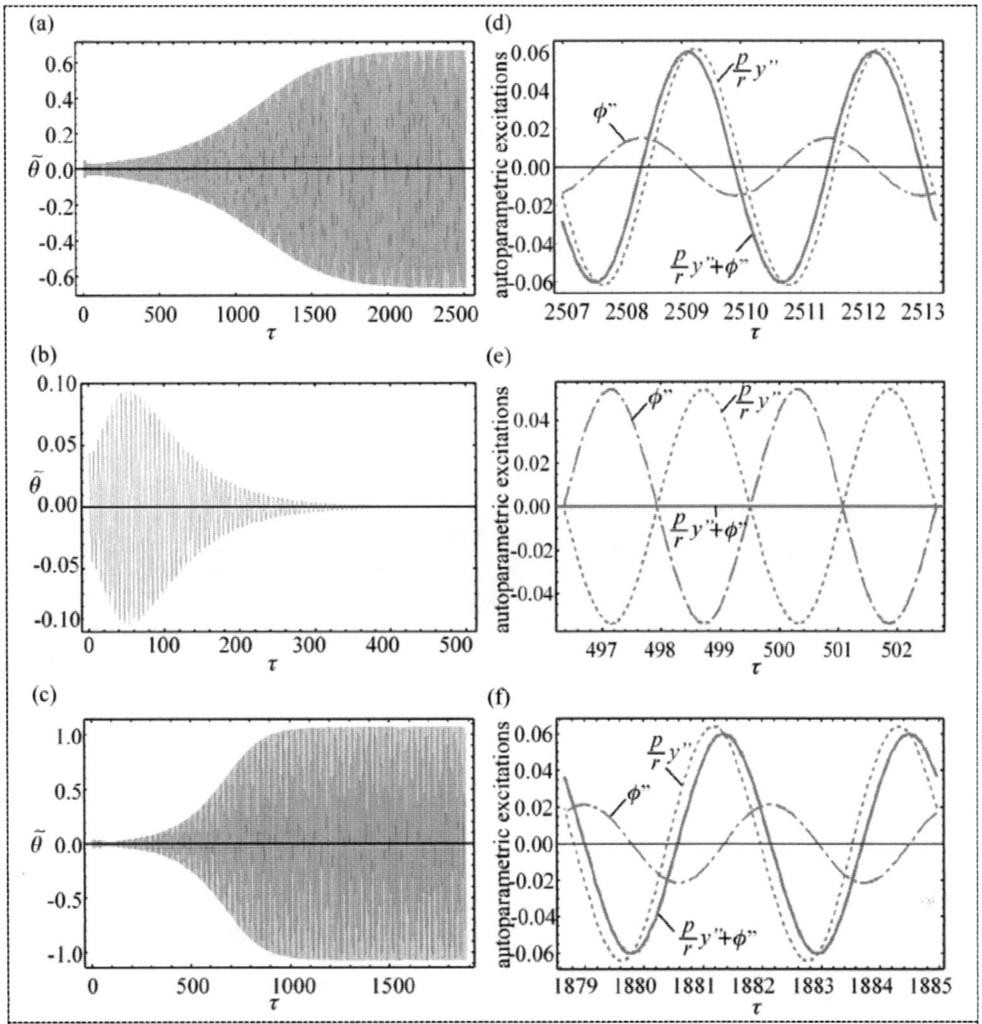

Figure 3. (a) - (c): Response of the right pendulum for the cases C1 – C3, respectively, labelled in Figure 2c; (d) - (e): Parametric excitations of the right pendulum from equation (2.3d) corresponding to C1 – C3, respectively.

5. References

[1] Matthews M R 2000 *Time for Science Education: How Teaching the History and Philosophy of Pendulum Motion can Contribute to Science Literacy* (New York: Kluwer Academic)

[2] Bennett M, Schatz M F, Rockwood H and Wiesenfeld K 2002 *Proc. R. Soc. Lond. A* **458** 563

[3] Pantaleone J 2002 Am. J. Phys. **70** 992

[4] http://www.youtube.com/watch?v=cauJViEl3KQ&feature=youtu.be

[5] Pikovsky A, Rosenblum M and Kurths J 2001 *Synchronization: A Universal Concept in Nonlinear Science* (New York: Cambridge Univ. Press)

[6] Pisarchik A N 2003 Phys. Lett. A **318** 65

[7] Cartmell M P, Kovacic I and Zukovic M 2012 *Proc IMechE, Part C: J Mechanical Engineering Science* (*In press* DOI: 10.1177/0954406212441748)

[8] Cartmell M P 1990 *Introduction to Linear, Parametric and Nonlinear Vibrations* (London: Chapman and Hall)

[9] Nayfeh A H 200 *Nonlinear Interactions: Analytical, Computational and Experimental Methods.* (New York: Wiley-Interscience)

Journal of Physics: Conference Series **382** (2012) 012056 doi:10.1088/1742-6596/382/1/012056

The fractional and nonlinear magneto-flexible rod

S A David [1,5], J M Balthazar [2], B H S Julio[3] and C Oliveira[4]

[1] University of São Paulo at Pirassununga, Pirassununga, SP, Brazil
[2] São Paulo State University at Rio Claro, Rio Claro, SP, Brazil
[3] University of São Paulo at Lorena, Lorena, SP, Brazil,
[4] Federal University of Dourados, Dourados, SP, Brazil

Abstract. This paper deals with a system involving a flexible rod subjected to magnetic forces that can bend it while simultaneously subjected to external excitations produces complex and nonlinear dynamic behavior, which may present different types of solutions for its different movement-related responses. This fact motivated us to analyze such a mechanical system based on modeling and numerical simulation involving both, integer order calculus (IOC) and fractional order calculus (FOC) approaches. The time responses, pseudo phase portraits and Fourier spectra have been presented. The results obtained can be used as a source for conduct experiments in order to obtain more realistic and more accurate results about fractional-order models when compared to the integer-order models.

1. Introduction

The theory of fractional calculus dates back to the birth of the theory of differential calculus, but its inherent complexity delayed the application of its associated concepts. In fact, fractional calculus is a natural extension of classical mathematics. Since the inception of the theory of differential and integral calculus, mathematicians such as Euler and Liouville developed their ideas about the calculation of non-integer order derivatives and integrals. Perhaps the subject would be more aptly called "integration and differentiation of arbitrary order."

The basic aspects of the theory of fractional calculus are outlined in [14]. Insofar as it concerns the application of its concepts, we can cite research in different areas such as viscoelastic damping [1], robotics and control [7-9], signal processing [15], electric circuits [10]. As for the adoption of this concept in other scientific areas, several researchers have been inspired to examine this new possibility. Some work has been carried out in the field of dynamical systems theory, but the proposed models and algorithms are still in the preliminary stage.

With these ideas in mind, this work introduces the fundamentals of fractional order calculus (FOC) in order to investigate, by means of numerical simulations, the modified motion equation of a flexible rod induced by two electromagnets and submitted to external periodic excitations.

2. Fundamentals of fractional calculus

Fractional order calculus can represent systems with high-order dynamics and complex nonlinear phenomena using few coefficients, since the arbitrary order of the derivatives provides an additional degree of freedom to fit a specific behavior.

Numerous mathematicians have contributed to the history of fractional calculus by attempting to solve a fundamental problem to the best of their understanding [14]. Each researcher sought a definition and therefore different approaches, which has led to various definitions of differentiation

[5] sergiodavid@usp.br

and antidifferentiation of non-integer orders that are provenly equivalent. Although all these definitions may be equivalent, from one specific standpoint, i.e., for a specific application, some definitions seem more attractive.

The three most commonly used definitions are the the Caputo, Riemann-Liouville and Laurent definitions. In this paper, we emphasize our simulations in Riemann-Liouville definitions, i.e.,

$$ {}_{c}D_{x}^{\nu} f(x) = \frac{d^{m}}{dx^{m}} \left[\frac{1}{\Gamma(\rho)} \int_{0}^{x} (x-t)^{\rho-1} f(t) dt \right] \tag{1} $$

where, for convenience, m is considered the smallest integer larger than υ and $0 < \rho \leq 1$.

3. Mechanical system-theoretical model

Many technical devices such as motors, generators, transformers, and fusion reactors are known to employ wide elastic plates in magnetic fields. A system involving a flexible rod subjected to magnetic forces that can bend it while simultaneously subjected to external excitations produces complex and nonlinear dynamic behavior [3-5], which may present different types of solutions for its different movement-related responses [2]. This fact motivated us to analyze such a mechanical system based on modeling and numerical simulation involving both, (IOC) and (FOC), approaches.

A continuum model based on linear elastic and nonlinear magnetic forces was developed, which can be reduced to an oscillator model with a single degree of freedom using the Lagrangian formalism [6] and the Galerkin's method [4-6].

The mechanical system whose theoretical model is developed is shown in Figure 1. A flexible rod is clamped in a rigid base. The electromagnets pull the beam in opposite directions and intensity of the field is strong enough to deflect the rod from one side to another.

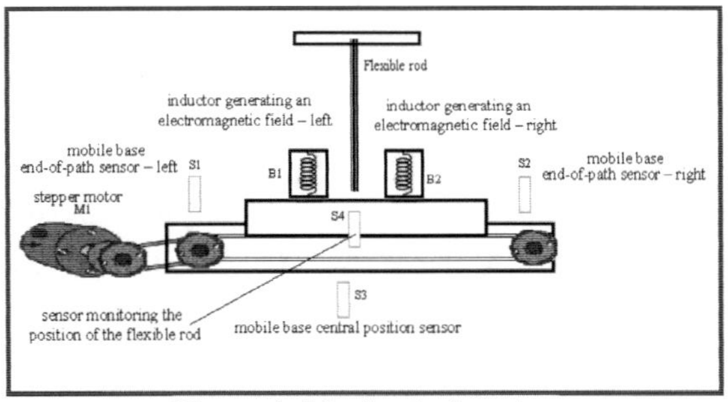

Figure 1: Mechanical system

The electromagnets generate a magnetic field that induces a magnetization **M** per unit volume in the solid. The rod can be modeled as a soft magnetic material where **M** is proportional to the local magnetic field in the solid [12]; i.e.,

$$ \vec{M} = \left[\frac{\chi}{(\chi+1)} \right] \frac{\vec{B}}{\mu_0} \tag{2} $$

μ_0 is the magnetic permeability of a vacuum; χ is the magnetic susceptibility

The field **B** can be written in terms of the field $\mathbf{B^0}$ produced by external magnets, and a field produced by the magnetization itself, $\mathbf{B^1}$. If self-forces on the rod are neglected then the external magnets produce both a force and moment distribution on the beam given by [12,13]

$$ \vec{F} = \vec{M} \cdot \nabla B^{0} \tag{3} $$

$$ \vec{C} = \vec{M} \times \vec{B}^{0} \tag{4} $$

These forces can be derived from a magnetic potential and therefore they are conservative. Thus,

$$W = -\frac{1}{2}\int \vec{M}\cdot\vec{B}^0\, dv \tag{5}$$

The non-linearities included in the analysis reflect the inhomogeneous nature of the magnetic field $\mathbf{B}^0$ [13] and the magnetic force and couple. If the x, y components of $\mathbf{B}_0$ are introduced, defined by $B_{0x}=B_0\cos\alpha$ and $B_{0y}=B_0\sin\alpha$ and the local slope of the rod with the x axis by θ, then the magnetic energy potential then takes the form

$$W = -\frac{\chi}{4\mu_0\mu_r}\int_0^L \left(B_1 + B_2\sin 2\theta + B_3\cos 2\theta\right)ds \tag{6}$$

with:

$B_1 = (\mu_r + 1)(B_{0x}^2 + B_{0y}^2)$; (7) $B_2 = 2(\mu_r - 1)B_{0x}B_{0y}$; (8) $B_3 = (\mu_r - 1)(B_{0x}^2 - B_{0y}^2)$. (9)

Here the integration is carried out over the original length of the rod and B_1, B_2 and B_3 are functions of the rod displacement. The nonlinear elastic forces are small even for the large displacements of the rod tip. Moreover, if a single mode approximation is made for the beam deformation, the elastic energy can be write in the form

$$P = \tfrac{1}{2} K y^2 + \text{(higher order terms)} \tag{10}$$

With the usage Lagrangian's, the Galerkin's method, and a typical choice for the ϕ_i in vibration problems are the normal modes of the associated linear problem, which when substitutes in equation of motion allow us write it with the dimensionless form [2],

$$\ddot{x} + \delta\dot{x} - x + x^3 = F\cos\omega t \tag{11}$$

where:

$\delta > 0$ is the damping constant, F is the forcing strength and ω is the forcing frequency.

In this article we investigate, by means of numerical simulations, the modified magneto-flexible rod dynamic equations (12) , i.e.,

$$x^{\lambda+1} + \delta x^{\lambda} - x + x^3 = F\cos\omega t \tag{12}$$

We presented the time responses of x and x^λ, pseudo phase portraits and Fourier spectra for some values of λ, ω, F and for $\delta = 0.2$.

4. Some numerical simulation results

We would like to illustrate some simulations results for non-integer order equation (12) in comparison to integer order for F = 0.18N , ω=1rad/s and λ = [0.8(red line);1.0(blue line); 1.2(green line)].

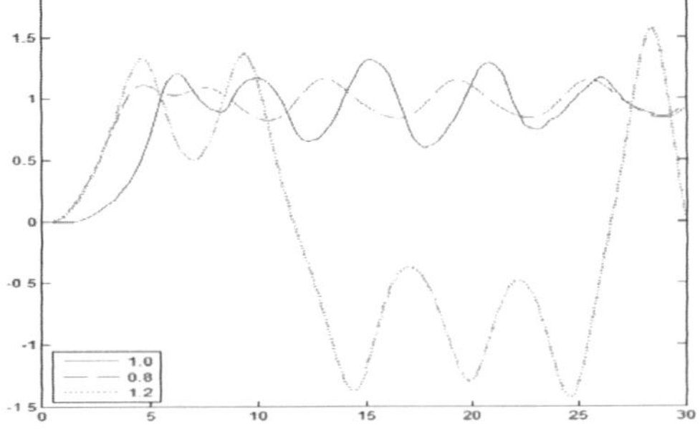

Figure 2 : x vs. t.

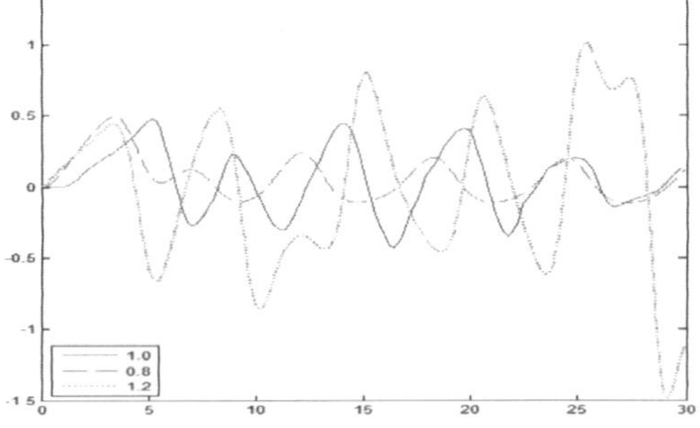

Figure 3: x^λ vs. t.

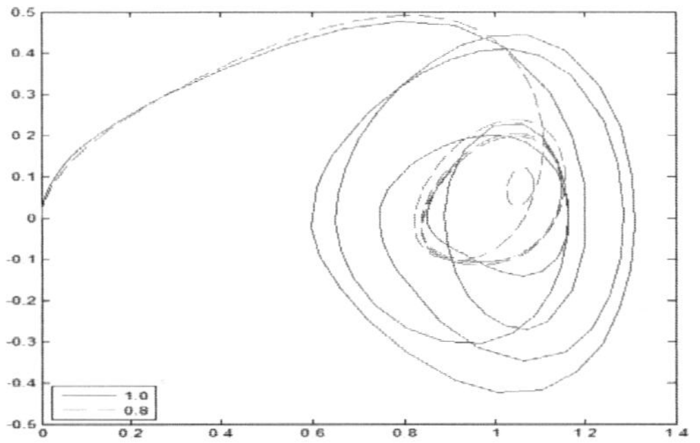

Figure 4 : Phase plane(x^{λ} vs. x), for $\lambda = \{0.8,1\}$.

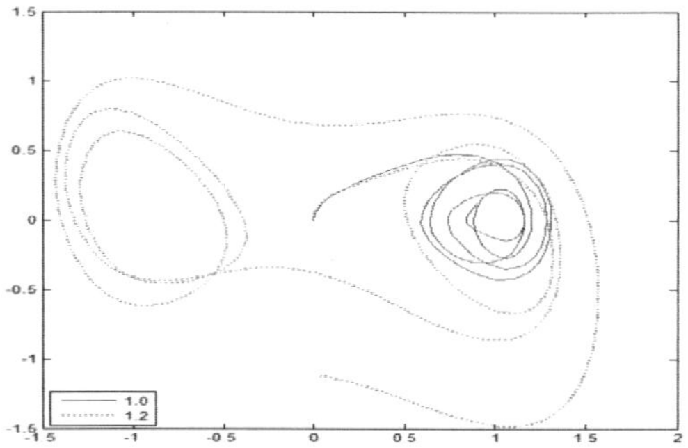

Figure 5: Phase plane (x^{λ} vs. x), for $\lambda = \{1, 1.2\}$.

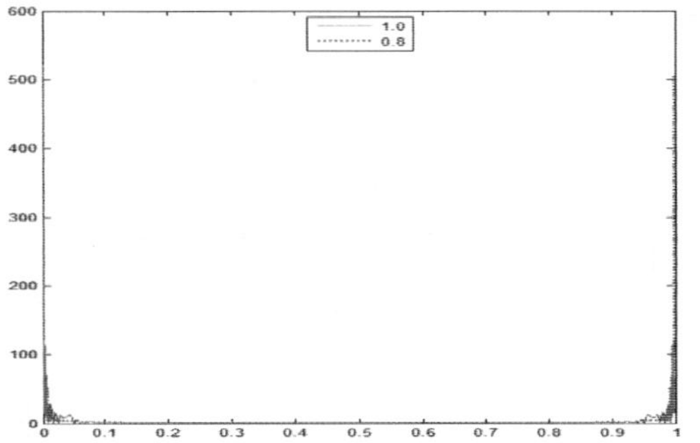

Figure 6: FFT for $\lambda = \{0.8, 1\}$.

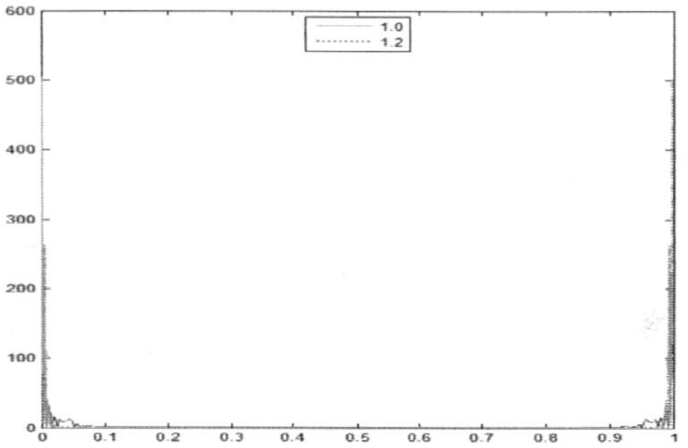

Figure 7: FFT for $\lambda = \{1, 1.2\}$.

We have illustrated too some simulations results for non-integer order equation (12) in comparison to integer order for F = 4N , ω=1.12 rad/s and λ = [0.8(red line); 1.0(blue line); 1.2(green line)].

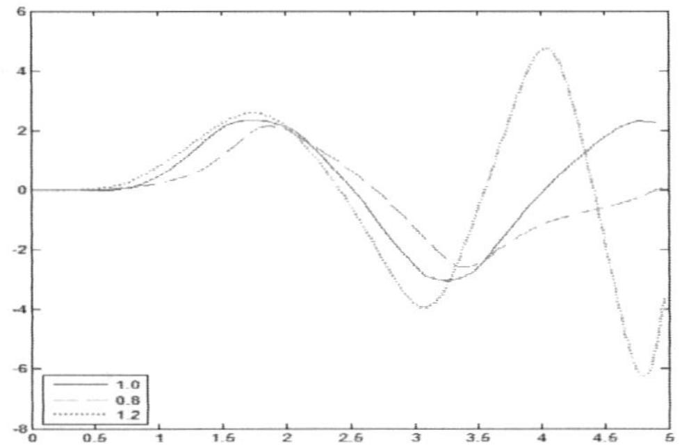

Figure 8: x vs.t.

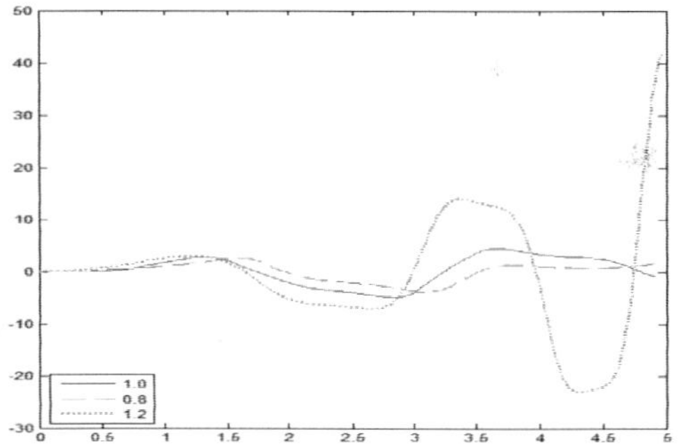

Figure 9: x^{λ} vs. t.

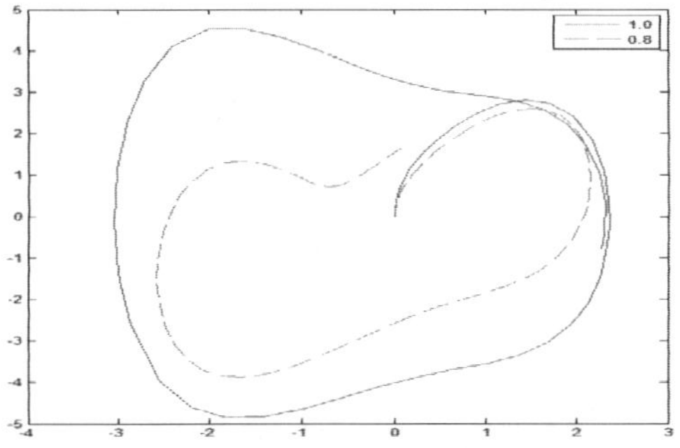

Figure 10: Phase plane, (x^{λ} vs. x), for $\lambda = \{0.8, 1\}$.

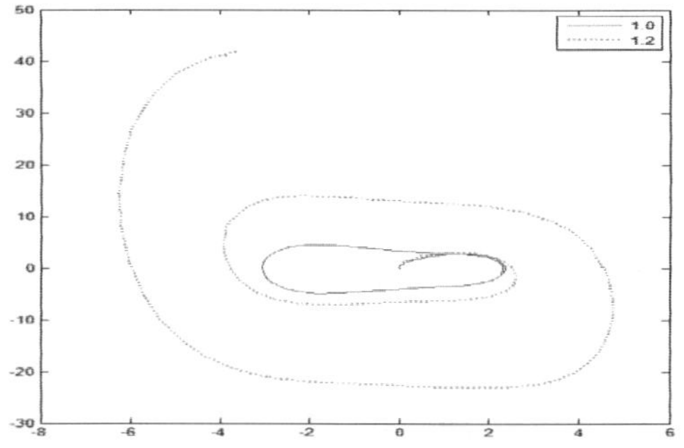

Figure 11: Phase plane , (x^{λ} vs. x), for $\lambda = \{1, 1.2\}$.

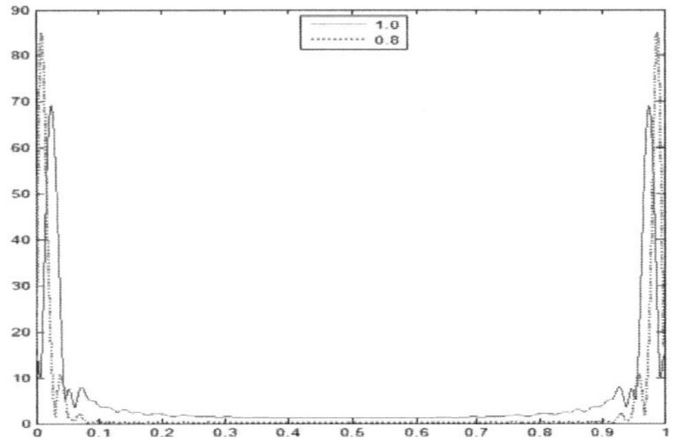

Figure 12: FFT for λ = {0.8, 1}.

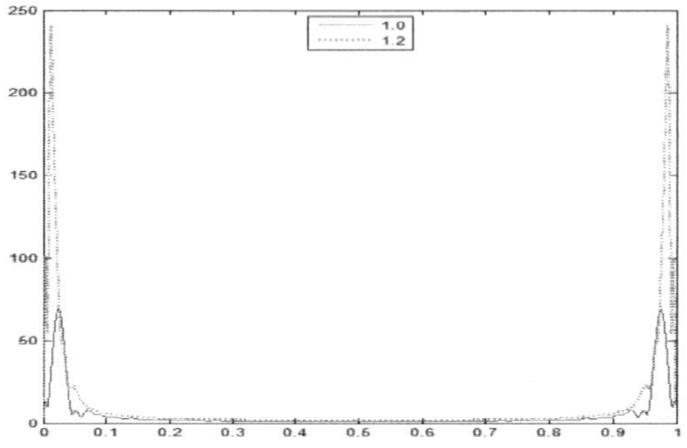

Figure 13: FFT for λ = {1, 1.2}.

We also have illustrated some simulations results for non-integer order equation (12) in comparison to integer order for F = 0.5N , ω=1.5 rad/s and λ = [0.8(red line); 1.0(blue line); 1.2(green line)].

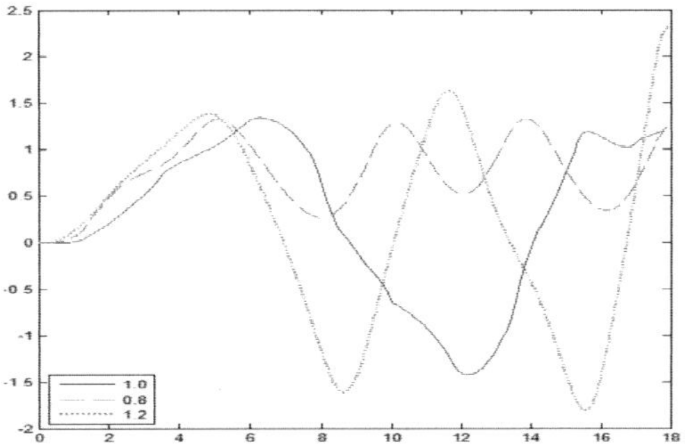

Figure 14: x vs.t.

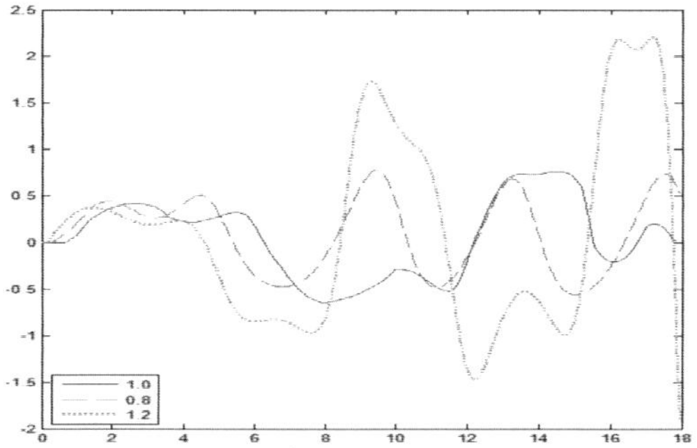

Figure 15: x^λ vs. t.

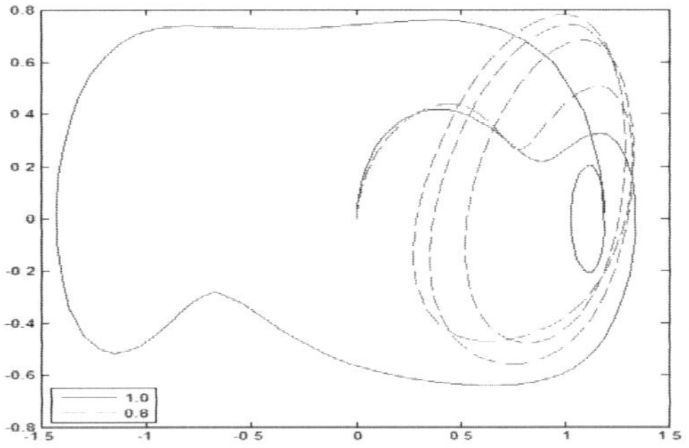

Figure 16: Phase plane, (x^λ vs. x), for λ = {0.8, 1}.

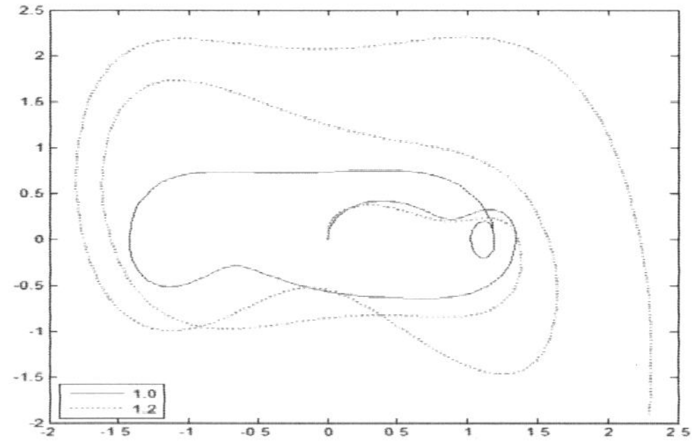

Figure 17: Phase plane , (x^λ vs. x), for λ= {1, 1.2}.

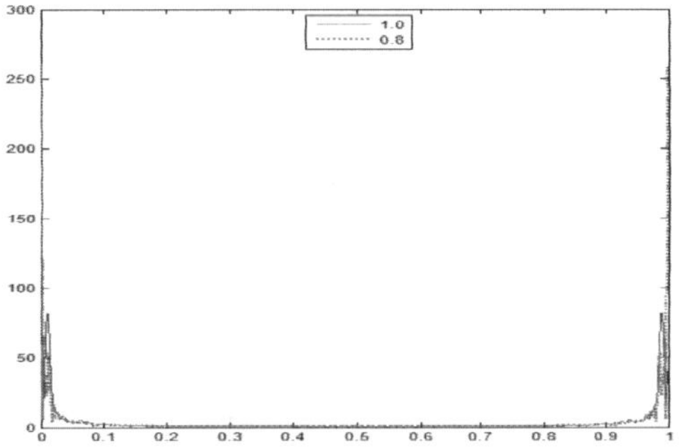

Figure 18: FFT for λ = {0.8, 1}.

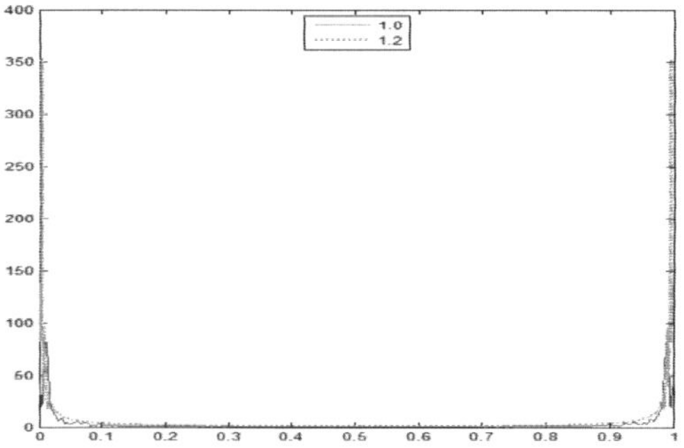

Figure 19: FFT for λ = {1, 1.2}.

5. Conclusions

In this paper we have proposed some versions of the modified magneto-elastic dynamic equation like presented by equation (12). Such modifications consisted in the introduction of a non-integer order time derivative in the standard equation (11). The results reveal that the non-integer order systems can exhibit different and curious behaviour from those obtained with the standard magneto-flexible rod dynamic system. The non-integer order can be useful for a better understanding and control of such nonlinear systems. Also, the results obtained can be used as a source for conduct experiments in order to obtain more realistic and more accurate results about fractional-order models when compared to the integer-order models.

Acknowledgments

The authors wish to acknowledge the financial support provided by the FAPESP, São Paulo Research Foundation (grant 2010/15824-0).

References

[1] Bagley R L and Torvik P J 1986 On the Fractional Calculus Model of Viscoelastic Behaviour, *Journal of Rheology,* **3**, 133-135.

[2] David S A, Rosario J M and Machado J 2001 Investigation about chaos in a magneto-elastic dynamical system. *Computational Fluid and Solid Mechanics - Solids and Structures*, Multi-Physics, vol 2, K.J.Bathe editor, (England: Elsevier Science Ltd) 1120-1123 ISBN 0-08-043944-6.

[3] Feynman R P, Leighton R, B and Sands M 1965 *The Feynman Lectures on Physics*, (Reading, MA: Addison-Wesley)

[4] Goldstein H 1981 *Classical Mechanics*, Reading Mass., Addison Wesley Publishing Company, 2nd Edition

[5] Guckenheimer J and Holmes P 1983 *Nonlinear Oscillations, Dynamical Systems, and Bifurcations of Vector Fields* (New York: Springer)

[6] Hagedorn P, 1984, *Oscilações Não Lineares*. (São Paulo: Editora Edgard Blucher LTDA)

[7] Machado J A T 2002 *System Modeling and Control Trough Fractional-Order Algorithms* Nonlinear Dynamics, Chaos, Control and Their Applications to Engineering Sciences, vol 4, pp 99-115

[8] Machado J A T 1987 *Analysis and Design of Fractional-Order Digital Control systems* SAMS – Journal Systems Analysis, Modelling, Simulation, vol. 27 , pp 107-122

[9] Machado J A T and Azenha A 1998 *Position/Force Fractional Control of Mechanical Manipulators* IEEE Int. Workshop on Advanced Motion Control, Coimbra, Portugal

[10] Mathieu B, Lay L L and Oustaloup A 1996 *Identifications of Non-Integer Order Systems in Time Domain*, IEEE-SMC/IMACS Symposium on Control, Optimization and Supervision, pp 952-956, Lille, France

[11] Moon F C 1992 *Chaotic and Fractal Dynamics: An Introduction for Applied Scientists and Engineers*" (New York: Wiley)

[12] Moon F C 1978 *Problems in Magneto-Solid Mechanics* chapter V in Mechanics Today - S. Nemat-Nasser Editor

[13] Nayfeh A H and Mook D T *1979 Nonlinear Oscillations* (New York: Wiley & Sons)

[14] Oldham K B and Spanier J 2006 *The Fractional Calculus: Theory and Applications of Differentiation and Integration to Arbitrary Order*, Dover Publication

[15] Ozaktas H M, Arikan O, Kutay M A and , Bozdagi G 1996 *Digital Computation of the Fractional Fourier Transform*, IEEE Trans. On signal Processing, vol 44, no 9, pp 2141-2150, Sept.

Aerofoil flutter: fluid-mechanical analysis and wind tunnel testing

A L Wensuslaus[1] and A J McMillan
Department of Aeronautical Engineering, University of Glamorgan, Wales, UK
E-mail: anto_aeen@yahoo.co.in

Abstract. This paper describes a three dimensional wing model, which has been developed for the purpose of studying flutter, both computationally and through wind tunnel testing. A three dimensional, laminar flow aerofoil wing, based on the NACA aerofoil has been designed. The natural frequencies for this aerofoil were obtained through modal analysis. A scale model wing, without taper was manufactured in the laboratory and tested in a wind tunnel. The pressure data was obtained from fluid flow analysis and the deformation results obtained through structural analysis. The analysis was performed in the ANSYS Workbench Environment, accessing FLUENT CFX for the computational fluid dynamics analysis and the ANSYS FEA package for the mechanical analysis. The computational results obtained are compared with the experimental data obtained in the wind tunnel. Comparison of the analysis and test results provides further understanding of the flutter characteristics.

1. Introduction

Flutter is an example of unstable self-excited vibration, and can occur under conditions of steady-state airflow. In the design of aircraft and aerospace components, design for aero-elastic performance is of fundamental importance: where flutter vibration amplitude cannot be controlled, catastrophic structural failure can result. The advent of computational flow analysis capability and high performance computing opens up the challenge of aero-elasticity assessment [1], and this work explores suitable methodologies. Computational fluid dynamics (CFD) and finite element method (FEM) provide the basic tools for predicting flutter, buffeting and limit cycle oscillation, hence computational aero-elasticity is expected to play a vital role in numerical modelling of combined solid-fluid interaction in the context of aerospace component and structure design. In this project, CFD analysis predicts the flow characteristics of the aerofoil, including turbulence and flow separation prediction. FEM analysis is used to predict the mode shapes and frequencies that may be excited in flutter [2, 3].

An additional objective of the project was to build a scale model of an aerofoil with a light-weight internal structure similar to that used in aircraft wing design. This aerofoil would be used in a wind tunnel, as an undergraduate laboratory demonstration of aero-elastic phenomena. With low airflow velocity and such a flexible structure, the intention was to find angle of attack conditions under which the aerofoil could be induced to flutter at controlled low amplitude. Under conditions of controlled amplitude, the energy supplied to the vibration from the air flow is exactly balanced by the vibration

[1] To whom any correspondence should be addressed

damping losses. Low amplitude vibration should give rise to low stress levels, and it is anticipate that the demonstrator aerofoil will have several years fatigue life in careful use.

2. Methodology

This paper describes a three dimensional wing model, which has been developed for the purpose of studying flutter, both computationally and through wind tunnel testing [4]. A three dimensional laminar flow aerofoil wing was designed in SolidWorks. The external aerofoil design was based on the scaled NACA aerofoil without taper, but the internal construction was hollow, and comprised of five subsections (Figure 1). The purpose of this light-weight structure was to make the aerofoil more flexible, and therefore more susceptible to aero-elastic phenomena. The natural frequencies for this aerofoil were obtained through modal analysis [4-7]. The scale model wing was manufactured in the laboratory and tested in a wind tunnel. The analysis was performed in the ANSYS Workbench environment, accessing FLUENT CFX for the computational fluid dynamics analysis, and ANSYS FEA for the mechanical analysis. The computational results obtained are compared with the experimental data obtained in the wind tunnel [2, 4]. Comparison of the analysis and test results provides further understanding of the flutter characteristics.

3. Wing design & Manufacturing

The wing has five aerofoil sections, joined by two stringers, and with a centre rod along the longitudinal direction of wing. The centre rod provides the mounting point into the wind tunnel. The aerofoil skeleton structure is then covered with a thin surface skin of aluminium foil. The wing structure is shown in the diagram below: length of the wing – 280 mm, chord – 160 mm, and the maximum thickness – 17 mm.

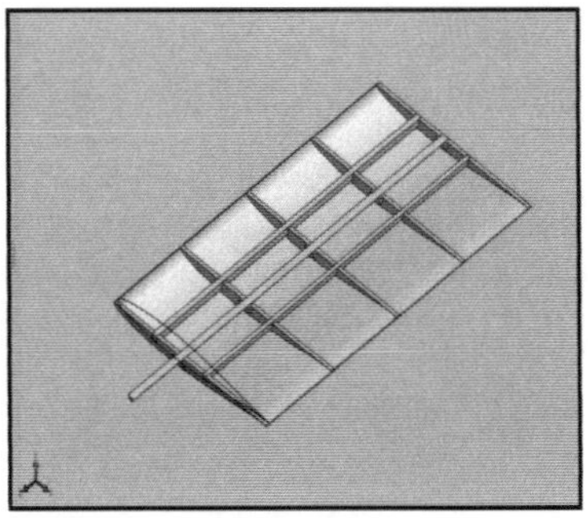

Figure 1. Cross-section through the wing in SolidWorks.

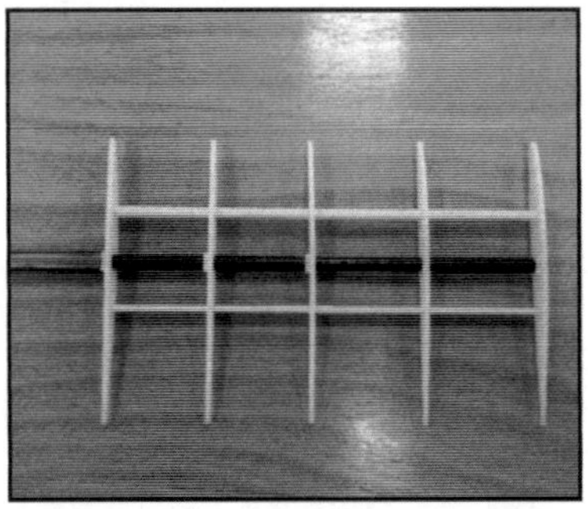

Figure 2. Assembled Wing structure without skin.

Table 1. Materials used in wing component manufacture.

No	Materials	Young's Modulus (GPa)	Poisson Ratio	Density (Kg/m^3)
1	ABS Plastic	2.62	0.35	1005
2	Bright steel	210	0.3	7850
3	Aluminium	70	0.35	2700

With the exception of the centre rod and the skin, all parts of the aerofoil were manufactured in ABS plastic using a rapid prototyping 3D printer. Initially all the aerofoil sections, the sheet and the rod were filed to remove any sharp edges. The two stringers were manufactured as four parts; and joined, the stringers were then assembled through bossed holes in the wing sections and glued for more support. The central rod was then passed through the aligned holes in the sections, and glued using araldite. The whole setup was then allowed to cure for 8 hours, after which the aluminium sheet was formed to the aerofoil shape and bonded around the structure, to form the final completed wing. It was then clamped for 24 hours to allow the glue to set.

4. Wind tunnel testing

Air enters the wind tunnel through the front cone shaped structure which increases the airspeed linearly before it reaches the test section. The air then leaves the test section and passes through the grill before reaching the fan and diffuser. The speed of the axial fan is controlled by an instrument unit attached separately to the wind tunnel. In this study the airspeed was varied from 5 m/s to 40 m/s and aerofoil incidence angles from 0° to 15°. The test model is fixed within the mount before starting the wind tunnel. The velocity and pressure values were measured using the Pitot-static tube fitted in the test section. All the results obtained are accessed by the computer connected to the electronic system [2, 4 and 8].

Figure 3.Wing mounted in the wind tunnel for testing.

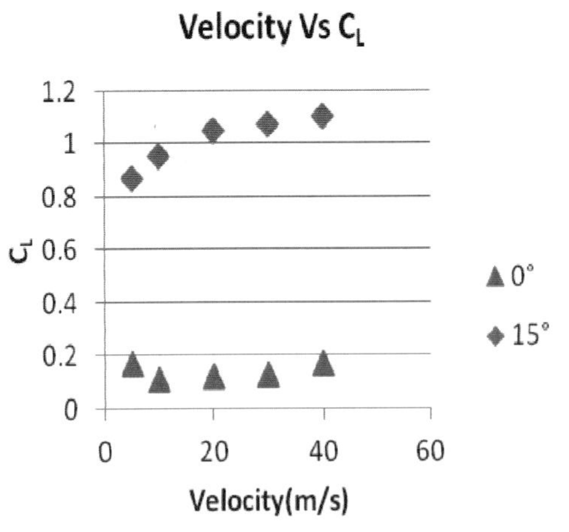

Figure 4. Lift coefficient wind tunnel results.

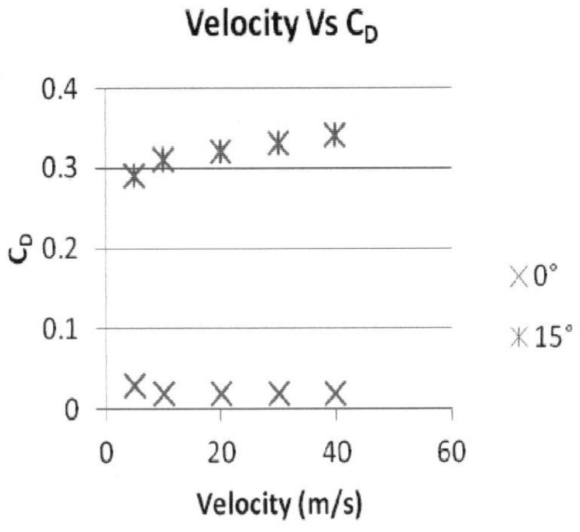

Figure 5. Drag coefficient wind tunnel results.

5. ANSYS analysis

In order to perform finite element analysis, meshing is important and a fine mesh gives a potentially more precise result. The total number of nodes used here was ~200, 000, and elements ~184,000. The fluid domain around the wing was also meshed. The two types of 3D cell types used here are

Journal of Physics: Conference Series **382** (2012) 012057 doi:10.1088/1742-6596/382/1/012057

tetrahedron and hexahedron. The structural analysis was performed to study the linear elasticity behaviour of the wing.

Different meshes were tried in order to confirm that they give the same modal frequencies [9, 10]. The first four mode shapes are shown in the Table 2. The flow analysis was performed with the intention of predicting insipient flutter. The Reynolds number for this scaled model was 963,280 for an airspeed of 40 m/s. When operating the wind tunnel at this speed, and with the aerofoil angle of attack at 15°, the wing will vibrate. We believe this to be the airspeed for incipient flutter. Analysis performed in ANSYS for airspeed at 40 m/s predicts the lift value obtained in the wind tunnel to within 5%.

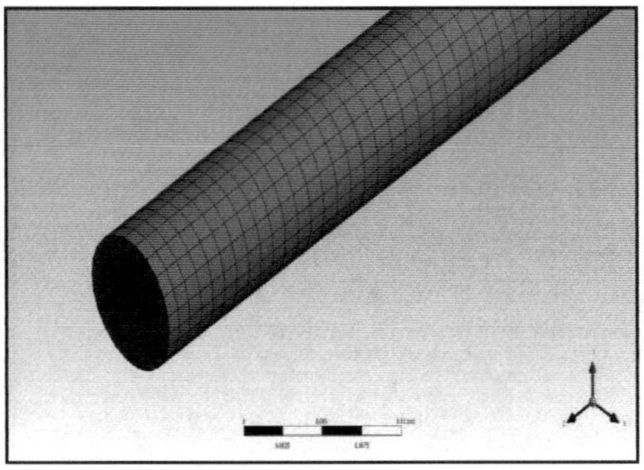

Figure 6. Mesh on the mounting rod.

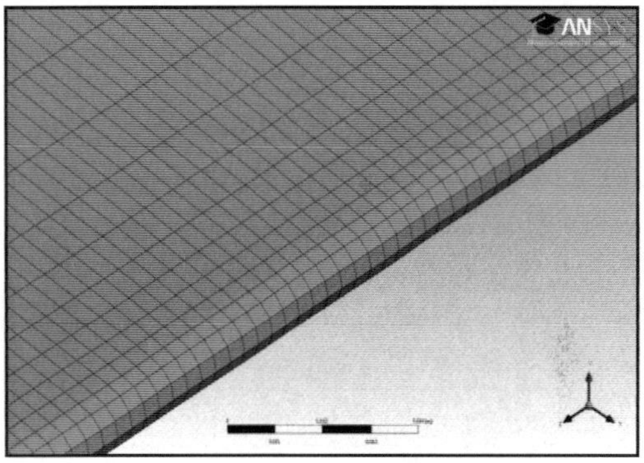

Figure 7. Mesh on the trailing edge.

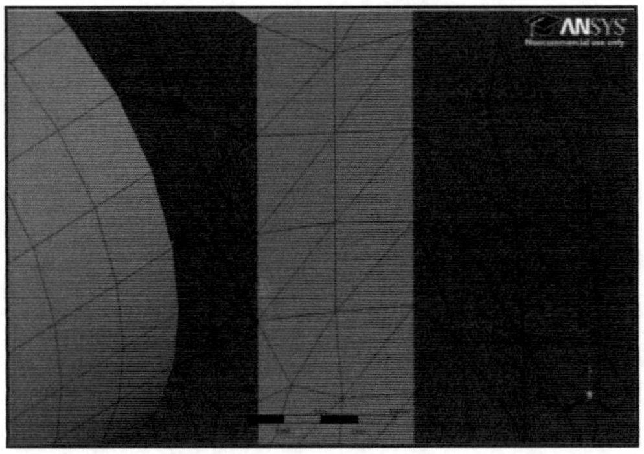

Figure 8. Tetrahedron and Hexahedron mesh.

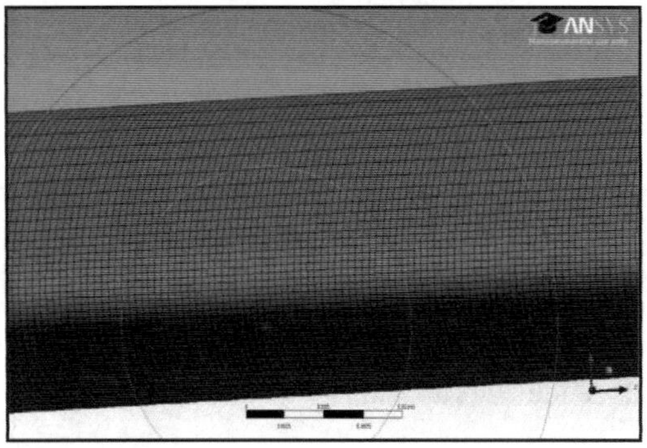

Figure 9. Mesh on the leading edge.

Table 2. Natural frequencies of the wing.

Mode	Frequency (Hz)
100	23.6
150	34.6
200	23.8
250	27.9

6. Discussion

The measured and predicted lift and drag coefficients compared at the airspeed of 40 m/s are within 5%. In the wind tunnel tests, the lift increases rapidly up to a certain level, after some time it stabilises and then gradually increases. The drag also increases, initially to a maximum, and then decreases to a constant stable value. In view of this, it would seem that transient analysis is pertinent. It was also noted that while the wing can only be made to start vibrating at the speed of 40 m/s, it will continue as the speed is slowly reduced, and stops vibrating at 30 m/s. The flutter of the wing was only seen when the angle of attack is maximum at 15°.

There is some discrepancy between the wind tunnel result for pressure and the CFD prediction. The difference is thought to result from the roughness and irregularities of the wing surface, which was compromised by the limited bending and shaping tools available for the aerofoil construction. At the speed of 40 m/s the turbulence observed in the CFD result is quite normal and hence no boundary layer separation. The modal frequencies were designed to be high, by designing the wing to be light but stiff, using five aerofoil sections. This was for the purpose of allowing the wing to flutter at low speed.

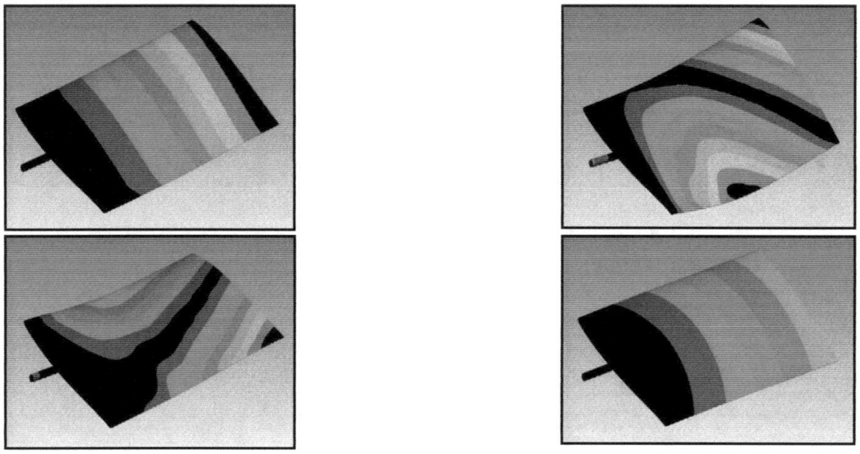

Figure 10. The first four mode shapes of the wing, (FEA results).

7. Conclusion

An aerofoil wing was designed, for which the flutter phenomenon was observed in the wind tunnel. It is well known that when the flutter exists at high velocity, any decrease in velocity of the aircraft will reduce the amplitude of flutter. This was demonstrated in the wind tunnel test, as flutter was observed at 40 m/s, and as the velocity was reduced to 30 m/s, the amplitude of vibration decreased until the wing stopped vibrating. Flutter was only observed at high angle of attack (15°), equating to flight conditions of cross wind or climb.

References

[1] Blazek J. *2001 Principles of Solution of the Governing Equations Computational Fluid Dynamics: Principles and Applications* (Switzerland: Elsevier) pp 29-57

[2] Yates Jr E C and Bland S R 1963 *Comparative Evaluation of Methods for Predicting Flutter and Divergence of Un-swept wings of Finite Span L R C* **64** 11902:55

[3] Cebeci T, Shao J P, Kafyeke F and Laurendeau E. 2005 *Turbulence Models Computational Fluid Dynamics for Engineers* (Germany: Horizons Publishing) pp 81-93

[4] Schwaz J B, Dowell E H, Thomas J P and Hall K C 2009 *Improved Flutter Boundary Prediction for an Isolated Two-Degree-of-Freedom Airfoil J. Aircraft* **46** 2069-76

[5] Rublin C L, Maynard C, Sandford and Yates Jr E C 1964 *Wind-Tunnel Flutter Studies of the Sweptback T-Tail of a Large Multi-jet Cargo Airplane at Mach Numbers to 0.90 L R C NASA* TN D-2179

[6] Date A W. 2005 Phase Change *Introduction to Computational Fluid Dynamics* (Cambridge University Press) pp 214-32

[7] Dewey H, Hodges G and Pierce A G 2002 *Introduction to structural dynamics and aeroelasticity* (Cambridge University press) pp 119-50

[8] Duraisamy K, McCroskey W J and Baeder J D 2007 *Analysis of Wind Tunnel Wall Interference Effects on a Subsonic Unsteady Airfoil Flows J. Aircraft.* **44** 1683-90

[9] Farhat C 2004 Encyclopedia of Computational Mechanics *CFD-based Nonlinear Computational Aeroelasticity* (Vol 3 Fluids) (John Wiley & Sons Ltd) pp 460-77

[10] Megson T H G 2007 *Aircraft structures for engineering students* 4th ed. (Elsevier Ltd) pp757-64.

The application of thermoelastic stress analysis to full-scale aerospace structures

R K Fruehmann[1], **J M Dulieu-Barton**[1*], **S Quinn**[1],

[1] Faculty of Engineering and the Environment, University of Southampton, Highfield, SO17 1BJ, UK

[*] E-mail: jance@soton.ac.uk

J Peton-Walter[2] **and P A N Mousty**[2]

[2] Institut Français de Méchanique Avancée, Campus de Clermont-Ferrand, BP365, France

Abstract. Non-destructive evaluation (NDE) techniques that can be applied in-situ are particularly relevant to the testing of large scale structures that cannot easily be taken into a laboratory for inspection. The application of established laboratory based techniques to the inspection of such structures therefore brings with it a new set of challenges associated with the change in operating environment between the laboratory and 'the field'. The current work investigates the use of thermoelastic stress analysis (TSA) to inspect carbon fibre composite aerospace components for manufacturing defects and in-service damage. An initial study using single transient loads to obtain a measureable change in temperature that can be related to the change in the sum of the principal stresses showed a good agreement with the traditional methodology. However, for large structures, the energy required to obtain a sufficiently large stress change to obtain a resolvable measurement may require an actuator that is not easily portable. Hence a number of ideas have been proposed to reduce the power requirement and deal with small signal to noise ratios. This paper describes the use of natural frequency vibration modes to enable large stress changes to be generated with minimal power input. Established signal processing in the form of a lock-in amplifier and Fourier signal analysis is applied. Tests on a laboratory scale flat plate and full-scale representative wing skin and stringer specimen are presented.

1. Introduction

The inspiration for the current work arises from the demand for improved inspection techniques for assessing a wide range of defect or damage types in high value, often large, composite structures. A key question that needs to be addressed is the serviceability, or remnant strength, of the component under inspection. To answer with confidence it is necessary to understand how a defect influences the stress or strain field, and how it will develop under the loading conditions. For this reason, it is of interest to use a stress or strain based metric in the inspection process. The need for stress or strain based inspection forms the basis for the Matera+ EU action AMUSED (**A**dvanced **Mu**lti-scale **S**train Based Non **D**estructive Evaluation) project, of which the work described in this paper is a part.

The current paper focuses on some of the solutions developed in the project to enable thermoelastic stress analysis (TSA) [1] to be applied outside of controlled laboratory conditions. Initial work has demonstrated the feasibility of using single transient loads to inspect components [2]. This was aimed at addressing (in part) the lack of a servo-hydraulic test machine to apply a controlled sinusoidal load. However, the magnitude of the applied stress change required to generate a temperature change exceeding the noise threshold of the infrared (IR) detector may prove to be a limiting factor. Also, the lack of information relating to non-adiabatic effects, obtained through a phase analysis of the periodic signal, reduces the amount of information available with which to characterise the defect.

A simple solution to the loading problem is to use service loads [3,4]. Often these include semi-periodic dynamic loads, e.g. road traffic or vibrating machinery. However, many composite structures operate in inaccessible environments, e.g. aircraft. Here it is necessary to have a portable system that can apply a suitable load to structure / component without requiring any disassembly of the structure. The requirement for portability, limits the energy / power available for imparting a change in stress. The proposed solution is therefore to excite the natural frequency vibration modes, using a versatile shaker / connector concept. Previously [5] it has been shown that a thermoelastic response can be obtained from excitation at resonant frequencies. The present paper describes the initial work assessing the use of vibrational modes in a structure with a view to applying TSA as a tool for strain based non-destructive evaluation outside the laboratory.

2. Initial experimental work

Initial work was conducted on a flat rectangular plate, fully clamped at the edges, representing a section of a typical grid reinforced plate structure. A shaker, driven by a signal generator, was used to excite the plate via a bonded connector and stinger, as shown in Figure 1. The reference signal used for the lock-in amplifier in the TSA processing was obtained directly from the signal generator. Two material types were used: a 0.9 mm thick aluminium sheet and a 0.75 mm thick E-glass / epoxy laminate comprising 4 layers of 210 gm^{-2} plain woven textile. The experiment was conducted with the stinger connected in different locations, to investigate the optimal location with regard to exciting different frequency modes.

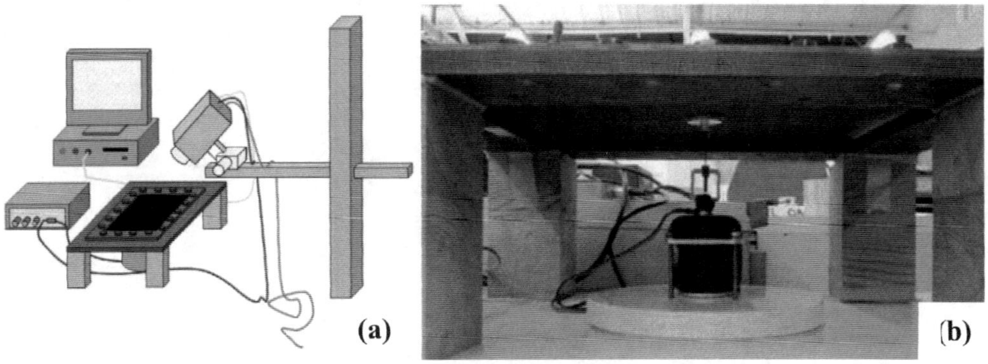

Figure 1: (a) schematic of the experimental setup, (b) image showing the bonded connection of the stinger to plate

Data is presented for the composite plate with the stinger located in the centre of the plate, showing the amplitude data from the TSA at the first and third vibrational modes (Figure 2). The stress distribution related to the mode shape is clearly visible, as are the stress concentrations associated with the bonded stinger connection. A significant difference in the location of the highest stress is clear between the first and third modes. The highest stress regions are above and below the connector (as viewed in Figure 2(a)). In contrast, in the third mode the same stress raiser results in stress concentrations to the left and right (Figure 2(b)). This highlights the significance of the loading configuration on how stress raisers (defects and damage) influence the stress field. The results also highlight the issue of the connector influencing the stress field. Clearly, a bonded connector is not an ideal solution as it causes local reinforcement, changing the mode shapes (especially mode 2), and may not be considered non-destructive as the removal results in damage to the specimen.

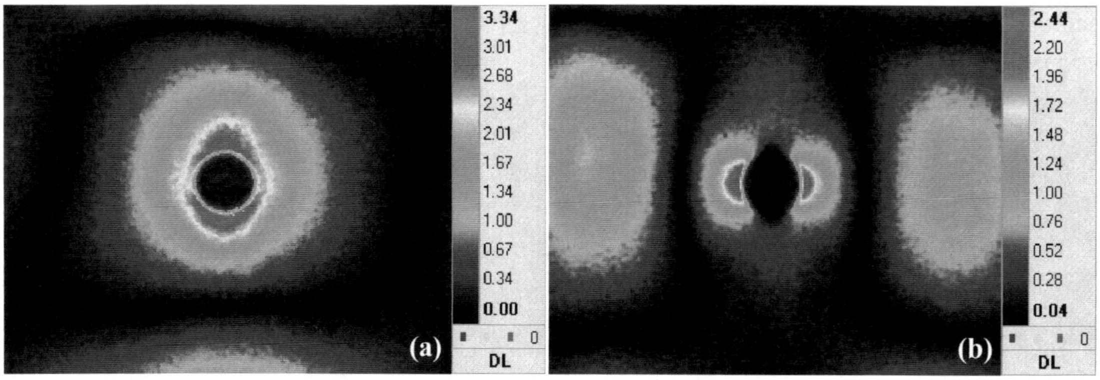

Figure 2: TSA amplitude data from the composite plate for (a) mode 1 and (b) mode 3

A test was therefore conducted with the stinger connector not bonded to the specimen. Instead, a thin layer of resin was applied between the connector and the plate to form a moulded fit. The excitation force was then applied as contact force only. Data from the aluminium plate excited in mode 3 with the stinger connector located 60 mm off-centre is shown in Figure 3. The influence of the connector on the stress field is clearly reduced, by comparison with Figure 2(b)), while still enabling the excitation to be effectively transferred into the specimen. Two areas of damage were introduced with a hemispherical hammer head with a radius of 3 mm. These are clearly visible on the right hand side of the specimen in Figure 3 as areas with reduced load carrying capacity and associated low thermoelastic response.

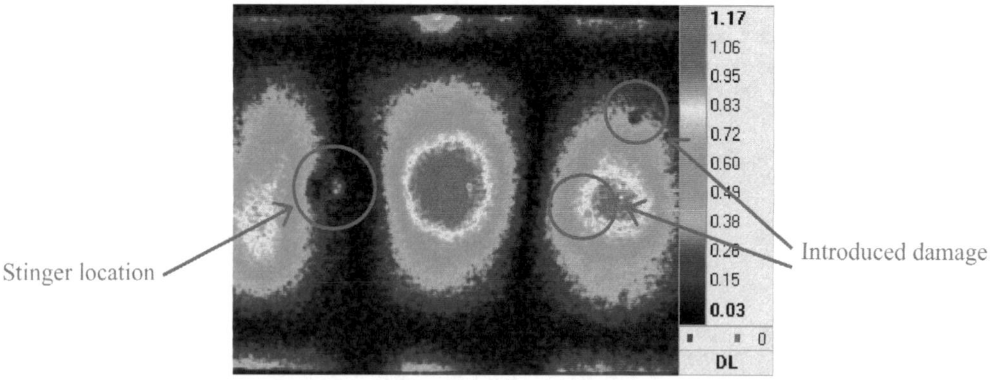

Figure 3: TSA data of mode 3 excitation for aluminium plate with contact connection only.

3. FEA of full-scale components

The following part of the project has been to apply the same methodology to full-scale carbon fibre / epoxy specimens comprising a bonded wing stringer and skin, 2270 x 240 x 110 mm, with a *quasi*-isotropic laminate 9.3 mm thick for the wing skin and 12 mm thick (tapering to 6 mm towards the middle of the specimen) for the stringer. Ultrasound inspection of the specimen revealed three regions with defects, either delaminations or void rich regions resulting from the manufacturing process. The location of these 'known' defects is shown in Figure 4 together with the main specimen dimensions. Since the specimen is not a 'real' component, the boundary conditions for the test were selected to best fit existing equipment. The specimen has therefore been simply supported at either end to simplify the rig and to enable the specimen to be loaded using both a shaker to excite the natural frequencies and a hydraulic ram to impart the mode I deformation in a three point bend setup. This enables results from the natural frequency test to be compared with a more standard approach.

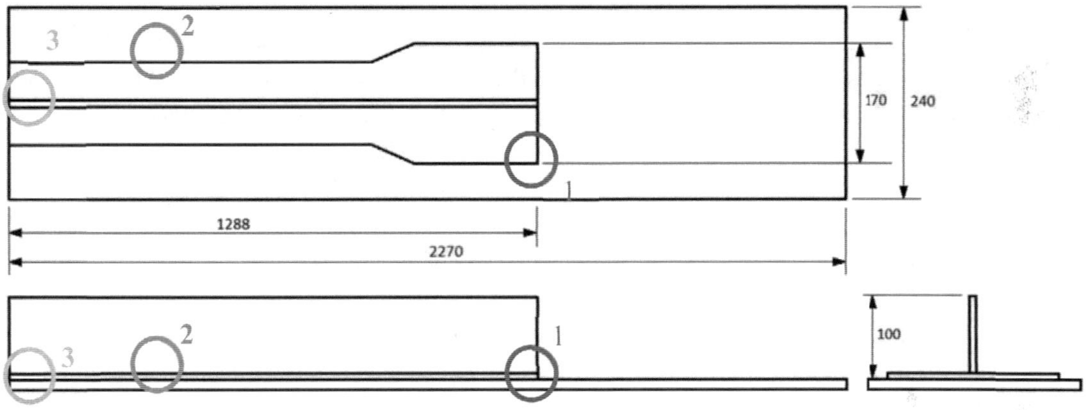

Figure 4: Geometry of wing-stringer specimens and location of defects, identified using an ultra sound C-scan (dimensions in mm)

A simplified finite element (FE) model was created using Ansys Workbench to estimate the natural frequencies and stress distributions associated with the mode shapes with and without damage. The principle aim of the modelling was to guide the experimental setup, regarding the preferred excitation mode to enable the known defects to be identified and to dimension and locate the shaker appropriately. The material was simplified as an isotropic solid with homogenised properties for a *quasi*-isotropic layup rather than using layered elements since many details of the up to 60 ply thick laminate were not known, e.g. the location of dropped plies in the stringer layup. Damage was simulated by removing a portion of material from an area corresponding to the area identified by the ultrasound C-scan. A convergence study was conducted using a structured mesh with SOLID 186 elements. Convergence of the 4th vibrational mode was shown to occur with 4821 elements using the structured mesh.

The model was verified experimentally using a tap test. This provided the natural frequencies for comparison with those obtained using the FE approach, as well as damping coefficients with which to assess the power required to achieve the desired vibrational modes and amplitudes. The natural frequencies from the FE model and the experiment are provided in Table 1. The first and fourth modes match well, considering the simplified material properties used in the modelling. However, the second and third modes are significantly different. An explanation to this discrepancy has not been forthcoming. The tap test was conducted at 7 different positions along the specimen length. The amplitudes of the different frequency components were assessed for each test, giving an indication of the best location at which to excite a given mode. This information can be used to optimise the shaker location for the experiment.

Table 1: Comparison of natural frequencies between FE model and experiment

Vibrational Modes	Frequency (Hz)		
	Model (undamaged)	Model (damaged)	Experiment
I	17.9	17.9	20.0
II	111.1	111.1	77.3
III	137.9	137.9	87.7
IV	145.7	145.7	143.6

The model results were also used to assess the effect of the defects on the surface stress field and thereby decide which natural frequency modes were best suited to identifying particular defects. The results from the 4th mode are presented in Figure 5, showing that defects 1 and 2 lie closely to regions of maximum stress and are therefore most likely to have a visible influence on the thermoelastic response. A line plot is presented in Figure 6, showing the stress sum along the specimen length, normalised against the maximum value in the undamaged state (as indicated by the dotted line in Figure 5) with and without the simulated damage. A line plot is shown on the secondary axis showing the percentage difference between the model with and without the defect, indicating the influence of the defect on the surface stress state.

To enable the thermoelastic effect to occur under *quasi*-adiabatic conditions, a minimum frequency of the order of 10 Hz is typically required. This means that mode I should provide a sufficiently fast oscillation to accurately capture the stress field. The infrared detector used in this work (a cooled Silver 480M InSb array detector manufactured by Cedip Infrared Systems) operates at a maximum

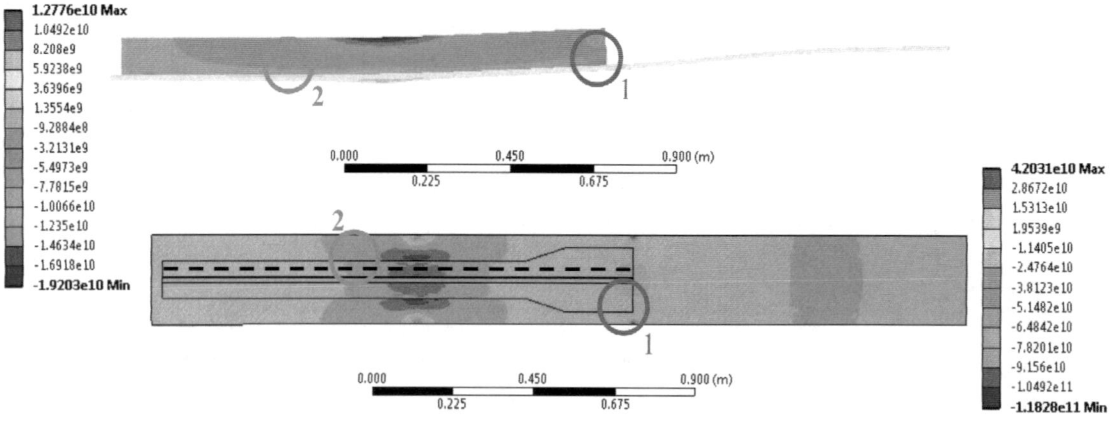

Figure 5: FE modal analysis mode 4 deformation and stress field

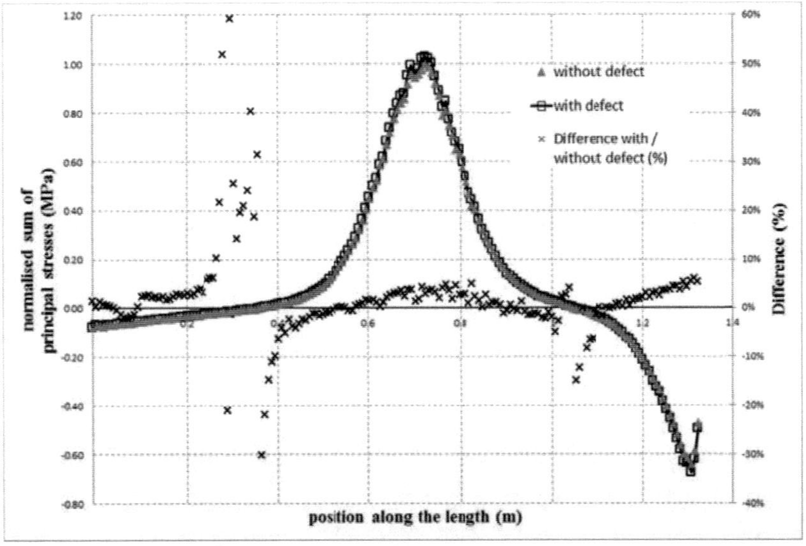

Figure 6: line plot of surface stress sum along the specimen length

frame rate of 383 Hz at full frame. Considering the Nyquist condition, that the recording rate must be at least twice that of the signal being measured, the fourth mode approaches the upper limit of the natural frequencies that can be analysed. In future work, experiments will be conducted for modes I and IV, with the camera focused on regions 1 and 2, as identified in the C-scan. Forced sinusoidal loading in three point bending will also be investigated for comparison. The results from the experiment will be presented at the conference.

4. Conclusions

The work present forms the first part of an on-going investigation into non-destructive inspection using stress or strain based measurands. The first part highlights the importance of a suitable connection between the shaker and the specimen. A simple contact connection with a distributed load introduction has been shown to have a minimal effect, even for very thin specimens. The selection of an appropriate excitation mode clearly has a significant influence on the ability to identify and assess defects in different parts of the structure. The initial results provide an encouraging proof of concept that will be further described in the aural presentation.

Acknowledgements
The authors would like to acknowledge Airbus UK for provision of specimens.

References

[1] Dulieu-Barton J M and Stanley P 1998 *Journal of Strain Analysis* **33** 93-104
[2] Fruehmann R K, Dulieu-Barton J M and Quinn S 2010 *Experimental Mechanics* **50** 1075-86
[3] Galietti U, Modugno D and Spagnolo L 2005 *Measurement Sci. and Technology* **16** 2251-60
[4] Boyce B, Zuraski P and Miles J 1993 *Proceedings of the SEM Spring Conference on Experimental Mechanics* 749-58
[5] Phan T S, Dulieu-Barton J M and Temarel P 2006 *Experimental Mechanics* **46** 463-72

Vibration tests on dismounted bridge beams and effects of deterioration

A. Quattrone[1,2], E. Matta[1], L. Zanotti Fragonara[1], R. Ceravolo[1] & A. De Stefano[1]

[1] Department of Structural, Geotechnical and Building Engineering, Politecnico di Torino, corso Duca degli Abruzzi 24 – 10129 Turin, Italy

[2]E-mail: antonino.quattrone@polito.it

Abstract. This paper aims at describing the tests campaign carried out on five precast bonded post-tensioned concrete bridge beams, recently dismounted after a service life of 50 years. The girders were part of the deck of a recently dismounted viaduct of an Italian motorway. The beams showed different deterioration levels, mainly due to the different exposure to corrosive agents. The test campaign were designed for evaluating the residual load bearing capacity of the members. Dynamic measurements were acquired before and after the static tests by using different excitation sources. This experimental research highlights that the natural frequencies of the beams can be used as a predictor of the ultimate bending moment of an existing structure.

1. Introduction

The network of infrastructures of a nation constitutes an asset of strategic importance to the life of the community. Its management requires periodic monitoring, through both short-term and long-term programs. In particular, the actual serviceability conditions, the assessment of the safety conditions and the vulnerability to natural catastrophic events must be checked, also resorting to experimental techniques. A thorough knowledge of the effective operational conditions and the residual safety margins of an infrastructure, achievable implementing a monitoring system, may lead to a rational scheduling of the interventions, and hence to an optimal management of economic resources.

The reduction of structural safety during the service life is a widely debated issue [1 - 5]. Several factors, such as corrosion, rheological effects, increasing of service loads, cracking propagation due to cyclic loads, may act simultaneously on a structure over its lifetime and lead to a substantial reduction in its capability to meet the required performances.

On this light, performing experimental campaigns on real structures with different deterioration levels should represent a great source of information to be used in developing predictive models and calibrating monitoring techniques.

Structural diagnostics is often associated to the analysis of vibration measurements, since modal quantities reflects the global behavior of the monitored object. The variation of these quantities can be used as indicator of occurring damages.

In this paper, the experimental vibration tests performed on five beams dismounted after 50 years of service life are illustrated and the results of the dynamic identifications and their evolution as a function of the ultimate bearing capacity are presented. Experimental modal analysis was performed and the results are presented and compared. In particular, the connection between the deterioration levels (expressed in terms of ultimate bearing moment) and the identified modal parameters is highlighted.

2. Description of the tests

The experimental testing campaign was commissioned by the Italian motorway company *ATS autostrada TO-SV S.p.A.*, at the aim of evaluating the residual strength of a set of bridge beams removed from *Pesio* viaduct after fifty years of service life during the modernization works of *A6 Torino-Savona* motorway.

The tested beams had all the same geometrical characteristics and presented different degradation levels. Each deck of the viaduct was constituted by five 35 m long simply supported beams. The section of the beams had a trapezoidal shape with linearly varying height from 1.55 m at the supports to 2.95 m at the midspan. The beams were constituted by a bottom precast part completed by a 0.1 m thick, cast-in-place slab. The post-tensioning reinforcement consisted of 5+5 six-strand tendons, placed in the two sloping lateral walls. The design characteristics of the employed materials were:

- Post-tensioning tendons steel - ultimate tensile strength: 1600 MPa;
- Concrete – cube compressive strength: 45 MPa;

Figure 1. A view of the *Pesio* Viaduct deck (a), the midspan cross section (b) and a beam placed on the test bench (c)

The tested beams presented different deterioration levels, mainly related to the position occupied in the deck. The health state of the tested beams was classified in three levels (good, intermediate and bad condition) on the basis of a visual inspection (see Table 1, second column). The estimated ultimate bending moment $M_{d,u}$ is 23950 kNm.

Table 1. Visual inspection rating and ultimate bending moments in midspan section

Beam	Condition Rating	Midspan ultimate moment M_u [kNm]	Percentage ratio $M_u / M_{d,u}$	Tests
B01	Bad	11172	46.7%	Static
B02	Intermediate	22581	94.3%	Static + Dynamic
B03	Good	20142	84.1%	Static + Dynamic
B04	Good	18336	76.6%	Static + Dynamic
B05	Intermediate	19321	80.7%	Static + Dynamic
B06	Bad	12087	50.5%	Static
B07	Bad	11242	46.9%	Static + Dynamic
B08	Good	14960	62.5%	Static
B09	Good	14924	62.3%	Static

2.1. Static tests

A four-point static bending test was chosen to determine the ultimate load capacity. The forces, applied by means of four hydraulic jacks, were progressively incremented performing loading and unloading cycles until reaching the ultimate load. The static tests have been designed in order to investigate the whole behavior of the post-tensioned beams, determining both the decompression moments and the ultimate bending strengths. Table 1 lists the condition rating of the beams and their ultimate bending moments, Figure 2 shows the test configuration.

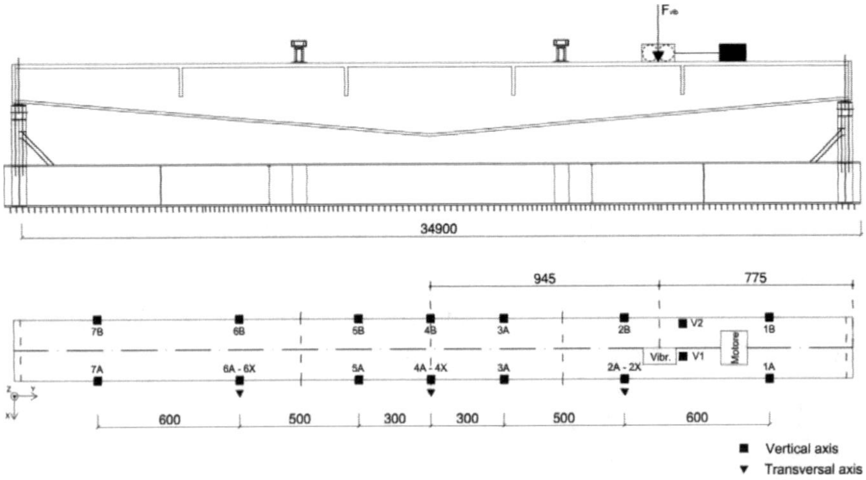

Figure 2. The dynamic tests configuration

2.2. *Dynamic tests*

The beams were tested after the dismantlement and, in three cases, after the execution of the static tests (Table 1).

A set of 18 unidirectional capacitive accelerometers was used to acquire the vibration response under both ambient noise and impact hammer excitations (Figure 2). Furthermore the beams B04, B05 and B07 have been tested by using an electro-mechanical vibrodyne placed on the upper slab. A sine sweep excitation with frequency varying in the range of the first two flexural modes have been applied. The vibrodyne has been positioned to maximize the number of excitable modes in its operative range. Thus it was possible to excite a range of frequencies including the first two flexural modes and the first torsional mode of the beams.

The accelerometers have been placed along the longitudinal axis of the beam on the slab, with a closer spacing in the middle part of the span, in order to get an adequate spatial resolution of the vertical flexural modes. Three accelerometers were placed along transversal axis to discriminate among torsional and transversal bending modes.

The acceleration signals were acquired at a 400 Hz sampling frequency for both ambient noise and impact hammer excitations. The impact tests were carried out by hitting the beams along the slab with a sledge hammer both in vertical and transversal directions.

3. Experimental modal analysis

Among all the available methods which can be employed to identify the modal parameters, it was decided to adopt two techniques working in the time domain. The Eigenvalue Realization Algorithm (ERA) was used to analyse the free decay responses [7], whilst ambient vibration signals called for a Stochastic Subspace Identification (SSI) [8].

After a preliminary analysis of the spectral content, the signals acquired were pretreated by filtering out of the bandwidth [0 - 50 Hz], removing the trends and reducing sampling frequency to 200 Hz. Several modal identification sessions were performed. Some tolerance criteria were adopted to discard computational modes, based on modal shape assurance coherence criterion (MAC) and fixing a maximum value of damping ratio. The chosen parameters are:

- MAC > 85%
- $0 \leq \zeta \leq 10\%$

A data cleansing process was further introduced in order to neglect those results which differed from the mean values of identified frequencies more than the computed standard deviation.

4. Discussion of the experimental results

The following graphs show the variation of the first four bending modes as a function of the ultimate bending moments in the midspan section.

The results of the analysis, reported in Figure 4, highlight a correlation between the dynamic characteristics of the beams and the ultimate bending strengths. Both static and dynamic tests showed the different evolution in time of the beams after 50 years in service, the former in terms of residual load-bearing capacity, the latter in terms of changes in modal frequencies.

The ultimate bending strength and the modal parameters are both influenced by the different levels of deterioration of the beams. These differences can be correlated mainly with the different position occupied in the deck. The edge beams are usually more exposed to weather, chemical attacks of chlorides contained in de-icing salts and pollutants emitted by vehicle traffic. Moreover, they typically undergo higher stress levels, resulting in the years in wider cracking patterns.

This phenomenon is pointed out by the behavior of beam B07. In this particular case, an important role was played by the lack of care in the design of the non-structural details, as it is possible to notice in Figure 5. The residual bending strength has been possibly affected by the localized damage induced in the section next to the drainpipes (Figure 5), as partially highlighted by the different failure behavior. The beams B02 and B04 showed mostly brittle failure due to the crushing of the upper slabs, whilst in the case of the beam B07 has been caused by the rupture of the corroded reinforcement.

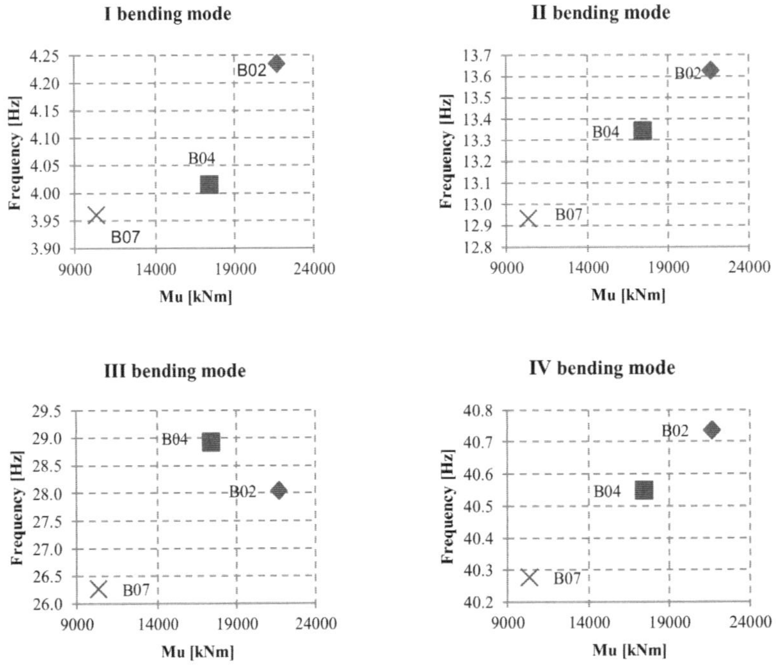

Figure 4. The bending modal frequencies identified in ambient vibration tests

Figure 5. Beam B07: corrosion of the post-tensioned tendons and ordinary reinforcement. The deterioration is particularly advanced in the sections next to the drainpipes

On the other hand, the ultimate bending moment of the interior beams (i.e. B02) is comparable to the ultimate bending moment $M_{d,u}$ calculated on the basis of the mean values of materials properties and the geometrical measures reported in the original design documentation.

5. Conclusions

This paper presents the experimental campaign carried out at the purpose of evaluating the residual strength of nine bridge beams with different levels of degradation, recently dismounted after 50 years of service life.

Dynamic tests were carried out on five beams and the modal parameters were identified. The evolution of the dynamic characteristics as a function of the ultimate flexural strength has been highlighted.

The identified modal parameters, in particular the natural frequencies, showed to be sensitive features of the strength reduction. This capability makes them suitable symptoms to be monitored in order to estimate the reliability of the structure.

6. References

[1] Enright MP and Frangopol DM 1998 Service-life prediction of deteriorating concrete bridges *Journal of Structural Engineering ASCE* 3 **124** 309-317

[2] Thoft-Christensen P 1998 November Assessment of the reliability profiles for concrete bridges *Engineering Structures* 20 **11** 1004-1009

[3] Stewart MG and RosowskY DV 1998 Structural safety and serviceability of concrete bridges subject to corrosion *ASCE Journal of Infrastructure Systems* **4** 146-155.

[4] Nowak, A. S., Collins, K. R., 2000. Reliability of Structures. Mc Graw Hill International Editions

[5] Ceravolo R, Pescatore M and De Stefano A 2009 Symptom based reliability and generalized repairing cost in monitored bridges *Reliability Engineering & System Safety* **94** (8) 1331-1339

[6] Ewins DJ. Modal testing: Research Studies Press Ltd; 2000.

[7] Juang JN and Pappa RS 1984 An eigensystem realisation algorithm (ERA) for modal parameter identification and modal reduction *NASA/JPL Workshop on identification and control of flexible space structures*

[8] Van Overschee P and De Moor B. Subspace Identification for Linear Systems: Theory and Implementation - Applications: Kluwer Academic Press Dordrecht; 1996.

[9] Friswell MI 2007 Damage identification using inverse methods *Philosophical Transaction of the Royal Society A* **365** 393-410

[10] Peeters B and DeRoeck G 1999 Reference-based stochastic subspace identification for output-only modal analysis *Mechanical Systems and Signal Processing* **13** 855-878

[11] Doebling SW, Farrar CR, Prime MB and Shevitz DW 1996 Damage Identification and Health Monitoring of Structural and Mechanical Systems from Changes in Their Vibration Characteristics: A Literature Review *Technical Report LA-13070-MS* Los Alamos National Laboratory, Los Alamos, NM.

Modern Practice in Stress and Vibration Analysis 2012 (MPSVA 2012) IOP Publishing
Journal of Physics: Conference Series **382** (2012) 012060 doi:10.1088/1742-6596/382/1/012060

Time-Frequency Analysis of Long Range Ultrasonic Signals

Keith Thornicroft[1,2] Cristinel Mares[1] Peter Mudge[2]

[1]School of Engineering and Design, Brunel University, Middlesex, UB8 3PH

[2] NDT Technology Group, TWI Ltd., Granta Park, Cambridge, CB21 6AL

E-mail: keith.thornicroft@twi.co.uk

Abstract. Long range ultrasonic testing (LRUT) is a relatively new development within the non-destructive testing sector. Traditionally, conventional ultrasonic testing (UT) is performed at high frequencies, in the MHz range, and is capable of detecting small flaws within a range of millimetres; whereas long range ultrasonic inspection is carried out at lower frequencies, typically between 20 and 100kHz, and is capable of highlighting structural detail and discontinuities tens of metres from a test position. Conventional ultrasonic testing relies on the transmission of bulk waves, the velocities of which are independent of frequency and can usually be predicted easily if the elastic properties of the material under test are known. The dynamics of guided waves, however, are dependent upon frequency making the analysis of received data from a specimen complex. This paper will serve as an introduction to time-frequency representation and may allow a clearer understanding of the non-stationary raw signals produced by this inspection process. Currently, LRUT data are assessed in the time or distance domain using the amplitude vs. time 'A-Scan', therefore structural features and potential flaws are highlighted on a time-of-flight basis. However, as the data obtained are dynamic in time and frequency (non-stationary), time-frequency distributions could provide a mode identification or de-noising process to deal with the problem of coherent noise.

1. Introduction

The main objective of this paper is to excite many vibrational modes observed within a tubular structure experimentally and compare the time-frequency results with analytical dispersion curves generated using the modelling software DISPERSE [1]. The class of structure examined here is carbon steel pipe; a common material used within industrial refineries and petrochemical plants worldwide to convey a variety of hazardous and non-hazardous materials. Assessing the condition of assets such as pipework and associated pipelines is therefore an essential task. The importance is highlighted further by the huge expense of a shutdown period due to a failure or loss of contents and the potential environmental impact of such a failure. LRUT is used to good effect commercially to screen for corrosion, erosion and other types of wall loss such as pitting. During normal inspections on a wave guide such as a pipe, an axi-symmetric ring of sound or vibrational mode is produced by a bracelet of dry coupled shear transducers wrapped around the circumference of the pipe. Each transducer can be viewed as a point source possessing an output and the ring of sound produced is the result of the superposition of each output. As the number of transducers in this array increases, the purer the axi-symmetric mode becomes. However, at the test frequency range for this technique there are many vibrational modes possible which separate shortly after transmission. Each mode has a unique frequency dependent group and phase velocity curve; these curves display the dispersion characteristics for a specific waveguide. This dispersion effect produces a temporal and therefore spatial resolution problem within the received data. The excitation signals used for LRUT are usually modulated tonebursts which are limited to a frequency range where the desired mode used is relatively non-dispersive. The modulation used generates a narrowband pulse in the frequency domain in attempt to minimise this effect. The two most common functions used for this amplitude modulation are the Hann window and the Gaussian window.

As full vibrational mode control using a finite number of transducers in the array is not possible the received data is multimodal. Moreover, as the excitation signal possesses a bandwidth within which a frequency dependent response is possible, the data becomes non-stationary *i.e.* it contains information in terms of time and frequency. By using only one transducer, it is possible to move away from the idea of controlling a true axi-symmetric ring and therefore excite many modes at once. This method can be used to experimentally validate a set of dispersion curves for any given structure. Traditionally, this has been achieved by using the time consuming two-dimensional Fourier transform. However by using a broadband input signal and displaying the result of a single scan in terms of time and frequency, the dispersive nature of the structure can be revealed.

A collective review of time-frequency can be found in [2]. Here Debnath includes a paper by Boashash and Sucic which demonstrates some practical applications of time-frequency distributions (TFDs) and introduces the idea of instantaneous frequency and reviews some of the desired criteria for TFDs. Chapter 8 in [2] focuses on time-frequency / time-scale reassignment. Chassande-Mottin, Auger and Flandrin discuss the trade-offs between several TFDs including the main alternative to the Fourier based spectrogram, the Wigner-Ville distribution. Results displayed in this section highlight the main disadvantage when using the Wigner-Ville; localised interference caused by superposition of negative and positive values within the domain. It was concluded that the reassignment principle applied to a spectrogram produced the closest to ideal results. In [3] Flandrin discusses the reassignment principle further and explains how phase calculated by the short-time Fourier transform (STFT) can be 'reassigned' to closely follow an ideal instantaneous frequency. Note that all of the TFDs suffer from resolution problems related to Heisenberg's notion that a specific frequency cannot be expressed exactly at a specific time. Prosser et al [4] carried out a set of acoustic emission experiments on a 6.35mm aluminium plate and displayed promising results. It was clear that the plate modes A0 and S0 were successfully generated and were dominant in the time-frequency domain. An analytical representation of S0 was overlaid onto the TFD, in this case a Pseudo-Wigner-Ville distribution (PWVD), showing good correlation. Documents [5] and [6] appear to contain the same results with [5] being a very concise report on a particular experiment and [6] providing an excellent review of TFDs including a scalogram using the continuous wavelet transform (CWT). The results show good contributions from the S0 and A0 plate modes using laser induced ultrasound in a 0.93mm thick aluminium plate. By using the reassigned spectrogram some of the higher order modes have been tracked successfully. However, where the modes intersect the solution becomes undefined. There are less vibrational modes present in a plate structure than in a hollow cylinder (*e.g.* steel pipe), so the problem under investigation in this paper is more complex. It is important to establish initially which modes can be excited effectively during this study.

2. Theoretical background

A transient signal such as those seen in LRUT can be displayed in the time domain as an A-Scan and also by using a Fourier transform, frequency content can be obtained. These display mechanisms are currently independent of each other in this field and do not therefore provide an adequate description of the multi-domain nature of this type of wave propagation. The first time-frequency distribution developed was the short-time Fourier transform (STFT). This method reduces a time domain signal into a series of windows of equal length. The Fourier transform for each window, also known as a Frequency bin, is returned yielding information about the signal in both time and frequency. Each

window is tiled to create an image, the energy density function of this process is known as the spectrogram.
The STFT of a signal *s(t)* is expressed as:

$$STFT(\omega,t)=\frac{1}{2\pi}\int_{-\infty}^{\infty}e^{-i\omega t}s(\tau)h(\tau-t)d\tau \tag{1.1}$$

Here *h(t)* is a window function and τ represents the length of the signal. The spectrogram suffers from a lack of resolution at higher frequencies, this is due to the window length (frequency bin) being fixed across the length of the function. This limits the ability of the STFT to display rapidly varying information accurately throughout the entire signal.

An alternative outlook on the STFT came in the form of the Wigner-Ville distribution. This distribution uses the time domain signal *s(t)* and its complex conjugate $s^*(t)$ as:

$$s(t+\frac{\tau}{2})s^*(t-\frac{\tau}{2}) \tag{1.2}$$

The TFD is then calculated by taking the Fourier transform of equation *(1.2)* with respect to the duration τ:

$$WVD(\omega,t)=\int_{-\infty}^{\infty}s(t+\frac{\tau}{2})s^*(t-\frac{\tau}{2})e^{-i\omega t}d\tau \tag{1.3}$$

The WVD is capable of achieving better time and frequency resolution through good local time representation but suffers from additional interference terms which can produce a noisy TFD. This effect can be improved upon by applying a two dimensional filter. A bi-product of the filtering process is a blurred result which cancels out the original high resolution advantage of the WVD.

2.1. Dispersion

Figures 1 and 2 below show the first five fundamental modes for the pipe used in this experiment. All of these modes have frequency dependent velocities which in this case have been calculated analytically using the modeling software DISPERSE. It is clear from these plots that in order to control an axi-symmetric test using longitudinal excitation it is essential to use a narrow bandwidth as dispersion of the input signal as it propagates along the wave guide will impair the resolution of the time domain signal. In this investigation it is essential to apply the opposite notion. A broadband signal will in theory generate all modes displayed in figures 1 and 2 and the TFD can be manipulated to give accurate mode velocities for the test piece experimentally and with further refinement yield a set of experimentally validated dispersion curves. From knowing that many modes can exist at the same frequency and similar amplitudes (depending on excitation conditions), the LRUT technique can suffer from input mode impurity. Time domain scans (A-Scans) can become unpredictable when more than one mode is generated upon input.

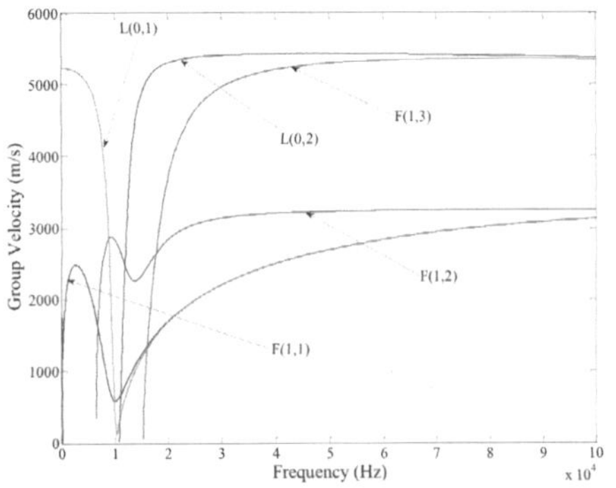

Figure 1. Group velocity curves for the test piece

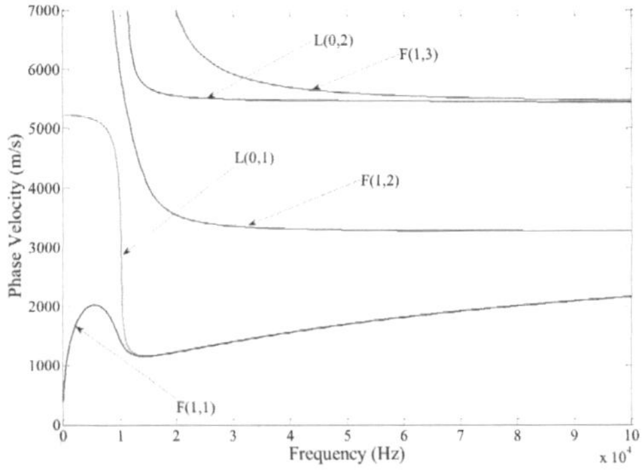

Figure 2. Phase velocity curves for the test piece

3. Experimental Measurements

To produce accurate arrival times and prevent any other scatter, it was important to maintain directionality during the experiment. This was achieved by setting up a through-transmission or 'pitch-catch' between the macro-fibre composite (MFC) transducer (Tx) seen in figure 3 and a Polytec PSV-400 1D scanning laser vibrometer (Rx). The MFC was dry-coupled to the outer surface of the pipe with a force normal to the pipe surface of approximately 100N. The total active area of the transducer was experienced a uniformly distributed coupling load. The major axis of the transducer was aligned with the major axis of the pipe; this ensured that the modes that propagated along the waveguide would be longitudinal. The transducer was also located at the very end of the pipe so that any backwards leakage of sound would not appear on the scans. The pulser/receiver used for the tests was a Teletest TT2. This unit has an external trigger which initiated synchronisation with the vibrometer during data collection.

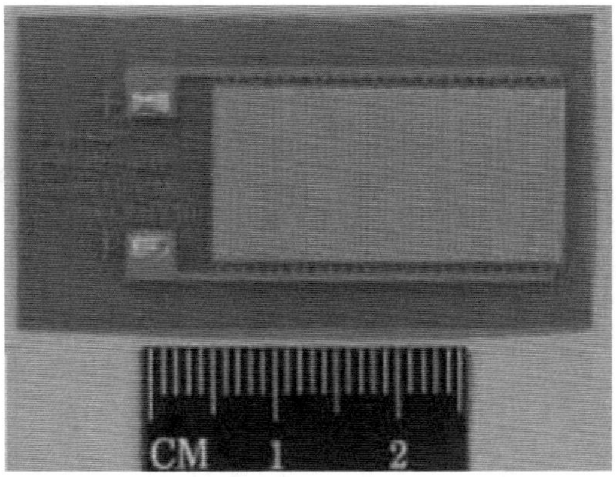

Figure 3. An example of the MFC transducer used in this investigation

In order to generate scans possessing a good signal to noise ratio (SNR), the collection averages on the vibrometer was set to 512. Random environmental noise is generally subdued by applying averages. The vibrometer was focused at a distance of 2.895m away and in line with the MFC delivering the input signal. A digital bandpass filter was also set up in the Polytec collection software in order to improve SNR further. The lower cut off frequency was set to 2kHz and the upper cut off frequency was set to 275kHz. This input is broadband with respect to normal LRUT range (20kHz to 100kHz), but it was viewed that any additional data outside of normal range could provide some additional information.

Once the data was collected a set of Matlab I/O routines were implemented and all files were set to a format which could be used as input to a C source code based time-frequency toolbox. The sampling frequency of the received data was set to 1MHz. With the collection length being nearly 3m and the sampling frequency being high, the data sets processed by Matlab were large. This caused problems during processing and the issue will need to be solved in order to make the process more efficient.

Journal of Physics: Conference Series **382** (2012) 012060 doi:10.1088/1742-6596/382/1/012060

4. Results

The processed results displayed were obtained from laboratory trials on a 6m long 6 inch schedule 40 carbon steel pipe (Outside diameter = 168.28mm, wall thickness = 7.112mm).

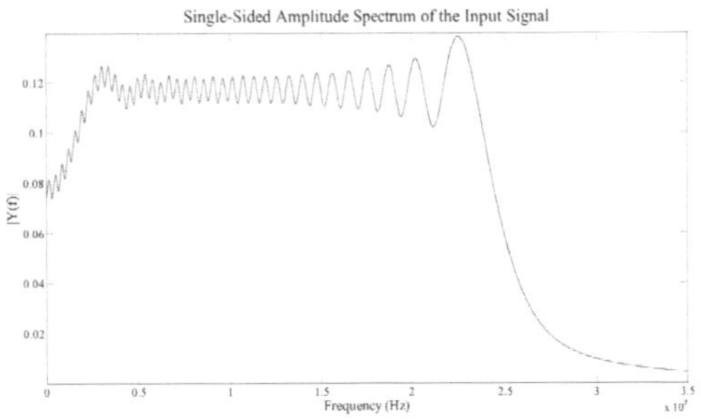

Figure 4. The Fourier spectrum of the input signal

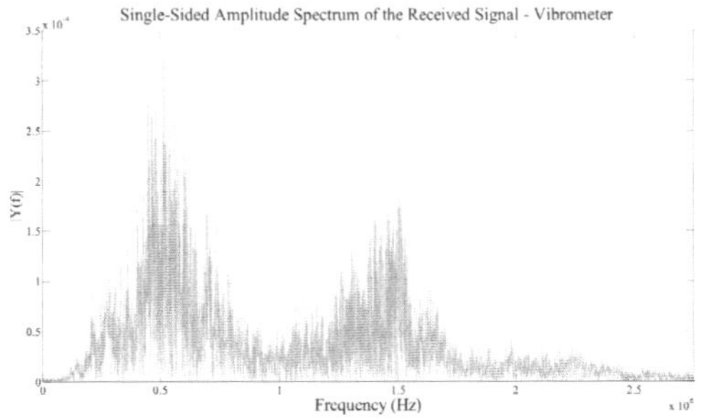

Figure 5. The Fourier spectrum of the received signal

In figure 4 a broadband, relatively flat (ignoring the Fresnel ripples) is observed before the signal enters the transducer. However, upon transmission, the transfer function of the MFC appears to distort the received spectrum seen in figure 5. The region between 50kHz and 150kHz is subdued in such a way that is becomes a trough in the data. It also appears that the data above 150kHz lies close to the noise floor where in fact the transmission spectrum in figure 4 shows good energy levels up to 250kHz. It is possible that the transducer is acting as a hardware filter and therefore providing unwanted amplitude modulation. Generally, the frequency response of a vibrometer of such specification is not usually thought to cause large distortion as it is flat. The twin spectrum peaks seen in figure 5 carry forward into the TFD as darker regions (displayed in figure 7). In an ideal case the spectrum would be flat and there would not be any dominant regions displayed in the TFD.

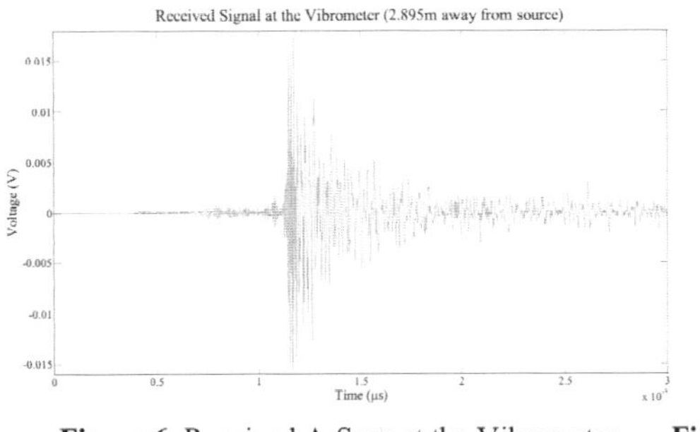

Figure 6. Received A-Scan at the Vibrometer

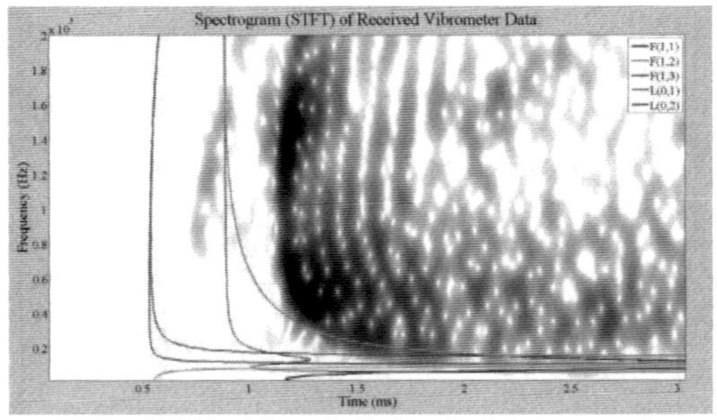

Figure 7. The STFT with analytical arrival times superimposed

Figure 6 shows the received A-Scan at the vibrometer. The scan seems surprisingly clean which suggests that either the MFC was not capable of generating all of the fundamental modes given the set up or that the vibrometer was not sensitive to some of the modes at the focal point. A wider scan of the received area will confirm this. The beginning of the A-Scan was padded with zeros as inter-channel leakage was observed and could affect the result of the TFD. There is modal information in front of the main reflection which upon comparing the TFD and the A-Scan appears to be either L(0,2) or F(1,3) (the 'F' is used to denote 'flexural'). Figure 7 shows the spectrogram of the data with the analytical arrival times for each fundamental mode superimposed for the purpose of identification. The data is not in synch with the analytical curves as there is a group delay associated with the processing and input signal parameters. However, again it is clear that L(0,1) and F(1,2) are dominant and that there is a weak presence of L(0,2) and F(1,3) arriving in front of the main reflection. The five fundamental modes shown in the dispersion curves and superimposed in figure 6 are not the only solutions in existence. The equally spaced ridges trailing behind the main reflection possessing similar intensity are likely to be higher order modes of the same family as L(0,1) and F(1,2).

5. Conclusions

In conclusion, this initial study has successfully generated and displayed experimental dispersion relationships for some of the fundamental vibrational modes that exist within a 6Inch schedule 40 steel pipe. It is clear that the experimental set up needs refining to promote some of the weaker modes present in the current data. Transducer selection is also a key factor in displaying uniform results across the desired bandwidth. The nature of contact transducers adds another dimension of complexity contained in coupling conditions, material surface condition and associated resonances.

Although the STFT was capable of displaying the multi-modal nature of these signals, refining the results to gain a more accurate image is needed. This may be achieved by exploring the attributes of other TFDs and perhaps include the CWT for decomposition. The modal analysis of pipes is more complex than plates as a large family of flexural modes exist. With this in mind, it is important to find a very resolute time-frequency technique to display or decompose this information accurately.

References

[1] B.N. Pavlakovic, M.J.S. Lowe, D.N. Alleyne, P. Cawley, "Disperse: a general purpose program for creating dispersion curves", in: D.O. Thompson, D.E. Chimenti (Eds.), *Review of Progress in Quantitative NDE, Vol. 16*, Plenum Press, New York, 1997, pp. 185–192.

[2] Debnath, L., "Wavelets and Signal Processing", *Applied and Numerical Harmonic Analysis*, Birkhäuser, Boston, (2003).

[3] Flandrin, P., "Time-frequency / Time-scale Analysis", *Wavelet Analysis and its Applications Volume 10*, Academic Press, London, (1999).

[4] Prosser, W. H., Seale, M. D. and Smith, B. T., "Time-frequency Analysis of the Dispersion of Lamb Modes", *Journal of the Acoustical Society of America*, 105, 2669-2676, (1999).

[5] Niethammer, M., Jacobs, L. J., Qu, J. and Jarzynski, J., "Time-frequency Representation of Lamb Waves using the Reassigned Spectrogram", *Journal of the Acoustical Society of America*, 107, L19-L24, (2000).

[6] Niethammer, M., Jacobs, L. J., Qu, J. and Jarzynski, J., "Time-frequency Representations of Lamb Waves", *Journal of the Acoustical Society of America*, 109, 1841-1847, (2001).

An analytical model of a longitudinal-torsional ultrasonic transducer

Hassan Al-Budairi, Margaret Lucas

School of Engineering, University of Glasgow, Glasgow, G12 8QQ, UK

h.al-budairi.1@research.gla.ac.uk, Margaret.Lucas@glasgow.ac.uk

Abstract. The combination of longitudinal and torsional (LT) vibrations at high frequencies finds many applications such as ultrasonic drilling, ultrasonic welding, and ultrasonic motors. The LT mode can be obtained by modifications to the design of a standard bolted Langevin ultrasonic transducer driven by an axially poled piezoceramic stack, by a technique that degenerates the longitudinal mode to an LT motion by a geometrical alteration of the wave path. The transducer design is developed and optimised through numerical modelling which can represent the geometry and mechanical properties of the transducer and its vibration response to an electrical input applied across the piezoceramic stack. However, although these models can allow accurate descriptions of the mechanical behaviour, they do not generally provide adequate insights into the electrical characteristics of the transducer. In this work, an analytical model is developed to present the LT transducer based on the equivalent circuit method. This model can represent both the mechanical and electrical aspects and is used to extract many of the design parameters, such as resonance and anti-resonance frequencies, the impedance spectra and the coupling coefficient of the transducer. The validity of the analytical model is demonstrated by close agreement with experimental results.

1. Introduction

Improvement in ultrasonic transducer performance requires a detailed investigation of the mechanical and electrical characteristics for different operating conditions. It is also important to simulate the effects of the mechanical, dielectric and piezoelectric losses which are associated with the transducer, prior to fabrication. For instance, numerical modelling based on the finite element (FE) method allows for an accurate description of the mechanical behaviour where the stresses and strains can be calculated accurately, but does not generally provide adequate insights into the electrical characteristics. Numerical modelling also requires large computational power and sometimes, due to the numerical characterisation, the physical meaning of the model may be lost [1]. Therefore, involving the optimized numerical model directly in the fabrication stage can lead to a mismatch between the actual and the expected performance.

The analytical model of wave motion is used as an alternative method to design piezoelectric transducers which is based on the solution of the wave equation along the transducer. It is generally used in one-dimensional vibration theory where the longitudinal dimension of the transducer is greater than the lateral dimension and therefore only the longitudinal modes are considered [2]. Analytical approaches can also be applied in two-dimensional and three-dimensional models to consider flexural, radial and longitudinal modes of vibration [3], requiring more advanced mathematics for solution. In the current work, a 1D model is modified to represent an optimised design of an LT transducer where the design optimisation has been presented in previous work [4]. The developed 1D model has the

Published under licence by IOP Publishing Ltd

following advantages: it is simple to model; it can be used to investigate the electrical characteristics of the transducer by calculating the electrical impedance, the phase diagram, and the coupling coefficient; it is possible to model the effects of mechanical and electrical losses of the piezoelectric and acoustic elements of the transducer [5]; it can be used to calculate the input electrical power and the output mechanical response for different excitation levels; and, finally, it is possible to create a mode shape by calculating the currents of the equivalent network branches, the mode shape being important for locating the nodal plane of the transducer.

2. Mason equivalent circuit model

Mason's model provides a powerful tool for the analysis and simulation of piezoelectric transducer elements. This model simulates both the coupling between the mechanical and electrical systems and the coupling between the mechanical and acoustical systems [6]. The equivalent circuits of piezoelectric and acoustic materials are shown in Figure 1, where the force, f_i, and velocity, v_i, are represented by an electric potential, V_{mi}, and current, I_{mi}, respectively, the acoustic impedance of these materials is represented by electrical capacitance, Z_C, and inductance, Z_L, elements, and these elements are calculated from the material properties as follows:

$$Z_L = j\,\rho\,c\,S\,tan(\frac{k\,l}{2}) \tag{1}$$

$$Z_C = \frac{-j\,\rho\,c\,S}{sin(kl)} \tag{2}$$

Where ρ, c, S, k and l are the material density, the sound velocity, the cross-sectional area, the wave number and the wave path length in the direction of wave propagation respectively.

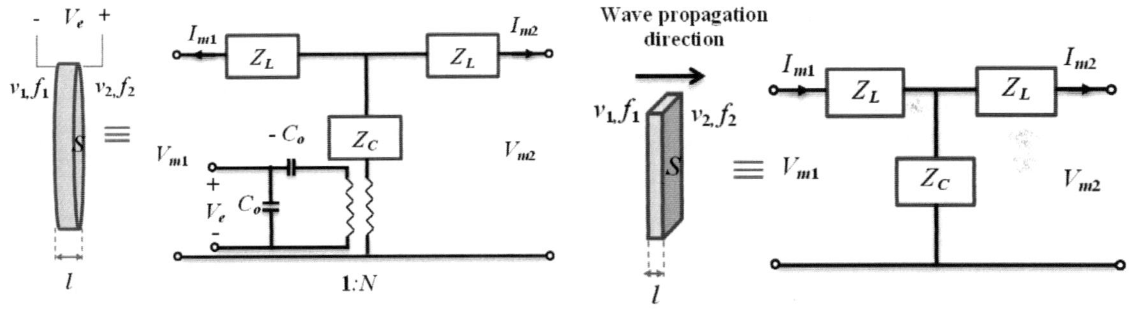

Figure 1. Mason equivalent circuit of a piezoelectric disc (left) and an acoustic solid (right).

3. Analytical model

The schematic of an LT transducer model is shown in Figure 2. The construction of the equivalent circuit is carried out by dividing the model into six regions where there are changes in material, cross-sectional dimensions, or both between these regions. After defining the material properties and the geometrical dimensions, equations (1) and (2) are used to calculate the inductance and capacitance impedances as functions of frequency for each region.

As shown in Figure 2, regions 2 and 3 are composite structures of different parts, region 2 has a hollow back mass and a bolt through the back mass and region 3 has piezoceramic disks, electrodes, and the bolt through the disks. In order to consider the acoustical effect of the bolt, the average values of the material properties of these regions are considered [7]. Region 5 represents the slotted helical geometry of the front mass, where this complex structure has an inner uniform solid core and an outer exponential slotted profile as shown in Figure 3(a). The force, F, generated at the piezoelectric disks region remains constant along the uniform section of region 4 and then divides at the slotted region 5 into two components: longitudinal force, F_L, and shear force, F_T, which can be expressed as;

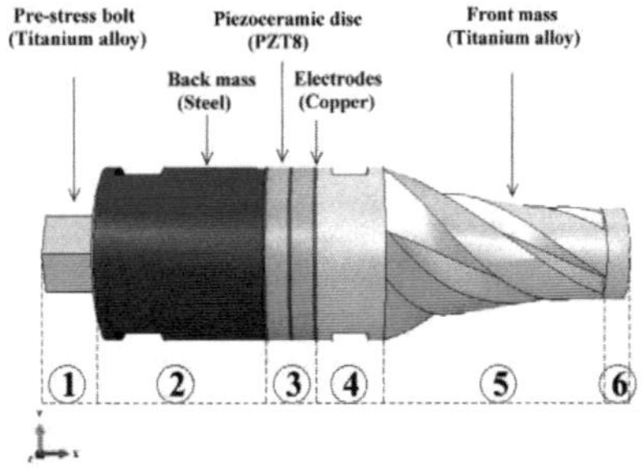

Figure 2. Exploded schematic of the LT transducer.

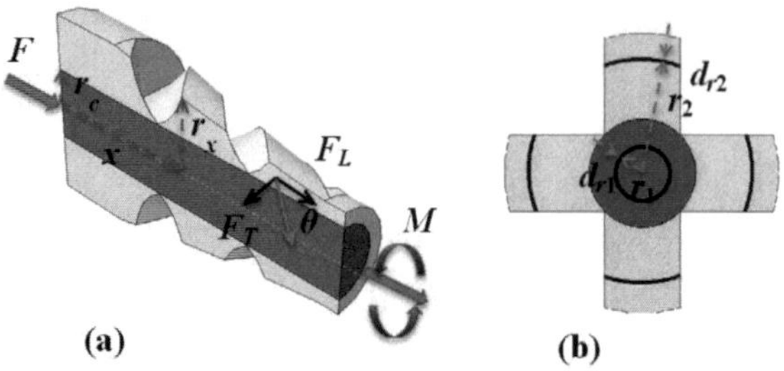

Figure 3. (a) Section of the exponential slotted front mass **(b)** cross-section normal to the wave motion.

$$F_L = F\,cos(\theta) \tag{3}$$

$$F_T = F\,sin(\theta) \tag{4}$$

where θ is the helix angle of the slots.

Based on the theory of longitudinal and flexural vibration, the longitudinal force component will create longitudinal vibration, while the shearing force component will create torsional vibration [8]. The shearing component produces a torsional moment, *M*, along the axis of the horn which can be expressed as;

$$M = \int r\,f\,dS \tag{5}$$

Where *f* is the shearing force acting on a unit area, *dS*, at radius, *r*.

The cross-sectional area of this region is a combination of slotted and solid core areas, as shown in Figure 3(b), which can be found as:

$$S_1 = \pi\,r_1^2 \qquad for \quad 0 < r_1 < r_c \tag{6}$$

$$S_2 = \pi\,r_2^2 - \pi(r_2 - r_c)^2 \quad for \quad r_c < r_2 < r_x \tag{7}$$

Differentiating equations (6) and (7), so that;

$$dS_1 = 2\pi\,r_1 dr_1 \tag{8}$$

$$dS_2 = 2\pi\,r_c dr_2 \tag{9}$$

Applying equations (8) and (9) into equation (5);

$$M = \int_0^{r_c} r_1\,\frac{F\,sin(\theta)}{\pi\,r_1^2}\,2\pi r_1 dr_1 + \int_{r_c}^{r_x} r_2\,\frac{F\,sin(\theta)}{2\pi\,r_c r_2 - \pi r_c^2}\,2\pi\,r_c dr_2 \tag{10}$$

The simplified equation is then;

$$M = 2F\,sin(\theta)\,[\int_0^{r_c} dr_1 + \int_{r_c}^{r_x} \frac{r_2 dr_2}{2r_2 - r_c}] \tag{11}$$

We have;

$$\int \frac{x\,dx}{ax+b} = \frac{x}{a} - \frac{b}{a^2} lin|ax+b| \tag{12}$$

From equation (11), it could be found that (a = 2) and (b = -r_c), by applying the integration solution, the shear moment equation will be;

Journal of Physics: Conference Series **382** (2012) 012061 doi:10.1088/1742-6596/382/1/012061

$$M = 2F\,sin(\theta)\,[\frac{r_c}{2} - \frac{r_c}{4}\,lin|r_c| + \frac{r_x}{2} + \frac{r_c}{4}\,lin|2r_x - r_c|] \tag{13}$$

The equivalent circuit of region 5 can then be obtained by two separate circuits of the 1D model, the first circuit representing the longitudinal vibration component and the other circuit representing the torsional vibration component. Each circuit interacts with region 4 through a conversion coefficient which is the ratio of degenerated vibration to the generated vibration at the dividing surface (output of region 4). The conversion coefficient of longitudinal vibration, n_L, and torsional vibration, n_T, are given by;

$$n_L = \frac{F_L}{F} \tag{14}$$

$$n_T = \frac{M}{F} \tag{15}$$

Applying equations (3) and (13) in equations (14) and (15) respectively, gives;

$$n_L = cos(\theta) \tag{16}$$

$$n_T = 2\,sin(\theta)\,[\frac{r_c}{2} - \frac{r_c}{4}\,lin|r_c| + \frac{r_x}{2} + \frac{r_c}{4}\,lin|2r_x - r_c|] \tag{17}$$

The complete equivalent network of the transducer is shown in Figure 4, where two more impedances are added at both end sides of the circuit which represent the front, $Z_{front\,L}$ and $Z_{front\,T}$, and back, Z_{back}, external loads. These impedances are neglected when the transducer is tested under free conditions (unloaded) because they represent the surrounding space impedance which has a small value, but when the transducer is tested under operational load, their values should be considered.

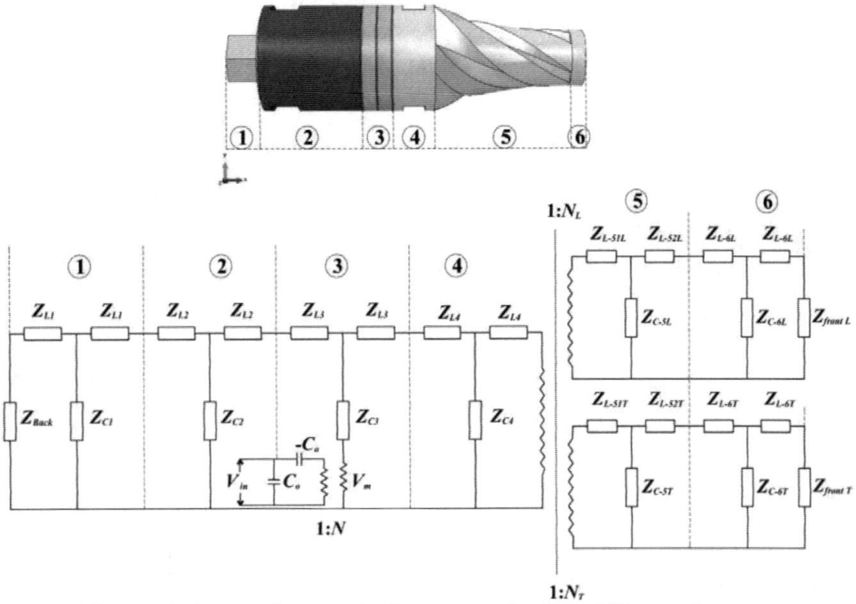

Figure 4. A complete equivalent network of the LT transducer.

4. Results

A Mathematica program is created to solve the complete equivalent circuit and extract the resonance and anti-resonance frequencies, the coupling coefficient and the ratio of torsional to longitudinal response at the output surface. The results are compared to experimental results as shown in Table 1; There is an excellent match between the frequency values and a good agreement of the response ratio, but, the difference in the coupling coefficient values may relate to a difference in the piezoelectric constant, d_{33}, due to the application of pre-stress, which is not considered in the analytical model.

Table 1. Comparison of analytical and experimental results.

	Resonance f_r (Hz)	Anti-resonance f_a (Hz)	Coupling Coefficient k_{eff}	T/L response
Analytical	19,451	19,706	0.16	0.64
Experimental	19,426	19,886	0.20	0.59
% diff.	0.1%	0.9%	20%	8%

The electrical impedance and the phase diagram are calculated and compared with the experimental results as shown in Figure 5. The resonance values are very well matched with a slight difference at the anti-resonance impedance which may relate again to the lack of pre-stress in the analytical model. The second peak, near 23 kHz, is the torsional mode of vibration which is extracted from the experimental analysis whilst the analytical model can only predict the longitudinal modes of vibration.

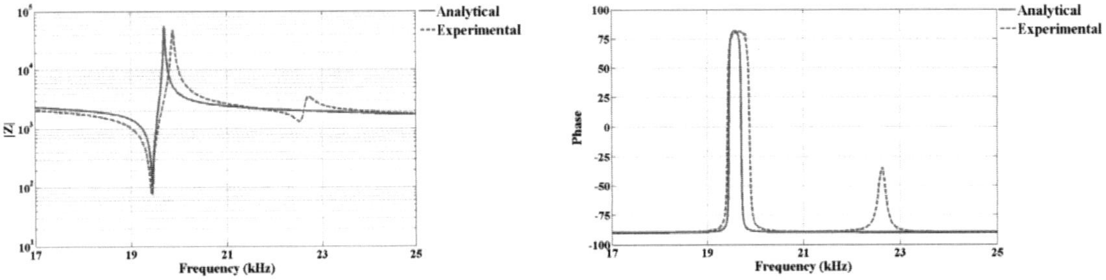

Figure 5. Analytical and experimental impedance spectra (left) and phase-frequency diagram (right).

The electrical currents in the network branches are equivalent to the velocities at the boundaries of the regions, therefore solving for these currents and employing interpolation can allow for calculating the L and T responses along the model as shown in Figures 6(a) and 6(b).

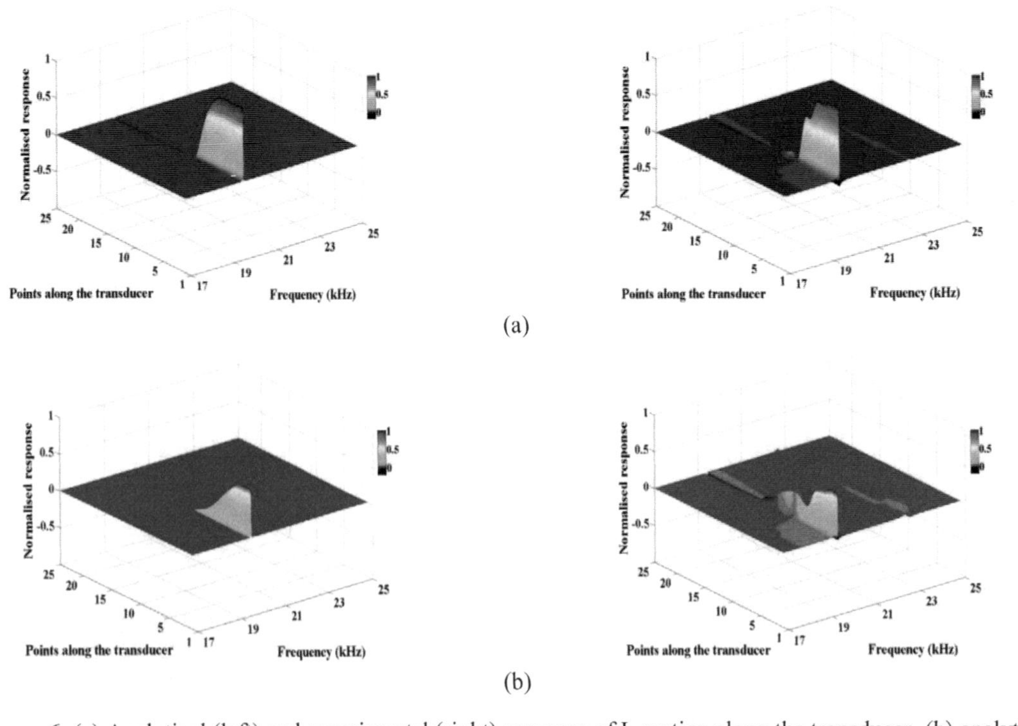

Figure 6. (a) Analytical (left) and experimental (right) response of L motion along the transducer, (b) analytical (left) and experimental (right) response of T motion along the transducer.

The location of the nodal planes can be found in these figures as the point of change from a positive to negative response. For L responses, the analytical model can predict the location accurately, however, there is a difficulty in locating the T nodal plane because the analytical model does not consider the T wave reflection which produces LT vibration in the piezoelectric and back regions of the transducer model.

Finally, the transducer L and T responses and the input electrical power for different excitation levels are calculated and compared with the experimental results as shown in Figure 7. Good agreement is obtained at lower excitations, however the difference increases at higher excitations as a result of the non-linear increase of mechanical and dielectric loss at higher excitation due to change in the properties of the piezoceramic at elevated temperatures [2].

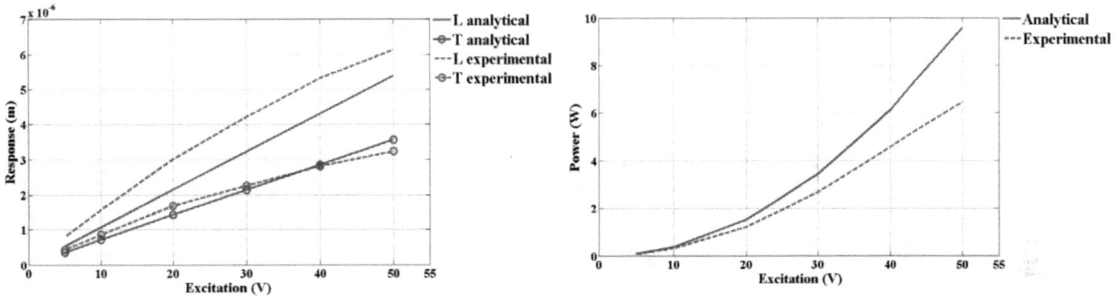

Figure 7. Analytical and experimental L and T responses (left) and electrical input power (right).

5. Conclusions

An analytical model based on the equivalent circuit approach is created for an LT transducer and is used to calculate different electro-mechanical parameters such as resonance and anti-resonance frequencies, coupling coefficient, and the electrical impedance spectrum, the model is also used to locate the nodal plane and to calculate the responses and the input power for different excitation levels. The results show good agreement with experimental findings, with some deficiencies at higher excitation levels.

References

[1] F. Arnold, 2008, Resonance Frequencies of the Multilayered Piezotransducers, *Journal of the Acoustical Society of America,* **123**(5) 3641.

[2] L. Shuyu and T. Hua, 2008, Study on the Sandwich Piezoelectric Ceramic Ultrasonic Transducer in Thickness Vibration, *Journal of Smart Materials and Structures,* **17**(1) 015034.

[3] D. Dragan, D. Mančić and G. Stančić, 2010, New Three-dimensional Matrix Models of the Ultrasonic Sandwich Transducers, *Journal of Sandwich Structures and Materials,* **12**(1) 63-80.

[4] H. Al-Budairi, P. Harkness and M. Lucas, 2011, A Strategy for Delivering High Torsionality in Longitudinal-Torsional Ultrasonic Devices, *Applied Mechanics and Materials,* **70** 339-344.

[5] S. Sherrit, B. Dolgin and Y. Bar-Cohen, 1999, Modeling of Horns for Sonic/Ultrasonic Applications, *Proceedings of the IEEE Ultrasonics Symposium,* **1** 647-651.

[6] W. Mason, 1948, Electromechanical transducers and wave filters, D. Van Nostrand Co., New York.

[7] T. Li, Y. Chen and J. Ma, 2009, Development of a Miniaturized Piezoelectric Ultrasonic Transducer, *IEEE Transactions of Ultrasonics, Ferroelectrics and Frequency Control,* **56**(3) 649-659.

[8] L. Shuyu, 1999, Study on the Longitudinal–Torsional Compound Transducer with Slanting Slots, *Journal of the Acoustical Society of America,* **105**(3) 1643-1650.

Numerical analysis and noise detection for design optimisation of an ultrasonic transducer

Farrukh Makhdum, Anish Roy and Vadim V Silberschmidt
Wolfson School of Mechanical and Manufacturing Engineering,
Loughborough University, Leicestershire, LE11 3TU, UK

E-mail: f.makhdum@lboro.ac.uk

Abstract. The characteristics of a Langevin transducer are studied using a combination of numerical and experimental techniques, which reveal the effect of minor design changes on its performance. The experiments were performed using a microphone and voice-recording software capable of measuring frequencies up to 41 kHz; the obtained signal was analysed in MATLAB. A three-dimensional finite element model of the analysed transducer was also developed in a commercial finite element software ABAQUS/Standard and used for numerical simulations of its response to different excitation conditions. The transducer system was optimised using the results of noise detection and FEA.

1. Introduction

Langevin transducers are widely used in high-power ultrasonics for applications ranging from cleaning to advanced machining and cutting of intractable materials [1-3]. In ultrasonically assisted machining, piezoelectric ceramic discs are used to excite the horn. Different types of cutting tools are attached to the horns; the tool attachment to the horn is achieved by means of screws, bolts or by other assembly procedures. In ultrasonically assisted drilling (UAD) studied here the drill bits can be attached to the transducer in different ways. Several authors presented the setup for conducting the ultrasonic machining detailed [4-7] but in certain studies the information about the setup, especially the tool holding procedure to the transducer, is kept confidential [8-10].

The studies revealed that the attachment of a tool to the horn plays a critical role in vibration modes; also the resonance frequency changes by changing the position of the free length of the tool with respect to the tool attachment [11]. When the transducer is excited the system produces sound/noise. The variation sound emanating by the system can be used as an indication of changes in the system. But such variations cannot be differentiated solely with human audibility.

Techniques such as acoustic emission and vibration measurement systems are used in the industry for different condition monitoring purposes e.g. for valve leak detection, bearing maintenance and detecting fluid flow through pipes [12-14]. In such applications the noise and vibration data is collected, then compared against the standard values and faults are monitored. Once the errors are detected, maintenance of the system is carried out often by replacing or repairing the individual parts of the assembly.

Finite element method (FE method) is broadly used by the engineering community in various tasks, from design and development to manufacturing and optimisation. Design of ultrasonic transducer with the aid of FE approaches are typically used to determine the vibration modes and amplitude, voltage required for piezoelectric ceramic materials and natural frequency of the systems [11, 15-16]. But optimisation of the ultrasonic transducers, especially used for UAD, has not been based on a combination of noise detection and FEA analysis.

Here, we study a behaviour of different waveguides for a horn as well as an attachment-systems for drill bit, connected to the ultrasonic transducer. The noise, beyond the audible range, emanated by the systems was recorded using a microphone and processed in Matlab™. The difference in the ultrasonic signal, recorded in the frequency range between 20-41 kHz, was analysed by this technique. A pre-validated FEA model was used for further analysis of each part of the attachment system, assessment of errors and optimisation of the transducer system.

2. Problem formulation

Ultrasonic transducers used for drilling are mostly manufactured for a single drill bit i.e. for each drill bit a special transducer was designed [4-7]. A special adapter which consists of a collet chuck for drill bits of different size was designed and manufactured so that the same pre-manufactured transducer can be used with different bits. The adapter was attached to the pre-manufactured transducer providing a capability to compensate the different size of drill bits with the help of collet-chucks of different sizes (Figure 1).

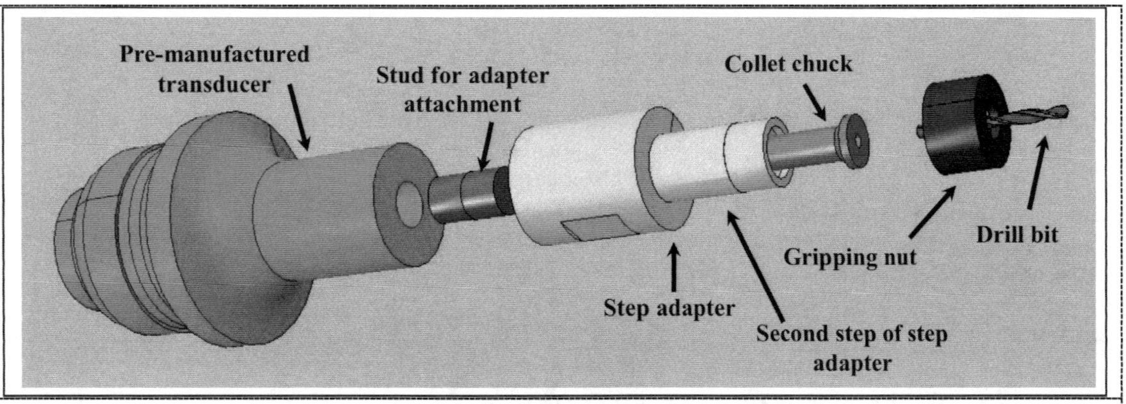

Figure 1. Transducer, adapter and collet-chuck for rapid tool change

Initial drilling experiments were conducted on carbon fibre reinforced plastic using the transducer assembly shown in Figure 1. However, with an increase in drilling depth the thrust force reduction in UAD was completely lost. It was around 50% in the beginning of complete drill bit engagement with the workpiece but the ultrasonic effect depleted completely through the half depth. The respective evolution of thrust force is presented in Figure 2.

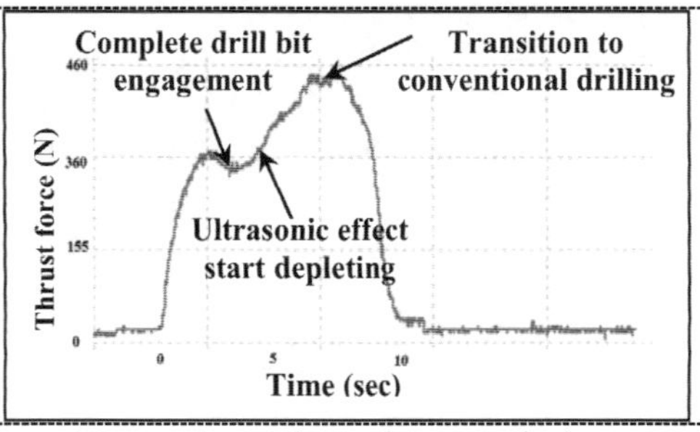

Figure 2. Thrust force evolution: ultrasonic effect depletion

Journal of Physics: Conference Series **382** (2012) 012062 doi:10.1088/1742-6596/382/1/012062

2.1. Noise detection

To elucidate the cause for ultrasonic effect-depletion, free-vibration tests were conducted on the transducer with the whole transducer assembly vibrating at the resonance frequency and produced sound being recorded. To achieve this, the microphone was placed near the ultrasonic-transducer system and noise/sound was recorded for 15 seconds. The acquired signal was processed using Matlab™; the obtained results are presented in Figure 3. The signals showed that the sound was not continuous, demonstrating that some of the transducer parts were not vibrating in harmony with the system. At that stage it was understood that the whole assembly should be analysed and finite element analysis was carried out to achieve this.

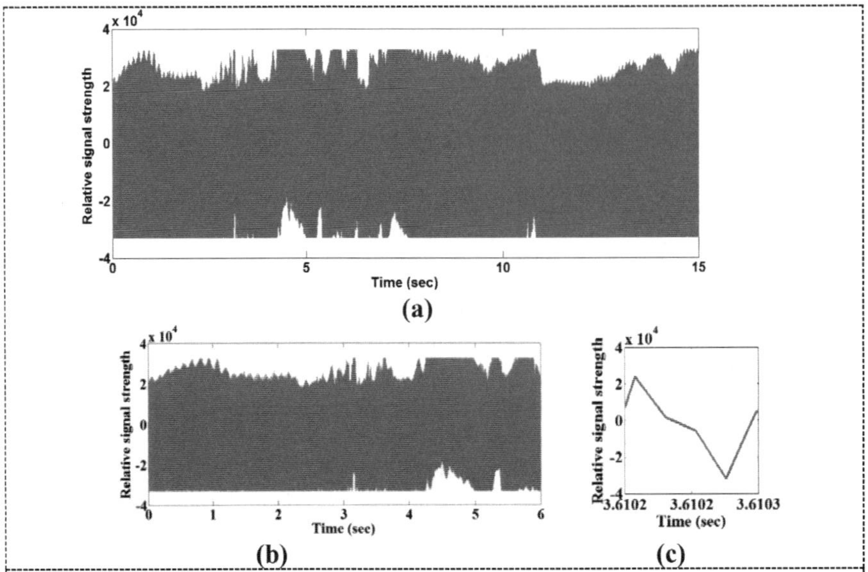

Figure 3. Sound evolution for transducer vibrating at resonance frequency
(a) For 15 Sec (b) For 6 Sec (c) For 0.0001 Sec from 3.6102 to 3.6103 Sec

2.2. Finite element analysis

A pre-validated 3-D model was developed in FEA software Abaqus/Standard and used to analysis the complete assembly. In the FEA model the Linear Perturbation Analysis was carried out for eigen-values, vibration modes and natural frequency analysis using the AMS solver. In our free-vibration experiments, the transducer vibrated at frequency of 24.6 kHz whereas in the simulation the same mode was observed at 23.6 kHz. The analysis revealed that under vibration the collet-chuck was getting dis-engaged with the adapter-assembly (Figure 4). This was considered to be the primary cause of loss of ultrasonic effects during drilling.

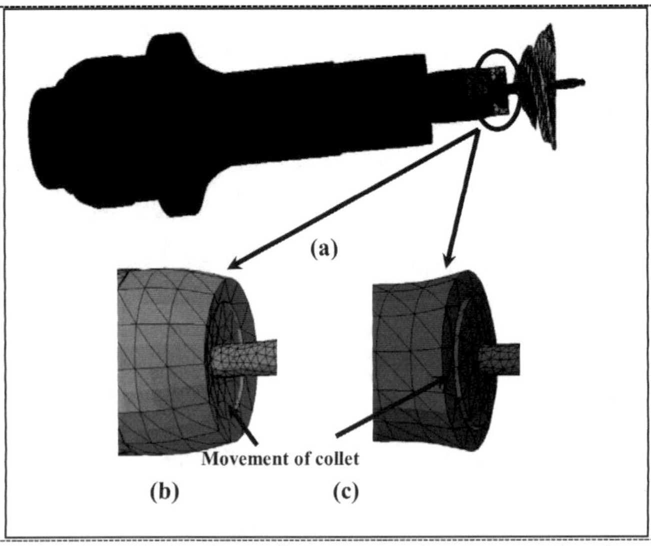

Figure 4. FEA of transducer system (resonance vibration – longitudinal mode)
(a) Complete transducer system (b) In compression (c) In expansion

Journal of Physics: Conference Series **382** (2012) 012062 doi:10.1088/1742-6596/382/1/012062

3. New adapter and grip screws

To solve the problem a new conical adapter was analysed using the developed FE model. The adapter holds the drill bit with the help of grip-screws. The FEA analysis revealed that the new system was more reliable compared to the collet-chuck system. After manufacturing free vibration tests were carried out. The experimental value of resonance was observed at 25.5 kHz whereas the FEA resonance value was found at 24.9 kHz. The longitudinal vibration mode is shown in Figure 6 (a) and the manufactured system in Figure 6 (b).

The sound test was carried out and a smooth sound signal was measured for the new design (see Figure 7). This conical adapter system also allows the use the same transducer for different sizes of the drill bits efficiently.

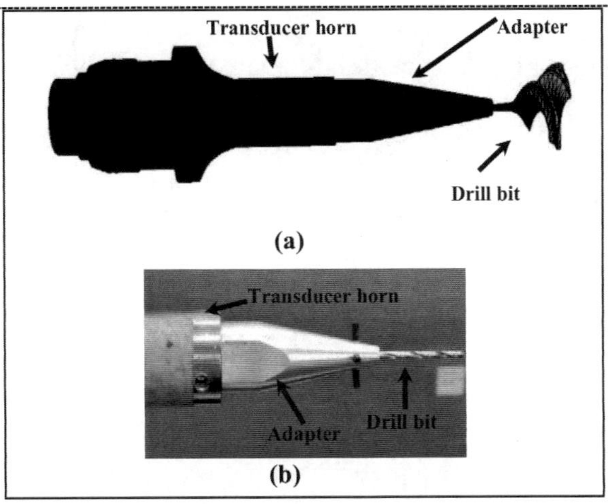

Figure 6. (a) FEA analysis (b) Manufactured system

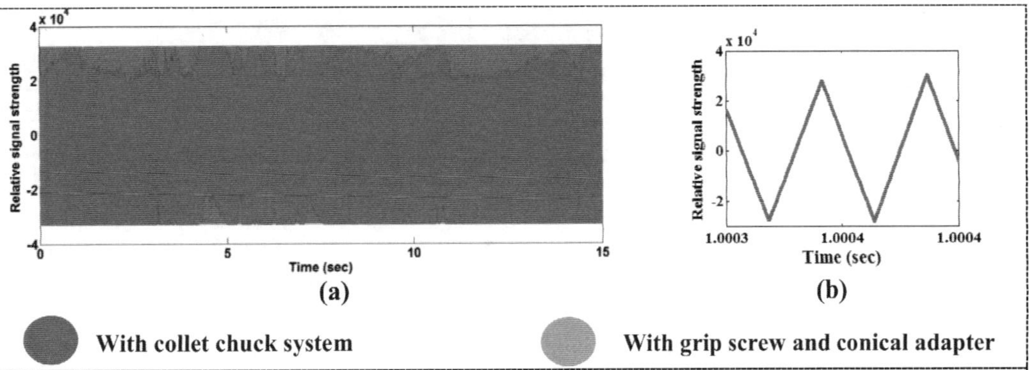

Figure 7: Noise evolution comparison for transducer vibrating at resonance frequency
(a) Comparison of systems noise (b) Zoomed noise trace for 0.0001 Sec of grip screw system

4. Drilling experiments

The drilling experiments were conducted using new transducer system and consistent results were observed. The system resonance frequency was found as 25.5 kHz with the vibration amplitude of 12 μm (peak to peak). The experiments were conducted on the same drilling parameters and same workpiece. The thrust force trance is presented in Figure 8.

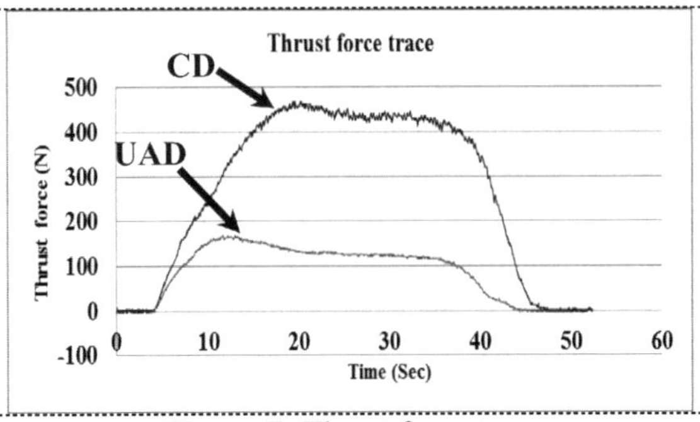

Figure 7: Thrust force trace

Conclusions

The following conclusions were achieved based on the undertaken study:

- A combination of experimental and numerical studies allowed to design a new transducer
- A combined approach for design optimisation of ultrasonic transducer was developed
- A successful series of drilling experiments revealed the suitability of optimised transducer system for drilling
- The noise detection allows to elucidate the design shortcomings as well as optimises the ultrasonic transducer
- The developed FEA model is an effective tool in designing and optimisation of the transducer

References

[1] Chen Y-C. 2011. A comparative assessment of classification methods for resonance frequency prediction of Langevin piezoelectric transducers. Applied Mathematical Modelling 35:3334-44

[2] Yan Z, Fang Q, Huang J, He B, Lin Z. 1997. Considerations and guides of the wattmeter method for measuring output acoustical power of Langevin-type transducer systems — II: experiment. Ultrasonics 35:543-6

[3] Philip WL. 2007. Piezoelectric ultrasonic transducers and waveguide finite elements. IEEE Transactions on Ultrasonics, Ferroelectrics, and Frequency Control 54:2045-51

[4] Singh R, Khamba JS. 2008. Comparison of slurry effect on machining characteristics of titanium in ultrasonic drilling. Journal of Materials Processing Technology 197:200-5

[5] Alam K, Mitrofanov AV, Silberschmidt VV. 2011. Experimental investigations of forces and torque in conventional and ultrasonically-assisted drilling of cortical bone. Medical Engineering and Physics 33:234-9

[6] Thomas PNH, Babitsky VI. 2007. Experiments and simulations on ultrasonically assisted drilling. Journal of Sound and Vibration 308:815-30

[7] Azarhoushang B, Akbari J. 2007. Ultrasonic-assisted drilling of Inconel 738-LC. International Journal of Machine Tools and Manufacture 47:1027-33

[8] Chang SSF, Bone GM. 2009. Thrust force model for vibration-assisted drilling of aluminum 6061-T6. International Journal of Machine Tools and Manufacture 49:1070-6

[9] Chang SSF, Bone GM. 2005. Burr size reduction in drilling by ultrasonic assistance. Robotics and Computer-Integrated Manufacturing 21:442-50

[10] Takeyama H, Kato S. 1991. Burrless drilling by means of ultrasonic vibration. CIRP Annals - Manufacturing Technology 40:83-6

[11] Cardoni A, Lucas M, Cartmell M, Lim F. 2004. A novel multiple blade ultrasonic cutting device. Ultrasonics 42:69-74

[12] Valve leak detection using acoustic emissions (AE). http://www.score-group.com/valve-leak-detection.php. 2012

[13] Puttmer A, Rajaraman V. 2007. P3F-6 Acoustic emission based online valve leak detection and testing. Ultrasonics Symposium. IEEE, pp. 1854-7

[14] Bloch HP, Geitner FK. 1997. Practical Machinery Management for Process Plants. Houston, TX, USA: Gulf Publishing Company

[15] Ramesh R, Prasad CD, Kumar TKV, Gavane LA, Vishnubhatla RMR. 2006. Experimental and finite element modelling studies on single-layer and multi-layer 1–3 piezocomposite transducers. Ultrasonics 44(4):341-9

[16] Wilm M, Reinhardt A, Laude V, Armati R, Daniau W, Ballandras S. 2005 Three-dimensional modelling of micromachined-ultrasonic-transducer arrays operating in water. Ultrasonics 457-65.

Journal of Physics: Conference Series **382** (2012) 012063 doi:10.1088/1742-6596/382/1/012063

Vibration characterisation of cymbal transducers for power ultrasonic applications

F Bejarano, A Feeney and M Lucas

School of Engineering, University of Glasgow, Glasgow G12 8QQ, UK

E-mail: f.bejarano-duran.1@research.gla.ac.uk

Abstract. A Class V cymbal flextensional transducer is composed of a piezoceramic disc or ring sandwiched between two cymbal-shaped shell end-caps. These end-caps act as mechanical transformers to convert high impedance, low radial displacement of the piezoceramic into low impedance, large axial motion of the end-cap. The cymbal transducer was developed in the early 1990's at Penn State University, and is an improvement of the moonie transducer which has been in use since the 1980's. Despite the fact that cymbal transducers have been used in many fields, both as sensors and actuators, due to its physical limitations its use has been mainly at low power intensities. It is only very recently that its suitability for high amplitude and high power applications has been studied, and consequently implementation in this area of research remains undeveloped. This paper employs experimental modal analysis (EMA), vibration response measurements and electrical impedance measurements to characterise two variations of the cymbal transducer design, both aimed at incorporation in ultrasonic cutting devices. The transducers are fabricated using the commercial Eccobond 45LV epoxy adhesive as the bonding agent. The first cymbal transducer is of the classic design where the piezoceramic disc is bonded directly to the end-caps. The second cymbal transducer includes a metal ring bonded to the outer edge of the piezoceramic disc. The reason for the inclusion of this metal ring is to improve the mechanical coupling with the end-caps. This would therefore make this design particularly suitable for power ultrasonic applications, reducing the possibility of debonding at the higher ultrasonic amplitudes. The experimental results demonstrate that the second cymbal design is a significant improvement on the more classic design, allowing the transducer to operate at higher voltages and higher amplitudes, exhibiting a linear response over a practical power ultrasonic device driving voltage range. The results also show that the device can be accurately tuned using finite element modelling and that the cymbal exhibits a modal response as predicted by the finite element models.

1. Introduction

Flextensional transducers have been in existence for a number of years, primarily being used in underwater and sonar applications since the 1920's [1]. Cymbal transducers are a variation of the flextensional 'moonie' transducer design, and were developed in the early 1990's by Newnham et al. at the Pennsylvania State University Materials Research Laboratory [2]. Although there are many applications for this transducer in low power applications, the incorporation of cymbal transducers into high-power ultrasonic technology is underdeveloped. One of the major reasons for this is that the bonding material imposes a natural limit at which the cymbal transducer can be driven.

The two most critical aspects of the cymbal design, from a dynamic and vibration response perspective, which directly impinge on the transducer performance, are the cavity dimensions and the thickness of the end-caps [3]. The end-cap of the transducer is used as a mechanical transformer, to convert high impedance, low displacement radial motion into low impedance, large axial-flexural motion [1]. A cymbal transducer has a high output vibration displacement amplitude and is sensitive to acceleration changes [4]. However, for the same materials and design, when the devices are used at high power, critical problems occur that can reduce the reliability of the cymbal due to high stress and electrical concentration points leading to degradations or debonding phenomena that can reduce drastically the operating life of the device [5].

The geometrical dimensions of the end-caps greatly affect the frequency response of the cymbal transducer and, due to this, the vast majority of cymbal transducers present a double resonance peak in the frequency response, as even small asymmetries in the epoxy layer or in the end-caps result in each metal cap being resonant at a different frequency. This effect reduces the efficiency of the whole transducer.

Since epoxy is commonly used as a bonding agent, there exists a finite limit for the performance of the transducer based on the components in the system, and therefore debonding will occur if the excitation level is too high. Because of the limitation imposed by the components of the system, such as the epoxy layer and the fact that radial motion is converted into large-amplitude flexural motion, strong mechanical coupling must exist between all components for its use in high power applications.

In 2010 Shuyu Lin developed a new design of the cymbal transducer suitable for high power applications in which the driver part consist of a ceramic disc bonded to a metal ring [4]. The end-cap, with a larger flange, is then attached directly to the metal ring through screws, increasing the mechanical coupling and reducing the stress in the epoxy layer.

The literature presents many studies, such as the work of Naidu et al., which describe the results of numerically-computed modal analyses of cymbal transducers [6], but this investigation utilises experimental modal analysis (EMA) to complement the numerical models in order to identify the differences in the operational mode of these devices. In this paper, two cymbal transducer designs are studied. The first is of the traditional design developed by Newnham et al., and the other is based on a design proposed by Shuyu Lin in 2010. The results of the experimental studies are presented along with finite element analysis (FEA) results obtained using Abaqus finite element software.

2. Cymbal transducer fabrication

2.1. Cymbal end-cap manufacture

The traditional and new designs used in the experiments are shown in Figure 1 and Figure 2.

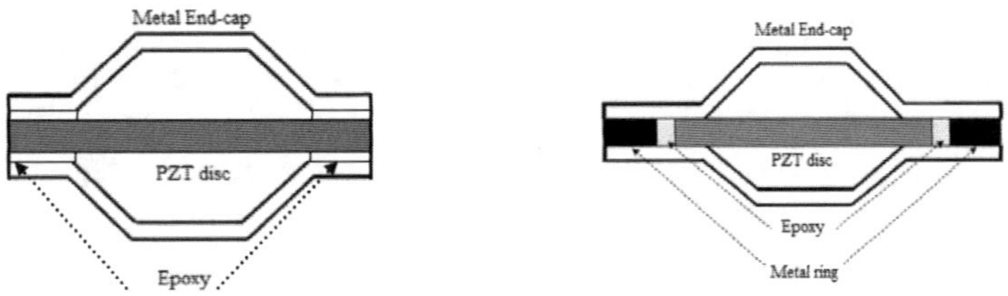

Figure 1. Schematics of the traditional device (left) and new device (right).

Figure 2. Traditional device (left), new device (centre), new device with end-caps removed (right).

Each cymbal end-cap was cut from a 0.25mm thick brass sheet and the piezoelectric elements were hard PZT (PZT-402) discs of 12.7mm diameter and 1mm thickness. Table 1 shows the dimensions of the cymbal transducer components. The new cymbal transducer incorporates a brass ring around the PZT disc which increases the mechanical coupling in the device and decreases the stress in the epoxy layer, thereby allowing operation at higher amplitudes. The dimensions of the brass end-caps are modified to allow the incorporation of the brass ring, as shown in Table 1. The cavity dimensions for the new cymbal transducer design remained the same as those in the traditional configuration, since the resonance frequency is largely influenced by this geometry.

TABLE 1. Cymbal transducer dimensions.

Dimension	Traditional design (mm)	New design (mm)
PZT Thickness	1.0	1.0
End-Cap Thickness	0.25	0.25
Total Ø	12.7	16.7
Base Cavity Ø	9.0	9.0
Apex Cavity Ø	4.5	4.5
Max. Cavity Depth	0.3	0.3
Ring Inner Ø	N/A	14.7
Ring Outer Ø	N/A	16.7
Ring Thickness	N/A	1.0

2.2. Cymbal transducer assembly

For the traditional cymbal transducer a layer of insulating epoxy resin, *Eccobond®* from Emerson & Cuming, at a ratio of three parts 45LV epoxy resin to one part 15LV resin hardener, was applied to the flange of each end-cap and left to cure at room temperature for 24 hours. The layers were approximately 40μm thick. The reason that an insulating and not a conductive epoxy was chosen is that conductive epoxies tend to have lower bonding strength [7]. To ensure that electrical contact could be achieved between the end-cap flanges and the piezoceramic disc, a small solder spot was applied to each surface of the PZT. The end-caps of the traditional cymbal transducer were bonded directly to the piezoceramic disc, but the new cymbal transducer includes a metal ring bonded to the outer edge of the piezoceramic disc. The new cymbal assembly required the use of threaded screws, with a head diameter of 0.50mm and thread diameter of 0.35mm, to fix the metal ring to both end-cap flanges. The gap between the PZT and the metal ring was filled with epoxy resin at a thickness of 1mm. In the device constructed by Shuyu Lin, there was no epoxy resin present between the end-caps and the ceramic. Because of that, the device had a lower resonance frequency [4]. In this investigation, epoxy resin has been included between the end-caps and the ceramic in the new design to enable vibration at a similar frequency to that of the traditional design. The inclusion of the bolts adds strength at the points of highest stress concentration in the epoxy layer. Two traditional cymbal transducers and one new device were assembled.

3. Device characterisation

The devices were driven over a range of voltage increments at resonance to characterise the response with the aim of identifying whether the transducer could be excited for high power applications in a range where a linear response was exhibited. The characterisation also aimed to determine the output displacement amplitude achievable before debonding occurred. A 1D laser Doppler vibrometer (LDV) was used to measure the output amplitude response. A burst-sine wave was applied with 2-4s between bursts depending on the voltage level, and all stated voltages are peak-to-peak. EMA was then conducted using the 3D LDV and DataPhysics SignalCalc acquisition software.

For the numerical analysis, Abaqus/CAE v6.10-2 was used. For the FEA model, steady-state dynamics analyses were performed using different input voltages as boundary conditions. The model is composed of elements of approximately 0.4 mm in size to ensure that optimal results are returned.

4. Experimental results

4.1. Dynamics characterisation

The displacements for both cymbal transducers are shown in Figure 3. The resonance frequency should match, but any discrepancy can come from the epoxy bonding layer and the tolerances in the fabrication process. Also the effect of the bolts improving the mechanical coupling of the system in the new device enables a more efficient transfer of energy from the driver to the metal end-cap. Also, differences in the strength of the coupling can lead in a variation in the resonance frequency.

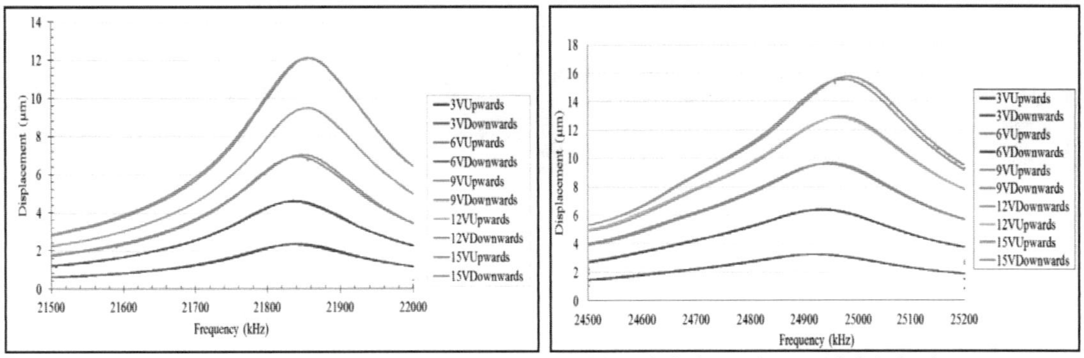

Figure 3. Vibration responses of the traditional (left) and new (right) devices.

In order to observe the differences in the behaviour of both devices at high displacements, the cymbal transducers were driven at different voltages to demonstrate that the new design can operate at higher amplitudes. In Figure 4 the results of the experiment are shown.

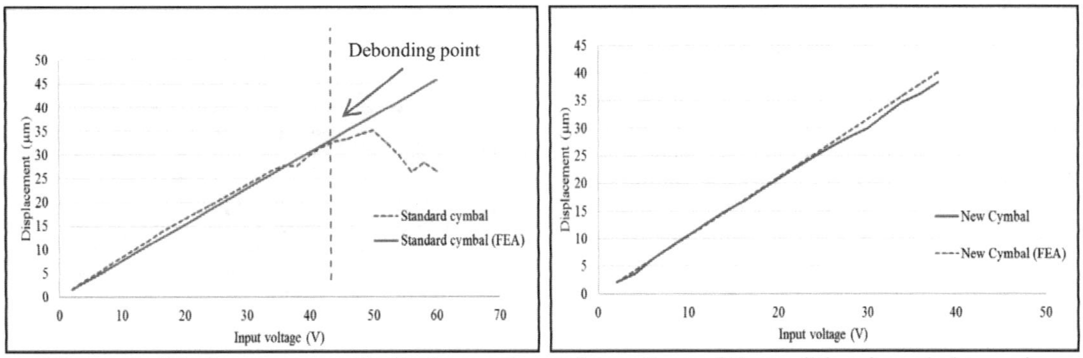

Figure 4. Displacement amplitude (pre and post debonding) of the traditional cymbal transducer (left) and new cymbal (right).

In the traditional cymbal transducer the maximum displacement is 32μm. Above this point, the device lost its performance due to debonding. However the new device reached 38μm and had the potential to achieve even higher displacements.

4.2. Experimental modal analysis

A frequency response function from the experimental modal analysis is shown in Figure 5. For the traditional design, it is very difficult to control the thickness and uniformity of the epoxy layer. One consequence of this loss of symmetry is the appearance of a double-peak in the response spectrum, as seen in Figure 5, meaning each end-cap vibrates at a different resonance frequency. The total energy of this mode is shared between these two frequencies. The new cymbal design advances the assembly procedure by improving the controllability of the epoxy layer. Also, the bolts allow a correct alignment of the end-caps with the ceramic surface. The results of the EMA of the new cymbal design show a single resonance peak, indicating considerably improved symmetry and increasing the output response.

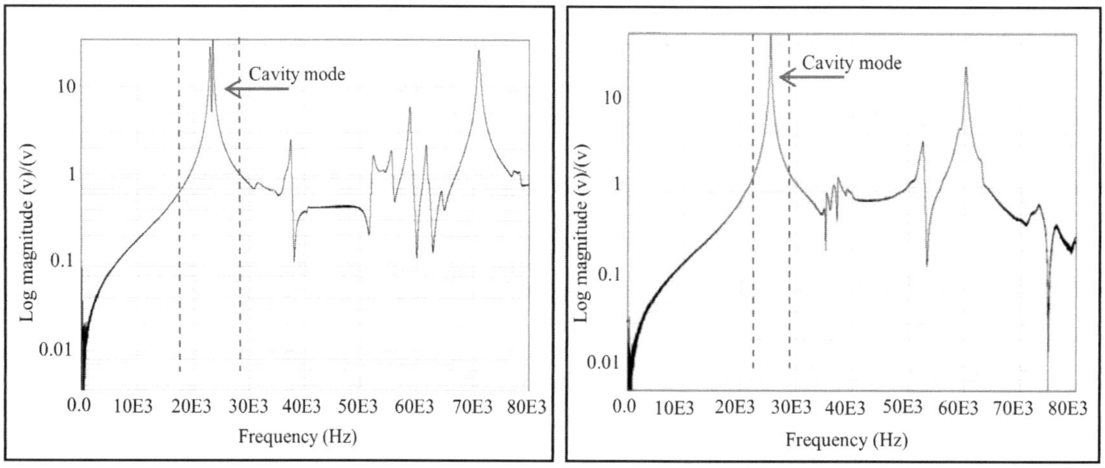

Figure 5. Frequency response of the traditional cymbal transducer (left) and new cymbal (right).

The mode shapes and associated frequencies from the EMA and finite element modal analysis are shown in Figures 6 and 7. It can be seen that there is a good agreement between the numerical and experimental results for frequencies and mode shapes, where the small differences can be attributed to the fabrication process of the different transducers.

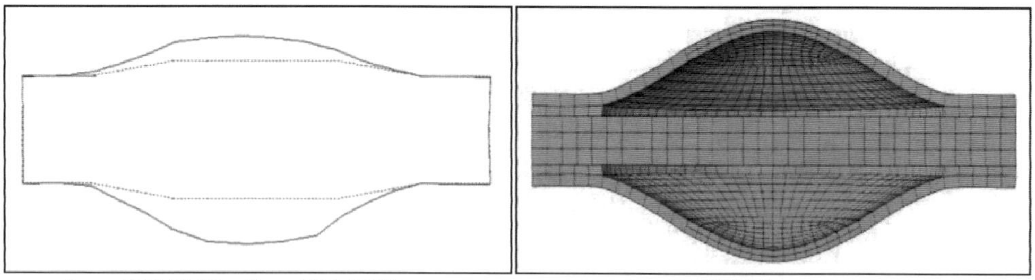

$f_{r,1}$ = 22945 Hz, $f_{r,2}$ = 23566 Hz f_r = 24280 Hz

Figure 6. EMA (left) and FEA (right) of the traditional cymbal transducer.

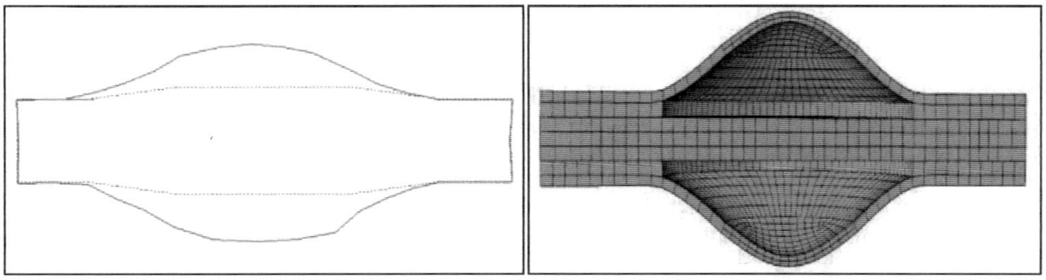

Figure 7. EMA (left) and FEA (right) of the new cymbal transducer.

5. Conclusions

In this paper two different designs of the cymbal transducer have been studied. In the new design, the mechanical coupling of the end-caps with the driver part of the transducer enables the device to be driven at higher excitation levels and displacements, and reduces the stress on the end-caps due to bending modes of the cavity. It is clear that even though the new design incorporates additional components of bolts and metal ring, the device exhibits the same cavity resonance mode as the traditional design.

Experimental results have demonstrated that the performance of the new cymbal transducer is similar to the traditional cymbal transducer, preserves the advantages of this design and solves the drawbacks in the power limitation imposed by debonding of the epoxy layer. The improvement in control of the fabrication of the new transducer, that allows significantly improved symmetry to be attained, removes the appearance of a double peak in the cavity mode frequency response, and therefore enables considerably improved resonance performance of the transducer.

Both the increased displacement amplitude capabilities and improved symmetry control of this cymbal transducer configuration mean that it can now be incorporated in high power ultrasonic applications for the first time.

6. Acknowledgments

This work has been funded by EPSRC Grants EP/P50418X/1 and EP/G046948/1.

References

[1] Zhang J, Hughes W J, Bouchilloux P, Meyer Jr. R J, Uchino K, Newnham R E 1999, A class V flextensional transducer: The cymbal, *Ultrasonics* **37** pp.387-393

[2] Zhang J, Hughes W J, Meyer Jr. R J, Uchino K, Newnham R E 2000, Cymbal array: A broadband sound projector, *Ultrasonics* **37** pp.523 – 529

[3] Sun C, Gou S S, Li W P, Xing Z B, Liu G C, Zhao X Z 2005, Displacement amplification and resonance characteristics of the cymbal transducers, *Sensors and Actuators A: Physical* **121**(1) pp.213-220

[4] Lin S 2010, An improved cymbal transducer with combined piezoelectric ceramic ring and metal ring, *Sensors and Actuators A: Physical* **163**(1) pp.266-276

[5] Ochoa P, Pons J L, Villegas M, Fernandez J F 2006, Advantages and limitations of cymbals for sensor and actuator applications, *Sensors and Actuators A: Physical* **132**(1) pp.63-69

[6] Naidu B P, Rao A B, Prasad N S, Trinath K 2009, Modal analysis of class V flextensional transducer, *Ocean Electronics (SYMPOL) International Symposium*, pp.122-127

[7] Zhang J, Hladky-Hennion A C, Hughes W J, Newnham R E 2001, Modeling and underwater characterization of cymbal transducers and arrays, *IEEE Transactions on Ultrasonics, Ferroelectrics and Frequency Control* **48**(2) pp.560-568

AUTHOR INDEX

AUTHOR INDEX